Vertebrate Evolution

Vertebrate Evolution

From Origins to Dinosaurs and Beyond

Donald R. Prothero

Illustrations by Nobumichi Tamura

CRC Press
Taylor & Francis Group
Boca Raton London

CRC Press is an imprint of the
Taylor & Francis Group, an **informa** business

First edition published 2022
by CRC Press
6000 Broken Sound Parkway NW, Suite 300, Boca Raton, FL 33487–2742

and by CRC Press
2 Park Square, Milton Park, Abingdon, Oxon, OX14 4RN

CRC Press is an imprint of Taylor & Francis Group, LLC

© 2022 Taylor & Francis Group, LLC

ISBN: 978-0-367-65176-3 (hbk)
ISBN: 978-0-367-47316-7 (pbk)
ISBN: 978-1-003-12820-5 (ebk)

DOI: 10.1201/9781003128205

Typeset in ITC Leawood
by Apex CoVantage, LLC

This book is dedicated to my students who have gone on to great careers in paleontology or are still preparing for their future careers. These include such paleontologists as John Foster, Jonathan Hoffman, Karen Whittlesey, Jingmai O'Connor, Joshua Ludtke, Val Syverson, Katherine Long, Patrick Gillespy, Thein Htun, Daniella Balassa, Sara Olson, Casey Cleaveland, Kristin Watmore, and Katherine Marriott. They have encouraged and supported me and inspired me throughout my teaching career, and I am so proud of their accomplishments. They are the future of our profession.

CONTENTS

PREFACE

I've already written several books about fossils and paleontology, including a college textbook in paleontology for upper-level undergraduates (*Bringing Fossils to Life*, 2nd ed., 2013, Columbia University Press) and a popular book about fossil collecting (*Fantastic Fossils*, 2020, Columbia University Press), but I've never found a book that gives a good general introduction to vertebrate history for the general reader and fossil enthusiast with high-quality color illustrations, so this book is intended to fill that need. Most of the books on this topic for the general reader are picture books with minimal information content. Given the rapid increase in knowledge about fossil vertebrates, and the many changes in old notions about vertebrates and how they lived, all such books are already grossly out of date. Thus, I wrote this book to reach the general reader and especially the fossil enthusiasts and collectors out there, who may have some knowledge about science, but not necessarily a college-level background in geology or paleontology. There is a real need to go beyond the pretty picture books, and present the latest information about extinct vertebrates at a level that the general reader can follow, but also enough information for college students in paleontology to learn about the topic as well.

For this reason, the book is written at an intermediate level. I do not assume any background in vertebrate anatomy, or the methods of systematics and classification. Of course, the concepts are completely modern and in line with the current thinking in cladistics, but I try to avoid jargon and excessive technical terminology as much as possible. I have tried to use familiar anatomical terms wherever possible, so it can be read and understood not only by the amateur fossil enthusiast, but also by students taking an undergraduate course in vertebrate evolution that does not require training in anatomy or systematics. This is a very challenging task, because most of the topics discussed in the book require a more advanced understanding of systematics or anatomy, but I have done my best. I hope the reader will find the book comprehensible and yet up to date and accurate, incorporating all the latest thinking and discoveries of these amazing animals.

ACKNOWLEDGMENTS

I thank my longtime friend Dr. Charles R. Crumly for encouraging me to write this book and Ana Lucia Eberhart at CRC Press/Taylor & Francis for editorial assistance. I thank Sathish Mohan of Apex CoVantage for supervising the book design and production. I thank the following scientists for their helpful reviews on various chapters: A. Henrici, D. Grossnickle, B. MacFadden, K.L. Marriott, S. Modesto, S.J. Nesbitt, J. O'Connor, K. Padian, S. Persons, T. Stidham, X. Wang, and several anonymous reviewers for the comments and corrections when my presentation needed updating. I thank K.L. Marriott for a lot of editorial help.

I thank the great paleontologists who taught and inspired me, from my early career contacts with Drs. Dave Whistler and J. Reid Macdonald, to my formal education in paleontology with Drs. Michael Woodburne, Michael Murphy, Malcolm McKenna, Gene Gaffney, and Bobb Schaeffer. Without their instruction and guidance, I would have never had the career in paleontology that I sought since I first got hooked on dinosaurs at age 4. From that age until today (over 63 years now), I never gave up, despite the difficult challenges of finding a career in this crowded profession.

Finally, I thank my amazing family: my incredible wife, Dr. Teresa LeVelle, and my sons, Erik, Zachary, and especially Gabriel, who also may become a paleontologist someday. They put up with my long months at the computer writing this book.

INTRODUCTION

FINDING, DATING, AND CLASSIFYING FOSSILS

1

Being a paleontologist is like being a coroner except that all the witnesses are dead and all the evidence has been left out in the rain for 65 million years.

—Mike Brett-Surman, 1994

HOW DO YOU FIND FOSSILS?

Many kids grow up fascinated with dinosaurs and other prehistoric creatures. Some even start digging holes in their backyards or driveways looking for dinosaur bones. Most give up and become discouraged, because fossil bones are extremely rare, and found only in a few places on earth.

If you wanted to find fossils, where would you look? Why are certain rocks and certain places on earth good for finding fossils, while others have none at all? First, nearly all fossils are primarily found in one kind of rock, known as sedimentary rocks. These are rocks that are made from the loose grains of sand, gravel, or mud, or other particles that weather out of the hard bedrock and are deposited in rivers or floodplains or in the bottom of the ocean. When animals and plants die, their hard parts (bones, shells, wood) can be buried (**Figure 1.1**). If the conditions are right, these hard parts will be deeply buried and covered by loose sediments. Over time, the sands are cemented together by minerals in the groundwater to become sandstone, or the soft mud grains are squeezed and compressed until they become a hard splintery rock called shale.

These sedimentary rocks might then be deeply buried in the earth's crust. Millions of years later, these ancient rocks might be uplifted to the surface by immense tectonic forces, and crumpled upward by the collision of continents to form a mountain belt. Or they might be tilted on their side and erosion will expose the ancient sediment. There they might be brought to the surface by erosion, and rain and frost and wind will break down the fossils as well as the rock surrounding them. This has been happening for millions of years, and most fossils that were once buried over millions of years have already been exposed by erosion and destroyed when no human was around to collect them. Only in the past 200 years have humans (especially paleontologists) been actively looking for and collecting and preserving fossils before they are lost forever. Fossil collectors have only visited small parts of the earth more than a few times. Even today, large areas of unexplored land remain in remote regions, and most fossils there are lost before any human sees them in time to save them.

DOI: 10.1201/9781003128205-1

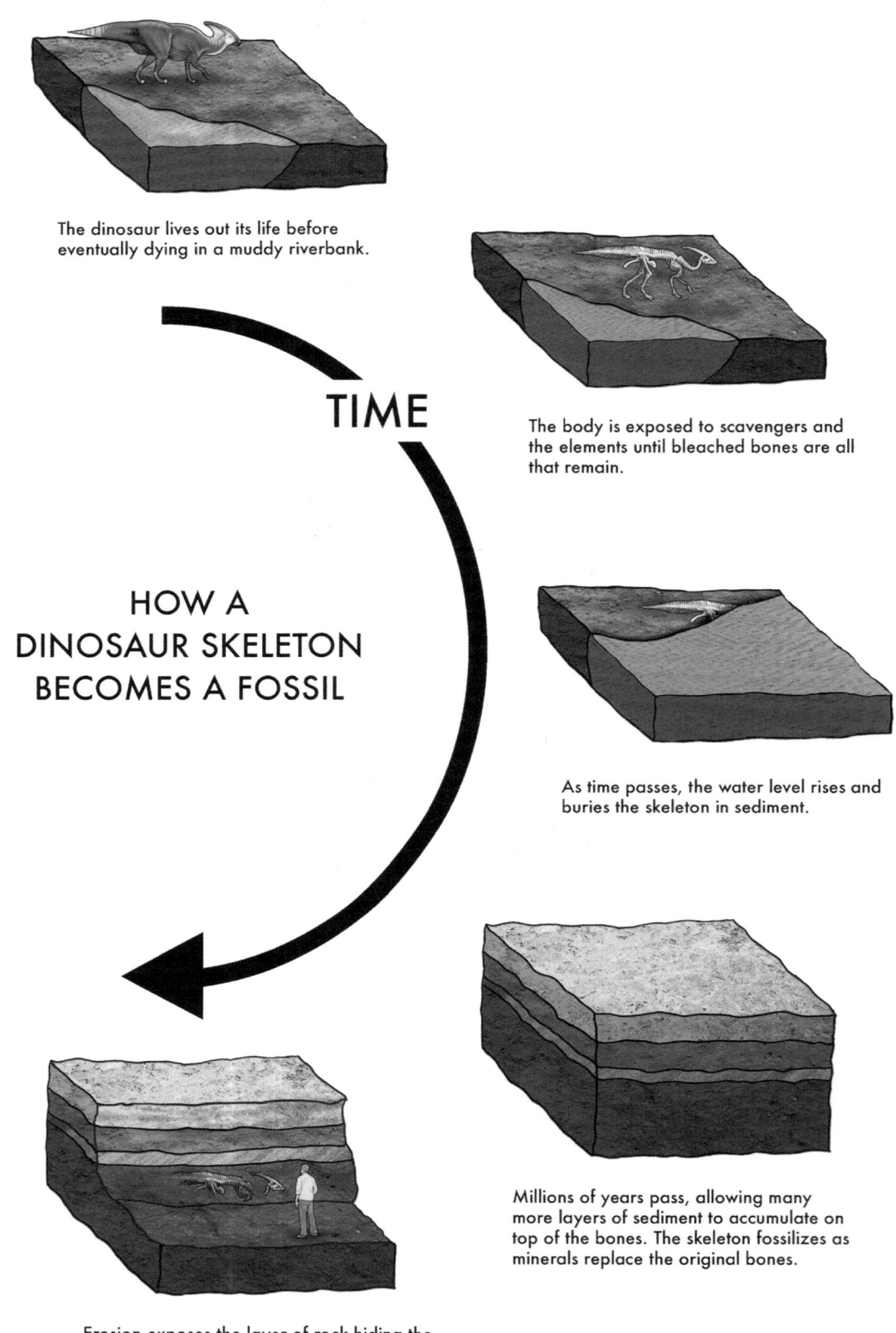

The dinosaur lives out its life before eventually dying in a muddy riverbank.

TIME

HOW A DINOSAUR SKELETON BECOMES A FOSSIL

The body is exposed to scavengers and the elements until bleached bones are all that remain.

As time passes, the water level rises and buries the skeleton in sediment.

Millions of years pass, allowing many more layers of sediment to accumulate on top of the bones. The skeleton fossilizes as minerals replace the original bones.

Erosion exposes the layer of rock hiding the skeleton. Lucky fossil hunters may now dig it out and begin their studies of the dinosaur.

Figure 1.1 How an animal, such as a dinosaur, becomes a fossil. As soon as the animal dies, its bones must survive being destroyed by scavengers, and it needs to be quickly buried in sand or mud. It then turns to stone and becomes a fossil as it is buried, as the groundwater seeps through it and deposits minerals that replace the actual bone. Finally, it must survive millions of years of heating and compression in the earth's crust, and then the rocks containing it must be uplifted and eroded and exposed in a dry badlands area over the last 200 years, where, if it's lucky, a collector might find it before it erodes away and is destroyed.

In addition to sands and gravels and muds, another common kind of sedimentary rock is known as limestone, and it is literally made of fossils—mostly the broken fragments of shells of sea creatures that lived millions of years ago. So, if you happen to be collecting in an area where limestones are common in the bedrock, fossils are everywhere. However, most may be highly fragmentary and not worth collecting.

There are two other classes of rock. Igneous rocks are formed by the cooling of magma, or molten rock, that comes up from the hot deep interior of the earth. This can happen when a volcano explodes and scatters volcanic ash across the landscape (as happened with Mt. St. Helens in 1980), or when lava flows out of an erupting volcano (as happens on Kilauea on the Big Island of Hawaii nearly every year). The magma might remain underground without ever erupting from a volcano, but instead cool in a deep magma chamber until it is a hard crystalline rock like granite. Either way, igneous rocks almost never preserve fossils. If the soft tissues of an animal or plant encounter hot magma, it usually incinerates or vaporizes without leaving a trace. Only in a few cases do volcanic ashes blown from long distance bury a creature and actually preserve it in some way.

The third class of rocks is known as metamorphic rocks. They are formed when igneous or sedimentary rocks descend deep into the earth's crust and are put under immense pressures and extremely hot temperatures. These conditions transform the original rock into a new rock with new minerals and a new fabric. Any remains of plants or animals are destroyed in this process, so there are no fossils to be found in metamorphic rocks (unless the rocks are just barely metamorphosed).

DATING FOSSILS

How old is your fossil? This is a question that is fundamental not only to identifying it, but also to knowing where to look. If you're looking in beds of the wrong age, you won't find the right kinds of fossils—or maybe no fossils at all.

There are two fundamental ways in which geologists and paleontologists determine the age of rocks and geologic events (**Figure 1.2**). The first method is by relative dating or relative age. In other words,

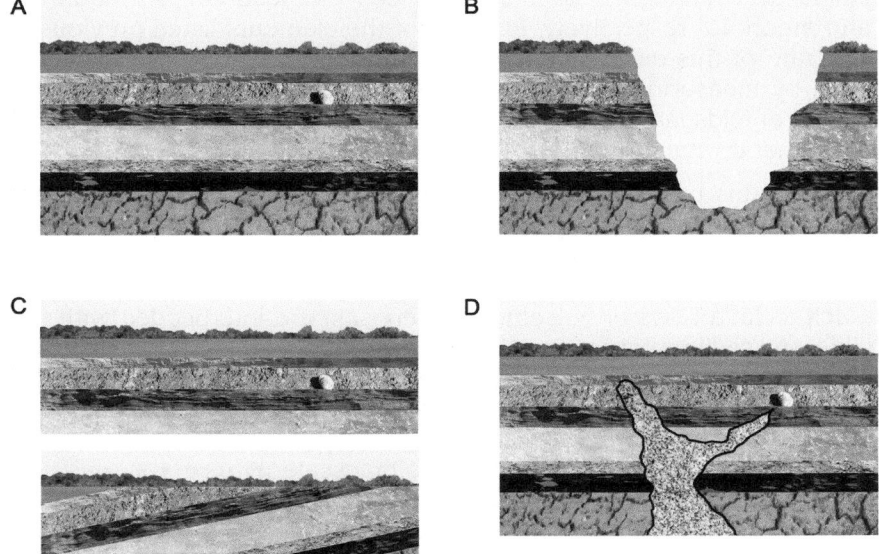

Figure 1.2 Steno's laws are used to determine the relative age of one rock body compared to another. (A) The principle of superposition says that the rocks near the top of a stack of layered sediments or lava flows are younger than those at the bottom of the stack. Thus, the top layer is the youngest and the bottom layer is the oldest. (B) The principle of original continuity says that rocks that match from one outcrop to another once connected, and have since been carved away by erosion. (C) The principle of original horizontality is based on the fact that rocks form in horizontal layers, so if you find them tilted or folded or faulted, then the deformation is younger than the rocks it deforms. (D) The principle of cross-cutting relationship says that when rock body (such as a dike of molten lava) or a fault cuts through another rock, then the material that cuts through is older than whatever it cuts.

geological event A is younger or older in relation to geologic event B. The primary way geologists do this is by using the principle of super-position first proposed by the Danish scholar Nicholas Steno in 1669. In any layered sequence of rocks (usually layered sedimentary rocks, although it applies to layered lava flows as well), the oldest rocks are at the bottom of the stack, and each layer above it is progressively younger (**Figure 1.2[A]**). In other words, the stack of rocks goes from older at the bottom to younger at the top. You can't stack one layer on top of another if the lower layer isn't already there first. A good analogy is the stack of papers on a messy desk or table. If they just keep accumulating through time without being turned over or sorted out, then the oldest papers will be at the bottom of the stack and the most recent ones will be at the top. Thus, if you are looking at the impressive pile of layers in the Grand Canyon, the oldest ones are always at the bottom and each layer above it is younger. They are like the pages in a book, with the first page at the bottom of the stack and the last at the top.

Another useful concept is the principle of cross-cutting relationships (**Figure 1.2[D]**). If a molten igneous rock intrudes into another rock (such an intrusion is usually called a "dike"), then the rock that does the intruding must be younger that the rocks that it cuts through. You can't cut through something if it isn't already there. Likewise, if a fault cuts through rocks, it must be younger than the rocks it cuts. The principles of relative dating not only go back to 1669, but also were in wide use when modern geology was born in about 1800–1830, and the geologic times-cale was born. The various names for the eras and periods and epochs of the geologic timescale are relative ages.

The other fundamental way to date rocks is known as numerical dating (formerly but incorrectly called "absolute dating" in older books). In other words, the date is given in number of years, or thousands of years or millions of years. Numerical dating is a young technique, only developed in the early twentieth century, and the most popular method, potassium-argon dating, has only been around since the 1950s.

Numerical dating is done by measuring the ticks of the radioactive "clock" in certain minerals. As minerals crystallize out of a magma, they trap radioactive elements such as uranium-238, uranium-235, rubidium-87, or potassium-40. These radioactive elements are naturally unstable, and spontaneously decay into different elements. As this decay proceeds over millions of years, the unstable radioactive parent atoms decay into a stable known daughter atom, such as lead-206, lead-207, strontium-87, and argon-40 (respectively, for each of the elements listed previously). The rate of this decay is known precisely for each of these elements; thus, by measuring the ratio of parent atoms to daughter atoms in a crystal of feldspar or mica or zircon, you can obtain the numerical date since that crystal formed.

Because this process only occurs in crystals that form from a molten rock, you can only date igneous rocks directly. What about sedimentary rocks, which contain the fossils? You cannot directly date them by radioactive minerals. Instead, we need to find places where igneous rocks (such as lava flows or volcanic ash deposits) are interbedded with fossiliferous sedimentary rocks. If a bed has Oligocene fossils ("Oligocene" is a relative age term), and the bottom of the bed has an ash dated 34 million years old, and the top of the bed has a lava flow dated 23 million years old, then we bracket the age of the Oligocene between 23 and 34 million years old. The entire geologic timescale (**Figure 1.3**) was constructed this way by finding fossiliferous sequences with fossil that gave well-determined relative ages and then using any and all available igneous rocks that are in right position to tell us the age.

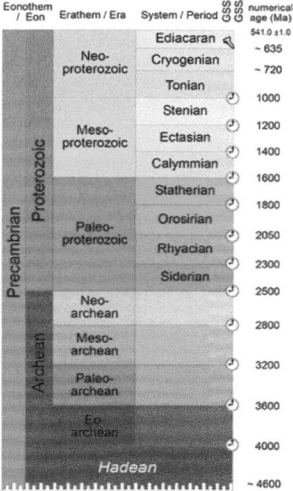
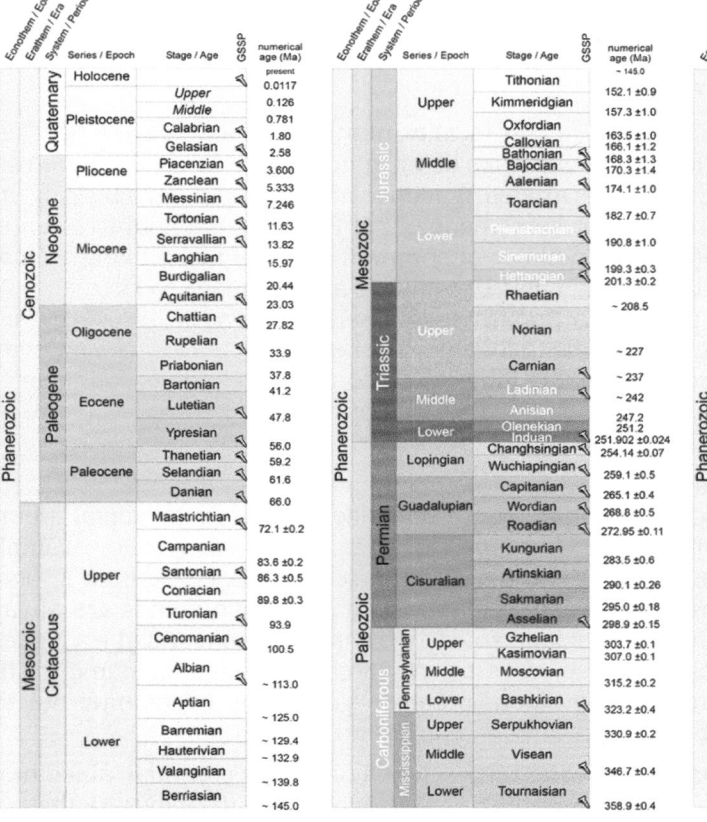
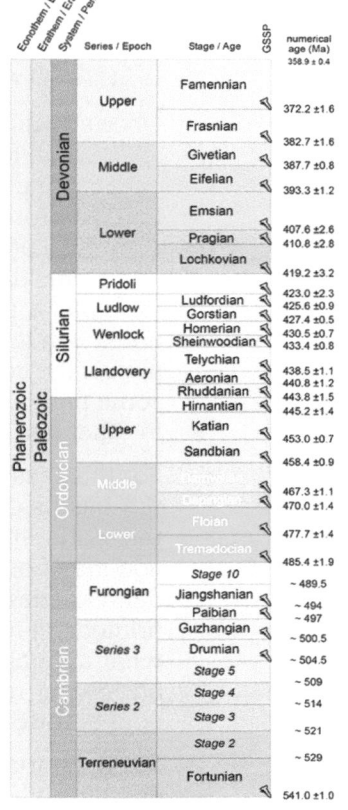

Figure 1.3 The modern geologic timescale.

There is one other radiometric system, known as radiocarbon dating, or carbon-14 dating. Unlike the other methods, you can date the fossil bones or shells or wood or any carbon-bearing substance directly, since you are measuring the decay of unstable radioactive carbon-14 incorporated into the organism before it died. However, the main drawback of this technique is that radiocarbon decays very rapidly. Half of the original carbon-14 parent atoms are gone in just over 5000 years, so the entire clock "runs down" in 60,000 to 80,000 years, and anything older than this cannot be dated by radiocarbon. The method is primarily used by archeologists who are interested in dating human bones and artifacts and by paleontologists studying the last Ice Age, which spanned the interval from 80,000 to 10,000 years ago. It is useless for anyone studying older fossils, since the clock has stopped for them. For this reason, you could never date a Mesozoic dinosaur bone using radiocarbon.

The principles of numerical dating have told us that the earth is immensely old. There are numerous meteorites and moon rocks that give ages of 4.6 billion years old, so that is how we estimate the age of the solar system. So far, the oldest earth rocks are only 4.28 billion years old, and the oldest earth minerals are 4.4 billion years old, so we have no earth rocks as old as the moon rocks or meteorites. But this is not surprising, because the earth's crust is constantly being mobilized and remodeled by plate tectonics, so we do not expect that any crust from the oldest earth could survive. The oldest known fossils are bacteria from South Africa and Australia about 3.4 to 3.5 billion years old, and there is organic carbon in rocks 3.8 billion years old that most scientists think is evidence of ancient life.

From these principles, we can reconstruct the sequence of events in any part of the earth. But to do this, we need to know the exact position in the sequence of rocks where each fossil was found. And if we are using lava flows or volcanic ash beds, they must be interbedded with fossils whose position is precisely known.

NAMING FOSSILS

Once a fossil is found, and its locality information recorded, the next step is determining what it was. Like living animals and plants, we assign scientific names to fossils as well. Scientific names may seem a bit long and hard to pronounce for some people, but they are essential to scientific communication. The popular or common name of many living animals and plants differs from culture to culture and language to language. For example, a peccary to English speakers is a *javelina* in Latin America, and a lion to us is *simba* to Swahili speakers. Even within the same language, the common name may not be consistent. If you say "gopher" in some parts of the United States, it means a small burrowing rodent, but in other parts it means a gopher tortoise.

For this reason, every organism (plant, animal, fungus, and even bacteria) has its own scientific name. Scientific names are universal around the world, no matter what language the scientist speaks. For example, you may not be able to read much of a scientific paper written in Mandarin Chinese or Russian Cyrillic, but the scientific names are always printed out in Roman script, so anyone can read them and at least guess what animal is the subject of the research. The scientific name for the burrowing rodent some people call a "gopher" is *Thomomys*, but the gopher tortoise is *Gopherus*, so there is no confusion.

For most fossils, knowing their scientific name is essential, since most don't even have a common name. You may know the saber-toothed cat by its English name, but it's different in other languages—yet to all scientists, it is *Smilodon*. Mammoths are familiar to us by that name, but in other languages they could be *mamut* in Spanish and *mammouth* in French, they are *Mammuthus* to a scientist. But nearly all other fossil animals and plants have no common name whatsoever, so there is no choice but to use their scientific names. You already know quite a few scientific names of prehistoric and living creatures. For example, everyone knows *Tyrannosaurus rex*, but that is its proper scientific name, and no other popular name exists. Nearly all the other dinosaur names you know, from *Brontosaurus* to *Velociraptor* to *Stegosaurus* to *Triceratops*, are scientific names as well.

All organisms on earth actually have a two-part (binomial) scientific name. The first part is the genus name (the plural is "genera", not "genuses"). It is always capitalized and either italicized (in print) or underlined (when handwritten). The names *Tyrannosaurus*, *Brontosaurus*, *Velociraptor*, *Stegosaurus*, and *Triceratops* in the preceding paragraph are all genus-level names or "generic names". But a genus typically includes a number of species. The species name (or "trivial name") is never capitalized (even if it came from a proper noun), but it is always underlined or italicized. Thus, *Tyrannosaurus rex* is a genus and species name; so is *Velociraptor mongoliensis*. Your scientific binomen is *Homo sapiens*, but there are other species of *Homo*, such as *Homo neanderthalensis*, *Homo erectus*, and *Homo habilis*.

Generic names are never used more than once in the animal kingdom, although there are a few cases of the same genus in plants and animals, but there is no likelihood of confusion between a plant and an animal. Species names, however, are used over and over again, so they cannot stand alone in a scientific paper. Thus, you can say *Tyrannosaurus rex* or

Homo sapiens, but not "rex" or "sapiens". You can abbreviate the genus name, so *T. rex* (but not "T-rex", as has become popular recently) or *H. sapiens* is acceptable.

Scientific names were originally based on Latin or Greek words, since in the early days of natural history, all scholars read and wrote in Latin or Greek as an international form of communication. Thus, most scientific names can be broken down to their original meaning. *Tyrannosaurus rex* means "king of the tyrant lizards" and *Homo sapiens* means "thinking human".

The criterion of Greek or Latin roots and Latinization of names has become more relaxed as fewer and fewer scientists learn the classical languages (the standard languages for all scholars less than a century ago), and much work is now being done in China, Japan, Russia, India, Latin America, and other less Western European-influenced scientific communities. Scientists have gotten more and more creative with their names, often to the point of silliness, or erecting names that are hard for others to use. For example, in 1963 mammalian paleontologist J. Reid Macdonald gave names based on the Lakota language to a number of specimens recovered from the Lakota Sioux reservation land near the old site of the Wounded Knee Massacre in South Dakota. Most non-Lakotans find them difficult to pronounce or spell. Try wrapping your tongue around *Ekgmowechashala* (iggi-moo-we-CHA-she-la), which means "little cat man" in Lakota. It is a very important specimen of one of the last fossil primates in North America, so it has gotten a lot of attention, and many people have struggled to pronounce its name. In the same publication, Macdonald also named *Kukusepasatanka*, a hippo-like anthracothere, *Sunkahetanka* for a primitive dog, and *Ekgmoiteptecela*, a saber-toothed carnivore. Then there is the transitional fossil between seals and their ancestors known as *Puijila*, which comes from the Inuktitut language of Greenland; you need to click on the website button (http://nature.ca/puijila/fb_an_e.cfm) to hear the correct pronunciation.

In Australia, there are many fossils that have names with Aboriginal roots, such as *Djalgaringa, Yingabalanaridae, Pilkipildridae, Yalkparidontidaem, Djarthia, Ekaltadeta, Yurlunggur, Namilamadeta, Ngapakaldia,* and *Djaludjiangi yadjana*. Some others include *Culmacanthus* ("culma" is Aboriginal for "spiny fish"), *Barameda* (Aboriginal for "fish trap"), and *Onychodus jandamarrai* after the Jandamarra Aboriginal freedom fighters. *Barwickia downunda* is named after Australian paleontologist Dick Barwick. *Wakiewakie* is an Australian fossil marsupial, supposedly named from the Australian way of waking up sleepy field crews in the morning.

About a century ago, an entomologist named George Willis Kirkaldy got a bit too creative naming different genera of "true bugs", or Hemiptera. He published the names *Peggichisme* (pronounced "peggy-KISS-me"), *Polychisme* for a group of stainer bugs, *Ochisme* and *Dolichisme* for two bedbugs, *Florichisme* for a plant hopper bug, *Marichisme, Nanichisme,* and *Elachisme* for seed bugs. For leaf hoppers and assassin bugs, Kirkaldy used male names such as *Alchisme, Zanchisme,* and *Isachisme*. In 1912, the Zoological Society of London officially condemned his naming practices, although so long as they were valid taxa, they could not abolish the names.

An entire website devoted to weird names (www.curioustaxonomy.net/) lists the gamut of odd inspirations, from puns to wordplay to palindromes that read the same way forward and backward. Some of the more clever names include the clams *Abra cadabra* and *Hunkydora*, the beetle *Agra vation*, the snails *Ba humbugi* and *Ittibittium* (related to the larger snail *Bittium*), the flies *Meomyia, Aha ha,* and *Pieza pi*, the wasps *Heerz tooya* and *Verae peculya*, the trilobite *Cindarella*, the Devonian fossil *Gluteus minimus*, the fossil carnivore *Daphoenus* (pronounced Da-FEE-nus) *demilo*, the fossil snake *Montypythonoides*, the Julius Caesar-influenced extinct lorikeet *Vini vidivici* and the water beetle *Ytu brutus*, and the Australian dinosaur *Ozraptor* (known as the "Lizard of Aus"). After a few too many beers, paleontologist Nicholas Longrich named a horned dinosaur *Mojoceratops*, because it had an elaborate heart-shaped frill that might have improved its "mojo" and its ability to attract mates. There is a Cretaceous lizard named *Cuttysarkus* (revealing the namer's preference for that brand of Scotch whisky), and a dog-like fossil mammal known as *Arfia*. The oldest known primate fossil is known as *Purgatorius*, not because the namer had some sort of religious point to make about humans, but because it was found in Purgatory Hill in the Hell Creek beds of Montana (suitably hellish in the summer time with their hot temperature and dangerous slopes). There are also fossils named after characters in *Star Wars* and *Lord of the Rings* and the *Harry Potter* series. Despite the musty reputation of taxonomists working away in dim museum basements, they certainly have a sense of humor!

Although taxonomic names sometimes attempt to describe the creature or give some idea of its main features, if the name becomes inappropriate it is still valid so long as no other senior synonyms are known. For example, the earliest known fossil whales were originally mistaken for large marine reptiles and named *Basilosaurus*, or "emperor lizard". Only later did scientists realize they were whales and mammals, not lizards, but the name is still valid even if it is inappropriate. In the 1920s scientists retrieved material of a bizarre predatory dinosaur from the Cretaceous of Mongolia and named it *Oviraptor* ("egg thief") from its proximity to nests of eggs they thought belonged to the most common dinosaur there, the horned dinosaur *Protoceratops*. But in the 1980s and 1990s, expeditions returned to Mongolia and found fossil skeletons of *Oviraptor* mothers brooding those same eggs, and the bones of unborn *Oviraptors* inside the eggs. The "egg thief" was actually the *parent* of the eggs, not a thief at all—but this slander to *Oviraptor* cannot be changed just because it's now inappropriate.

The species is the fundamental unit in nature, since it is species that evolve due to natural selection on populations within the species. The genus is a bit more arbitrary, depending upon the scientists' judgments as to which species cluster together. Genera are clustered into larger groups known as families. For example, our genus *Homo* belongs to the family Hominidae, along with other genera such as *Sahelanthropus*, *Ardipithecus*, *Paranthropus*, *Australopithecus*, and others. The dogs are all members of the family Canidae, the cats are Felidae, and the rhinoceroses are in the Rhinocerotidae. In the animal kingdom, all family names end with the suffix -idae, which is a quick clue when you encounter an unfamiliar name. (In the plant kingdom, families end in *-aceae*, so Rosaceae is the plant family that includes roses.)

Families are clustered into a larger group called an order (**Figure 1.4**). Humans, apes, monkeys, lemurs, and their relatives form the order Primates, while the order Carnivora includes most of the flesh-eating mammals including cats, dogs, bears, hyenas, weasels, raccoons, seals, and walruses. The rodents are an order (Rodentia), as are the rabbits (Lagomorpha), and most of the larger groups of mammals are orders. Orders are clustered into classes. Within the backboned animals, the families of mammals are clumped into class Mammalia, while the birds (class Aves), the Reptilia, the Amphibia, and so on are classes. Classes are clustered into a larger group called a phylum (plural is phyla). Vertebrates (animals with backbones) are members of the phylum Chordata, but there is a phylum Mollusca (molluscs, including clams, snails, squids,

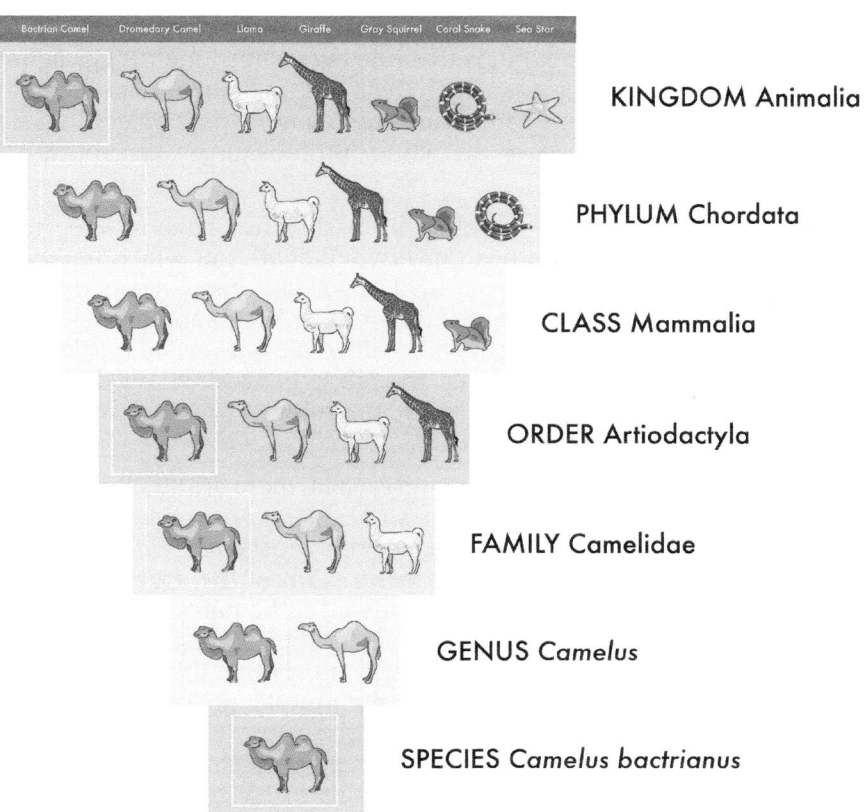

KINGDOM Animalia

PHYLUM Chordata

CLASS Mammalia

ORDER Artiodactyla

FAMILY Camelidae

GENUS *Camelus*

SPECIES *Camelus bactrianus*

Figure 1.4 The hierarchy of classification, showing how each rank or group is nested within a larger one. In this example, the genera of camels are all clustered into the family Camelidae, the order Artiodactyla, the class Mammalia, and the phylum Chordata.

and their relatives), the phylum Arthropoda (jointed segmented animals, including insects, spiders, scorpions, crustaceans, millipedes, trilobites, and many others), and so on. The highest rank of all is kingdom. We are members of the kingdom Animalia, but there are also kingdoms for the plants, the fungi, and so on.

Here is an example of how their hierarchy of groups within groups looks:

KINGDOM	Animalia	Animalia	Animalia
PHYLUM	Chordata	Chordata	Chordata
CLASS	Mammalia	Reptilia	Sarcopterygia
ORDER	Primates	Theropoda	Coelacanthiformes
FAMILY	Hominidae	Tyrannosauridae	Latimeriidae
GENUS	*Homo*	*Tyrannosaurus*	*Latimeria*
SPECIES	*Sapiens*	*rex*	*chalumnae*

There are strict rules for how organisms are named. They are specified in official rule books, such as the *International Code of Zoological Nomenclature* (ICZN), which used to exist only in bound printed copies, but now can be read online (http://iczn.org/code). There are similar codes for plants, fungi, and bacteria and protists. Most rules are important only to specialists who are about to give a new scientific name, but others are commonly encountered by anyone who follows fossils or biology, and are worth mentioning.

Most important is the Rule of Priority. The first name given to an organism is the only valid name (unless there are problems), no matter how unfamiliar or inappropriate it is. For example, many paleontologists regard the name "Brontosaurus" as invalid, because the same paleontologist, O.C. Marsh, who named that fossil had already given the name *Apatosaurus* to another specimen of the same animal earlier. Thus,

Apatosaurus is the proper senior synonym of "Brontosaurus", and paleontologists have been bound by this rule ever since Elmer Riggs figured it out in 1903. No matter how much the public is familiar with the name "Brontosaurus", scientists cannot use that name. (Some paleontologists have recently tried to revive the name "Brontosaurus", but this is still controversial.)

When paleontologists are working on fossils, they have to keep track of all the names that have been given, and figure out which name has priority; the other later names are known as junior synonyms. This is true, even if the senior name turns out to be inappropriate. As we already mentioned, the first known fossil whale was called *Basilosaurus* ("emperor lizard" in Greek), even though later work showed it was a whale and mammal, not a reptile. By the rules of priority, *Basilosaurus* must stand no matter what it means.

In addition to rules about which name is valid, there are strict rules about creating new names for new species or genera. For the last century, a new scientific name must include a clear diagnosis of how to tell it apart from other similar species, a good description of the specimens, good illustrations, a list of specimens considered to be part of the species, a type specimen that is the physical basis for the species, the geographic range and time range of the species, and many other things. All of these must be published in a reputable scientific journal, not on a web page or unpublished dissertation or somewhere else. Otherwise, the new name of a genus or species is not valid, and other scientists will not recognize or use it. A scientist cannot name a genus or species after himself or herself, but they can name it after someone else, and have that person return the favor on a different fossil.

These rules may seem boring and excessively legalistic, but they are essential to maintain order and stability in scientific names. Scientists agreed to these rules over a century ago to prevent pointless arguments about whose name for an organism is right. All other scientists (and especially the scientific journals) follow these rules, and will not publish any work that violates them. It's like knowing the rules of the road before you take your driving test. The Department of Motor Vehicles, and all other drivers, must assume that you know the proper rules for driving, because they don't want to get into a deadly accident if you suddenly break the rules. Thus, we have many cases where amateur fossil collectors try to create new names, or even publish them in books and websites, but without following the rules properly. The rule book allows the professional scientists to quickly determine who is right, and who is not, and whose work deserves attention and whose work should be ignored.

HOW DO WE CLASSIFY ANIMALS?

We've established how we give names to animals, but how do we decide how to classify them? There are lots of ways that people might classify things. We could sort them into categories like "good to eat" and "toxic and bad tasting", or "dangerous to humans" and "not dangerous" (as some cultures do). We could cluster them by color patterns, or where they live, or how they behave. The science of classification is called taxonomy, and any group of organisms (a genus, a species, a family) can be called a taxon.

Before the time of Linnaeus, many natural historians realized that the best way to classify creatures was by unique anatomical specializations that distinguish them from other similar creatures. Some classifications clustered creatures like fish and dolphins together because they were aquatic, or turtles and armadillos because they had a hard shell. By Linnaeus' time, scholars began to realize that the overwhelming number of anatomical specializations clump some animals together and not others.

For example, fish and whales have superficial similarities because they are both swimmers with streamlined bodies and a tail fin, but if you look past these ecological overprints, you find that every other anatomical feature of fish and whales are completely different. This is called evolutionary convergence, and it has occurred often in the history of life. As we shall discuss in Chapter 21, the pouched marsupial mammals of Australia have converged in body form on many of the placental mammals of the rest of the world, since they evolved in isolation in Australia and did not encounter competition from placentals. Saber-toothed predators evolved at least four independent times in mammalian history, including once in pouched mammals, once in an extinct group called creodonts, and twice in the order Carnivora (once in the "false cats" or nimravids, and once in the true cats, or felids).

To get away from this problem of convergence, we try to find characteristics that are unique specializations for the group of animals we are classifying, not features left over from their remote past. These are known as shared derived characters, or synapomorphies. For example, the order Primates is distinguished by having grasping hands and feet, nails instead of claws, forward-pointing eyes with binocular vision, and good color vision. However, groups within the Primates are defined by their own specializations, so apes and humans (family Hominidae) share anatomical features such as the loss of a tail, complex nasal sinuses, five or six vertebrae in the hip, elongated middle finger, and another dozen features in just the skeleton. We would not use the occurrence of grasping hands to define the apes and humans, because for them it is a shared primitive character, or symplesiomorphy; it is only useful to distinguish primates from other mammals, but not groups within the primates. Nor would we use a very primitive feature, such as the occurrence of four limbs. That is a shared primitive feature that apes and humans inherited from the first amphibians, and not useful to defining the Hominidae. Nor would we bother to diagnose hominids by the presence of a backbone, which we inherited from the earliest vertebrates.

This emphasis on basing classification on shared derived features, or shared evolutionary novelties, makes classification a reflection of the evolutionary branching history of life. This was apparent when Linnaeus' 1758 classification showed branching pattern like the bushy "tree of life", but it was a century later that Charles Darwin pointed out the pattern of classification was evidence for evolution. Since these days, biologists have argued over the best way to classify organisms. Traditionally, some groups of animals were clustered together based on a mixture of both shared evolutionary novelties plus primitive features. For example, the category "fish" is useful to anglers and grocery stores and restaurants and fishmongers, but it has no meaning as a taxonomic group of animals. Some "fish", such as lungfish and coelacanths, are actually part of the lineage that leads to land vertebrates. Others, such as lampreys and sharks are very primitive creatures not at all closely related to the bony fish, such as a tuna or a goldfish. Some people talk about "jellyfish" or "starfish" or "shellfish" or "crayfish", using the name "fish" broadly to include *any* animal that lives in water, no matter what they're related to. Thus, over the past 40 years, biologists have been trying to avoid groupings of animals that are unnatural wastebaskets based on primitive similarities (like the aquatic body form of "fish").

This has long been a problem in the vertebrates. For example, many of the orders of mammals were once huge wastebaskets of creatures united only by shared primitive similarity. For a long time, the "order Insectivora" was such a wastebasket, including three kinds of mammals which were actually interrelated (shrews, moles, hedgehogs) and a wide spectrum of other creatures that were insect-eaters but not closely related, including tree shrews, elephant shrews, golden moles, tenrecs, and several others. As we shall discuss in Chapters 21 and 22, this wastebasket has long since been broken up, and only the original three groups are clustered any more. Another wastebasket was the "Condylarthra", long used for any extinct hoofed mammal that wasn't a member of a living group of hoofed mammals (see Chapter 23). Such "wastebasket groups" are not only unnatural, but as my coauthors and I found when we sorted out the true relationships of their members, "Condylarthra" was covering up and obscuring what we didn't know and hid the problems that needed to be solved. Once the "condylarths" were abandoned as a meaningless group, we made great strides in figuring out how all the hoofed mammals were interrelated.

Another problem with classification as we tease out the branching points of the tree is that now there are more splits that need names than we have names for. As an example, if we cluster mammals into class Mammalia, and treat each order as a separate group, we find that there are lots of branching points between them. The living Mammalia first splits into three groups, the monotremes (platypus and echidnas), marsupials (pouched mammals like kangaroos and opossums), and placentals (mammals that give birth to live, developed young). Are those subclasses? Then we have numerous splits within the subclass Eutheria (placentals) before we get down to the rank of order. We can use ranks such as "infraclass" and "superorder", but quickly we run out of ranks between superorder and order, and between subclass and infraclass. Consequently, the traditional ranks of classification are receiving less and less emphasis now, and there are lots of ways of showing the branching pattern of evolution without creating formal ranks for each evolutionary branch point.

During the 1960s through the 1980s, there were a number of breakthroughs in thinking about classification and how to do it. Most taxonomists came to agree with the emphasis on shared evolutionary specializations and avoiding wastebasket groups based on shared primitive similarity, and the idea that classification should reflect the evolutionary branching sequence and nothing else. A group that includes all the descendants of a common ancestor is known as a monophyletic group, or a natural group. This has been difficult for people accustomed to the traditional groups of animals. One of the main ideas is that if classification reflects evolutionary branching history and nothing else, then each monophyletic group should include all its descendants within it. Otherwise, it is an unnatural, arbitrary paraphyletic group (**Figure 1.5**). Thus, some biologists were scandalized when it became clear that birds evolved from a subgroup of dinosaurs resembling *Velociraptor*, and thus birds are descended from dinosaurs. To the modern taxonomist, birds are a group within dinosaurs, not a separate class Aves distinguished from the class Reptilia as two parallel groups of equal rank. If you don't put birds within Reptilia, then "reptiles" become a wastebasket group for all land vertebrates that are not birds. Likewise, we use "Amphibia" to talk about salamanders and frogs, but Reptilia are all descended from lineage within the "Amphibia" and thus a subgroup of them—or "Amphibia" becomes another unnatural wastebasket. Modern classification is gradually abandoning these ancient "wastebasket" categories that are well known but not natural by using a new set of names that are defined only by their shared evolutionary specializations. Thus, the Tetrapoda (four-legged vertebrates) includes all amphibians, reptiles, and their descendants. The Amniota

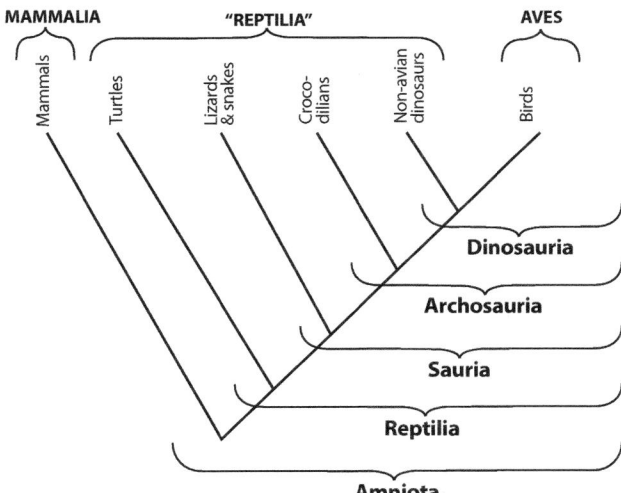

Figure 1.5 Different ways of classifying the same group of organisms. Traditional classifications (top) prefer to emphasize the tremendous evolutionary radiation of birds and mammals and place them in their own classes, equal in rank to the rest of the amniotes, lumped in the paraphyletic "Reptilia". A cladistic classification (bottom) does not permit mixing of phylogeny and other factors such as evolutionary divergence. Instead, every group is monophyletic and defined strictly by evolutionary branching. In this view, birds are a subgroup of dinosaurs, archosaurs, saurians, and reptiles.

(tetrapods that lay land eggs) includes all reptiles, birds, and mammals. Gradually, obsolete concepts of the old names "Reptilia" and "Amphibia" and "invertebrates", which are clumps of creatures defined on shared primitive specializations and not including all their descendants, are vanishing from biology (even if the public still doesn't know it).

But how are all the major orders of vertebrates interrelated? Since the beginning, the primary line of evidence was their anatomy, especially the bones and teeth, although the entire anatomy of every biological system (muscles, nerves, brains, and every other soft tissue) can be used. This evidence was first studied in detail by natural historians in the 1800s, who noticed many anatomical features that we still use to classify vertebrates. Yet there were still lots of problems, and issues with groups of mammals that had little clear evidence as to who their closest relatives were.

In the 1980s and 1990s, however, the emphasis on shared evolutionary novelties and only natural monophyletic groups led to great strides in understanding vertebrate relationships. The phylogeny of vertebrates based on anatomical features had more or less reached consensus.

At the same time, however, a new data set emerged from material that was not known to Linnaeus or Darwin or any biologists until the 1960s. These were data from molecular biology, especially the detailed sequence of biochemicals (amino acids in proteins, nucleotides in DNA) making up the genes and proteins found in any living animal. Sure enough, these data also produced a branching sequence that closely matched the evolutionary pattern deciphered from the external anatomy, confirming that both kinds of data carried an original signal. In most cases, the branching sequences produced by anatomy closely matched the molecular branching sequence. We will talk about some of the exceptions in later chapters.

FURTHER READING

Ager, D.V. 1981. *The Nature of the Stratigraphical Record* (2nd ed.). Wiley, New York.

Behrensmeyer, A.K.; Hill, A., eds. 1980. *Fossils in the Making*. University of Chicago Press, Chicago.

Cracraft, J.; Donoghue, M.J., eds. 2004. *Assembling the Tree of Life*. Oxford University Press, New York.

Donovan, S.K., ed. 1991. *The Processes of Fossilization*. Columbia University Press, New York.

Donovan, S.K.; Paul, C.R.C. 1998. *The Adequacy of the Fossil Record*. Wiley, New York.

Faul, H.; Faul, C. 1983. *It Began with a Stone*. Wiley, New York.

Hennig, W. 1966. *Phylogenetic Systematics*. University Illinois Press, Urbana, IL.

Prothero, D.R. 1990. *Interpreting the Stratigraphic Record*. Freeman, New York.

Prothero, D.R. 2013. *Bringing Fossils to Life: An Introduction to Paleobiology* (3rd ed.). Columbia University Press, New York, 704 pp.

Prothero, D.R. 2020. *Fantastic Fossils! A Paleontologist's Companion*. Columbia University Press, New York, 321 pp.

Prothero, D.R.; Schwab, F. 2003. *Sedimentary Geology* (2nd ed.). W.H. Freeman, New York, 557 pp.

Prothero, D.R.; Schoch, R.M., eds. 1994. *Major Features of Vertebrate Evolution*. Paleontological Society Short Courses in Paleontology, no. 7. Paleontological Society, Lawrence, KS, 270 pp.

Rudwick, M.J.S. 1972. *The Meaning of Fossils: Episodes in the History of Palaeontology*. Macdonald, London.

Shipman, P. 1981. *Life History of a Fossil*. Harvard University Press, Cambridge.

Wiley, E.O.; Liebermann, B.S. 2011. *Phylogenetics: Theory and Practice of Phylogenetic Systematics*. Wiley-Blackwell, New York.

Wiley, E.O.; Siegel-Causey, D.; Brooks D.; Funk, D. 1991. *The Compleat Cladist, a Primer of Phylogenetic Procedures*. University of Kansas Museum of Natural History Special Publication 19, Lawrence, KS.

THE ORIGIN OF VERTEBRATES

2

Most species do their own evolving, making it up as they go along, which is the way Nature intended. And this is all very natural and organic and in tune with mysterious cycles of the cosmos, which believes that there's nothing like millions of years of really frustrating trial and error to give a species moral fiber and, in some cases, backbone.

—Terry Pratchett (1991) *Reaper Man*

WHAT IS A VERTEBRATE?

What do we mean when we say an animal is a "vertebrate"? Why should we be interested in them? Certainly, vertebrate paleontology is in the news all the time thanks to dinosaurs. Dinosaur movies and TV shows and paraphernalia are a huge business, and this aspect of paleontology has a high public profile. Many paleontologists first got hooked (as did I) when they succumbed to dinomania as children and parlayed that childhood interest into their life's work. Sadly, of the millions of dollars of profits made on dinosaur paraphernalia every year, none of it actually supports the research that makes it all possible.

Nevertheless, an understanding of the broad features of all of vertebrate evolution is a worthwhile goal. For one thing, in addition to cool creatures like dinosaurs, we humans are also vertebrates, and we want to understand our roots and where we came from. Although vertebrates are not as diverse or numerically abundant as molluscs or arthropods (there are about 45,000 living species of vertebrates, only a few percent of all the animal species on Earth, compared to millions of species of insects), the larger body sizes and sophisticated adaptations of vertebrates give them a dominant ecological role on both the land and sea, especially in the higher levels of the food pyramid. With their incredible ecological diversity, vertebrates occupy the deepest oceanic waters, cover the land, and reign in the air. Today, one species of vertebrate (*Homo sapiens*) has completely changed the face of the planet, wiping out thousands of other species of vertebrates and invertebrates and plants, while causing the proliferation of certain other vertebrate species, such as cattle, pigs, chickens, rats, and pigeons. Humans might not survive much longer on this planet, but some vertebrates, such as rats, will probably persist as long as the cockroaches and bacteria.

Where do vertebrates come from? First, we must define what we mean by "chordate" or "vertebrate". Vertebrates are a group of animals that have a number of unique specializations, including a tissue called bone, plus red blood cells, a thyroid gland, and a backbone made of numerous bony or cartilaginous segments (the vertebrae). All mammals, birds, reptiles, amphibians, bony fish, sharks, and lampreys are vertebrates. The phylum Chordata includes vertebrates and several related groups that have certain unique features in common (**Figure 2.1**). The name "chordate"

DOI: 10.1201/9781003128205-2

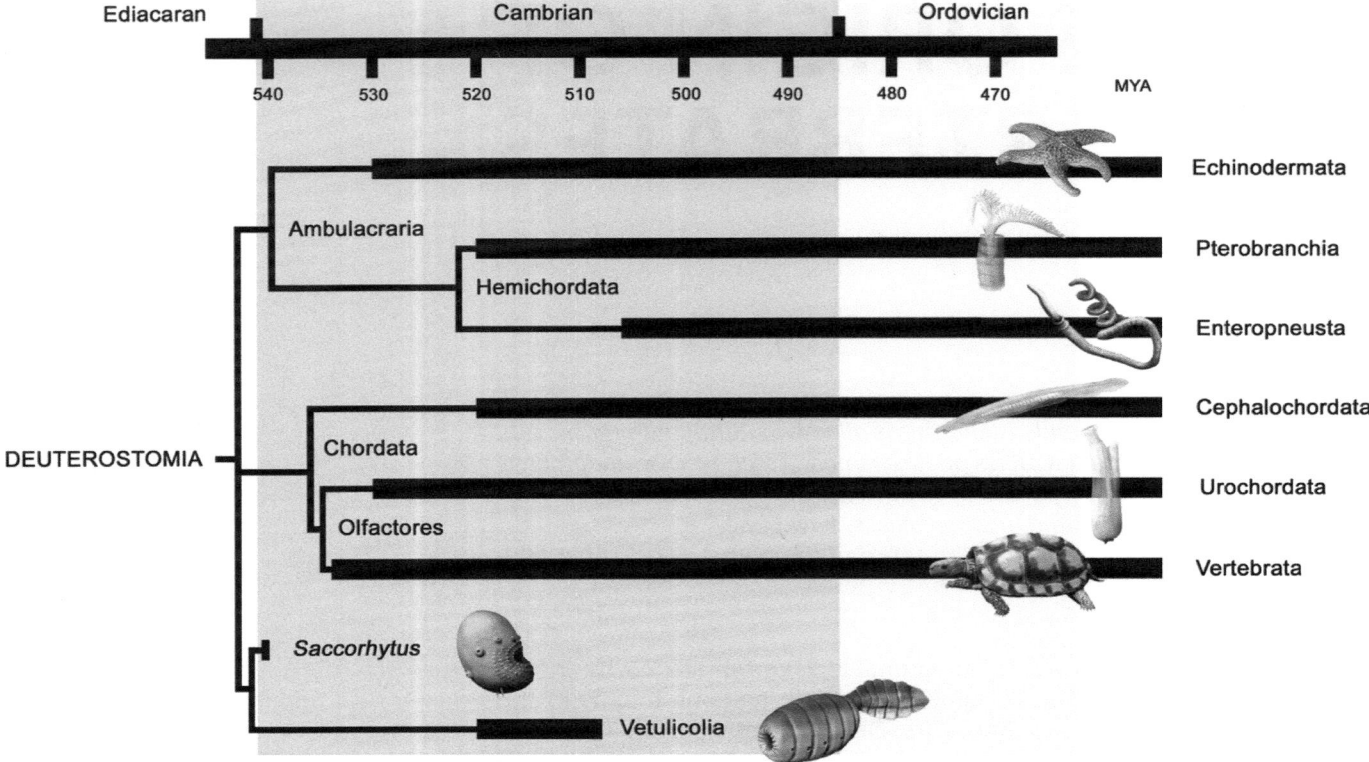

Figure 2.1 Relationships of the chordates and their relatives among the deuterostomes, such as the echinoderms, hemichordates, urochordates, cephalochordates, and other close relatives.

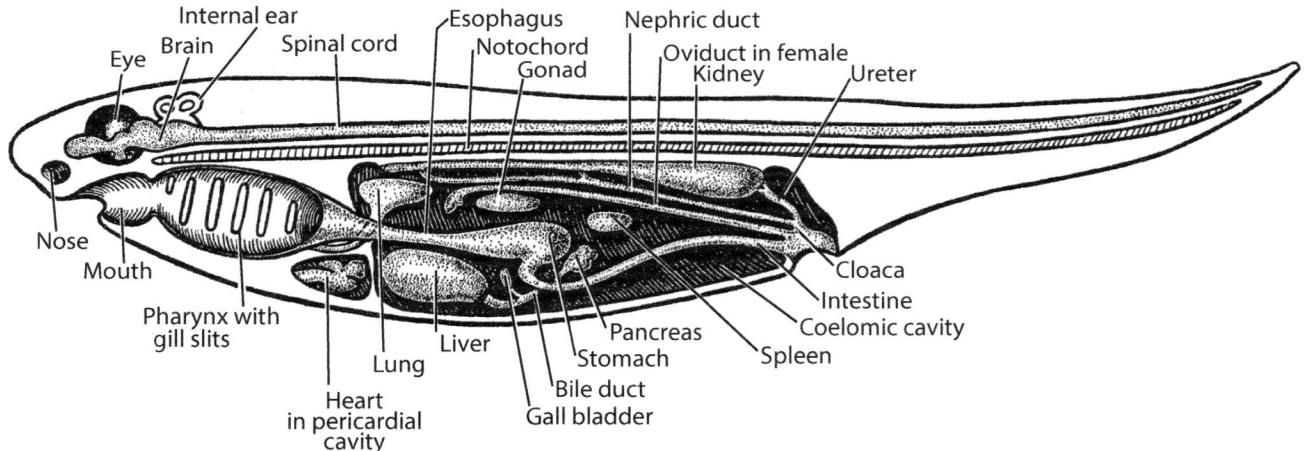

Figure 2.2 Diagrammatic section through the idealized chordate body plan, showing the relative position of the major organs. (Redrawn from several sources.)

refers to the flexible rod of cartilage, or notochord, along the back that serves for support (**Figure 2.2**). This rod transforms into the backbone of bony vertebrae in the adults of most vertebrates. However, the primitive relatives of vertebrates have no bony spine, only a notochord, and all vertebrate embryos (including you) have a notochord in the early stages of development before it was replaced by the bony spinal column. This stiffening rod is important not only to support the elongate body but also for swimming. When the muscles along the body contract, they pull on the notochord and cause it to flex from side to side, resulting in an efficient swimming motion. If there were no notochord, the contraction of these muscles would telescope the body and cause it to collapse.

Just above the notochord is a nerve cord in the dorsal (along the back) position. By contrast, annelids, arthropods, and molluscs have a nerve cord that runs in a ventral (along the belly) position (**Figure 2.2**). The front end (anterior) of most chordates usually has a concentration of sense organs surrounding a cartilaginous or bony structure, the braincase, that encloses and protects the brain. The mouth is in the front, and passes into a "throat" region (the pharynx). In primitive vertebrates, the pharynx is muscular and serves to pump food particles through the digestive tract and to pump oxygenated water past the gills. The rest of the digestive tract may have a differentiated stomach, intestines, or other organs, and ends in an anus that is behind the midpoint of the body. Behind the anus, the back end (posterior) of the body is usually composed of a muscular tail, composed mostly of V-shaped muscle masses known as myomeres, which propel the body in a tadpole-like fashion. By contrast, many worms and other creatures have their anus at the very end of their bodies, and do not have a muscular tail behind the anus. The main vessels of the circulatory system run ventrally (along the belly) of the chordate body plan, in contrast to the dorsal heart and circulatory system of annelids, arthropods, and molluscs. The internal body cavity of the chordate contains not only the digestive and respiratory system, but also excretory and reproductive organs, and frequently, many other organ systems as well.

OUR KINFOLK IN THE SEA

Although most vertebrates are characterized by a hard, bony skeleton and have a decent chance of fossilization, the earliest vertebrates and their ancestors were soft-bodied, and therefore rarely fossilized. If they had any rigid tissues at all, it was cartilage, which is also difficult to fossilize. For this reason, the primary approach to understanding vertebrate origins has been to look at all the closest living relatives of the vertebrates and their embryonic history, and then to try to place the few available fossils in this context. Three important groups of living animals give us examples of the steps in the evolution of the basic chordate body plan (**Figure 2.1**). They are the hemichordates (pterobranchs, "acorn worms", and the extinct graptolites), the urochordates (the tunicates or "sea squirts"), and the cephalochordates (the amphioxus or lancelets). All of these groups, plus the echinoderms (sea stars, sea urchins, brittle stars, sea cucumbers) are known as deuterostomes because they have several distinctive features of their embryology found in no other animal group. Recent analysis of the molecular sequences of animals has confirmed that all deuterostomes (echinoderms, hemichordates, and chordates) are very closely related to each other.

Of all the animals in the sea, some of our closest relatives are the echinoderms, plus a group called the phylum Hemichordata (Greek: "half-chordates"). Hemichordates are in a different phylum from the Chordata, because they do not have the diagnostic notochord. Instead, they have a tubular structure, the stomochord, in the position of the notochord that is actually derived from a pouch off the digestive tract and is thought to be equivalent to the embryonic precursor of the notochord. Hemichordates do have a few specialized chordate features, including a pharynx with multiple gill openings, a dorsal nerve cord, and ventral blood vessels.

Today, the hemichordates are represented by about 90 species in two classes of invertebrates that look as different from vertebrates as could possibly be imagined. One group, class Enteropneusta (about 80 species), is known as the "acorn worms" (**Figure 2.3[E,F]**), because they have a worm-like body, with a proboscis on the front that is used for burrowing and for trapping food particles. Mucus flows along the proboscis, capturing the food and transporting it to a collar-like structure that ingests the good stuff and rejects the rest. Most acorn worms are just a few centimeters long, although some are as long as 2.5 m. Acorn worms live in

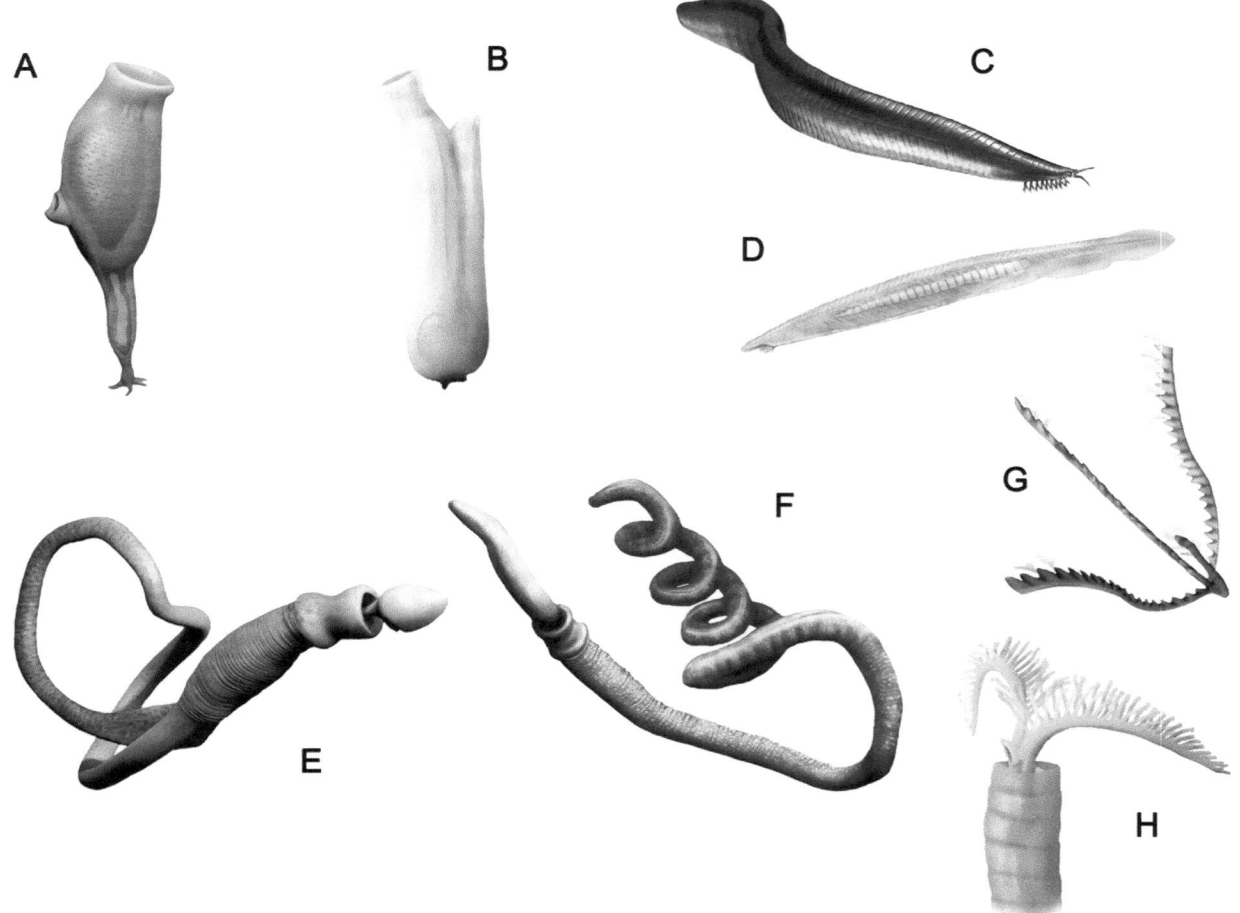

Figure 2.3 Early deuterostomes (chordates, echinoderms, and relatives) are represented here with their modern counterparts. (A) *Shankouclava anningense* from the Lower Cambrian Maotianshan Shale, China, is the earliest known definite tunicate to which the sea squirts (B) are modern representatives. (C) *Pikaia gracilens* from the Middle Cambrian Burgess Shale was possibly a cephalochordate, of which the lancelets (D) are the only modern surviving representatives. (E) *Spartobranchus tenuis*, also from the Middle Cambrian Burgess Shale, is the oldest known fossil of the acorn worms (enteropneusts) (F). A modern acorn worm. (G) The extinct graptolites are very similar is the structure of their branches and cup-like structures which held living animal, very much like the modern pterobranch *Rhabdopleura* (H).

U-shaped burrows in shallow marine waters, but they are not common members of the seafloor community. Most people would never be able to distinguish them from any other marine worm, but to the astute eye, there is an important clue: multiple pharyngeal gill openings just behind the collar, showing that they are relatives of the chordates. In addition, they have a dorsal nerve cord and ventral blood vessels, also chordate features. Their elongate, bilaterally symmetrical body is also similar to that of the chordate body plan and very different from the radial symmetry of the other main group of deuterostomes, the echinoderms.

As hard as it is to imagine an acorn worm as our cousin, the other group of hemichordates, the class Pterobranchia (about 10 living species in 3 genera) is even less like vertebrates. As adults (**Figure 2.3H**), they are tiny colonial filter-feeding animals, with a fan of tentacles for catching food particles, which are then processed by the small proboscis and collar before they enter the mouth. Pterobranchs still possess pharyngeal gill slits, but almost all other similarities to chordates have been lost. Living pterobranchs like *Rhabdopleura* form large colonies of multiple individuals, each secreting a long, segmented tube of organic matter with a very distinctive structure. Each individual animal has a long tube called a stolon that connects it with the rest of the animals in the colony.

Another important group related to pterobranchs were the graptolites, a mystery fossil that wasn't figured out for more than 150 years. First discovered in the late 1700s from flattened graphite specimens on lower Paleozoic deep-sea shales, they looked more like pencil marks on slates than real fossils (**Figure 2.4[A]**). For over a century no one could tell what kind animal they came from. Nevertheless, they proved to be one of the best index fossils of the Ordovician and Silurian and Early Devonian, because they evolved rapidly and are found in both shallow marine and deep marine sedimentary rocks, allowing correlation across oceans and continents. Eventually in the 1940s, scientists found uncrushed three-dimensional specimens preserved in limestones and especially cherts. When these were carefully sliced into numerous parallel section (like slices of a loaf of bread, but extremely thin), or etched out of the limestone using acid,

A

B

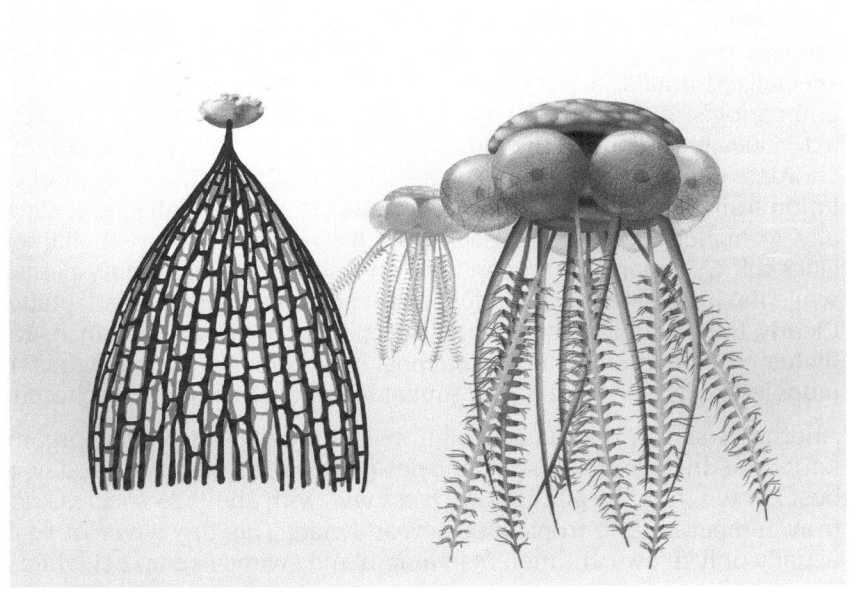

Figure 2.4 (A) Graptolite fossils tend to look like little graphite carbon films on black shales, and for a long time no one knew what they were related to or what they looked like in life. (B) A reconstruction of graptolites in life, with the filter-feeding organisms on the branches (stipes) hanging down from a flotation device above.
[(A) Courtesy Wikimedia Commons.
(B) Drawing by Mary Persis Williams.]

their 3D structure could be obtained. Lo and behold, the detailed structure of the little cups on each branch were built exactly like those of living pterobranchs! Apparently, graptolites lived in big colonies which floated across the surface of the Ordovician and Silurian seas, attached to floating objects or held up by air-filled floats on the colony (**Figure 2.4[B]**). They made a living trapping tiny planktonic prey with their tentacles, and kept growing and expanding the little branches of each colony. When they died, graptolites sank to the bottom of the sea in both shallow and deep water, and made it possible to correlate deep-marine shales with shallow-marine sandstones around the world's oceans during the Ordovician and Silurian.

Hemichordates such as the acorn worms, pterobranchs, and graptolites are not considered chordates, because they lack a notochord. The molecular evidence suggests that hemichordates are more closely related to echinoderms than chordates. However, there are several organisms in the ocean today that have the full complement of chordate features, yet they are still a long way from being vertebrates. These are known as the tunicates, ascidians, or "sea squirts", and they are placed within the phylum Chordata as the subphylum Urochordata. The name "sea squirt" is an undignified but apt description for our close relatives. As adults, they are soft little sacs of jelly that pump water in through a "chimney" at the top, filter it through a basket-like pharynx, into the surrounding cavity, and then expel it from a little tube on the side (**Figure 2.3**). They have a flexible outer body sheath called a tunic, whence they get the name "tunicates". Although this simple body plan may not seem impressive, they are a very successful group, with over 2000 living species. Some are so small and often translucent, however, so that most beachcombers and divers never even see them. Others, like the colonial pelagic group known as salps, form a translucent and bioluminescent colony of thousands of individuals that reach a length of several meters.

How could such a humble little sac of jelly be related to us, or any other chordate? As we have mentioned, the presence of the pharynx in adults is one clue. Better evidence, however, can be seen in their tadpole-like larvae (**Figure 2.5**), which have not only a pharyngeal basket, but also a notochord, dorsal nerve cord, and muscular tail with paired myomeric muscles. After the gametes form a fertilized egg and then a multicellular larva, this larva swims for a few hours to a few days, trying to find a hard surface. When it does, the adhesive papillae at the front end attach, and within 5 min the tail begins to degenerate and the notochord disappears. About 18 hours later, the metamorphosis is nearly complete, and the body has reduced to a simple sac filled with a pharyngeal basket.

This remarkable metamorphosis is a classic case of embryonic development providing important evolutionary clues that are lost in highly specialized adults, a trend that we will see again and again. In 1928, embryologist Walter Garstang first suggested that retention of larval characteristics into reproductive adulthood may have been very important in chordate evolution. He visualized a sequence (**Figure 2.5**) of steps of evolution from echinoderms to hemichordates to urochordates, and eventually, to higher chordates. In each step, the retention of larval characteristics allows organisms to continue on this main evolutionary pathway, while the highly specialized adults branched off in their own adaptations. Clearly, the odd adult tunicate body form could not have been ancestral to higher chordates, so the next step must have been the persistence of their tadpole-like larva, which is very similar to many other primitive chordates.

Another stage in chordate evolution is a small, soft-bodied organism known as the amphioxus or lancelet (**Figures 2.3** and **2.5**). Today, it is best known from the genus *Branchiostoma*, with about 25 species known from temperate and tropical seas worldwide. This tiny sliver of flesh is usually only a few centimeters in length and swims like an eel while it is

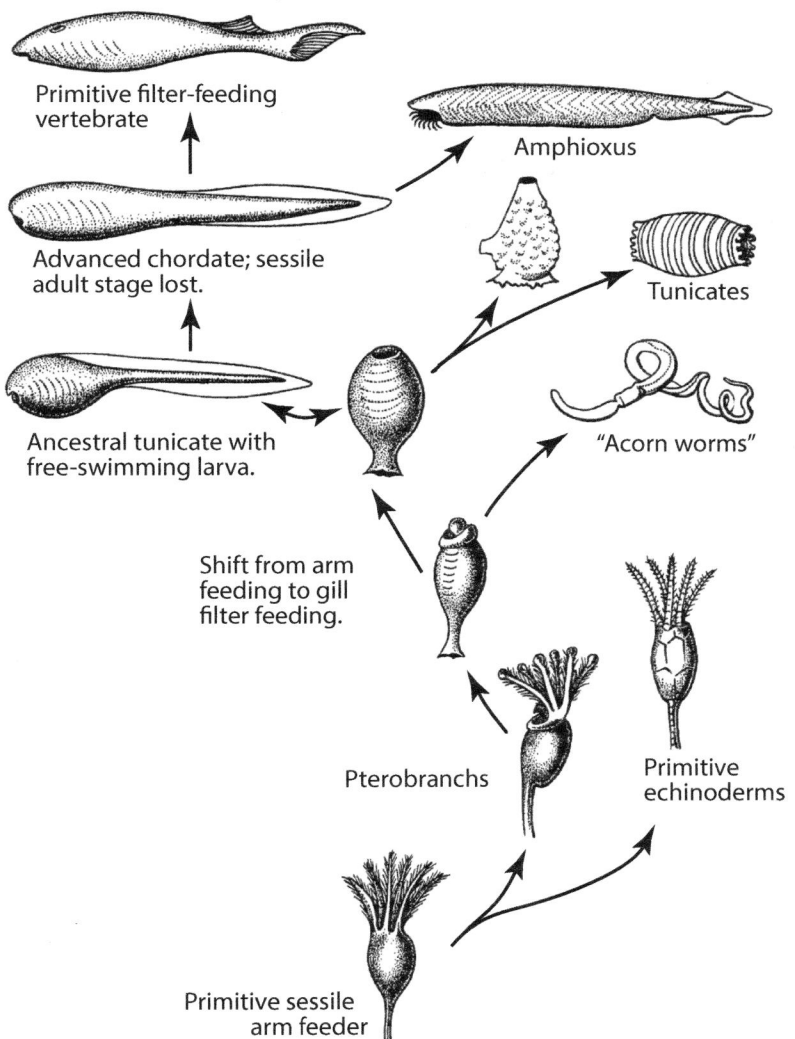

Primitive filter-feeding vertebrate

Amphioxus

Advanced chordate; sessile adult stage lost.

Tunicates

Ancestral tunicate with free-swimming larva.

"Acorn worms"

Shift from arm feeding to gill filter feeding.

Pterobranchs

Primitive echinoderms

Primitive sessile arm feeder

Figure 2.5 Garstang and Romer's scenario for the role of retention of juvenile features in the evolution of chordates. From a sessile arm feeder came both the echinoderms and the pterobranchs. A shift from arm feeding to gill feeding produces the acorn worms. Tunicates represent the culmination of this gill-feeding stage, but chordates escaped this adult specialization through retention of features of their free-swimming larvae. Eventually, the sessile adult stage is lost, producing amphioxus and finally a filter-feeding jawless vertebrate. (Redrawn from several sources.)

a larva. As adults, lancelets burrow in the sandy seafloor tail first, leaving only their heads protruding to filter feed via tentacles around their mouth. Their elongate, worm-like body has many chordate features: a well-defined notochord and dorsal nerve cord, a ventral blood vessel, a pharyngeal basket with over a hundred slit-like openings on the sides, and a post-anal tail. Unlike the larval tunicates, however, lancelets have well-developed myomeres that run the length of the body, rather than just in part of the tail. They also have other additional advanced chordate features, including a gut diverticulum structure homologous to the liver, a more advanced nervous system, and other genetic and molecular features that are unique to lancelets and higher chordates. However, the molecular evidence suggests that tunicates are closer to vertebrates than are the lancelets, so there is an interesting conflict between the anatomical and molecular evidence that has not yet been resolved.

Although lancelets definitely have a front end with a mouth, they do not yet have a well-defined head. They have no eyes, but they do have a photosensitive pigment spot in the front that detects light and darkness. Their mouths have no jaws, but instead they use tentacles with many tiny cilia to trap floating food particles. The food then passes through folded, cilia-lined tracts called the wheel organ (so-called because the beating cilia give the impression of a wheel in motion). On the right dorsal side of these tracts is Hatschek's pit, which secretes mucus to help collect the food particles and may be the equivalent of the pituitary gland in vertebrates. The endostyle produces more mucus to hold the food particles together as they pass through the pharynx. It is also found in the tunicates, and is homologous with the thyroid gland in vertebrates.

Such a small, soft-bodied animal would seem unlikely to fossilize, but fortunately there are specimens in extraordinary fossil deposits which preserve soft tissue. The Middle Cambrian Burgess Shale yields an animal known as *Pikaia* (**Figure 2.3**), which has a broad, lancelet-like body with visible notochord and myomeres and a distinct tail fin. Another Middle Cambrian deposit with extraordinary preservation in China has produced a similar animal known as *Yunnanozoon*. However, in detail both of these fossils have a lot of features not seen in the living cephalochordates, so there is some controversy as to whether they are truly members of the living group. If they are relatives of the lancelet, then we have a fossil record in the Cambrian not only of hemichordates (graptolites) but also of cephalochordates, and as we shall see later, even vertebrates are now known from the Cambrian. The evolutionary radiation of the chordates and their primitive relatives must have occurred rapidly entirely within the Early–Middle Cambrian or possibly earlier. Another fossil lancelet, *Palaeobranchiostoma*, is known from the Permian of South Africa.

GETTING A HEAD: THE VERTEBRATES

A lancelet has many features of vertebrates, but it does not have a true head region, nor does it have a true skeleton. But these features are found in the living jawless vertebrates and their extinct relatives (**Figure 2.6[A]**). These chordates have with a well-defined head region, including a brain at the front end of the nerve cord and cranial nerves connected to well-defined sense organs (eyes, nose, ears). Vertebrates are also the first chordates to have more than just a notochord for support, exhibiting a true internal skeleton made of cartilage. Vertebrates have many other specializations, including a much more sophisticated circulatory system with a two-chambered heart. During their embryology, they have a distinctive region developing along the spine called the neural crest, a feature found in all vertebrate embryos. Neural crest cells are very important in embryology, because they migrate from that region to form parts of the skeleton, skin, nervous system, sense organs, and other systems in vertebrates.

The only living members of these primitive vertebrates are two jawless eel-like forms, the lamprey and the hagfish (**Figure 2.6[A]**), known as the cyclostomes ("ring mouth" in Greek). Hagfishes (also known as "slime eels" or "slime hags") are a group of eel-like vertebrates with highly degenerate features, so the cartilaginous skeleton and the vertebrae are nearly gone. About 60 species in six genera are known, restricted to the deeper continental shelf, and often found to depths of 400–1000 meters. They burrow along the seafloor, seeking out dead and dying fish and marine worms, which they slurp up like strands of spaghetti. They use tooth-like ridges on their muscular tongue to rasp out pieces of living fish, although they prefer to burrow into the body cavity of a dead fish and eat it from the inside. When they are trying to tear a chunk off a fish, they grasp onto a protruding surface (like the gills or anus) and then tie themselves into a knot. They then force the knot against the surface of their victim until they rip a piece loose. This same kind of knotting behavior is useful for wriggling out of the grasp of a predator.

Hagfishes are known as "slime eels" because they produce copious amounts of mucus when they need to evade a predator. Their bodies are lined with 90 slime pores, and it is said that they can fill a 2-gallon bucket of water with slime in a matter of minutes. The mucus contains a special kind of protein that unfolds in water and gathers up the water molecules

A

B

Figure 2.6 (A) The hagfish (*Myxine*) is a slimy eel-like jawless vertebrate which spends most of its time slurping up worms from the sea bottom muds, or scavenging dead animals on the seafloor. (B) The lamprey is the only other jawless fish alive today, with a sucker mouth full of tiny teeth which helps it rasp a hole in the side of its victim and suck out its fluids. (Both photos courtesy Wikimedia Commons.)

to form a jelly-like substance that is almost impossible to wash off. There are several excellent video clips of this amazing process. If you do a search for "hagfish slime" videos online, you will find many fascinating demonstrations of this. Fishermen hate hagfish, because these creatures scavenge the fish caught in gill nets so quickly that the catch may be ruined. However, hagfish themselves are becoming overfished, because their tanned skins are used for a popular kind of leather sold as "eelskin". Their populations have been decimated in East Asian waters and off the west coast of North America, and are rapidly declining in the Atlantic.

Although it looks superficially similar to hagfishes, the lamprey (**Figure 2.6[B]**) is a more advanced animal, with most of the specializations of vertebrates (**Figure 2.1**). Lampreys usually live as parasites that attach to the side of another fish with their suction-cup mouth, armed with hundreds of tiny teeth and a rasping tongue. They clamp on and open a festering wound in the side of the host fish, using an anticoagulant in their saliva to keep the blood flowing, and then they suck out the body fluids. Like hagfish, lampreys live mostly in marine waters, but they swim up rivers to breed in lakes and streams. If they encounter rapids or a waterfall, they use their sucker-like mouth to cling to rocks and slowly creep up the cataracts. They have become established in many bodies of freshwater, such as the Great Lakes. In recent years, humans have built canals that connected the Great Lakes, aiding the spread of lampreys through the lakes and their tributary streams and rivers. Lampreys have become a major economic problem in the region, because they quickly decimate the native fish species. Alarmed officials have tried to use nets across the Saint Lawrence River to keep them out, and poisons and electrical barriers have been tried in other rivers and lakes. Although native fish populations are now beginning to recover, the effort was very expensive and cannot be relaxed, because humans have once again introduced an exotic species to a region that had no effective predators, and now we must spend the money to control it ourselves.

CONODONTS

If graptolites were one of paleontology's longest-running mysteries, the conodonts are a close second. First described in 1856, for over a century no one had a clue about their biology, yet this fact did not prevent them from becoming useful for biostratigraphy. In fact, conodonts were so ubiquitous during the Paleozoic and Triassic that practically any marine sedimentary rock (especially organic-rich sandstones and limestones dissolved in acid) of that age yields them. Conodonts evolved so rapidly, and they are so abundant, that they have become the index fossil of choice for most of the Paleozoic.

Like vertebrate bone, conodonts are made of calcium phosphate (the mineral known as apatite), so many scientists have guessed they were related to vertebrates (although some groups of worms also produce tooth-like phosphatic structures, and inarticulate brachiopods also use this chemical compound). Their tooth-like structure (**Figure 2.7[A]**) suggested they might be teeth of some sort of a minute vertebrate, but there are problems: most conodonts show no wear on the crowns of the "teeth", and the growth lines of these structures show that even the tips of the cusps were normally embedded in tissue. For a century, each distinctive conodont shape was described as a new taxon, but eventually specialists began to realize that many different conodonts were parts of the same animal. Specimens were found with a variety of different conodonts associated in a bilaterally symmetrical apparatus (**Figure 2.7[B]**), suggesting that they all supported some kind of structure in a bilaterally symmetrical animal. This created problems for conodont taxonomy, because many different named taxa were found to belong to the same apparatus and were apparently all synonyms. In the last 50 years, conodont workers have made many revisions and adjustments, so a large number of apparatuses and associations have been documented. Unfortunately, 99% of conodonts are still found as isolated elements in residues of rocks dissolved with acid, so synonymies have still not been determined for the majority of these taxa.

Yet, despite this handicap, conodont paleontologists have made enormous progress describing what is preserved. The earliest conodonts are simple coniform, cusp-shaped objects (**Figure 2.7[A]**), with a broad base and a long, curving tip. More advanced conodonts are multi-cusped, blade-like (ramiform) elements, and the multi-cusped, flattened platform elements, which have a broad base with a pit underneath, and many cusps and ridges on top. The terminology of each of these elements is relatively simple and straightforward, but nevertheless hundreds of different genera of conodonts have been described.

The earliest conodonts are simple coniform elements known as proto-conodonts. These are known from the Early Cambrian, but many authors doubt that they are related to more advanced conodonts, because there are important differences in their chemistry and ultrastructure, and because they were exposed to wear and may have actually been used as teeth. However, two groups of more advanced conodonts, the

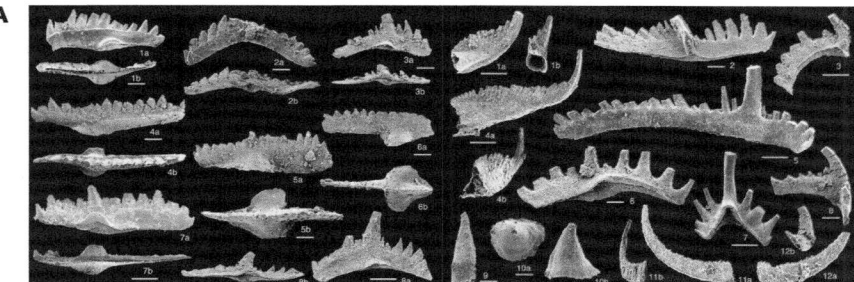

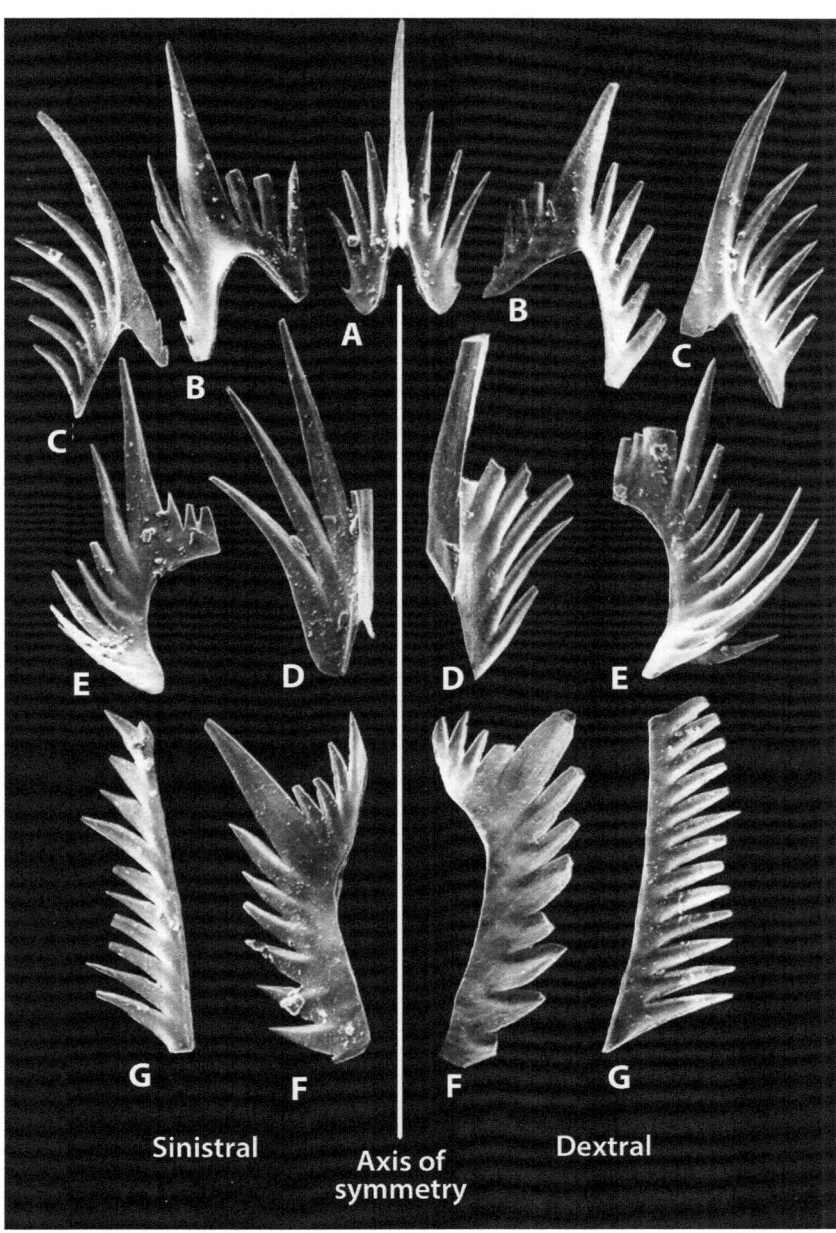

Figure 2.7 **(A) Series of typical conodont elements, some shaped like simple single cusps, others with platforms studded by rows of cusps. (B) Reconstruction of a conodont apparatus (*Ellisonia*, from the Permian of Texas), using associations of conodonts found in rocks. A bilaterally symmetrical element (A) probably lay across the midline of the body, and each of the other elements (B–G) apparently were symmetrically arranged on each side.** (Courtesy Wikimedia Commons.)

paraconodonts and euconodonts, both placed in the Conodontophorida. They radiated in the Early-Middle Cambrian, and there were at least 60 genera known by the Middle Ordovician. Conodont diversity crashed in the Middle Silurian, then recovered in the Devonian, when it reached a peak of about 30 genera. After the Late Devonian extinctions, conodonts were much less diverse through the later Paleozoic, but a few genera did survive the Permian catastrophe and lingered on through the entire Triassic before their final extinction.

So what animal made conodonts? In recent years, a number of specimens have been advanced as the "conodont animal", only to be discredited. Most of these candidates actually had conodonts in their stomach area, so they might better be described as conodont predators. In 1983, Derek Briggs and others described fossils from the Granton "Shrimp Bed" of the Lower Carboniferous of Scotland that are our best candidate for the conodont animal. The best specimen has a long eel-like body, with a complete conodont apparatus in what appears to be its mouth or throat region (**Figure 2.8**). The body appears to have had eyes, ears, a pharynx, fins supported by rays, a dorsal fin, and segmented muscles. Another specimen from the Lower Silurian rocks of Wisconsin is less well preserved, but shows similar structures. Some conodonts even have cellular

Figure 2.8 Reconstruction of the eel-like conodont animals corresponding to the conodont fossils *Promissum* (A) and *Clydagnathus* (B).

bone, like vertebrates. On the basis of this evidence, most specialists are now convinced that the conodont animal was a chordate, and probably a vertebrate as well, possibly related to the hagfish, but more primitive than lampreys and the extinct jawless fish. The conodont animal is now reconstructed much like a hagfish or lamprey, with a bristling array of conodont elements in its pharyngeal region. Although they are not true teeth, some conodont elements may have functioned as grasping organs to hold a prey item once it reached the pharynx (as the hagfish uses its tiny teeth). So conodonts, which had long been in the realm of micropaleontology and studied in biological isolation from nearly all other fossil groups, may turn out to be one of our closer relatives.

FURTHER READING

Conniff, R. 1991. The most disgusting fish in the sea. *Audubon.* 93: 100–118.

Delarbre, C.; Gallut, C.; Barriel, V.; Janvier, P.; Gachelin, G.; et al. 2002. Complete mitochondrial DNA of the hagfish, *Eptatretus burgeri*: The comparative analysis of mitochondrial DNA sequences strongly supports the cyclostome monophyly. *Molecular Phylogenetics and Evolution.* 22 (2): 184–192.

Donoghue, P.C.J.; Purnell, M.A. 2009. The evolutionary emergence of vertebrates from among their spineless relatives. *Evolution: Education and Outreach.* 2: 204–212.

Kinya, G.O.; Kuratani, S. 2007. Cyclostome embryology and early evolutionary history of vertebrates. *Integrative and Comparative Biology.* 47: 329–337.

Kuratani, S.; Kuraku, S.; Murakami, Y. 2002. Lamprey as an Evo-Devo model: Lessons from comparative embryology and molecular phylogenetics. *Genesis.* 34: 175–195.

Long, J.A. 2010. *The Rise of Fishes* (2nd ed.). John Hopkins University Press, Baltimore.

Maisey, J.G. 1996. *Discovering Fossil Fishes.* Henry Holt, New York.

JAWLESS FISH

3

Half my closet walls are covered with the peculiar fossils of the Lower Old Red Sandstone; and certainly a stranger assemblage of forms have rarely been grouped together; creatures whose very type is lost, fantastic and uncouth, and which puzzle the naturalist to assign them even their class;—boat-like animals, furnished with oars and a rudder;—fish plated over, like the tortoise, above and below, with a strong armor of bone, and furnished with but one solitary rudder-like fin; other fish less equivocal in their form, but with the membranes of their fins thickly covered with scales;—creatures bristling over with thorns; others glistening in an enamelled coat, as if beautifully japanned—the tail, in every instance among the less equivocal shapes, formed not equally, as in existing fish, on each side of the vertebral column, but chiefly on the lower side—the column sending out its diminished vertebrae to the extreme termination of the fin. All the forms testify of a remote antiquity—of a period whose "fashions have passed away".

—Hugh Miller, 1841, The Old Red Sandstone

FISH IN ARMOR

The lamprey and hagfish are the only living groups of jawless vertebrates, but they are not typical of the earliest stages of vertebrate evolution. They are but a tiny remnant of a huge radiation of jawless vertebrates that flourished in the early Paleozoic (**Figures 2.1** and **3.1**). These creatures include a number of soft-bodied boneless fossils from the Early Cambrian of China, including some complete specimens with the entire soft-bodied outline preserved, such as *Haikouella* and *Haikouichthys* (**Figures 3.1[A]** and **3.3[A]**). The earliest possible fossil bone material of jawless fish are small denticles of armor from the skin of *Anatolepis* from the Upper Cambrian Deadwood Sandstone of Wyoming, although there is still controversy about whether these fossils are truly bone or not. The plates have some of the detailed histological features of vertebrate bone, but in other details of the histology are not like modern bone. Isolated bony plates of early fish are also known from the Ordovician Harding Sandstone of Colorado, and the Ordovician of Australia yields a few partial specimens, such as the jawless fish *Arandaspis* (**Figures 3.1[C]** and **3.3[B]**). This fish had large curved plates around its front end, with a groove down the side that hinged the two plates, and provided room for the eyes and gill openings. The rest of its body was covered by small scales, which were articulated so that the tail could be flexed, and the tail apparently had a diamond-shaped fin on it. The oldest relatively complete fossil of a jawless fish is *Sacabambaspis* from the Upper Ordovician beds near the town of Sacabamba of Bolivia (**Figures 3.1[D]** and **3.3[C]**). The best complete specimens of jawless fish do not occur until the Silurian and especially from the Devonian, when they reached their peak of diversity.

Heterostracans

All these early jawless fish had an internal skeleton made mostly of cartilage (as do lampreys and sharks), but most had bony plates around the head region and bony scales covering much of the body. Their jawless

DOI: 10.1201/9781003128205-3

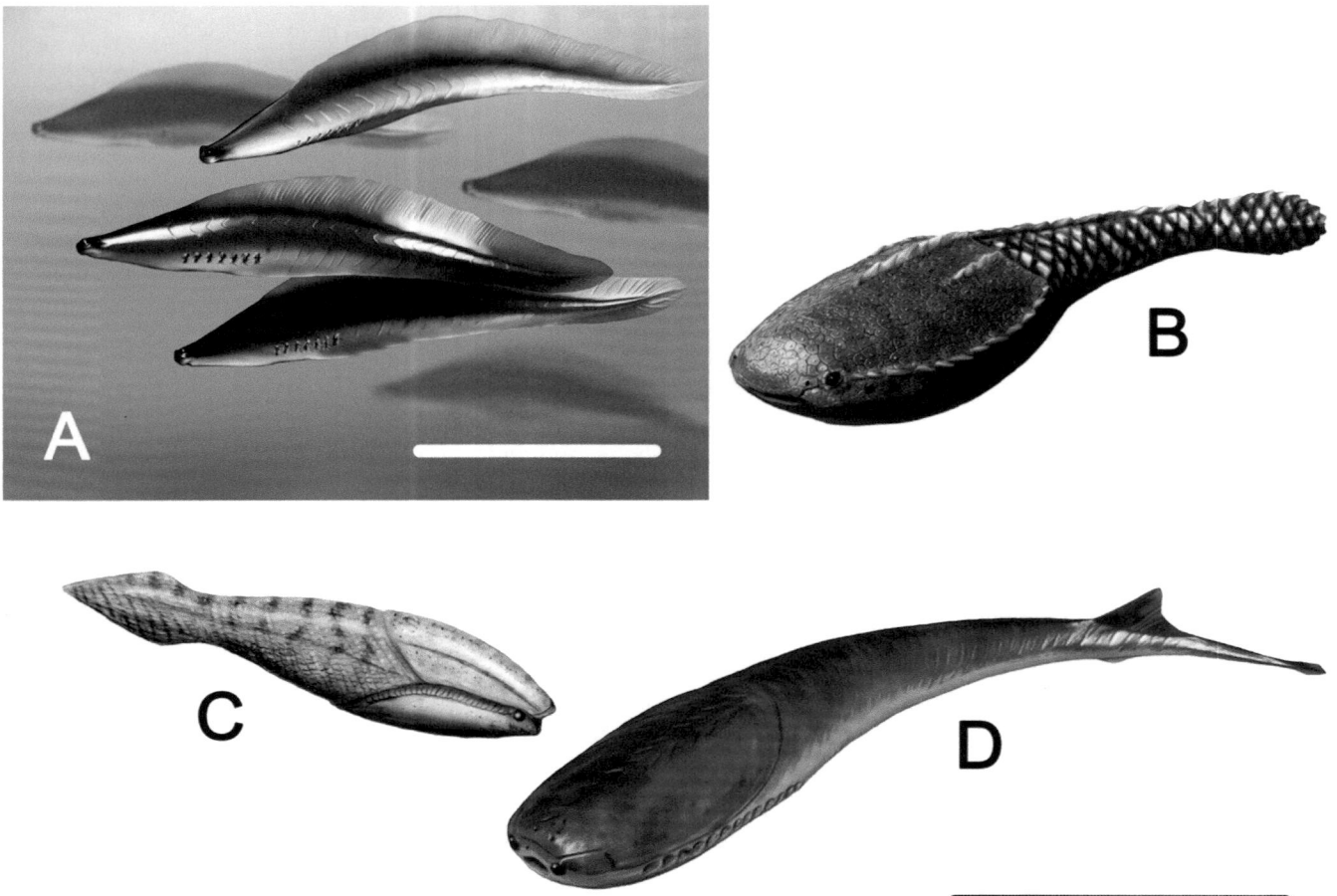

Figure 3.1 Reconstruction of some of the earliest chordate fossils. (A) *Haikouichthys* (scale = 1 cm). (B) *Astraspis*. (C) *Arandaspis*. (D) *Sacabambaspis*. Scale bar: 10 cm.

mouths were usually simple slits that could be used to feed on detritus or to filter feed by passing water through the pharynx. One group, the heterostracans, included many fish with a cylindrical or football-shaped body, with the head and thoracic region covered by solid bony armor (**Figures 3.2** and **3.3**). Early examples included *Astraspis*, *Arandaspis*, and *Sacabambaspis* (**Figures 3.1** and **3.3**), which had multiple gill openings along the side of their head. More advanced heterostracans had only a single gill opening along the side of their bodies between the top and bottom armor plates on the head. Most individuals were only a few centimeters long, although the largest ones were up to 2 meters long. Small bony scales covered the rest of the body, and the tail was asymmetrical, with its major lobe pointed downward (known as a reversed heterocercal or hypocercal tail). Some had small spines protruding from the head shield, or a dorsal fin spine. However, they apparently had no fully developed pectoral or pelvic fins, and thus no stabilizing mechanism for swimming, so they must have swum very erratically, like a large tadpole with armor.

There has been much debate about how they used the mouth plates in their jawless mouth—scooping up sediment, slurping up prey with suction, or other ideas—although their head shield did not allow them to create much suction in their mouth cavity. Detailed studies of the mouth plates show no wear whatsoever, so they did not grind anything with their plates, or pass lots of gritty sediment over the plates, but must have focused on filter feeding of plankton or microbial mats that passed through their mouth and into their oral cavity and gill chamber. Inside the head shield of some specimens, there are internal impressions for

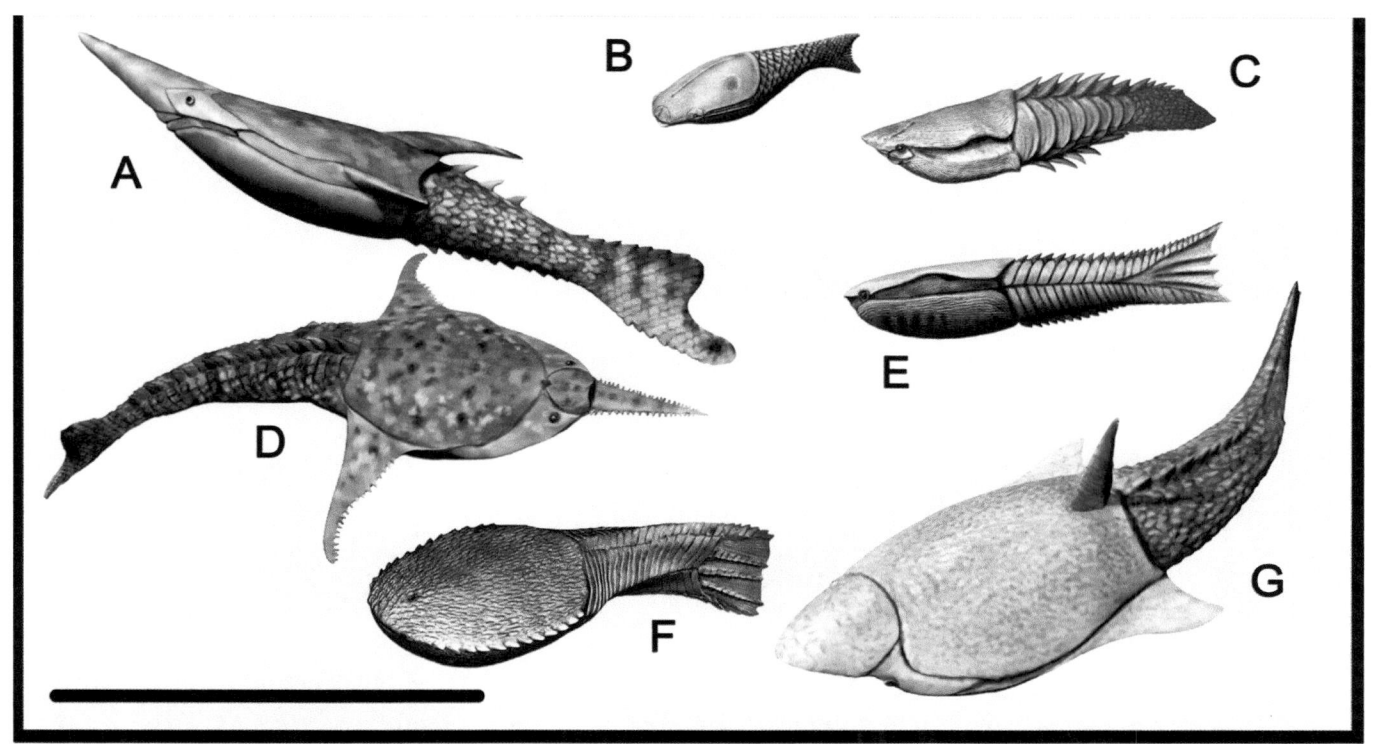

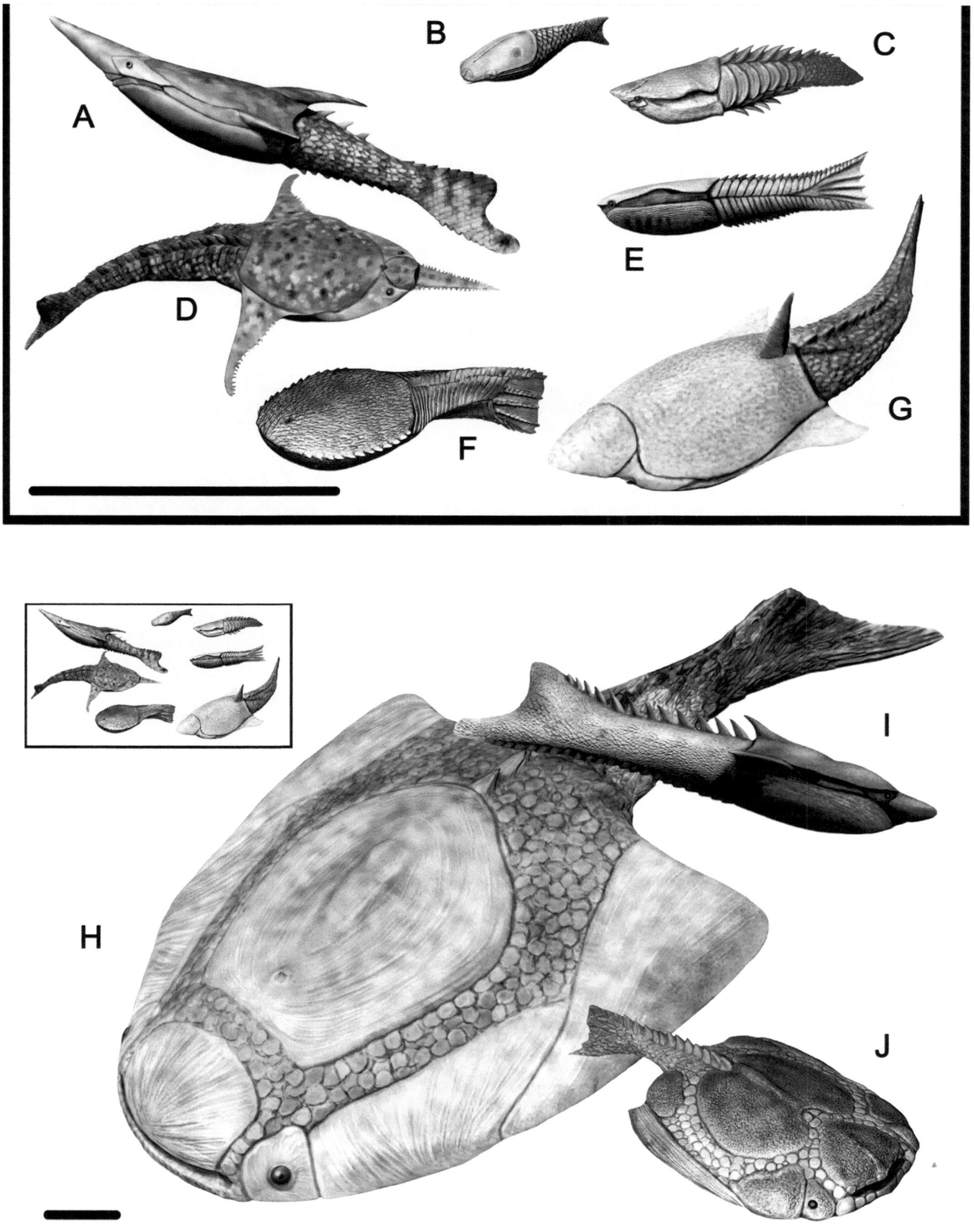

Figure 3.2 A variety of heterostracans: (A) *Pteraspis*, (B) *Athenaegis*, (C) *Anglaspis*, (D) *Doryaspis*, (E) *Poraspis*, (F) *Ctenaspis*, (G) *Larnovaspis*, (H) *Psammolepis*, (I) *Panamintaspis*, (J) *Drepanaspis*. Scale bar: 10 cm.

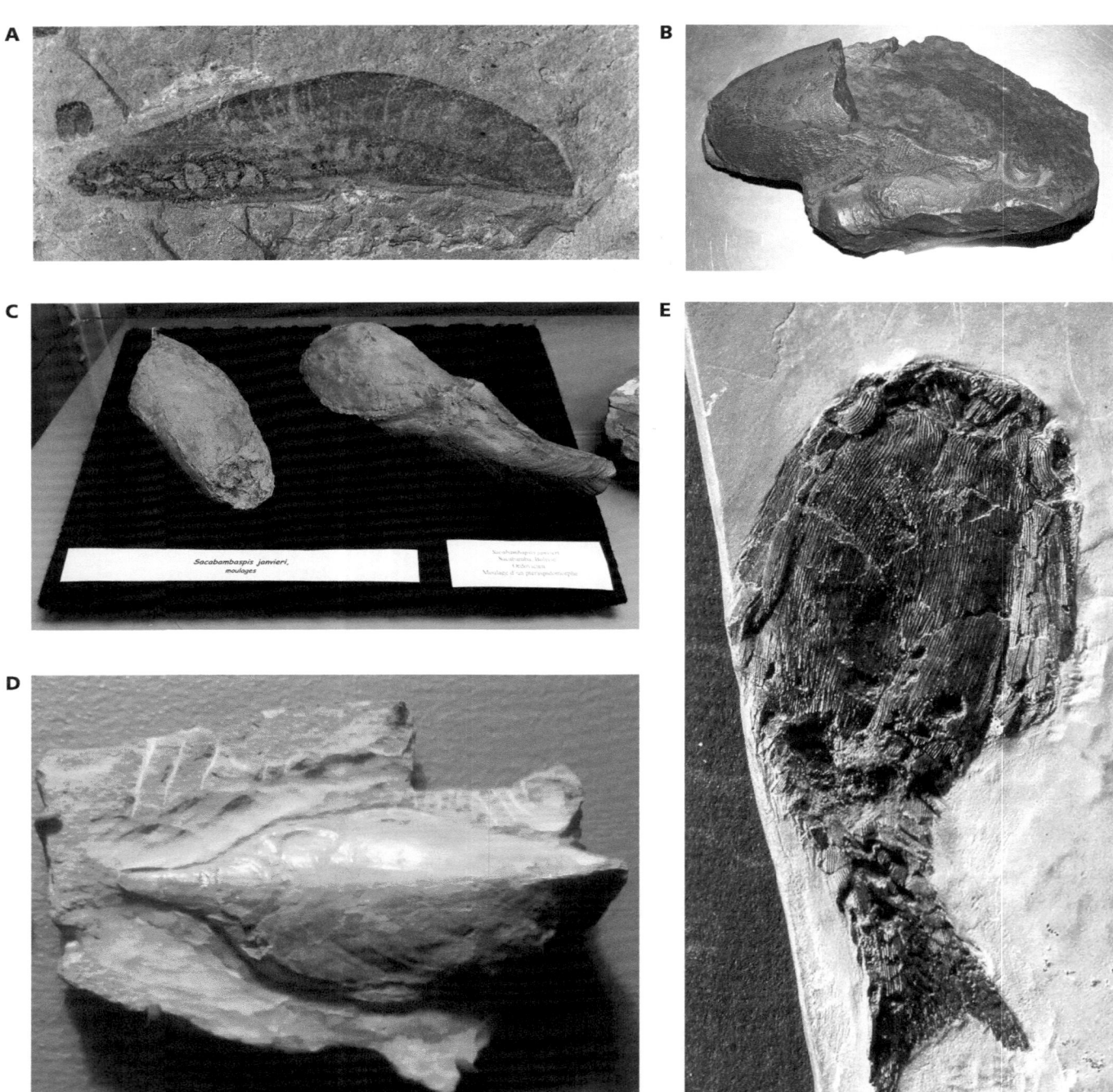

Figure 3.3 Some fossils of primitive jawless vertebrates. (A) Crushed specimen of *Haikouichthys*, the oldest known vertebrate, from the Lower Cambrian of China. (B) Fossil of *Arandaspis*, a primitive heterostracan from the Ordovidian of Australia. (C) Fossils of *Sacambaspis*, a primitive heterostracan from the Ordovician of Bolivia. (D) Fossil of the head shied of the heterostracan *Pteraspis*. (E) Fossil of the primitive heterostracan *Athenaegis* from the Devonian of Canada. [(A–D) Courtesy Wikimedia Commons. (E) Courtesy M.H.V. Wilson.]

the gills, the braincase, and other features, including large olfactory lobes of the brain that allowed them to smell differences in the water chemistry. Most heterostracans are known only from marine settings, although a few might have lived in fresh or brackish waters.

Thelodonts

Another group, the thelodonts, were covered by a "chain mail" armor of small interlocking dentin scales like those of sharks, some with protruding spines (**Figure 3.4**). Many had a very simple body with only weak

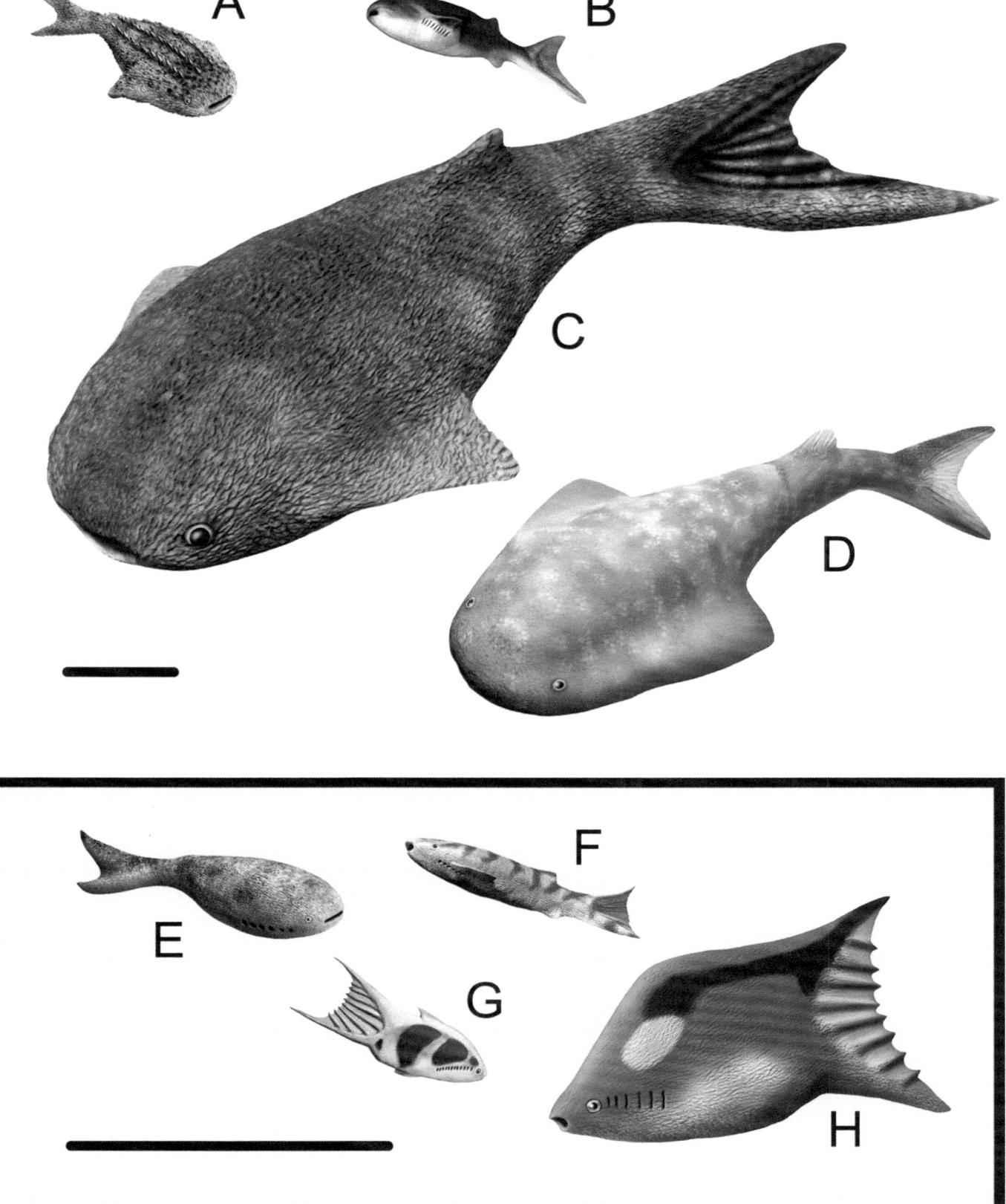

Figure 3.4 Reconstructions of typical thelodonts. (A) *Lanarkia*, (B) *Loganiella*, (C) *Thelodus*, (D) *Turinia*, (E) *Archipelepis*, (F) *Phlebolepis*, (G) *Furcacauda*, (H) *Sphenonectris*. Scale bar: 10 cm.

pectoral fins and a heterocercal tail, while some had well-developed pectoral fins, as well as a dorsal fin on the back and an anal fin near the bottom rear of the fish. Some, like *Furcacauda*, had a forked tail and resembled some of the living fish with forked tails, like angelfish. Others were broad and flattened, and apparently fed by slowly cruising along the bottom and sucking in prey, like the angel shark does today. Thelodonts are found in marine deposits. They first appeared in the Ordovician, are known from a number of Silurian rock units around the world, as numerous isolated scales (complete articulated specimens are very rare), and then became restricted to the Gondwana continents by the Middle Devonian. They vanished in the great extinction near the end of the Devonian. Their unique scales apparently helped them live in Devonian reef settings, since the spiny armor protected them from abrasion by reef rocks. The isolated scales of thelodonts are abundant and distinctive in many Silurian and Devonian deposits, so much so that they can be used to tell time in rocks of that age. The first known thelodonts were originally considered to be mud-grubbers who sucked in mud and strained out the food, but more recent research has shown that some of them were active swimmers, and probably fed on plankton that they could filter through their jawless mouth and gill apparatus.

Anaspids

A third group of jawless fish, the anaspids, are represented by specimens with fine rows of tile-like scales, rather than a bony body shield as in most other jawless fish. Instead, their scales were made of an acellular tissue called aspidine, which is mineralized, but not as complex in structure as true bone. Anaspids had a simple slit-like mouth, paired eyes, and a strongly reversed heterocercal tail (**Figures 3.5** and **3.6[A]**). They had a row of 6–15 paired gill openings through their external armor along the sides of the head. They had paired fins, or paired ridges along their sides, that would have stabilized their swimming, although they did often have an anal fin, and some had spikes or ridges of bony scales along their backs. Their lack of pectoral fins suggest that they were poor swimmers, and probably either filter-fed with their jawless mouth and gill apparatus, or maybe plowed through the bottom sediment with their mouth to strain out the food particles in the mud. Most were small, reaching about 15 cm in length, although there are huge scales from the Early Silurian of Canada suggesting much bigger anaspids existed. Anaspids were found exclusively on the Euramerican continent from Western Canada to Scotland to Norway to Estonia to Siberia during the Silurian and Devonian, and apparently never made it to the southern Gondwana continents. All anapsids were found in marine rocks. Some scientists think that the detailed anatomical structures of Late Devonian anaspids such as *Endeiolepis* and *Euphanerops* from Scaumenac Bay in Quebec, Canada, suggest that anaspids and lampreys are very closely related, so that lampreys are just anaspids that have lost their armor. This idea is still controversial. *Euphanerops* appears to have a ring of cartilage around its mouth, like that of the living lamprey. Nonetheless, the simple mouth of the anaspids is consistent with either a filter feeding lifestyle, or possibly as a lamprey-like parasitic existence.

Osteostracans

The best-known group of early jawless fishes is the cephalaspids, or osteostracans, which typically had a large flattened head shield (**Figures 3.6[B]** and **3.7**). They were formerly called "ostracoderms" but that name once was used for all armored jawless fish, so it is now obsolete. In most species, the head shield is made of a single plate of bone, with many polygonal subunits fused together. It had openings for two

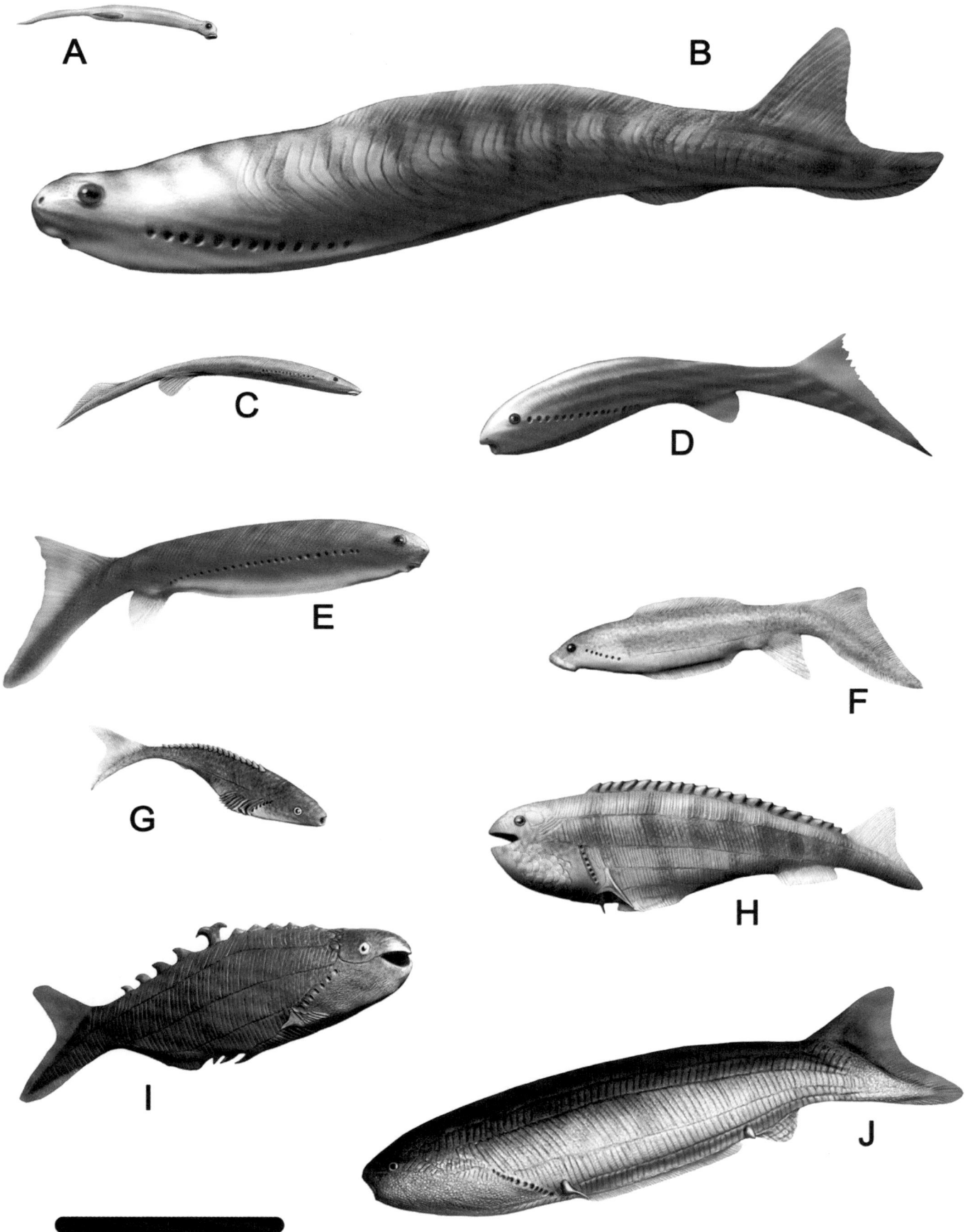

Figure 3.5 A variety of different genera of anaspids. (A) *Ciderius*, (B) *Jamoytius*, (C) *Achanarella*, (D) *Cornovichthys*, (E) *Euphanerops*, (F) *Endeiolepis*, (G) *Lasanius*, (H) *Cowielepis*, (I) *Birkenia*, (J) *Pharyngolepis*. Scale bar: 5 cm.

Figure 3.6 (A) Crushed specimen of the anaspid *Birkenia*. (B) Head shields of the osteostracan *Hemicyclaspis*. [(A and B) Courtesy Wikimedia Commons. (C) After Stensiö (1928).]

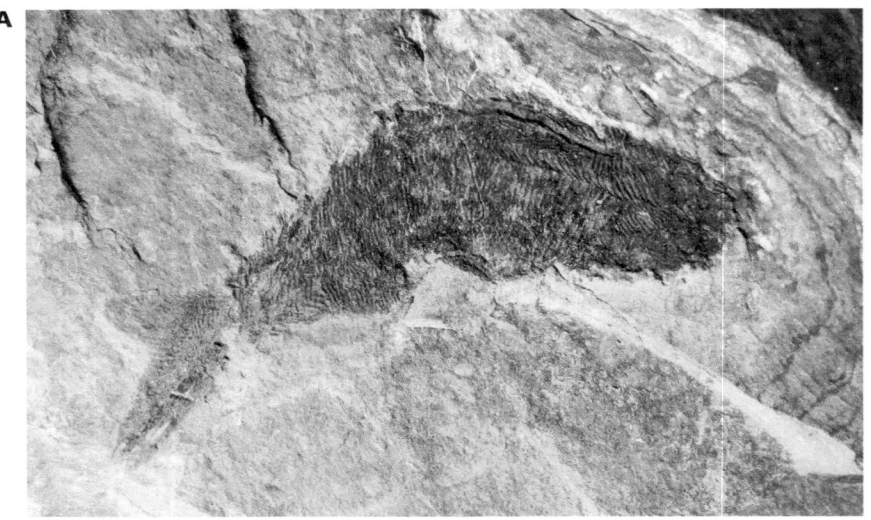

C

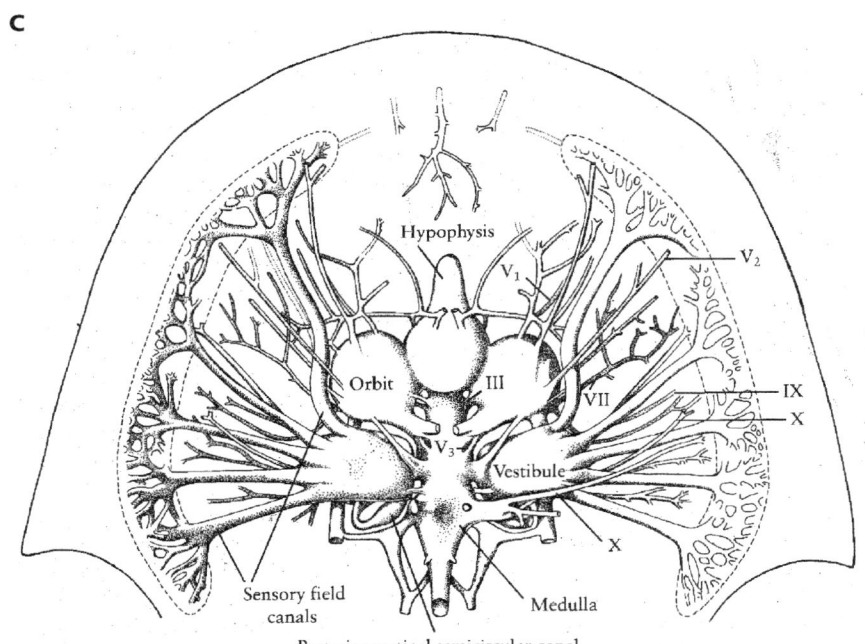

Hypophysis

V_1

V_2

Orbit   III

VII

IX

X

V_3

Vestibule

X

Sensory field
canals

Medulla

Posterior vertical semicircular canal

Figure 3.6 (Continued) (C) Reconstruction of the internal nerves and organs of the head shield of an osteostracan.

upward-directed eyes and another opening between them. This opening is thought to be the medial nasohypophyseal opening, which had the olfactory organ and the hypophysial pouch containing the pituitary gland. Osteostracans had a broad scale-covered area around the brim, apparently with features for sensing motion in the water. Many of the osteostracans had weird spines and spikes protruding from their head shields (**Figure 3.7**). The internal structure of these head shields has been exhaustively studied by scientists, who took numerous specimens of abundant cephalaspids, and sliced them into very thin slices like a loaf of bread. Using this method, they have traced the detailed course of all the nerves and blood vessels and other structures within these exquisitely preserved specimens (**Figure 3.6[C]**). Behind the head shield, cephalaspids had flap-like pectoral fins that would have helped control swimming, and some of these flaps apparently had cartilaginous supports. However, these were not true fins with bony supports that are used for paddling in most living fishes. Inside the head shield and the body was a calcified endoskeleton, where calcium carbonate has replaced the cartilage. The rest of the body was covered with fine bony scales, and the lobe of the asymmetrical tail bent upward, as in sharks (a heterocercal tail). The flattened bottom of the head shield, with its scoop-like mouth opening, and the overall flattened body with the heterocercal tail, suggests that cephalaspids were bottom-feeders who probably scooped up detritus and sifted out the food with their pharynx and gills.

Osteostracans appear to be much more advanced than other groups of fossil jawless fish. They are the first fish with true muscular pectoral fins, supported by an internal skeleton of cartilage. In addition, the details of the structure of their bones and scales are more advanced than other jawless fish, again putting them closer to the jawed vertebrates than any other group of jawless fish.

Many scientists argue that osteostracans are the closest relatives of the jawed fish among all the different jawless fish (**Figure 4.1**). Not only did they have muscular pectoral fins and the beginnings of a shoulder girdle to support those fins, plus details of the bone microstructure, and the bony sclerotic ring protecting the eyeball, and true gill slits over the gill

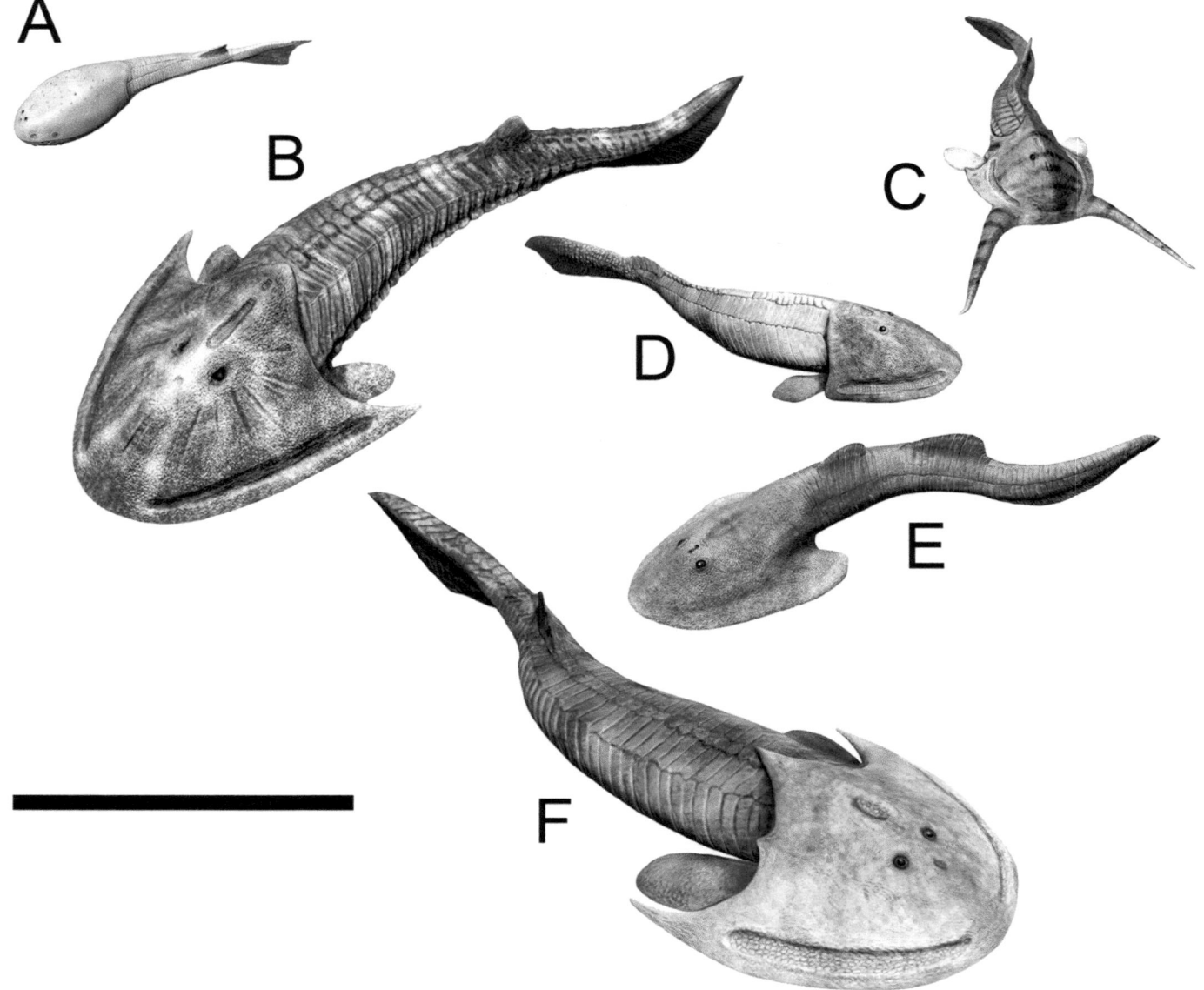

Figure 3.7 A range of different kinds of osteostracans. (A) *Tremataspis*, (B) *Cephalaspis*, (C) *Tauraspis*, (D) *Hemicyclaspis*, (E) *Ateleaspis*, (F) *Zenaspis*. Scale bar: 10 cm.

chamber (rather than a series of round openings), as well as two dorsal fins and a shark-like heterocercal tail. In fact, embryonic evidence shows that there is strong connection to the development of the eye region (especially the bony sclerotic ring) and the development of lower jaws. Once eyes with sclerotic rings developed, they may have triggered the embryonic pathways that allowed the jaw to develop next.

Galeaspida

Since the 1960s and 1970s, the world of jawless fish has been amplified by the discovery of yet another distinct group of fossils that don't fit into any of the previous categories. Found primarily from Silurian-Devonian rocks in China (as well as Tibet and Vietnam), they are known as the Galeaspida (**Figure 3.8**). Galeaspids are unique in that the opening to the pharynx and the gill chamber is on the top surface of the head shield, not on the sides (as in many jawless fish) or the bottom (as in osteostracans). This opening on the top center of the head for the intake of water resembles the position of the homologous nasopharyngeal duct of hagfishes. Galeaspids also

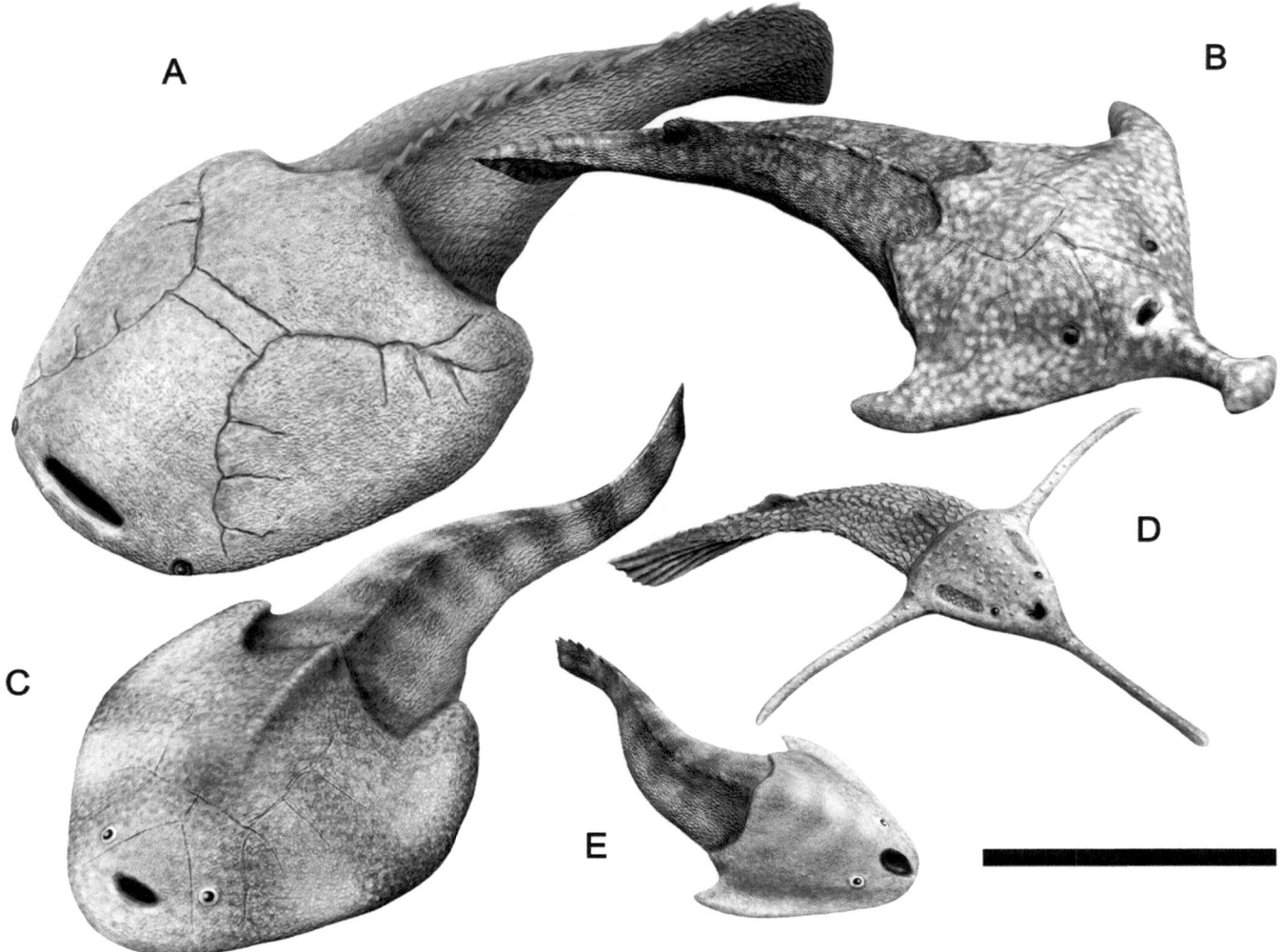

Figure 3.8 The odd Asian jawless fish known as galeaspids: (A) *Hanyangaspis*, (B) *Sanchaspis*, (C) *Laxaspis*, (D) *Lungmenshanaspis*, (E) *Shuyu*. Scale bar is 10 cm.

have a scalloped pattern on the sensory lines on the top surface of the head shield. Their head shields are also very thick and large, which gave them their names galeaspids (Latin for "helmet shields"). The head shields are often the only parts preserved on these animals, so the bodies are usually reconstructed like those of osteostracans, with a long trunk covered in small plates, and some sort of tail (usually heterocercal).

Despite these features and their limited geographic extent and time range, galeaspids came in an amazing array of different, often bizarre shapes. Some, like *Shuyu*, *Laxaspis*, and *Hanyangaspis*, had flattened oval-shaped head shields with a small slit mouth at the front. Others, like *Sanchapsis*, had a broad horseshoe-shaped head shield with an odd-looking bony tubular snout sticking out from the front. The weirdest of all was *Lungmenshanaspis*, which had a small triangular head shield with a long spike sticking out of the front (like that of a swordfish), and two long spikes or spines sticking out from each side, so it forms a cross pattern when viewed from above (**Figure 3.8**). Over 76 species in 53 genera of Galeaspida have been named, so they are very diverse as well.

The relationships of the Galeaspida are still a puzzle. They lack any sort of fin or flap on the sides of their heads, so in that respect they resemble heterostracans and other primitive jawless fish. However, there are details of the interior of the braincase that seem to ally them with the osteostracans, so that is currently considered their closest relatives.

WHERE DID THEY COME FROM? WHERE DID THEY GO?

A variety of group names have been created to classify the jawless vertebrates (**Figure 4.1**). The term Agnatha (Greek: "without jaws") is the most widely used, but unfortunately it is a "wastebasket" group for craniates without jaws, not a natural group. "Agnatha" are defined by the lack of the specialized jaws found in all other vertebrates, not by unique characters of their own. Indeed, some of them (like osteostracans) appear to be closer to jawed fish, while others (such as heteostracans) are much more primitive and appeared earliest, while still others (anaspids) probably belong with lampreys, so they do not form a group defined by any unique anatomical features. Likewise, some of the extinct osteostracans plus other unrelated fish were long called "ostracoderms", but that group is also a wastebasket, originally coined to lump all the armored jawless fish into one group (**Figure 4.1**). In recent years, these obsolete terms have gradually drifted out of use, and most scientists refer to natural groups defined by unique evolutionary specializations, like vertebrates or gnathostomes. If you want to talk about "agnathans", for example, "jawless vertebrates" conveys the same information content without the misleading implication that they are a natural group.

Another longstanding controversy was the issue of whether vertebrates originated in freshwater or in saltwater. When armored cephalaspids were first found in abundance in the freshwater river deposits of the Devonian Old Red Sandstone in Britain in the 1820s and 1830s, the freshwater environment was the most popular explanation. As older specimens of Silurian, then Ordovician, then Cambrian age, were found, however, it was apparent that all the Cambrian and Ordovician specimens came from normal marine deposits. In 1934, Romer and Grove argued that the complicated brain and nervous system of the earliest vertebrates suggested that they needed to swim in the turbulent waters of streams. In his provocative 1953 book *From Fish to Philosopher*, physiologist Homer Smith argued that the vertebrate kidney is a structure designed to pump out excess water. Because vertebrate body fluids are saltier than freshwater, if they lived in freshwater initially, they would have needed a strong kidney or they would have bloated up. Vertebrate body fluids are less concentrated in salts than seawater, so the kidney would not have been as useful in marine settings. In this case, the problem is getting rid of salts or gaining water, not losing it. However, Smith's arguments have since fallen out of favor, primarily because the kidney's primary function is getting rid of nitrogenous wastes (urine or ammonia), and it only secondarily became used for osmoregulation. Vertebrate kidneys work equally well in fresh, brackish, and salt water. More importantly, all the nearest relatives of vertebrates are salt-water organisms, as are the majority of early invertebrates, and all the earliest vertebrate fossils are marine, so it is a much less likely hypothesis to assume, against all the evidence, that freshwater origins are required by the kidney or brain structure.

What happened to all the armored jawless fish? After ruling the seas as the only common vertebrates in the Ordovician and Silurian, they were apparently overcome by the huge radiation of new kinds of fishes in the Early and Middle Devonian. They found themselves competing with much more advanced fish, like sharks, placoderms, acanthodians, and bony fish (see Chapters 4 and 5). But jawless fish began to decline as all this competition emerged, and by the Late Devonian, they were actually quite rare. Perhaps they were outcompeted by the more advanced fish like placoderms, who adopted many of the same lifestyles (especially bottom feeding) and would have been more efficient with their jawed mouth. For example, in the later Devonian of deposits of the Baltic Sea region (mostly in Estonia and Latvia and Russia), the flattened bottom-feeding psammosteid jawless fish were replaced in younger beds by a group of placoderms that were also flattened and even better bottom feeders, the phyllolepids. Whatever lineages were around near the end of the Devonian, the first of the two great extinct events at the beginning of the Late Devonian Fammenian Stage (one of the four biggest extinction events in earth history) was apparently the last straw for these archaic slow-moving armored forms. Only the lampreys and hagfish, unencumbered by armor, managed to survive past the Devonian and are still with us today.

FURTHER READING

Donoghue, P.C.J.; Purnell, M.A. 2009. The evolutionary emergence of vertebrates from among their spineless relatives. *Evolution: Education and Outreach*. 2: 204–212.

Forey, P.; Janvier, P. 1984. Evolution of the earliest vertebrates. *American Scientist*. 82: 554–565.

Forey, P.; Janvier, J. 1993. Agnathans and the origin of jawed vertebrates. *Nature*. 361: 129–134.

Gai, Z.; Donoghue, P.C.J.; Zhu, M.; Janvier, P.; Stampanoni, M. 2011. Fossil jawless fish from China foreshadows early jawed vertebrate anatomy. *Nature*. 476: 24–327.

Janvier, P. 1998. *Early Vertebrates*. Oxford University Press, Oxford.

Long, J.A. 2010. *The Rise of Fishes* (2nd ed.). John Hopkins University Press, Baltimore.

Maisey, J.G. 1986. Heads and tails: A chordate phylogeny. *Cladistics*. 2: 201–256.

Maisey, J.G. 1996. *Discovering Fossil Fishes*. Henry Holt, New York.

Moy-Thomas, J.; Miles, R.S. 1971. *Palaeozoic Fishes*. Saunders, Philadelphia.

Shu, D.-G.; Luo, H.-L.; Conway Morris, S.; Zhang, X.-L.; Hu, S.-X.; Chen, L.; Han, J.; Zhu, M.; Li, Y.; Chen, L.-Z. 1999. Lower Cambrian vertebrates from China. *Nature*. 402: 42–46.

Smith, H. 1953. *From Fish to Philosopher*. Little, Brown, Boston.

PRIMITIVE GNATHOSTOMES

4

It is hard to imagine life without jaws: giant killer sharks, carnivorous dinosaurs, saber-toothed tigers, and that talkative neighbor just would not be the same without them. The acquisition of jaws is perhaps the most profound and radical evolutionary step in craniate history, after the development of the head itself.

—John Maisey, *Discovering Fossil Fishes*, 1996

JAWS

When we hear the word "jaws", the first thing that comes to mind is the 1975 movie, the sinister foreboding music by John Williams warning of the shark's presence, or the terror and fear it inspired that kept people from swimming in the ocean for years after the movie was released. We seldom stop to think how important jaws are to vertebrates, or how limited vertebrates were without them. The jawless craniates could do little more than feed on suspended food or detritus, or in the case of lampreys and hagfish, suck the fluids out of their victims. Jaws are essential for grabbing a food item, and armed with teeth, they allow an animal to chop up the food to edible sizes. The evolution of jaws made it possible for vertebrates to exploit a wide variety of food sources, and with this skill, to evolve into many different habitats and body sizes, including the largest fish in the sea and the largest land animals. Jaws are also critical to many other functions, such as manipulation of objects. Vertebrates use their jaws for functions as different as digging holes, carrying pebbles or vegetation to build nests, grasping mates during courtship or copulation, carrying their young around, and making sounds or speech.

The presence of jaws is such an important innovation that it defines a group known as the *gnathostomes* (*gnathos* is "jaw" and *stoma* is "mouth" in Greek), which includes all vertebrates except the jawless fish (**Figure 4.1**). However, jaws are not the only evolutionary novelty of the group. Gnathostomes are also characterized by having their gills lying on the outside of their gill supports. By contrast, in jawless vertebrates such as the lamprey, the gills lie inside the gill arches. The gill arches in jawless vertebrates are a complex network or web of cartilage, but in gnathostomes they are completely separated, free, and segmented. In the ear region, gnathostomes have three semicircular canals that are used to detect motion and maintain balance in three different perpendicular planes; lampreys have only two semicircular canals. Finally, gnathostome fins have cartilaginous supports with muscles to allow them to move and flex; these are connected to the pectoral and pelvic girdles (shoulder bones and hip bones in mammals). This is just the beginning of a long list of the evolutionary novelties that define the gnathostomes. Several scientists have pointed out at least 30 more, making the gnathostomes one of the best-supported natural groups known.

DOI: 10.1201/9781003128205-4

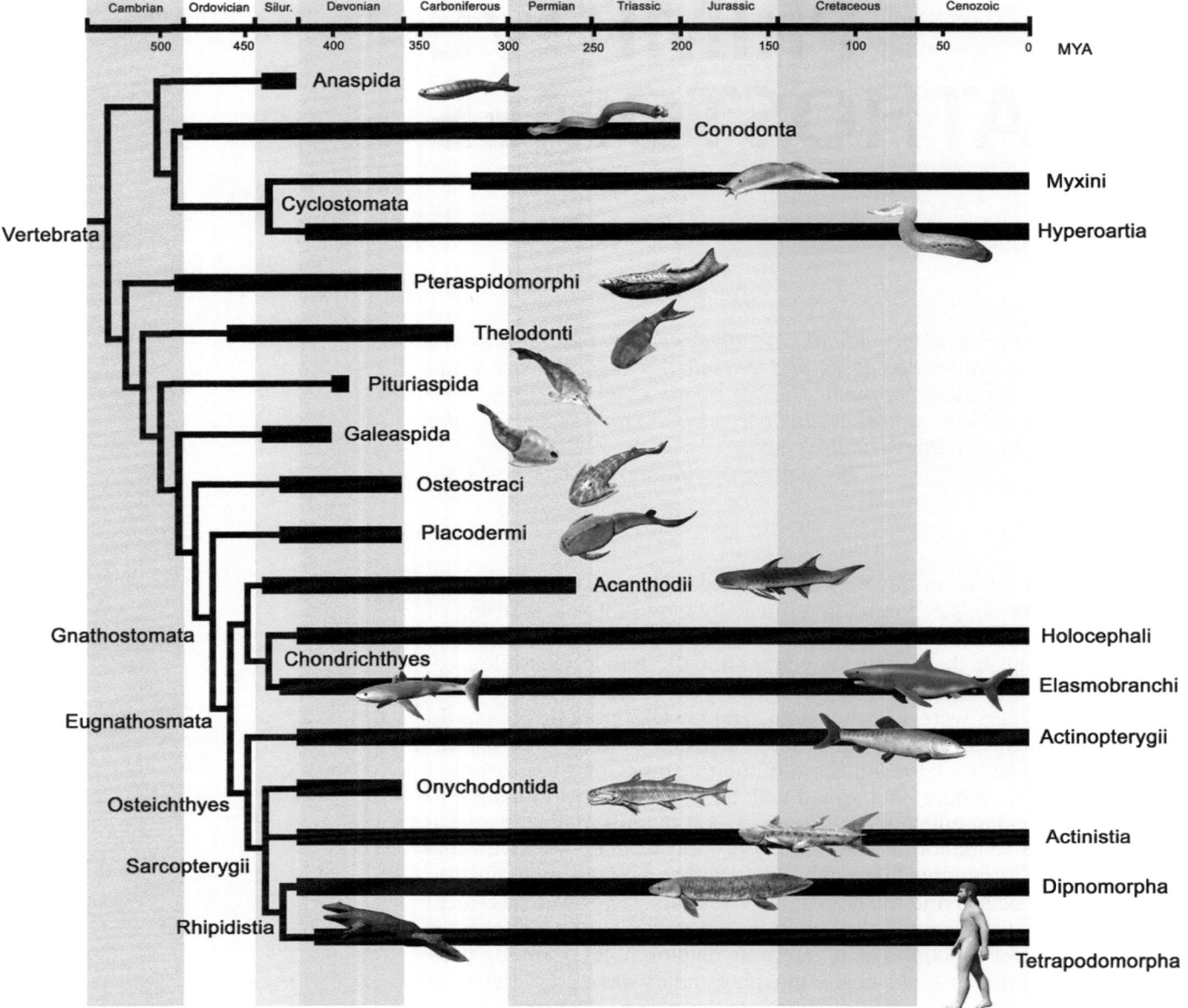

Figure 4.1 Family tree of early vertebrates, including the jawless fish, the jawed vertebrates (gnathostomes), and their tetrapod descendants.

In the Late Silurian and especially the Devonian, four groups of gnathostomes evolved and quickly radiated into a huge diversity of fish that dominated the seas (**Figure 4.1**). Indeed, the Devonian is often called the "Age of Fishes" for that reason. These four groups are: (1) the extinct armored fish called placoderms; (2) the cartilaginous sharks; (3) the acanthodians, or "spiny sharks"; (4) the bony fish, which comprises the overwhelming majority of fish species alive today.

PLACODERMS

Among the most common, most diverse, largest, and most impressive group of fish in the Devonian is the placoderms (**Figures 4.1–4.4**). Their name means "plate skin", because nearly all of them had bony armored plates surrounding their head and jaws, and sometimes down the rest of their body. Their internal skeleton, on the other hand, was supported by cartilage, not bone (as in sharks), although some specimens show perichondral bone encasing the cartilage. Most placoderm fossils consist of the plates of their head shields, with occasional specimens that show the outline of their soft tissues of the rest of their bodies. When there are complete specimens, we know that they had a long body with a heterocercal tail that bent upward, as in sharks, as well as a distinct pectoral fin with cartilaginous fin supports, so they had much better control when swimming (compared to the clumsy tadpole-like body form of most jawless fish with no pectoral or pelvic fins for control). Placoderms also had well-developed pelvic fins in the back, also supported by cartilage, and most had a long dorsal fin along their backs as well.

As shark-like the rear half of their body was, their bony armor was made of a pattern of plates that were once considered unique and non-homologous with other fish. But more recent analyses of primitive Chinese placoderms like *Entelognathus* and *Quilinyu* (**Figure 4.2[A]**) showed that there is a common pattern of plates shared between placoderms and the bony fish. Placoderms also have only a single gill slit at the back of their head armor, rather than the multiple gill slits found in sharks and some bony fish. Last, placoderms have both true teeth, as found in sharks and bony fish, but they also have sharpened edges of the armored plates on their jaw shields was used for biting and grabbing their food (**Figure 4.2[D]**). From their origin in the late Silurian with Chinese fish called *Entelognathus* (**Figure 4.2[A]**), *Silurolepis*, *Shimenolepis*,

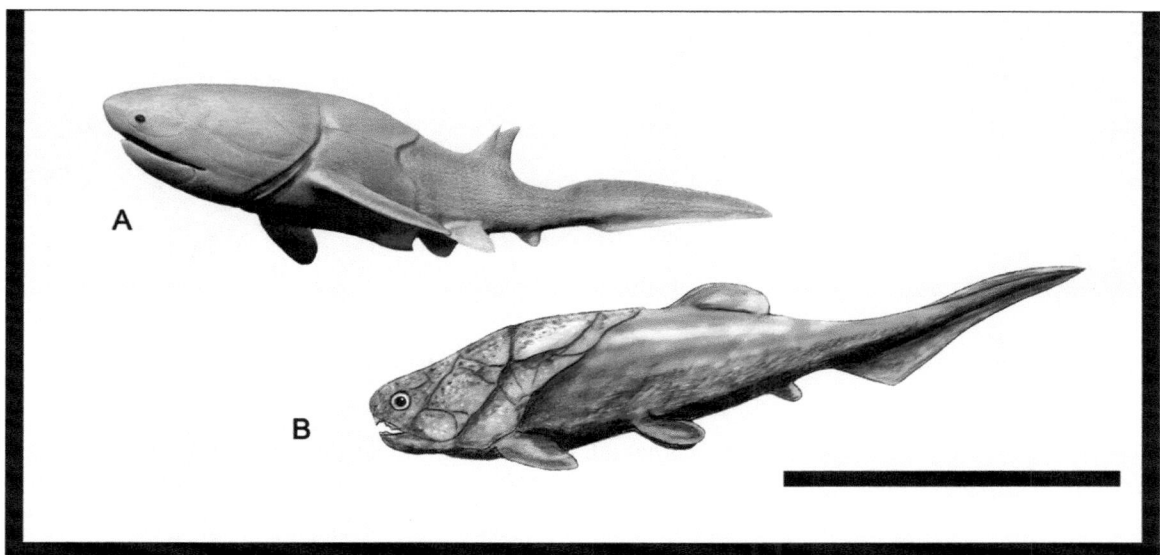

Figure 4.2 Placoderms. (A) An early placoderm from the late Silurian of China, *Entelognathus*. (B) The common small arthrodire *Coccosteus*. (Scale bar = 20 cm.) (C) The giant predatory arthrodire *Dunkleosteus*.

D

Figure 4.2 (Continued) (D) The head and body armor of *Dunkleosteus.* [(D) Courtesy Wikimedia Commons.]

and *Qilinyu* (all from China), they radiated rapidly in the Devonian. There are about 20 different families of placoderms known, with more than 240 genera named, so they were incredibly diverse and abundant.

Arthrodires

The anatomical features listed earlier are typical of most placoderms, but starting with their basic body plan, placoderms evolved into an amazing variety of different shapes and lifestyles. Perhaps the most spectacular were the ferocious predators known as arthrodires (Greek for "jointed neck"). They got this name because in the advanced forms, their head shield is connected to the thoracic armor by a distinctive ball-and-socket joint that allowed the head to raise upward when they opened their mouth for a big bite. More primitive arthrodires had a sliding joint between the head shield and thoracic shield. Their armored lower jaw could drop down sharply, so they had an enormous gape for grabbing big prey. Their biting edges on their jaw plates were like sharp self-sharpening scissor blades, so they were clearly voracious predators who could attack and swallow almost anything that shared the Devonian oceans with them. The cheek plates on the side of the face between the eyes and jaws were also hinged, giving their heads even greater flexibility.

Some arthrodires were tiny, and some were only a few meters long, but others were huge, like the famous predator *Dunkleosteus* (formerly called *Dinichthys*) (**Figure 4.2[C,D]**). At over 8 meters (30 feet) long, not only was it the largest creature in the oceans in the Devonian, but also no larger animal came along until the Age of Dinosaurs. There are specimens of large arthrodires with puncture wounds on their head plates from other arthrodires, so these were the nastiest predators of the oceans, attacking not only the largest sharks but also others of their own species. The first arthrodires are found in the Early Devonian in many places in the world from Canada to Scandinavia to Siberia to China to Antarctica. These early forms had large thoracic shields that encased their entire trunk in a cylinder of bone, almost like a barrel. A typical early arthrodire is the Australian fossil *Wuttagoonaspis*, which was a meter long, with a reduced head shield and weak lower jaws, suggesting that might have bottom feeders rather than active predators of large swimming prey.

As arthrodires evolved through the Devonian, the lower part of the thoracic shield was reduced and shortened, and their pectoral fins got larger, so their bodies became lighter and more flexible. This feature made them better swimmers, without losing the essential armor on their face and back. A common example from the Devonian of North America and Europe is *Coccosteus* (**Figure 4.2[B]**), a small form about 40 cm in length, which had an additional joint between its neck vertebrae and the skull, giving it an even bigger gape. Arthrodires were very diverse, with numerous families and genera, and they make up about 60% of all the named species of placoderms. Some arthrodires secondarily modified their biting plates into other purposes, like blunt plates for crushing molluscs, or plates shaped like picks for stabbing prey. In *Dunkleosteus* and many others, there is a bony ring around the eyeball (the sclerotic ring), which protected the eye when diving to deeper water with higher pressures (**Figure 4.2[D]**). Their eyes were relatively large, suggesting that they were highly visual predators. Recent discoveries of well-preserved arthrodires from the Gogo Formation of western Australia showed that the males had bones called claspers behind the pelvic fin (also found in sharks) to aid in copulation, and that females gave birth to well-developed live young (also something that sharks do). Some of these Australian Gogo arthrodires have strange tubular snouts (*Oxyosteus*), or jaws with a long "prow" overhanging their mouth (*Fallacosteus*). Yet other groups of arthrodires, the phyllolepids, became flattened bottom feeders, while the actinolepids had large curves spines in front of their pectoral fins.

Antiarchs

From the fast-swimming streamlined predatory arthrodires, we next look at the antiarchs, which were heavily armored bottom-feeding placoderms (**Figure 4.3**). In contrast to the arthrodires, antiarchs had then entire front of their body encased in a rigid box of armor, complete with jointed bony tubes that looked like crab legs encasing their pectoral fins (**Figure 4.3[B,C]**). Their head shield had single opening for both upward-facing eyes, plus the nostrils and pineal eye. The bulk of the body was enclosed in a barrel-like trunk shield, and only a short, naked tail stuck out the back to propel them along. Although some antiarchs reached over a meter in length, most were quite small (10–20 cm long). Clearly, such heavily armored fishes were not strong swimmers, and their flattened bodies suggested that they were mud grubbers, or perhaps ate smaller prey on the sea bottom. Specimens of *Bothriolepis* (**Figure 4.3[B]**) have been sliced into sections to show their internal structures, and they had a pair of lung-like organs, as well as a spiral intestine like sharks have. A spiral intestine is a long tubular part of the digestive tract, with a corkscrew-spiral divider down the middle, to promote absorption of food with the increased surface area of its walls.

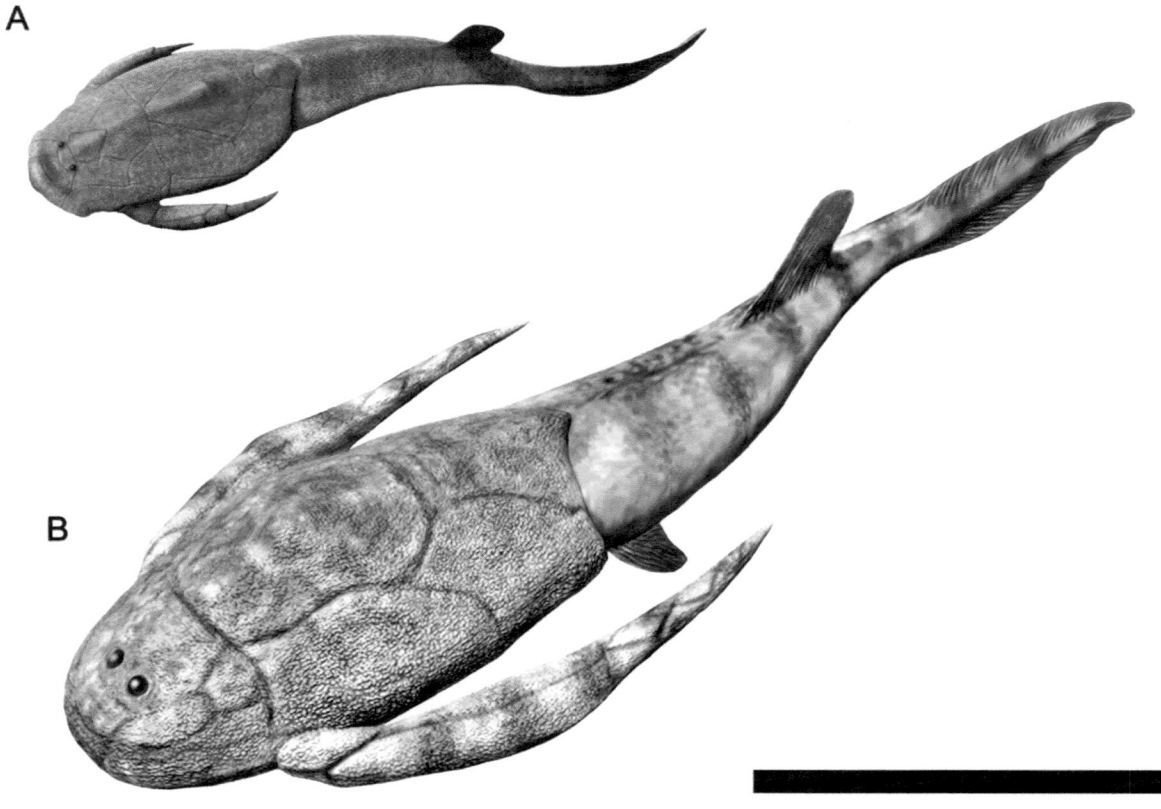

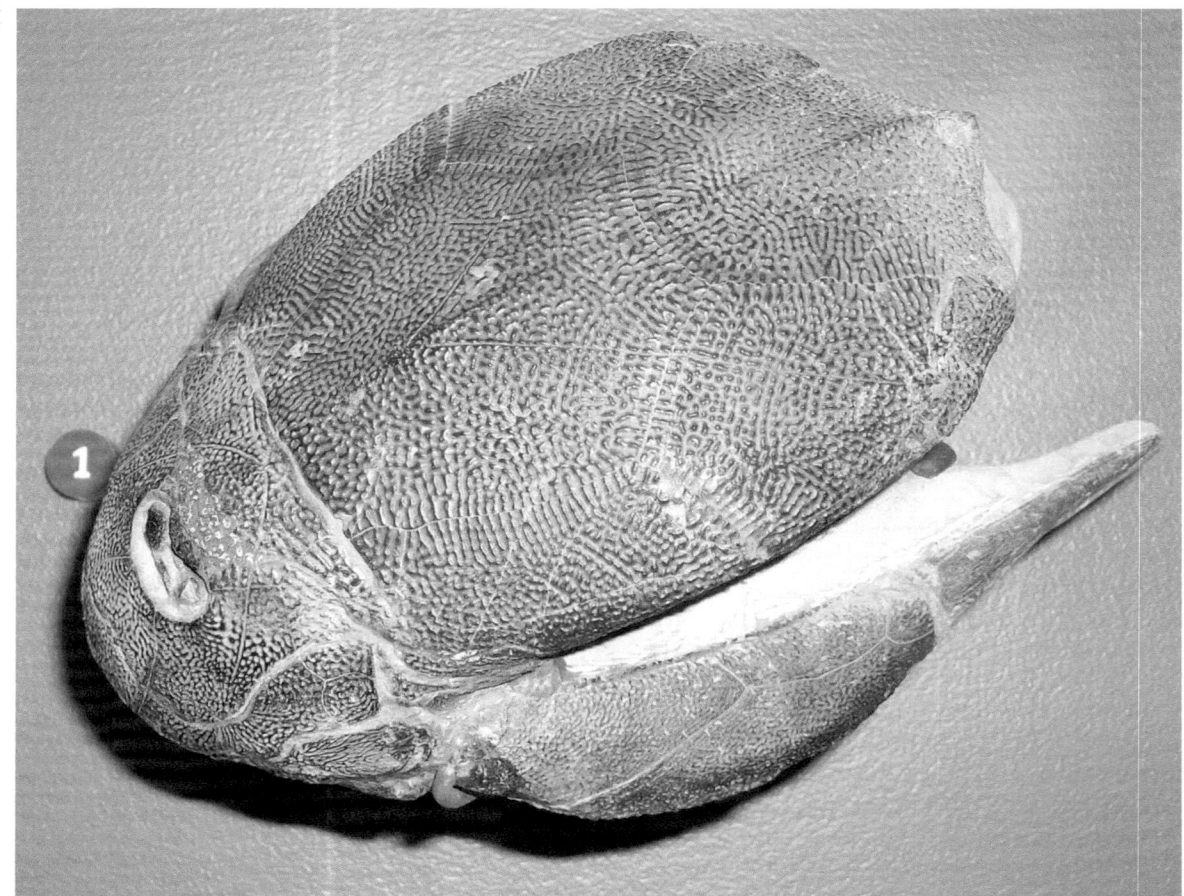

Figure 4.3 Typical antiarch placoderms. (A) *Yunnanolepis*. (B) *Bothriolepis*. (Scale bar = 20 cm.) (C) Fossil of *Bothriolepis* armor. (Scale bar = 20 cm.) [(C) Courtesy Wikimedia Commons.]

The earliest antiarchs (**Figure 4.3[A]**) from the Late Silurian of China (*Yunnanolepis*) had a single unjointed prop-like pectoral fin, but by the Middle Devonian, antiarchs were flourishing around the world, from China and Australia to Siberia to Estonia to Scotland and Canada. These early antiarchs were divided into two major groups, the bothriolepids and the asterolepids. By the Late Devonian, the genus *Bothriolepis* (**Figure 4.3[B,C]**) was extraordinarily abundant, known from over 100 different species in both freshwater and saltwater deposits, and found on every continent including Antarctica.

Besides the predatory arthrodires and the heavily armored antiarchs, placoderms came in many other shapes and sizes. One group, the rhenanids (**Figure 4.4[A,C]**), were extremely flattened with broad wing-like pectoral fins, very much like modern rays and skates. The skull was roofed by a mosaic of many unfused tile-like bony plates, and the eyes pointed upward. They had a short trunk shield, and plates along the top of the long and narrow tail, also like that of a skate or ray. Presumably these features suggest that they also lived like modern skates and rays, feeding on creatures that lived on the sea bottom, and crushing them with the plates in their mouth. By contrast to the abundant arthrodires and antiarchs, rhenanids were relatively rare, because the many small plates of their head shield fell apart when they died. Only five genera are known in the Early and Middle Devonian, but they were probably more diverse, but they were just harder to fossilize. Rhenanids were virtually worldwide in distribution, found everywhere from the Early-Middle Devonian of Germany (*Gemuendina*) to Saudi Arabia (*Nefudina*) to Bolivia (*Bolivosteus*) to Ohio (*Asterosteus*, *Ohioaspis*), and the last of them were found in the Late Devonian of Germany (*Jagorina*).

Yet another odd-shaped group of placoderms were the ptyctodonts (TIK-doh-donts), which had deep, narrow heads with greatly reduced head armor with only a tiling of small plates around the face, and robust crushing plates in their upper and lower jaws (**Figure 4.4[B]**). In this regard, they resemble the modern ratfish or chimaeras, which use the plates in their mouth to crush molluscs and other shelled prey. Also like sharks and ratfish (and certain arthrodires), ptyctodonts had long whip-like tails, a tall dorsal fin, and pelvic claspers in the males, to guide their sex organs to the reproductive tract in the female. The Australian ptyctodont *Materpiscis* (**Figure 4.4[B]**) was discovered with the skeletons of embryos inside of it. Unlike the arthrodires and antiarchs, which are well armored and commonly fossilized, ptyctodonts were like rhenanids in being lightly armored and rare, so only about a dozen genera are known.

Arthrodires, antiarchs, rhenanids, and ptyctodonts make up just four of the ten orders of placoderms known. Another group, the petalichthyids, were flat bodied but had long-curved spines sticking out of their shoulders, near the pectoral fins (**Figure 4.4[D]**). Even more peculiar are the long-snouted Brindabellapsida from the Early and Middle Devonian. Still others, such as the Stensioellida and Pseudopetalichthyida, are known from the Early Devonian, and represent archaic, primitive members of the placoderm radiation. Finally, at the end of the Devonian, during the climax of their diversification (especially among arthrodires), the placoderms vanished, as did most of the marine species of the time, since the Devonian mass extinctions were the third biggest in earth history. The first extinction event was at the beginning of the last stage of the Devonian (known as the Famennian), when 35 of 46 families of fish died out during this great crisis, including the last of the armored jawless fish, virtually all the placoderms, and 10 families of lobe-finned fish. The rest of the placoderms vanished in the second extinction pulse at the end of the Famennian Stage (and the end of the Devonian). The reasons for this extinction are still controversial, but whatever the cause, the largest fish in the sea (especially its largest predators) vanished for good, leaving the oceans to be inhabited by the two groups of fish that dominate

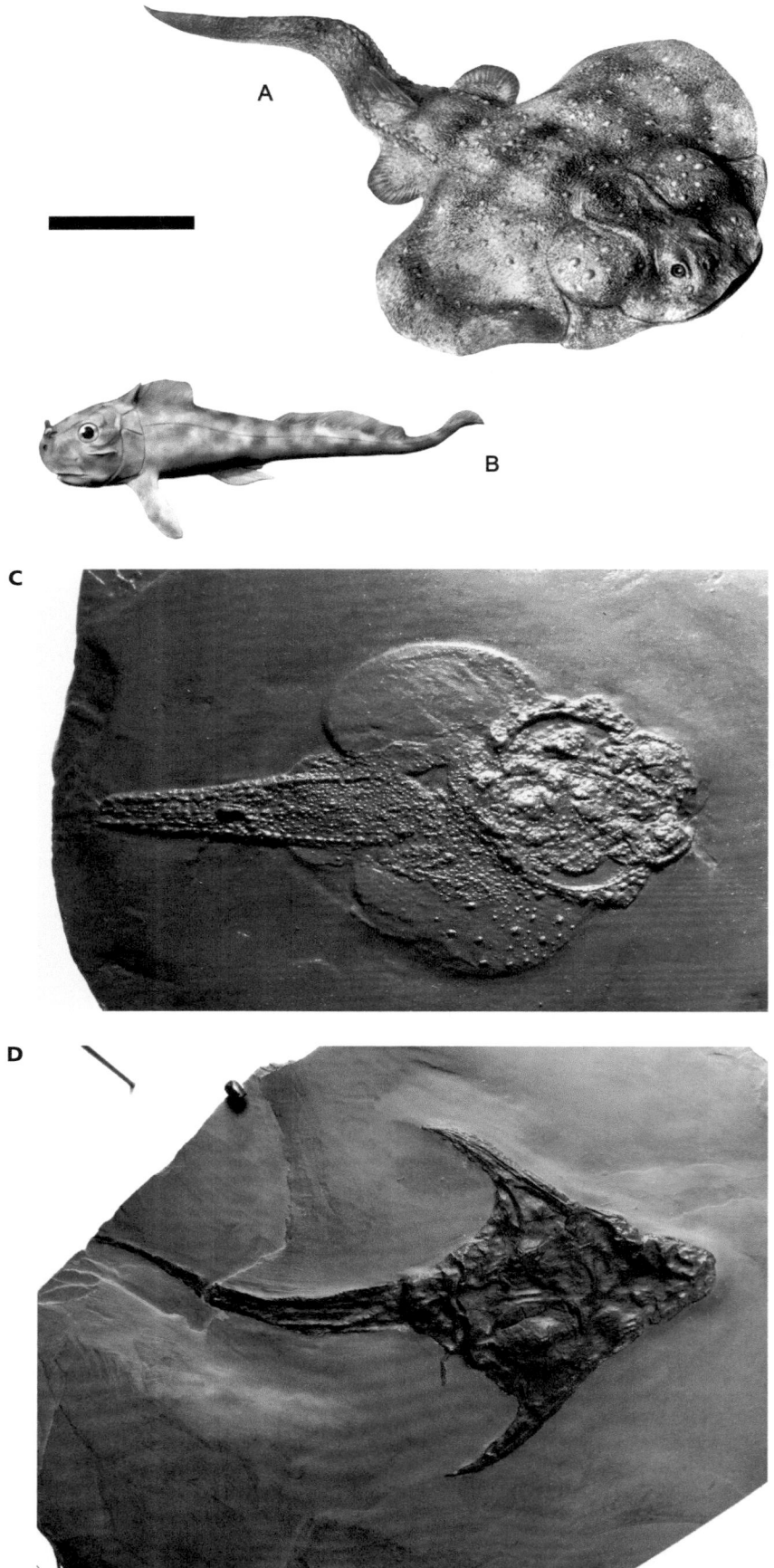

Figure 4.4 (A) The ray-like flattened rhenanid placoderm *Gemuendina*. (B) The ptyctodont *Materpiscis*. (Scale bar = 20 cm). (C) Fossil of *Gemuendina*. (D) A fossil of the petalichthyid *Lunaspis*. [(C–D) Courtesy Wikimedia Commons.]

today, the cartilaginous fish (sharks, skates, and rays), and the bony fish. These are discussed in the following sections.

CHONDRICHTHYANS: SHARKS, RAYS, SKATES, AND CHIMAERAS

The jawless fish and the extinct placoderms are relatively unknown to the public, but everyone knows sharks. They are the subject of fear in the movies and cable TV, and the Discovery Channel once a year devotes an entire week to "Shark Week", with videos about nothing but sharks. Yet sharks are not merely "menaces of the deep" or "eating machines with soulless eyes" as they portrayed in the media, but amazing animals that have been around since the Late Silurian, and have survived most of the earth's great mass extinction events. They are incredible creatures, and have adopted many ways of life besides the ones familiar on TV and movies. Sadly, they are now threatened in many parts of the world, since people fear them and hunt them, and they are slaughtered by the millions just for their dorsal fins, one of the chief ingredients of shark-fin soup. Humans are deathly afraid of getting bitten by sharks (an extremely rare occurrence), but humans are actually a much greater threat to sharks than they are to us. Humans kill about 73 million sharks each year, while sharks kill at most 20–30 people each year.

Despite their long history, sharks have a disadvantage in that most of their bodies is made of softer tissues, held up by a skeleton of cartilage, so they are harder to fossilize. The sharks, skates, rays, and chimaera are known as the Chondrichthyes, which literally means "cartilaginous fish". The only bony tissue in their bodies are the sharp denticles imbedded in their skin, and of course, their hundreds of teeth, which is often the only fossils we have of many sharks. The bony structure of the teeth and denticles are unique to chondrichthyans, and helps define them as a group. Another unique evolutionary novelty which defines the Chondrichthyes is a set of cartilages around their mouth known as labial cartilages.

Another unusual feature of sharks is a set of rod-like cartilages that trail behind the pelvic fin and anal opening in male chondrichthyans known as pelvic claspers. These facilitate internal fertilization (as mentioned earlier, they are also found in two kinds of placoderms as well). During copulation, the male wraps his body around the female and inserts one of the claspers into her urogenital opening, known as the cloaca. The claspers often have hooks and spines on them to prevent them from slipping out. Sperm is then injected into the female, and she then nurses the eggs inside her body until they are ready to be laid in their distinctive egg cases. Some sharks even give birth to live young. This kind of internal fertilization is very different from the external fertilization of most fish, in which the females lay their eggs in the water in clumps attached to a surface; the male then sprays them with sperm. Most sharks give birth to only a few well-developed young that have a good chance of survival, whereas fish without the internal fertilization mechanism lay hundreds of eggs, only a few of which survive and grow to maturity. This is why sharks reproduce so slowly, and are now threatened by excessive overfishing.

There are many other unusual features seen in the chondrichthyans. They do not have a swim bladder or any other air-filled sac, such as is found in bony fish, so sharks and rays have trouble staying neutrally buoyant. Many sharks have an oil-filled liver to reduce their density, but it is not as effective as an air pocket, so sharks must swim continuously to avoid sinking. Contrary to the popular myth, however, sharks do not need to swim to breathe—their gills have very effective pumping mechanisms that work fine when the fish is stationary. Many chondrichthyans, like rays and skates, are bottom dwellers that lie still and breathe just fine.

The body fluids of a shark are controlled by a complex mechanism of retaining urea in their blood to maintain osmotic balance between the salt water outside their bodies and their own body fluids. When they absorb too much salt, they have a special rectal salt gland to get rid of salts. A few sharks (sting rays, sawfish, and bull sharks) are capable of living in brackish or fresh water, changing the urea concentration in their bodies to maintain their osmotic balance in freshwater.

There are two main groups of Chondrichthyes: the familiar sharks and rays, or elasmobranchs, and the weird deep-water fish known as ratfish or chimaeras (discussed next). The teeth of sharks are found in rocks as far back the Early Silurian, so we know they are among the oldest jawed vertebrates to evolve. In the Early Devonian, we have the oldest articulated shark fossil, *Doliodus*, from Canada. The earliest elasmobranchs known from nearly complete fossils are the Devonian cladodont sharks. The best known of these is *Cladoselache* (clad-o-SELL-a-key), generally with a shark-like body form, but much more primitive than living sharks in many ways (**Figures 4.5[A]** and **4.6[A]**). It had a very wide, triangular pectoral fin with a broad base that clearly could not be flexed or rotated; this fin must have served as a stiff stabilizer. There were two small dorsal fins, each with a thick horn-like spine in front of it. The heterocercal tail fin had a well-supported lower lobe, making it almost symmetrical. These sharks are best known from the Upper Devonian Cleveland Shale of Ohio, where they reached about 2 meters (6–7 feet) in length. However, they were probably prey for their contemporaries, the giant arthrodires like *Dunkleosteus*, which were up to 4 times their size.

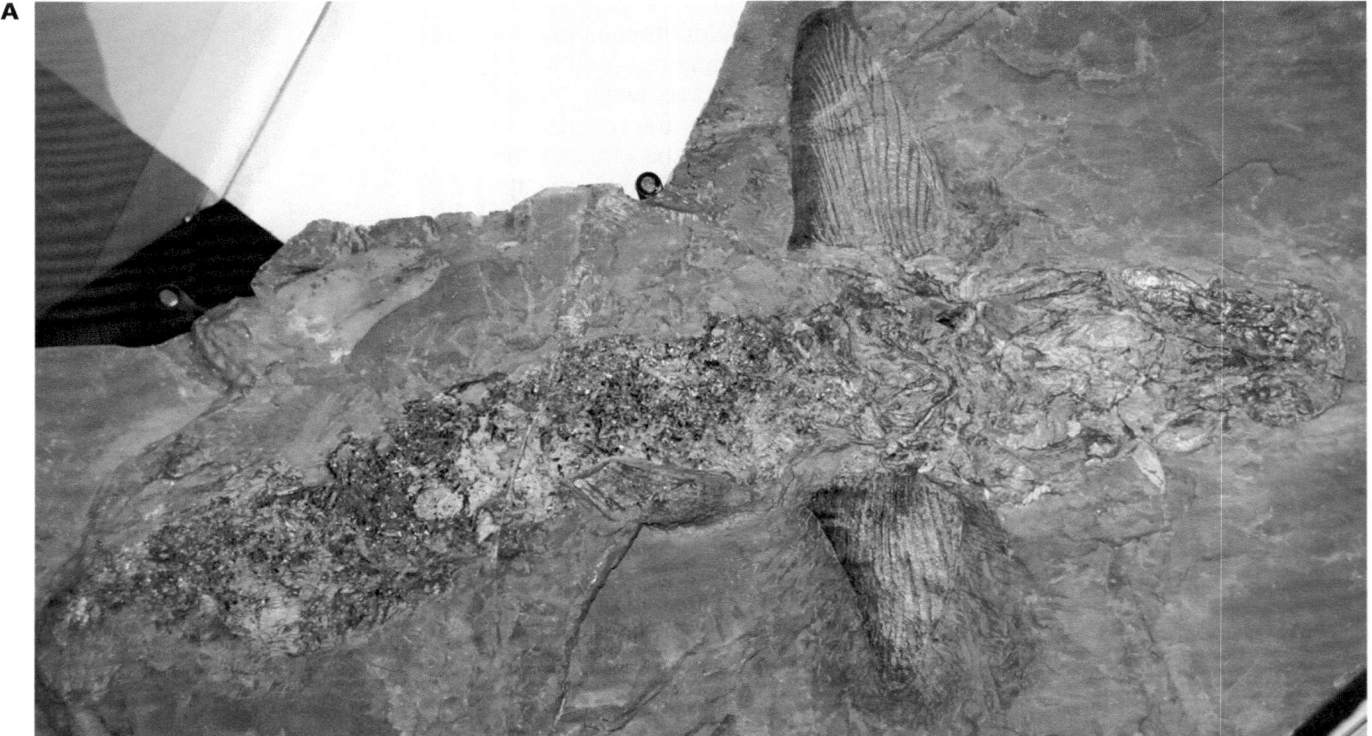

Figure 4.5 Fossils of various extinct sharks. (A) Nearly complete crushed specimen of the Late Devonian cladodont shark, *Cladoselache*.

Figure 4.5 (Continued) (B) Teeth of the gigantic *Otodus megalodon*, among a large number of mako shark teeth. (C) Life-sized reconstruction of *O. megalodon*.

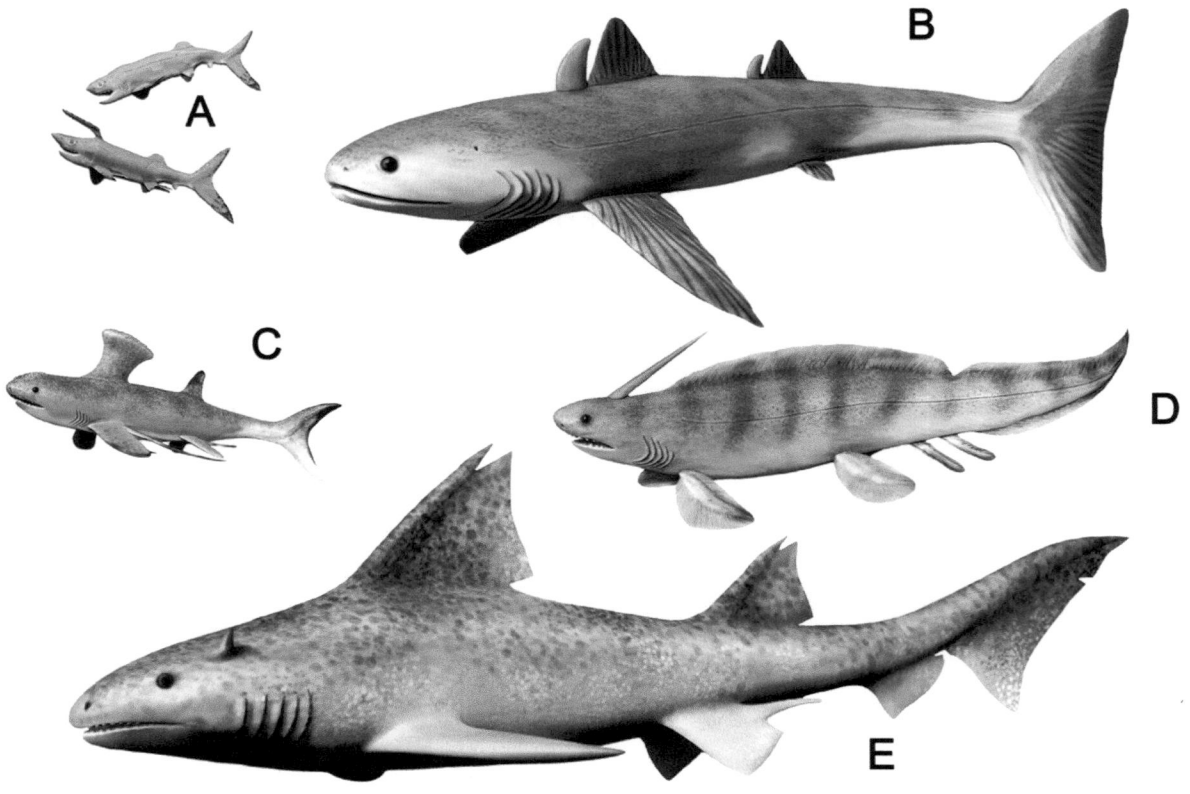

Figure 4.6 Primitive sharks. (A) Restoration of *Falcatus*. (B) *Cladoselache*. (C) *Sthethacanthus*. (D) *Xenacanthus*. (E) The Jurassic hybodont shark *Egertonodus*.

By the Carboniferous, sharks got more even weirder. Some, like *Falcatus*, from the Mississippian Bear Gulch Limestone of Montana, had an odd sickle-shaped spine curved up and forward over its snout (**Figure 4.6[A]**). *Akmonistion* from the Early Carboniferous of Scotland, and *Stethacanthus* (**Figure 4.6[C]**) from the Late Carboniferous of Eurasia and North America, had a strange anvil-shaped structure over their head and in front of the dorsal fin, called the "spine-brush" complex. The function of this weird structure has been debated for a long time. Scientists have suggested that it might be a feature used to advertise the maturity and status of males (like horns and antlers in some mammals), or a structure that might have served to make the shark look larger and scarier to predators (these sharks were only 1–2 meters total body length). These sharks also had long extensions on their pectoral fins called "fin whips", whose function is also unknown.

In the Permian, even weirder sharks evolved. The most common sharks were the odd-looking xenacanth or pleuracanth sharks (**Figure 4.6[D]**). Their distinctive double-pronged teeth are abundantly preserved in the freshwater and deltaic deposits of the Carboniferous coal swamps and the Permian floodplain red beds. They even survived the Permian catastrophe, although they disappeared for unknown reasons during the Triassic. Some xenacanths were big, with bodies over 3 meters (10 feet) long, and thus were among the largest freshwater fish the world had yet known (only the giant lobe-fins like *Rhizodus* of the Carboniferous were larger, reaching 7 m). Xenacanths are also one of the most anatomically peculiar sharks. Instead of the normal heterocercal tail, they were the only sharks with a symmetrical tail that came to a central point, something like the tail of an eel. The large pectoral and pelvic fins were supported by a central rod of cartilage and branching fin rays, a structure seen in no other shark or fish. The single long dorsal fin ran more than half the length of the body and merged with the tail fin. On top of their heads was a long bony spine that projected up and backward. No one knows what this was for, although extraneous spines are a common feature in many sharks. The earliest xenacanths look more like normal sharks, without the weird fins or tail. Their peculiar eel-like body shape of later xenacanths is thought to have been an adaptation for living in the sluggish waters of swamps and lakes and lunging at their prey from the murky water (in contrast to the slow, steady swimming of marine sharks).

Sharks went through yet another evolutionary radiation beginning in the late Paleozoic and continuing through the Mesozoic. Most early Mesozoic sharks are known as hybodonts (**Figure 4.6[E]**), a group that is the closest relative of the living sharks. Hybodonts had pectoral and pelvic fins with a much more advanced support structure, and narrow bases so their fins could be turned and flexed for good control of swimming. They had two dorsal fins, each with a thick spine in front. Their teeth were highly differentiated, with multi-cusped pointed teeth for piercing up in front and blunt-cusped teeth for crushing toward the back of the jaw. Hybodonts were common in both marine and fresh waters during the Mesozoic, but reached only about 2 meters in length, so they probably didn't prey on the larger dinosaurs.

In the Late Triassic and Jurassic, we find the first evidence of sharks with calcified vertebrae, a feature of the modern group of sharks, or neoselachians. Since then, sharks have radiated into hundreds of species, including not only familiar forms like the dogfish and great white shark and hammerhead (360 living species), but also 456 living species of skates and rays. The largest of the extinct sharks were the giant great white sharks (*Otodus megalodon*, formerly placed in the genus *Carcharocles* or *Carcharodon*), which reached almost 12 meters (40 feet) long (**Figure 4.4[B,C]**). Neoselachians have further modified their jaws so that the upper jaw cartilage (the palatoquadrate) is suspended from the hyomandibular cartilage (the upper element of the first gill arch). When the shark gapes, not only does it drop its lower jaw, but also it can protrude its upper jaw, increasing its bite capacity. The next time you see a video of a shark opening its mouth, watch the upper jaw bulge forward as the lower jaw drops downward.

Neoselachians exhibit a wide variety of lifestyles. Most of the familiar forms are voracious open-ocean predators, but the flat-bodied skates and rays swim with their huge pectoral fins and spend much of their time lying on the bottom buried in the sediment and feeding on molluscs, crabs, and other bottom-dwelling invertebrates. The largest living chondrichthyans, such as the manta ray, whale shark, and basking shark, however, are plankton feeders. They open their mouths as they swim and filter plankton, crustaceans, and fish from the water with their gill apparatus.

The other major branch of the chondrichthyans is the holocephalans, also known as the chimaeras, ratfish, or rabbit fish (**Figure 4.7**). These fish are alive today (50 species in 6 genera in four different families), but few people other than deep-water fishermen or ichthyologists ever see them, because they live in waters over 80 meters deep. They have large eyes (adapted for the dark waters in which they live), a long whip-like tail (hence the name "ratfish"), and very broad pectoral fins, and they are very spiny. Their short snout bears large pavement-shaped tooth plates for crushing molluscs, their principal prey. The name "holocephalan" means "whole head" in Greek, because unlike other living sharks, they have fused the upper jaw cartilage (palatoquadrate cartilage) to the braincase cartilage (chondrocranium), so their jaws cannot be protruded like in neoselachian sharks. Instead, the jaws are rigid and reinforced within the head for the stresses of shell crushing. Unlike the typical shark configuration

Figure 4.7 The chimaera or ratfish. (Courtesy Wikimedia Commons.)

of multiple gill slits, holocephalans have a single gill opening. As different as they are from typical elasmobranch sharks, they still have all the key chondrichthyan characteristics, including pelvic claspers, which are shaped like a medieval mace. Their place among the Chondrichthyes was once debated, but now it is clear from both molecular evidence and anatomy that holocephalans are the nearest relative of all the elasmobranchs. The earliest fossil holocephalans are known from the Carboniferous, although they are much rarer as fossils than are the elasmobranchs.

Perhaps the weirdest chondrichthyan of all was known only from the strange spirals of shark teeth long known as "tooth whorls" from a genus dubbed *Helicoprion* (**Figure 4.8**). Some of these tooth whorls suggested a shark 12 meters in length (almost 40 feet long), one of the largest fish of the Paleozoic. For a long time, nobody could visualize how such a whorl of teeth could fit into the mouth of any fish, let alone function in a shark. All sorts of weird suggestions were made, including a whorl of teeth curling back from the nose and over the head. But in 2013, nearly complete body specimens were found which showed that the tooth whorl fit into the middle of the lower jaw, forming a sort of flexible spiral saw blade that produced a vertical slice as it chopped into its prey. They would have had a very long narrow skull, with crushing teeth in the upper jaw, to complement the rotary saw blade of the lower jaw. These specimens also have features that suggest that *Helicoprion* is not a shark (as long thought) but a weird side branch of the holocephalans, without the specialized shell-crushing lifestyle of living species.

A

B

Figure 4.8 (A) *Helicoprion* **fossil. (B) Modern reconstruction of** *Helicoprion.* [(A) Courtesy Wikimedia Commons.]

ACANTHODIANS

After placoderms and chondrichthyans, the third great group of early and middle Paleozoic fish was the acanthodians (**Figures 4.1** and **4.9**). Their oldest known fossils are isolated spines from the Late Ordovician, so they are probably the earliest known jawed vertebrates (unless the reports of Ordovician shark teeth are true). Their nickname is the "spiny sharks", but they are not sharks or even closely related to any specific chondrichthyan group. For a long time, their relationships were controversial (**Figure 4.1**). Some thought that they were closely related to bony fish, while others placed them with sharks or with placoderms. The current consensus is that they are distantly related to sharks based on the similarity of their body, their cartilaginous skeleton, and their early occurrence. Some analyses point to the osteichthyan features of some specimens, such as the large ear bones and the details of their internal anatomy, gill structure, and braincases, and suggest that certain acanthodians are more closely related to bony fish, while others are closer to sharks and their kin.

The most striking feature of acanthodians is their multiple spines (hence the nickname "spiny sharks"; the Greek word *akanthos* means "spiny").

Figure 4.9 Reconstructions of a variety of acanthodians: (A) *Cheiracanthus*, (B) *Acanthodes*, (C) *Gyracanthus*, (D) *Diplacanthus*, (E) *Climatius*, (F) *Ishnacanthus*, (G) *Parexus*. Scale bar is 5 cm.

All their fins are supported by spines on the front edge, including the two dorsal fins, and in some species, there are paired rows of as many as six fins that run along the belly between the pectoral and pelvic fins. They even have a separate anal fin. These fins were probably not very mobile, however, so they did not make the fish more maneuverable, but probably served as stabilizing devices for such an active swimmer. The bony spines were sunk deep into the body so they did not flex or move, but the structure of isolated acanthodian spines is very distinctive. Most acanthodians are known only from a few spines, but in some Devonian deposits, there are rare impressions of complete fish. These fossils show that acanthodians had large eyes, a short snout, and very advanced jaws with a single row of unreplaced teeth embedded in them (not the multiple teeth of sharks, which are continuously shed). They did have some primitive, shark-like features, such as a heterocercal tail and five gill arches. The majority of acanthodians are known from freshwater deposits, although their spines are also common in marine rocks.

FURTHER READING

Ahlberg, P.E.; Trinajstic, K.; Johanson, Z.; Long, J.A. 2009. Pelvic claspers confirm chondrichthyan-like internal fertilization in arthrodires. *Nature*. 460 (7257): 888–889.

Amaral, C.R.L.; Pereira, F.; Silva, D.A.; Amorim, A.; de Carvalho, E.F. 2017. The mitogenomic phylogeny of the Elasmobranchii (Chondrichthyes). *Mitochondrial DNA. Part A, DNA Mapping, Sequencing, and Analysis*. 29 (6): 867–878.

Davis, S.P.; Finarelli, J.A.; Coates, M.I. 2012. *Acanthodes* and shark-like conditions in the last common ancestor of modern gnathostomes. *Nature*. 486 (7402): 247–250.

Janvier, P. 1998. *Early Vertebrates*. Oxford University Press, Oxford.

King, B.; Qiao, T.; Lee, M.S.Y.; Zhu, M.; Long, J.A. 2016. Bayesian morphological clock methods resurrect placoderm monophyly and reveal rapid early evolution in jawed vertebrates. *Systematic Biology*. 66 (4): 499–516.

Long, J.A. 2010. *The Rise of Fishes* (2nd ed.). Johns Hopkins University Press, Baltimore.

Long, J.A.; Trinajstic, K.; Johanson, Z. 2009. Devonian arthrodire embryos and the origin of internal fertilization in vertebrates. *Nature*. 457 (7233): 1124–1127.

Long, J.A.; Trinajstic, K.; Young, G.C.; Senden, T. 2008. Live birth in the Devonian. *Nature*. 453 (7195): 650–652.

Maisey, J.G. 1996. *Discovering Fossil Fishes*. Henry Holt, New York.

Moy-Thomas, J.; Miles, R.S. 1971. *Palaeozoic Fishes*. Saunders, Philadelphia.

Nelson, J.S. 1994. *Fishes of the World* (3rd ed.). Wiley, New York.

Norman, J.R.; Greenwood, P.H. 1975. *A History of Fishes* (3rd ed.). Ernest Benn, London.

Sallan, L.; Coates, M. 2010. End-Devonian extinction and a bottleneck in the early evolution of modern jawed vertebrates. *Proceedings of the National Academy of Sciences*. 107 (22): 10131–10135.

OSTEICHTHYES

THE BONY FISH

5

The term "fish" is of value on restaurant menus, to anglers and aquarists, to stratigraphers and in theological discussions of biblical symbolism. Many systematists use it advisedly and with caution. Fishes are gnathostomes that lack tetrapod characters; they have no unique derived characteristics. We can conceptualize fishes with relative ease because of the great evolutionary gaps between them and their closest living relatives, but that does not mean they comprise a natural group. The only way to make the fishes monophyletic would be to include tetrapods, and to regard the latter merely as a kind of fish. Even then, the term "fish" would be a redundant colloquial equivalent of "gnathostome" (or "craniate", depending upon how far down the phylogenetic ladder one wished to go).

—John Maisey, "Gnathostomes", 1994

FISH BONES

By far the largest group of fish alive today is the Osteichthyes, or bony fish. Other than the lamprey, hagfish, and the sharks and their kin, all the fish you encounter are bony fish, from the fish tank to the seafood restaurant to the lakes and oceans (**Figure 4.1**).

Bony fish are distinctive in a number of specializations, most of which involve using bone rather than cartilage for most of their skeleton. They have dermal bone (derived from the embryonic precursor of skin) in their shoulder girdle, palate, along the outside of the jaw, in the bones that cover the gill flap, and in their throat region—and often in many other places. The remaining bones of the skeleton, on the other hand, are derived from the embryonic cartilage precursors, such as the chondrocranium for the braincase, the palatoquadrate cartilage for the upper jaws, and Meckel's cartilage for the lower jaws. (Sharks still use these cartilages for their head skeleton, with no bone replacing it.) Bony fish usually have some sort of gas-filled chamber (a lung, which later evolved into a swim bladder) which evolved as a branch of the gut region (typically branching off the esophagus). Typically, bony fish have only three or fewer gill arches to breathe with, which are concealed behind the gill cover, or operculum. All of these features and more confirm that bony fish are a natural group.

The Osteichthyes are divided into two groups (**Figure 4.1**): the lobe-finned fish or sarcopterygians (*sarkos* is "flesh" and *pterygos* is "fin" in Greek) have a series of robust bones and fleshy muscles supporting their lobed fin. These include the lungfish, coelacanth, a number of extinct fish in the wastebasket group called "rhipidistians", and their descendants include all four-legged vertebrates (tetrapods). They are discussed in the next chapter.

The other group, the ray-finned fish or actinopterygians (*aktinos* is "ray" and *pterygos* is "fin" in Greek), support their fins with many thin parallel rods of bone. These ray fins have a unique characteristic in that the

DOI: 10.1201/9781003128205-5

fins are relatively long, and attach to relatively small bones or cartilages which articulate with the limb girdle. In addition to the ray fins, actinopterygians are defined by many other anatomical features, including a single dorsal fin (most other fish have two or more). Primitive actinopterygians have a unique type of body armor known as ganoid scale, which is composed of a thin layer of mineralized tissue called ganoine overlying layers of dentin and bone. There are also many soft-tissue and biochemical and molecular features that are unique to the actinopterygians that are found in no other group of vertebrates.

The earliest bony fish known is *Andreolepis*, from the Upper Silurian rocks of Sweden, as well as *Lophosteus* from Russia and Estonia. By the Devonian there was a big radiation of primitive ray-finned fish known as the palaeoniscoids, typified by the Devonian fossil *Cheirolepis* (**Figure 5.1[A]**). Their heavy skulls were completely encased in a roof of dermal bone (while modern bony fish have greatly reduced the bony armor). They had large eyes and a short snout. Their ray fins were triangular in shape, heavily built, and included paired pectoral and pelvic fins and an anal fin. Like sharks and acanthodians (but unlike later bony fish), they had a primitive heterocercal tail. Complete body fossils are rare, but their thick distinctive rhomboid scales (typically shaped like a parallelogram) are common fossils in the Devonian, and the group persisted through Cretaceous. These palaeoniscoids were not as diverse or abundant in the Devonian as the placoderms or sharks or jawless fish, but they survived the Late Devonian extinctions that wiped out most of the rest of these groups. Then they

A

B

Figure 5.1 (A) Reconstruction of *Cheirolepis*. (Scale = 2 cm.) (B) The Permian palaeoniscoid *Rhabdolepis*. [(B) Courtesy Wikimedia Commons.]

underwent a huge evolutionary radiation in the Carboniferous when the other groups had already vanished after the Devonian extinction.

These primitive ray-finned fish did not completely vanish, but still have three surviving groups living today (**Figure 5.2[A,B,D]**). They include the sturgeon *Acipenser* (the source of caviar), the paddlefish *Polyodon*, and a strange-looking African freshwater fish known as the bichir or reedfish (genus *Polypterus*). Some sturgeons can reach truly impressive sizes of 6 meters (20 feet) in length. Paddlefishes are equally distinctive, with their long flat prow on their nose and a throat region that expands out to filter-feed on plankton that have been stirred up out of the mud with their long noses. Although these fish (except *Polypterus*) have mostly lost the heavy dermal bones of the skull found in palaeoniscoids, they still have archaic jaw and fin configurations and (in sturgeons and paddlefishes) the heterocercal tail. These fish have been lumped until a wastebasket group (a grade of evolution, not a natural group) called "chondrosteans" ("cartilage bone" in Greek), because they have a mostly cartilaginous skeleton with only limited dermal bone except for bony ossicles in their skins.

The primitive grade of "chondrostean" ray-finned fishes was replaced in the Jurassic and Cretaceous by more advanced fishes known as the neopterygians (**Figures 5.2** and **5.4**). They are considerably more advanced than the "chondrostean" grade of fish evolution. Instead of shark-like heterocercal tail, they have a tail in which the supporting spine is shortened and sharply upturned, forming a homocercal tail, in which the top lobe is symmetrical with the bottom lobe. Even more striking is the change in their

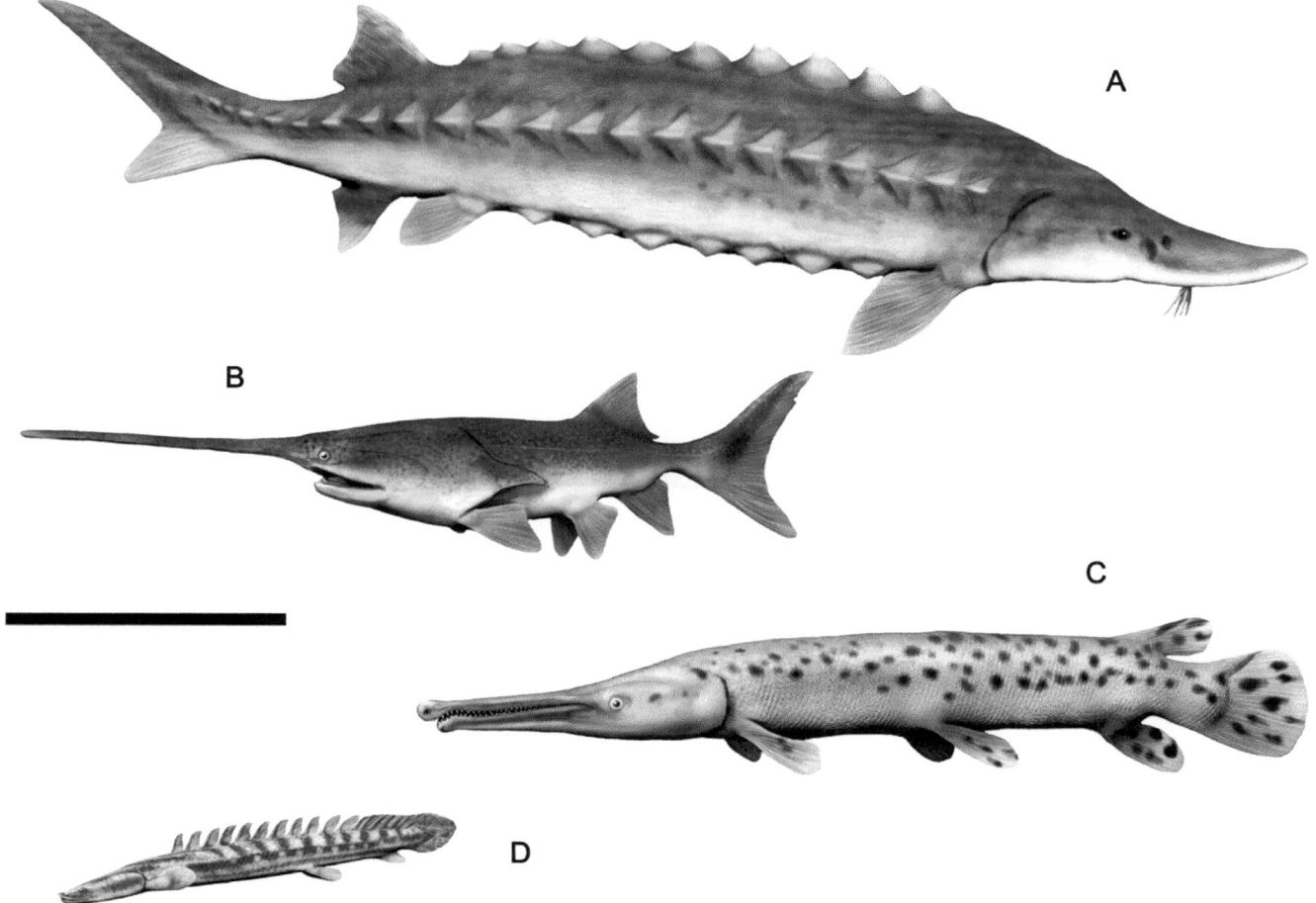

Figure 5.2 Examples of living archaic actinopterygians. (A) The sturgeon (*Acipenser*). (B) The paddlefish (*Polyodon*). (C) The garfish (*Lepisosteus*). (D) The bichir (*Polypteru*). (A), (B), and (C) used to be called "chondrosteans" and (C) was once called a "holostean". (Scale = 1 meter.)

jaw apparatus. Palaeoniscoids (**Figures 5.1** and **5.4**) had relatively robust, thick bony jaws, which were capable only of a simple snapping bite. Their jaw muscles were small and restricted to a narrow slot in the jaw hinge. However, in more advanced actinopterygians, the maxillary bone of the upper jaw is detached from the skull at the back end and swings on a hinge at the front end, restrained only by muscles; this allows the jaw to open much wider and permits the expansion of much stronger jaw muscles.

Neopterygians dominated the waters of the Mesozoic (**Figure 5.3**), and evolved to a wide variety of body forms that converged on many modern fish alive today. In addition to fishes with the standard fish-like body shape, early Neopterygii evolved into torpedo-shaped fast swimmers, eel-shaped fish, fish with long dorsal fins, and deep-bodied fish. Many of these body shapes were also seen among the palaeoniscoids, and are seen again among the living teleost fish, so clearly a lot of convergent evolution must have taken place. Although most of this "holostean" radiation was extinct by the end of the Cretaceous, two lineages are still alive today: the bowfin (*Amia*) and the gar (*Lepisosteus*) (**Figure 5.2[C]**). Despite the fact that these archaic fish are far outnumbered by their more diverse and successful teleost relatives, they are remarkably durable and versatile creatures. They are found in most bodies of freshwater in the southeastern United States, surviving as "living fossils", although their fossil record shows that both bowfins and gars once lived in many parts of the world. Bass fishermen

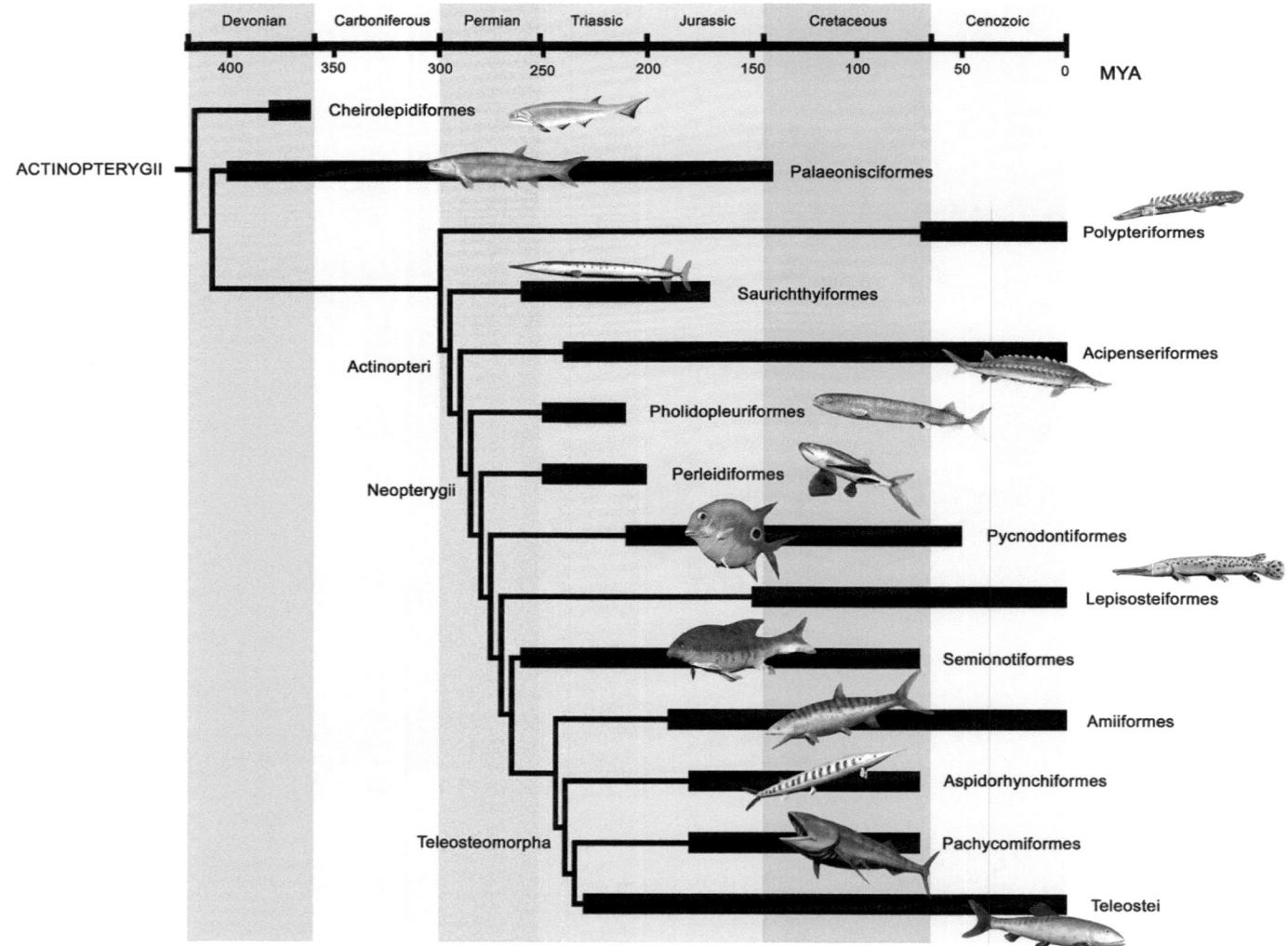

Figure 5.3 The radiation of primitive ray-finned fishes (actinopterygians), from the Mesozoic radiation of "chondrosteans" and "holosteans", followed by the explosion of teleosts in the Cretaceous and Cenozoic.

have repeatedly attempted to eradicate gars, because these fish are voracious predators of the bass and have spread to most lakes and rivers in the
southern United States. With their heavily armored bodies, long tubular
snouts with needle-like teeth, and distinctive heavy scales, gars are very
easily recognized, and their fossil scales are common in the dinosaur beds
of the West. Some alligator gars are known to reach 4 meters (12 feet) in
length. Gars even have adaptations for breathing air when the water in
their pond becomes too stagnant, so they are true survivors.

THE AGE OF TELEOSTS

The most advanced ray-finned fish are the teleosts ("completely bone"
in Greek), which originated in the Triassic, then underwent an explosive
adaptive radiation in the Cretaceous to completely take over the waters of
the world (**Figure 5.3**). With almost 30,000 species, they make up almost
99% of all the fish alive today. Nearly every fish you see in an aquarium, or
eat in a seafood restaurant, or catch in the ocean or a lake or a stream, is a
teleost. In fact, teleosts are more diverse than amphibians, reptiles, birds,
and mammals put together. Their diversification since the Cretaceous
dwarfs that of the radiation of early mammals or any other group. As
tetrapods and mammals, we like to think of the Mesozoic as the "Age of
Dinosaurs" and the Cenozoic as the "Age of Mammals", but if sheer numbers of individuals or species diversity or variety of shapes and lifestyles
mean anything, among vertebrates the Cretaceous and Cenozoic have
always been the "Age of Teleosts". (Of course, insects and other arthropods have all vertebrates beat in terms of either numbers of individuals
or taxonomic diversity.) If this book were to fairly represent the diversity
of vertebrates, more than half of it would be dedicated to teleosts. Most
paleontology books focus on dinosaurs, which have about 1000 known
species (excluding birds), but teleosts have 30 times as many species.

What makes a teleost distinctive? Their skulls (**Figure 5.4**) show the most
advanced stage in evolution from the simple "snap-trap" jaws of primitive
palaeoniscoids to more highly modified skulls of primitive neopterygians,
with their reduction in the dermal bone on their skulls that turned the
skull into a series of bony struts. Teleosts represent the ultimate step in
reduction of bones in the skull. Nearly every skull bone is reduced to long
thin rods which can flex or stretch using the tendons that connect them.
The premaxillary bone on the front of the upper jaw becomes enlarged
and is part of a pivoting framework that allows the entire mouth to open
wide quickly and produce great suction. If you watch a goldfish or most
other aquarium fish when they are feeding, you will see that most of them
do not bite their food, but suck it into their mouths by suddenly protruding and expanding their jaws and mouth cavity. Among the unique bones
that contribute to this unique suction mechanism are two supramaxillary
bones that are found in no other group of animals except teleosts. The rest
of the skull is also lightened and the bony components reduced, making
it a highly mobile, kinetic structure of bony struts that can flex and stretch
much more easily than the solid skulls of archaic fish.

Most teleosts have a swim bladder, a gas-filled chamber in their bodies
that makes them neutrally buoyant and much more adept at swimming
and hovering in the water. For this reason, their pectoral and pelvic fins
are no longer as critical for providing forward thrust, but instead can be
used for fine steering, hovering, and even backward swimming. Consequently, the pectoral fin in some teleosts shifts up to the middle of the
body, just behind the gill opening, and the pelvic fins shift forward, so
some teleosts can turn on a dime.

Even more diagnostic is the anatomy of their tail (**Figure 5.4**). The bones of
the symmetrical neopterygian homocercal tail have been further modified to
a distinctive series of radiating elements known as uroneurals, originating

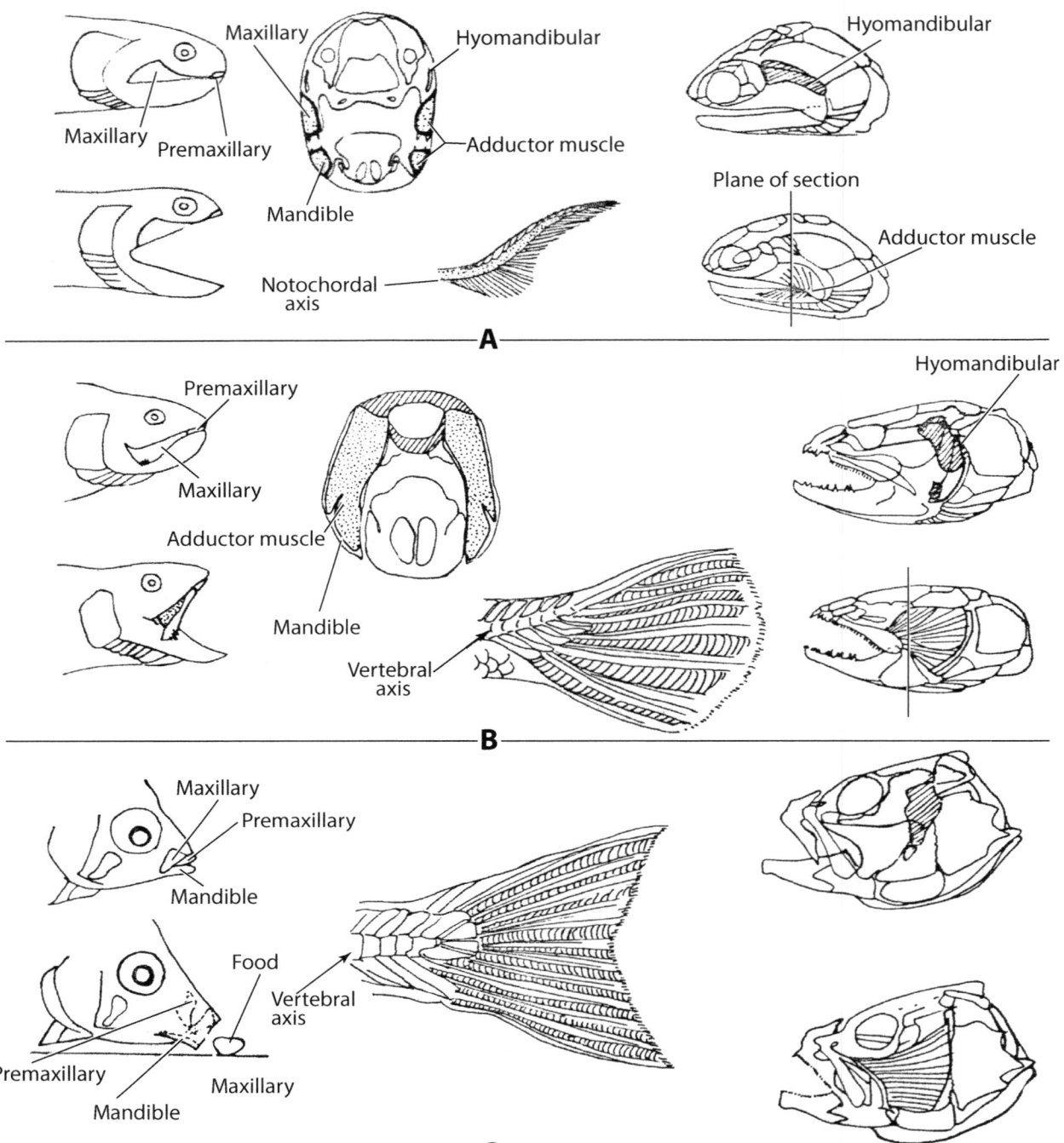

Figure 5.4 Stages in the modification of the actinopterygian skeleton, from the heavily armored bony "snap-trap" skull of palaeoniscoids (A), to the more open hinged skull of "holosteans" (B), and finally the skull of a teleost (C), with most of the skull reduced to bony struts which are hinged, and make the skull highly flexible. Most teleosts feed by expanding and opening their mouth cavity suddenly, so they suck their prey without having to bite down hard on it. Meanwhile, the tail changes from the simple heterocercal tail (found in many primitive fishes) to a homocercal tail with a reduced spinal column, and many more fin rays. The teleost tail has two distinctive bones called uroneurals, which are unique to the group. (Modified from several sources.)

from the upturned back end of the spinal column. The detailed configuration of the tail is one of the key features used in teleost classification.

Clearly, in a short chapter like this we cannot begin to describe the diversity of the 30,000 living species of teleost and all their fossil relatives. But we can mention a few outstanding examples. The earliest teleosts in the Cretaceous were generalized primitive fish without the extreme specializations we now see in everything from a seahorse to a tuna to an eel. But there were some spectacular examples, nonetheless. The largest and best known of these (**Figure 5.5**) was the huge (up to 6 meters or 20 feet long) marine fish *Xiphactinus* (once called *Portheus*). They were the largest fish predator of the Cretaceous inland seas that once flooded the American Plains region from the Gulf of Mexico to Hudson Bay, but were especially well preserved in the chalk beds of western Kansas (**Figure 5.5[B]**), as well as the Cretaceous beds of the southeastern United

A

B

C

Figure 5.5 The giant Cretaceous ichthyodectiform teleost *Xiphactinus*. (A) A complete skeleton with another fish inside, its last meal. (B) A photo of this enormous specimen collected in the field with George F. Sternberg, the collector (left), and his crew in the Cretaceous Smoky Hill chalk in 1926. (C) Reconstruction of *Xiphactinus*. [(A) Photo by the author. (B) Courtesy Wikimedia Commons.]

States. These voracious predators are often found with a smaller fish inside their stomach, causing them to choke and die and becoming fossilized shortly after swallowing their last meal (**Figure 5.5[A]**).

These Cretaceous teleosts like *Xiphactinus*, known as ichthyodectiforms, are now extinct, but the ancestors of many living groups first appeared in the Cretaceous and are still alive today. A short outline of the major groups of living teleosts would include the following (**Figure 5.6**):

- Osteoglossomorphs are often considered the most primitive group of living teleosts, many with strange eel-like shapes and long ridge-like fins that run the length of their bodies. They were common in the Cretaceous, but today there are only 217 species, found mostly in tropical freshwater lakes and rivers. They include *Arapaima*, a huge (4.5 meters or 14 foot) freshwater predator from the Amazon; the meter-long Amazonian predator *Osteoglossum*, along with the mudskippers, knifefish, elephant fish, mooneyes, and featherbacks, and some electric fish as well.
- Elopomorphs also considered a very primitive group of teleosts; some place them at the base of the teleost radiation (**Figure 5.6**), while others consider the osteoglossomorphs to be more primitive. They include about 350 species, including tarpons, eels, and many of the unusual fish of the deep ocean, like viperfish, gulpers, and others. All elopomorphs have a distinctive, ribbon-like larva known as the leptocephalus.
- Clupeomorphs are the third main group of primitive teleosts. They include the herrings, shad, sardines, anchovies, and their relatives. Of the 350 living species, most are specialized for living in large schools and feeding on plankton in the open ocean with their specially adapted gills. Clupeomorphs are of tremendous economic importance, as herring, sardines, and anchovies are among the most common food fish in many parts of the world.
- Euteleosts make up all the rest of the teleosts. One of the largest groups of euteleosts are the ostariophysans. About 6500 species of ostariophysans live today, and they are the dominant group of freshwater fish on this planet. Most freshwater fish in your fish tank, backyard pond, or in the pet store are ostariophysans, including the goldfish, carp, minnows, suckers, loaches, tetras, piranhas, and catfish. Ostariophysans have a unique series of bones attached to the front of the swim bladder known as the Weberian apparatus. These bones connect the swim bladder to the inner ear, allowing the fish to hear the sounds amplified by the hollow chamber of the swim bladder, and thus greatly enhancing their sensitivity to sound.
- Salmoniformes are another group of euteleosts. These include the salmon, trout, pike, pickerel, muskellunge, and lantern fish of the deep sea. Some of these fish are fast-swimming predators of the freshwaters and lakes and are very important to both sportsmen and commercial fishermen.
- A third group of euteleosts are the paracanthopterygii. These include about 1160 living species, including the cod and anglerfish and their relatives.
- The largest group of euteleosts are the acanthomorphs, or spiny teleosts. These include about 15,000 living species in 300 families. Many acanthomorphs bear hundreds of erectile spines on their bodies. When they are threatened, they bristle like a porcupine and become harder to swallow. Among the many familiar groups of acanthomorphs are the atherinomorphs (silversides, grunions, half-beaks, killifish, seahorse, barracuda, flying fish, guppies, mollies, swordtails, kissing fish). The atherinomorphs are distinguished by their extraordinarily protrusible mouthparts, which is why many of them can "kiss" or feed with a similar suction motion. Finally, the largest of all the acanthomorph groups are the percomorphs (about 9000 living species), including the

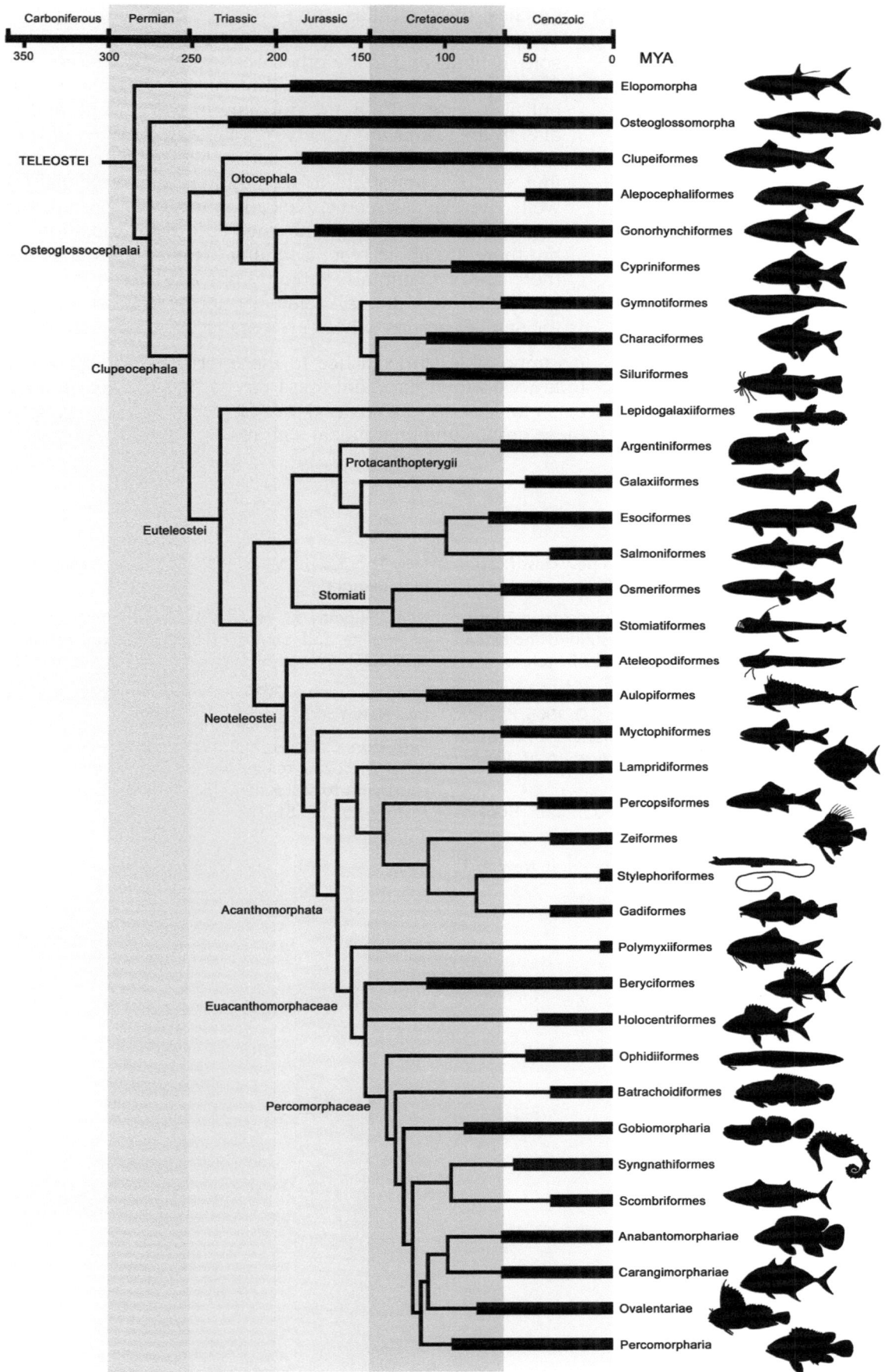

Figure 5.6 A modern family tree of teleost fishes, focusing on the living groups. [Modified from Betancur et al. (2013).]

perches, bass, cichlids, tuna, marlin, swordfish, angelfish, remoras, scorpionfish, stonefish, puffers, porcupine fish, sunfish, flounders, sole, halibut, and many other freshwater and marine forms. As the list of species indicates, percomorphs are not only incredibly diverse but also widely disparate, not only in body shape and habitat, but also in their amazing variety of adaptations. Many of them, such as the perch, tuna, sunfish, and flatfish like the flounder, sole, and halibut, are also important to recreational or commercial fishermen as well. The explosive adaptive radiation of percomorphs occurred in the early Cenozoic, about the same time as the radiation of mammals, but there are about twice as many living species of percomorphs as there are of mammals. Percomorphs are such a large and complex group that they are still relatively poorly understood, in contrast to the smaller groups, whose systematics have been studied since 1966.

It's impossible to do justice to the enormous variety of teleosts alive today, with their excellent fossil record, but the next time you look at a fish tank or go to a seafood market, see if you can recognize some of these groups and anatomical features.

FURTHER READING

Betancur, R.; et al. 2013. The tree of life and a new classification of bony fishes. *PLoS Currents Tree of Life*. 5 (Edition 1).

Clarke, J.T.; Friedman, M. 2018. Body-shape diversity in Triassic: Early Cretaceous neopterygian fishes: Sustained holostean disparity and predominantly gradual increases in teleost phenotypic variety. *Paleobiology*. 44 (3): 402–433.

Greenwood, P.; Rosen, D.; Weitzman, S.; Myers, G. 1966. Phyletic studies of teleostean fishes, with a provisional classification of living forms. *Bulletin of the American Museum of Natural History*. 131: 339–456.

Long, J.A. 2010. *The Rise of Fishes* (2nd ed.). John Hopkins University Press, Baltimore.

Maisey, J.G. 1996. *Discovering Fossil Fishes*. Henry Holt, New York.

Moy-Thomas, J.; Miles, R.S. 1971. *Palaeozoic Fishes*. Saunders, Philadelphia.

Near, Thomas J.; et al. 2012. Resolution of ray-finned fish phylogeny and timing of diversification. *Proceedings of the National Academy of Sciences*. 109 (34): 13698–13703.

Nelson, Joseph S. 2016. *Fishes of the World*. John Wiley & Sons, Inc., New York.

Patterson, C.; Rosen, D.E. 1977. Review of ichthyodectiform and other Mesozoic teleost fishes, and the theory and practice of classifying fossils. *Bulletin of the American Museum of Natural History*. 158 (2): 81–172.

THE TRANSITION TO LAND

THE TETRAPODS

6

In my day it was believed that the place for a fish was in the water. A perfectly sound idea, too. If we wanted fish, for one reason or another, we knew where to find it. And not up a tree. For many of us, fish are still associated quite definitely with water. Speaking for myself, they always will be, though certain fish seem to feel differently about it. Indeed, we hear so much these days about the climbing perch, the walking goby, and the galloping eel that a word in season appears to be needed. Times change, of course—and I only wish I could say for the better. I know all that, but you will never convince me that fish that is out on a limb, or strolling around in vacant lots, or hiking across the country, is getting a sane, normal view of life. I would go so far as to venture that such a fish is not a fish in its right mind.

—Will Cuppy, *How to Become Extinct*, 1941

LOBE-FINNED FISH

We have seen how bony fish have taken over the waters of the world, and became the most diverse group of vertebrates on the planet. But how did fish make the transition to land? What kinds of anatomical adaptations did they have to make when they moved from the watery realm to the brave new world of dry land? And what does the fossil record show us about that transition?

In the previous chapter, we have seen the great radiation of bony fish (Osteichthyes), and focused mainly on the most diverse branch, the ray-finned fish (Actinopterygii). The other main branch of the bony fish is the Sarcopterygii, or the lobe-finned fish. They get this name because their fins are supported not by a fan of thin bony rods (as in the ray-finned fish), but by a robust set of bones running down the axis of the fin. Their fins are fleshy with numerous muscles, and the fin rays splay out from this support. The fleshy muscular aspect of the fin gives them their name, since *sarkos* means "fleshy" and *pterygos* is "fin" in Greek. Many other features distinguish lobe-fins from ray-fins besides the fins themselves. Lobe-fin scales are composed of a thick layer of porous bone, and are known as cosmoid scales, very different from the thin acellular coating of ganoine on scales of ray-fins. Their teeth are also covered with cosmine, and have a highly infolded enamel surface; when they are sliced, they look like a labyrinth, so these "labyrinthodont" teeth not only define sarcopterygians, but are found in the earliest tetrapods as well (which used to be called "labyrinthodont amphibians"). In addition, there are many unique characters of the jaws, jaw-support mechanisms, gill arches, and shoulder girdle in lobe-fins that are not found in any other group of fish.

DOI: 10.1201/9781003128205-6

The sarcopterygians have reduced the teeth on the edge of the jaw, in contrast with the numerous teeth on the edge of the jaws of actinopterygians. Instead, the lobe-fins emphasized the teeth on the palate for holding prey as they swallow. Sarcopterygians have a sac off their gut used for respiration (which became the lung in lungfish and tetrapods).

The earliest known sarcopterygian was the Late Silurian Chinese fossil *Guiyu*, which is also the oldest bony fish fossil known as well (**Figure 6.1[A]**). Although it still had ganoid scales and other features that are primitive for the bony fish in general, it has the fin structure and features of the skull show that it was an extremely primitive lobe fin. In the Devonian, three main groups of sarcopterygians arose. The most primitive of these are the coelacanths, or Actinistia (**Figures 6.1** and **6.2**). They are notable in the great abundance of their lobed fins, including paired pectoral fins, pelvic fins, two tandem dorsal fins, an anal fin, and a symmetrical tail with a lobe in the center. Another distinctive feature is the triangular shape of the operculum, or the bone covering the gills. Coelacanths were as common as the palaeoniscoids in both marine and freshwater habitats in the late Paleozoic and Triassic, but their fossils long suggested that they became extinct at the end of the Cretaceous. For almost a century, there was no evidence to suggest that they survived into the Cenozoic (although Miocene coelacanth fossils are known from several places now).

Then, in 1938, a remarkable discovery was made at the mouth of the Chalumna River in South Africa. Fishermen trawled up a huge, shiny blue fish that looked like nothing they had ever seen before (**Figure 6.2**). The

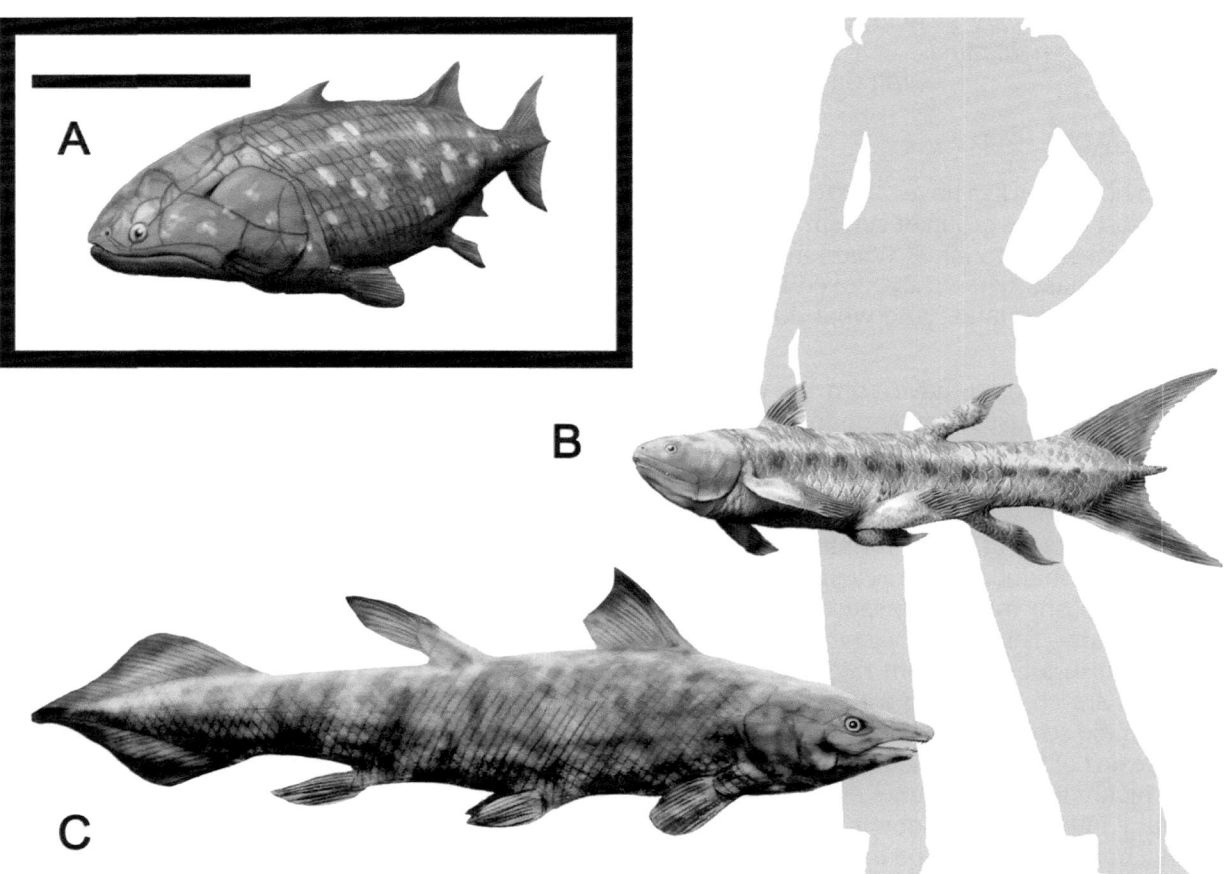

Figure 6.1 (A) The Late Silurian Chinese fish, *Guiyu*, the oldest bony fish known, and the oldest sarcopterygian. (B) The Triassic coelacanth *Rebellatrix* from British Columbia. (C) The coelacanth *Chinlea*, from the Triassic Chinle Formation of Arizona.

Figure 6.2 A specimen of the living coelacanth *Latimeria.* (Courtesy Wikimedia Commons.)

local museum curator, Marjorie Courtenay-Latimer, was summoned and immediately recognized its importance. Although the huge fish weighed 127 pounds, and was already dead and rotting rapidly, she did her best to preserve it, but eventually had to discard everything but the skin. A few weeks later, her letter and sketches of the fish reached the ichthyologist J. L. B. Smith. As he wrote later,

> Then I turned the page and saw the sketch, at which I stared and stared, at first in puzzlement, for I did not know any fish of our own or indeed of any seas like that; it looked more like a lizard. And then a bomb seemed to burst in my brain, and beyond that sketch and the paper of the letter I was looking at a series of fishy creatures flashed up as on a screen, fishes no longer here, fishes that had lived in dim past ages gone, and of which often only fragmentary remains in rocks are known. I told myself sternly not to be a fool, but there was something about the sketch that seized on my imagination and told me that this was something far beyond the usual run of fishes in our seas . . . I was afraid of this thing, for I could see something of what it would mean if it were true, and I also realized only too well what it would mean if I said it was it was not.
>
> (Smith, 1956, p. 62)

When he finally saw the specimen,

> Coelacanth—yes, God! Although I had come prepared, that first sight hit me like a white-hot blast and made me feel shaky and queer, my body tingled. I stood as if stricken to stone. Yes, there was not a shadow of doubt, scale by scale, bone by bone, fin by fin, it was a true Coelacanth. It could have been one of those creatures of 200 Ma come alive again. I forgot everything else and just looked and looked, and then almost fearfully went close up and touched and stroked.
>
> (Smith, 1956, p. 73)

Smith named the specimen *Latimeria* (after the discoverer) *chalumnae* (after the place it was found), and it caused a sensation in the scientific world in 1939. However, 13 years of diligent searching by Smith and many others failed to turn up another specimen in South African waters. They even put out "wanted" posters with a reward of £100 for another specimen and circulated them all over the African coast. Finally, in 1952, another specimen was found in the Comoros Islands north of Madagascar, and it was preserved soon enough that all of its internal organs remained intact. Scientists soon realized that the Comoros was the best place to find coelacanths. In the next 20 years, 83 specimens were hauled out of the deep waters off these steep volcanic islands. In 1997, a second species of coelacanth was found in the waters off Indonesia, and even more recently, another specimen was found off South Africa, showing that the original discovery was not a stray or a fluke. But coelacanths are still so rare and valuable that humans are driving them to extinction all over again, and there may be more specimens in museums now than there are in the waters of the Comoros. Fewer than 1000 are thought to be alive in the wild, so officially they are on the list of endangered species.

Even though *Latimeria* is an undoubted coelacanth, with the distinctive fins and triangular operculum, it has many peculiarities of its own. Fossil coelacanths have some bone, but *Latimeria* is mostly cartilaginous, and even retains a notochord in its spinal column. *Latimeria* has a peculiar spiral intestine, shaped like a long straight cylinder with a helical divider running down its length. A similar type of intestine is found elsewhere only in sharks. Instead of an air-filled lung, *Latimeria* has a swim bladder filled with fatty tissue that gives it some neutral buoyancy but allows it to dive to deeper waters where the pressure is much higher. When some preserved specimens were dissected, completely developed embryos were found inside, showing that *Latimeria* gives birth to live young, rather than laying eggs. *Latimeria* has phosphorescent eyes and a deep blue color consistent with the depths of 200 to 500 meters where it lives. It comes near the surface only at night, so it is hard to catch, and is almost never seen alive in its habitat. Using a miniature submarine, Hans Fricke and his colleagues were able to observe living specimens and found that it has a very peculiar "dance" when it swims, flexing its fins in surprising ways and even standing on its head. However, it does not seem to use its lobed fins to "walk" in a way that anticipates walking in tetrapods.

The second main group of sarcopterygians are the lungfish, or dipnoans (**Figure 6.3**). In the Devonian, lungfish were almost as common as coelacanths, and primitive Devonian members of both groups look remarkably similar. They were abundant and diverse in freshwater and marine deposits all over the world, with the distinctive lungfish tooth plates found on many continents during the Mesozoic. But today only three genera survive (**Figure 6.3**), all restricted to Gondwana continents: the African lungfish *Protopterus*; the Amazonian lungfish *Lepidosiren*; and the Australian lungfish *Neoceratodus*. The first two genera are highly specialized and degenerate, shaped more like eels than their ancestral lungfish, and the lobed fins are reduced to skinny ribbon-like fins that have little propulsive power. But the fins of the Australian lungfish are much like those of their ancestors, and in most respects *Neoceratodus* looks like many of the Paleozoic and Mesozoic fossils. Indeed, it got its name because it looks like a younger ("neo" or new in Greek) version of the Triassic lungfish *Ceratodus*.

Lungfish use their lungs to supplement their respiration by gills in stagnant lakes and ponds. When the lake dries up completely, they estivate by encasing themselves in a cocoon of dried mud and mucus, with only a breathing hole at one end, to protect themselves from drying up. When the rains come again, the soggy cocoon breaks down and the lungfish swims free. Unlike the highly flexible mobile skulls of actinopterygian fish, lungfish have fused the upper jaw (palatoquadrate) to the braincase, so their bite can crush their prey. They have no teeth on the edge of the mouth, but instead their palates are armed with peculiar, ridged

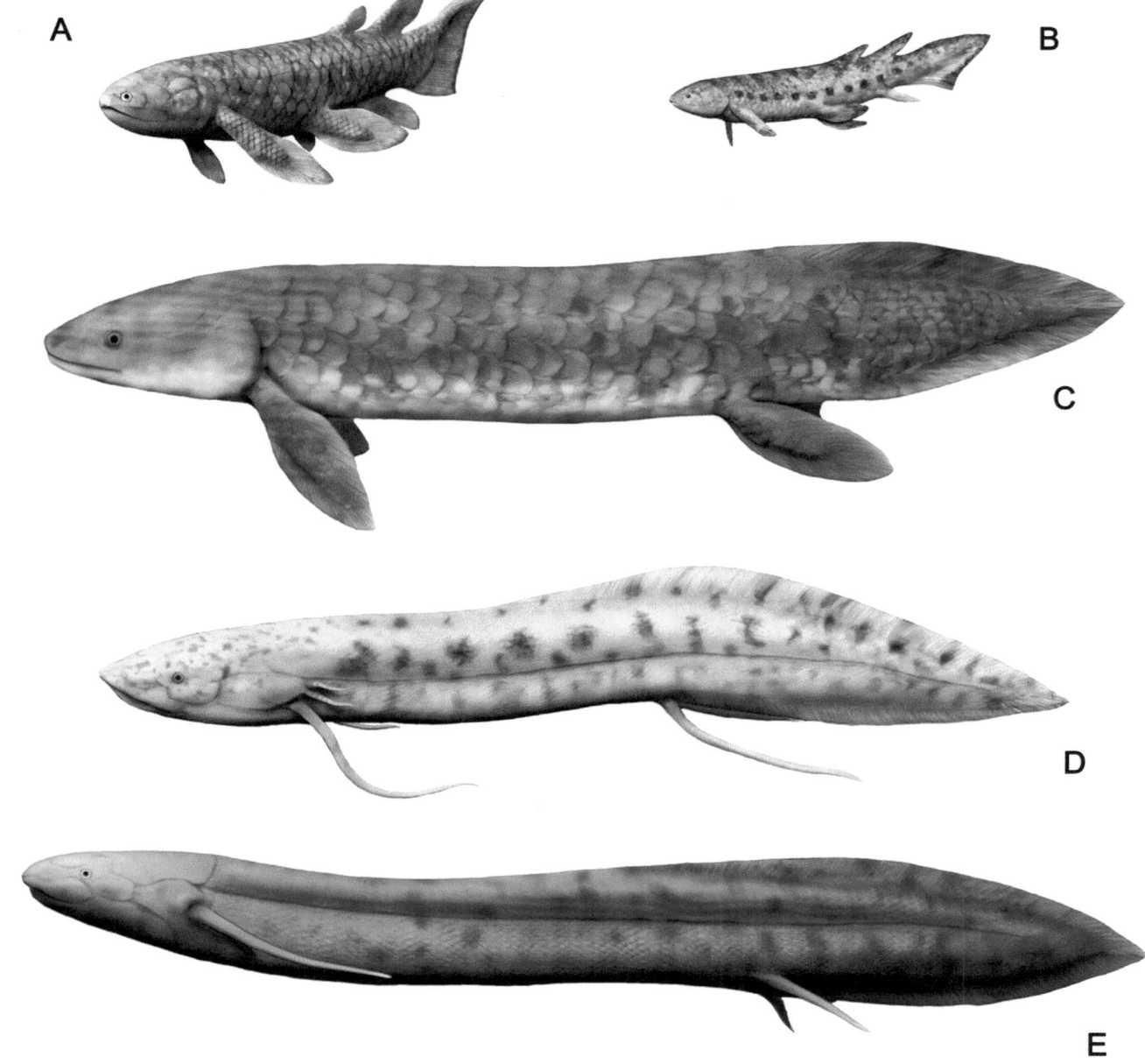

Figure 6.3 A selection of fossil and living lungfish. (A) The Devonian lungfish *Holoptychius*. (B) The Devonian lungfish *Dipterus*. (C) *Neoceratodus*, the Australian lungfish. (D) *Protopterus*, the African lungfish. (E) *Lepidosiren*, the South American lungfish. Scale bar = 10 cm.

tooth plates, which are their most commonly and distinctively fossilized remains. The dermal bones of the skull are also peculiar, with more numerous elements forming a completely different pattern from that found in other groups of bony fish.

The third group of lobe-fins are a wastebasket assemblage of all the fish that are more closely related to tetrapods than they are to other fish. They have been called "rhipidistians" but that paraphyletic group name only means they are lobe-fins closely related to tetrapods that have not developed all the unique features of tetrapods. Thus, "rhipidistia" is not a natural group. One of the best studied of these "rhipidistians" is *Eusthenopteron* (**Figure 6.4**). It was a big (up to 1.8 meters or 6.5 feet long) lobe-finned fish that was more like an amphibian than any living lungfish or coelacanth. It is now known from many beautiful specimens from a famous fossil locality called Miguasha, on Scaumenac Bay, Quebec. It had all the right bones to construct the tetrapod arm and leg from and all the right bones in the skull to be ancestral to tetrapod. Yet it still had the two dorsal fins typical of all lobe-fins; these were lost in the tetrapods, plus the symmetrical tail for swimming full-time; it was not able to crawl out of the water. Like other lobe-fins and also tetrapods, it had internal nostrils (choana) that ran from the outside through the nose and snout and to the mouth cavity. Even its teeth were constructed on the classic labyrinthodont pattern.

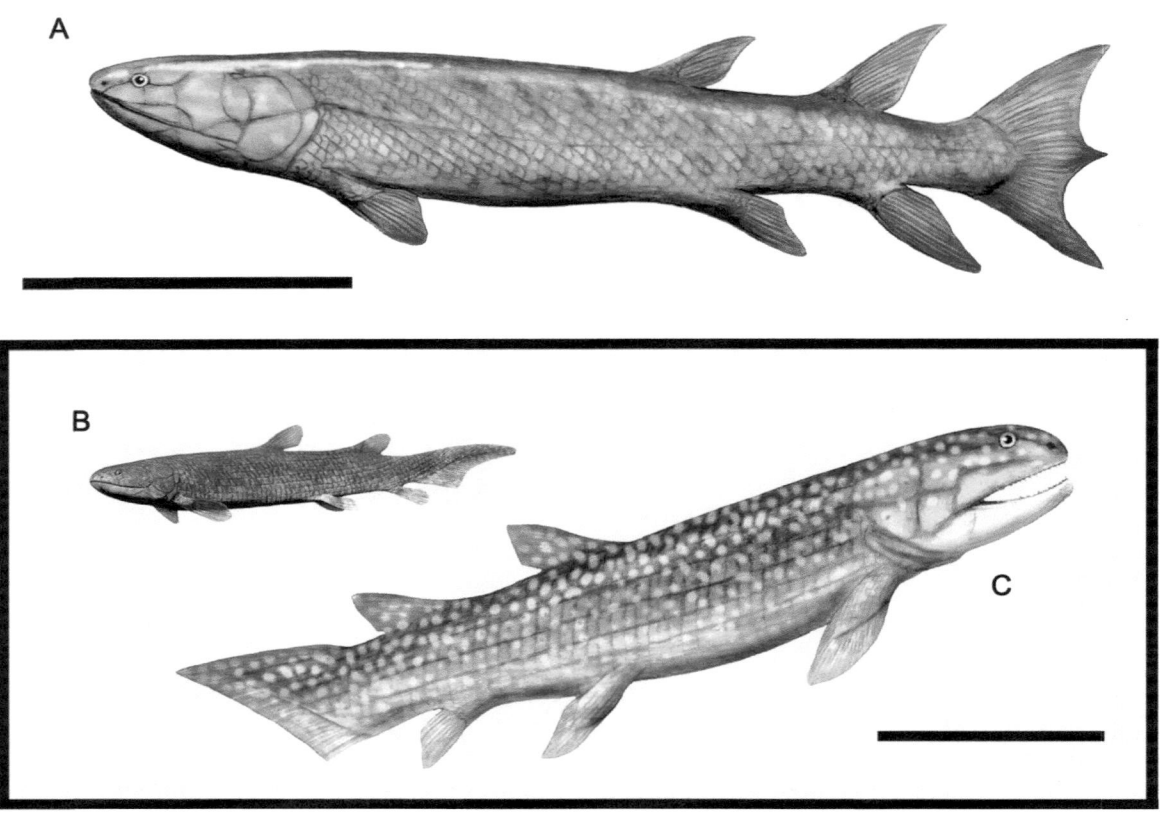

Figure 6.4 Reconstructions of typical "rhipidistians", including: (A) *Eusthenopteron* (scale bar: 50 cm), (B) *Osteolepis*, (C) *Gogonasus* (scale bar: 10 cm). (D) Fossil specimen of *Eusthenopteron*. [(D) Courtesy Wikimedia Commons.]

INVASION OF THE LAND: THE TETRAPODS

For generations, scientists puzzled and marveled about the daunting challenges for a fish to crawl out on land during the Late Devonian, around 370 Ma. These creatures were not the first on land, of course. In the Late Ordovician, over 100 million years earlier, there is evidence not only of the first land plants, but also of burrows made by millipedes, the first animals on land. By the Silurian, the earliest relatives of centipedes, scorpions, spiders, and eventually insects show up in the fossil record, but no vertebrates were on land during the Silurian or even most of the Devonian. Only near the end of the Devonian, when the first forests developed, do we get fossils of lobe-fins that are on their way to becoming tetrapods.

Yet as radical as this step seems, it is not as difficult as scientists used to think. The ray-finned fish (99% of living fish, including most of the fish you eat or have in your fish tank) have done it many times independently in many different groups. For example, mudskippers (**Figure 6.5[A]**) live permanently right on the boundary of land and water. They graze on algae on the surface of mudflats in mangrove swamps at low tide and prop themselves up by their ray fins. They use their stalked eyes to see out of the water when they are submerged. They can flee to water when predators threaten from land, and run to land when predators occur in the water. The so-called "walking catfish" can wriggle across the land from one pool to another when their home begins to dry up or the water becomes foul, or also to find a new pool with new food resources when the old pool is too crowded (**Figure 6.5[B]**). Climbing perch wriggle and crawl across dry land to find better pools; they can even crawl up trees, hence their name. Many tidepool fish, such as gobies and sculpins, spend much of the time at low tide crawling along the rocks with their hand-like fins, preying on animals trapped by the low tide. Spotted moray eels wriggle out of water during low tides to prey on crabs that are looking for smaller food to eat. A number of other fish have modified the fin-rays of their front fins into clumsy "fingers" that allow them to crawl across surfaces.

Yet these are all ray-finned fish, not closely related to lungfish or coelacanths or the other lobe-finned fishes that gave rise to amphibians. All of these examples of semi-terrestrial lifestyles in these fish evolved independently in multiple groups, all in different ways. Clearly there are strong

A

Figure 6.5 Two examples of teleost fish which have adapted to crawling out on land for extended periods of time. (A) The mudskipper, which lives on the edge of the water in tidal mudflats, typically in mangrove swamps. (Courtesy Wikimedia Commons.)

Figure 6.5 (Continued) (B) The "walking catfish", which is capable of wriggling across the ground to reach a new pond when its old pond becomes stagnant or dries up.

B

pressures for fish to exploit the land (at least spend short periods of time out of the water) to find new food or escape predators or crowding in the water. And it is also clear that it is no big deal for fish to do this, if it evolved many different times to different degrees in entirely unrelated groups of fish. Instead of the difficulties that scientists imagined just a few decades ago, it now seems like a trivial task if it was done so often by so many different unrelated groups of fish. As Neil Shubin wrote in *Your Inner Fish*:

> *What possessed fish to get out of the water or live in the margins? Think of this: virtually every fish swimming in these 375-million-year-old streams was a predator of some kind. Some were up to sixteen feet long, almost twice the size of the largest Tiktaalik. The most common fish species we find alongside Tiktaalik is seven feet long and has a head as wide as a basketball. The teeth are barbs the size of railroad spikes. Would you want to swim in these ancient streams?*

Finally, recent study showed just how easy it is for fish to become modified for at least some kind of land life. A group of scientists led by Emily Standen took a very primitive ray-finned fish, the bichir found in Africa (genus *Polypterus*) **(Figure 5.2[D])**, which is distantly related to sturgeons and paddlefish. Its fins are constructed like much more primitive ray-fins, and have some similarities to the earliest fossil lobe-fins. These bichirs were raised on land rather than water (since they are already good air breathers). In just a few generations of breeding, their fins became more robust and better for land crawling than those of their ancestors. Clearly, the genes for modifying fins into something else are easy to trigger, and this mechanism was employed by many of the land-living ray-finned fish we just listed.

But making a permanent life on land requires a number of adaptations that no teleost has completely developed. These include:

Respiration: As we have just seen, the lungfish (and probably the "rhipidistians") already have an air-breathing lung in place, which helps them survive when their ponds dry up. (This may have been one of the important factors that drove the fish onto the land, because the drying of ponds forces most fish to crawl out or die—except for lungfish, which can estivate during drought). An aquatic fish can gulp air and then force it back into the lungs by diving downward. But on land, a tetrapod needs a pumping mechanism to force the air backward. Salamanders and some other amphibians do this by expanding and contracting the rib cage, while frogs pump air with the sac on the base of the throat. The earliest tetrapods had broad flanges on their ribs, which might have been to help with pumping the air into the lungs. They also still had gills as well, so they had not completely given up on breathing in water.

Locomotion: Once again, sarcopterygian fins already have robust bony supports with lots of fleshy muscular tissue to support and manipulate their fin rays, so all that is needed for the lobed fin to become a tetrapod limb is further development of these robust supporting bones into the bones of the tetrapod arm or leg and the replacement of the fin rays with toes. Although fossils of these specimens are rare, the most advanced "rhipidistian" fins and the earliest tetrapod limbs are not very different (**Figures 6.6** and **6.7**). Even more surprising is the recent discovery that

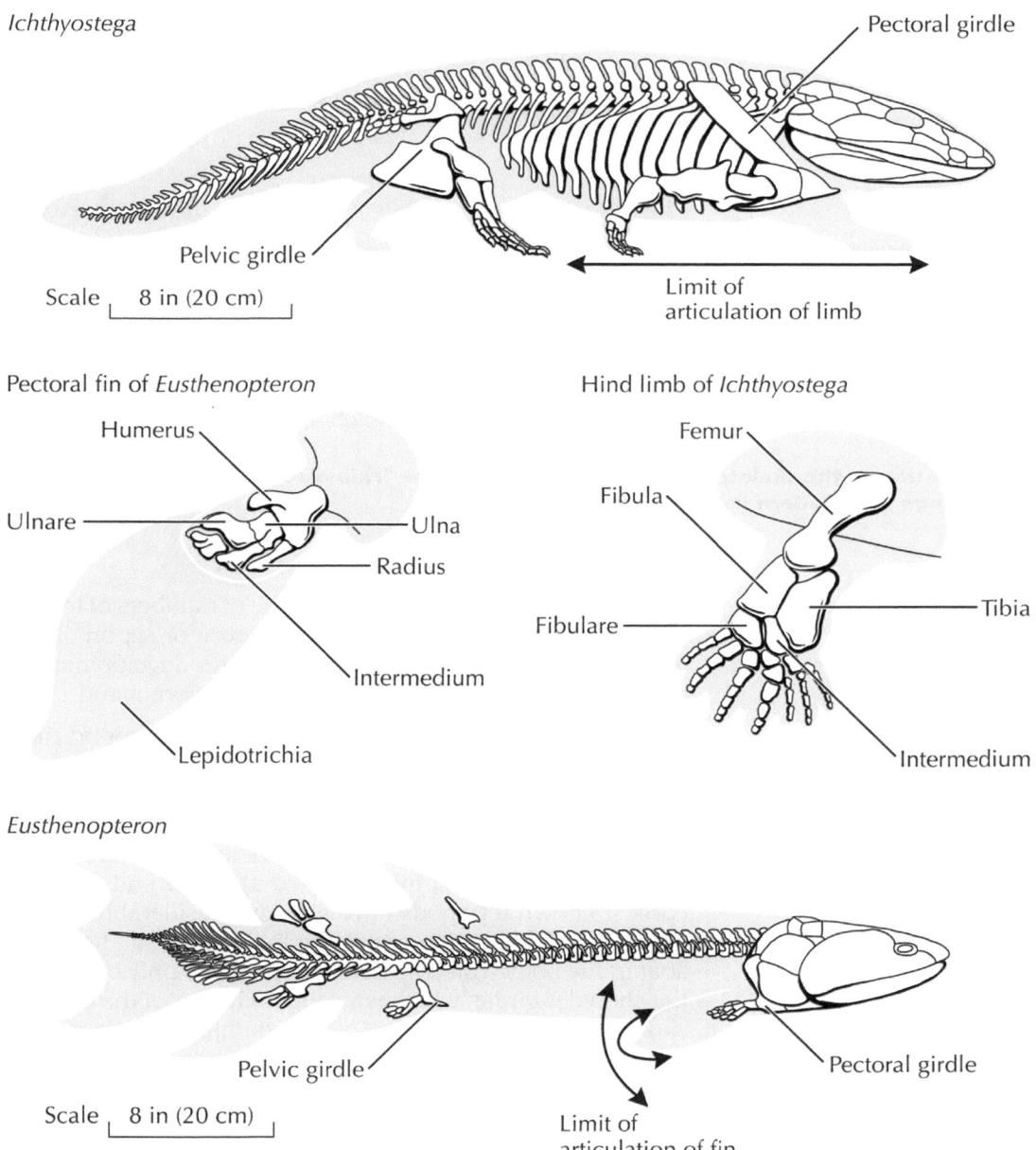

Figure 6.6 The evolutionary modifications of the skull, limbs, and the rest of the skeleton from a "rhipidistian" like *Eusthenopteron* to a tetrapod like *Ichythyostega*.

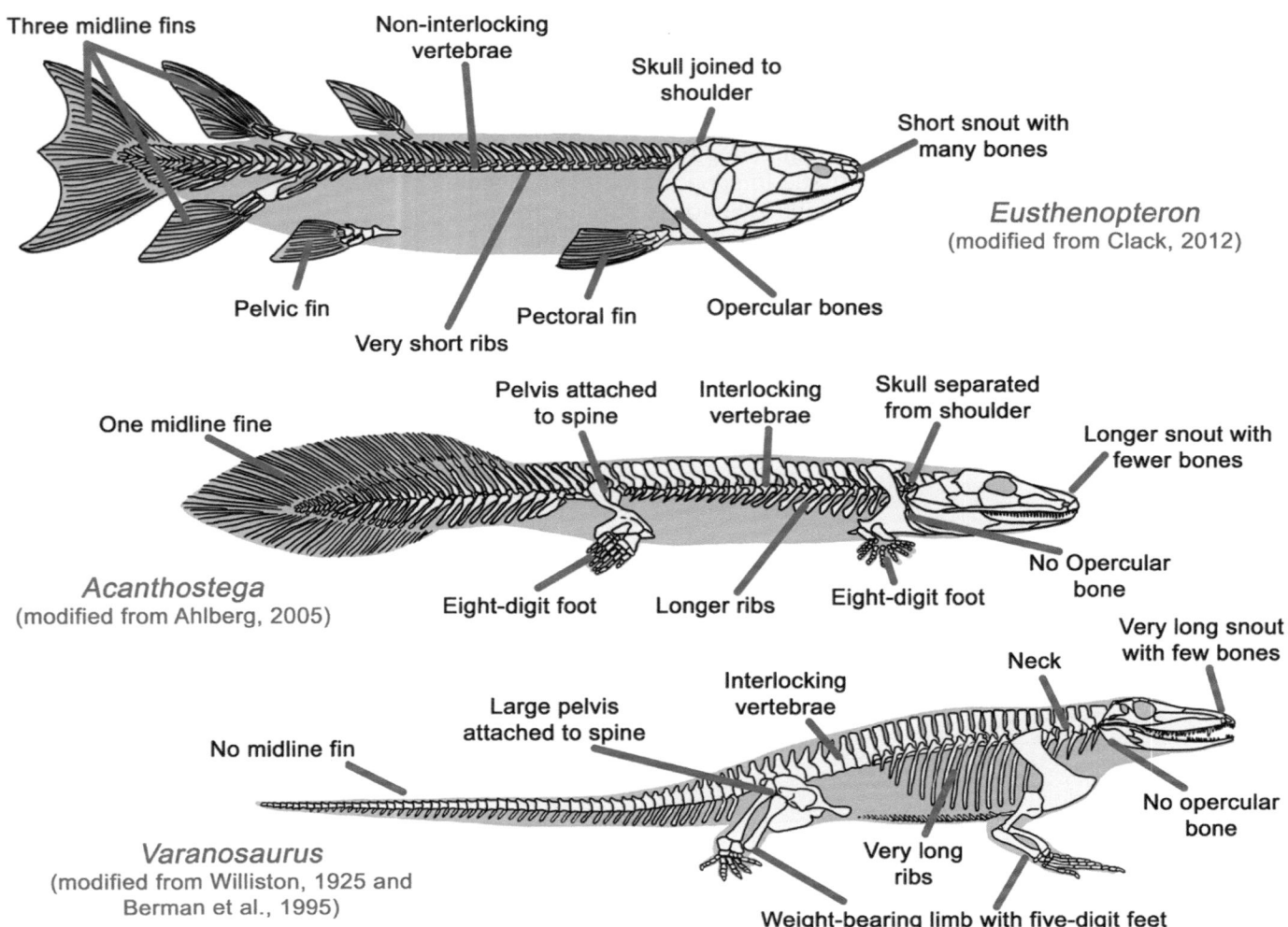

Figure 6.7 The evolution of the skeleton in the transition from the "rhipidistian" *Eusthenopteron* to the primitive tetrapod *Acanthostega* to a modern lizard.

the earliest tetrapods had a wide range of numbers of toes on their hands and feet—eight on some specimens, seven or six on others. The standard count of five toes and fingers is a later development that stabilized after the earliest tetrapods had crawled out onto land.

But strong limbs with toes are not enough. To support the body on land without the buoyancy of water, the limbs must have solid connections to the main axis of the spinal column. This is comparable to the engineering problem of a supporting the span of a bridge. To hold the long horizontal span of the bridge stable, at least two strong vertical support columns need to be firmly attached at either end. For this reason, the earliest known tetrapods have already considerably enlarged the pelvic elements and fused them to the spinal column (in fish, the pelvic bones float in the body wall connected to the spine only by muscles). Likewise, the shoulder girdle, which was originally part of the dermal bones of the skull roof and gill-covering elements, has become detached from the head, allowing the tetrapod to move its head independently of its front limbs. Clearly, the ability to freely twist and turn their head (which no fish has) allows a land predator to see better, and to turn its head quickly to fight or to catch prey.

Lack of Buoyancy: The vertebral column of fish is only loosely joined together, because the water supports most of their body weight.

However, when the spine no longer has the support of water, it must become a much stronger, more directly articulated structure. Consequently, the vertebrae of tetrapods developed a variety of bony spines and joints between them that lock the vertebrae together tightly, while permitting some movement between them, but the enlarged spines are also attachment points for much stronger bundles of muscles and tendons that run along and reinforce the spinal column. The details of how the various bony elements of tetrapod vertebrae are put together have long been used in classifying different primitive tetrapod groups.

Desiccation: Most fish have thin, soft scales that allow gases to pass through their skin, and many respire through their skin more than they do through their gills. However, a permeable skin is a disadvantage in the dry air, so most tetrapods have developed an impermeable protein called keratin in their scales that protects them from drying up. (Keratin is the same protein found in your hair and fingernails and also in bird feathers.) Keratinous scales are relatively rare in primitive tetrapods, but are found in all reptiles, birds, and mammals, most of which live permanently out of the water. Among living amphibians (frogs and salamanders), the skin is permeable and helps with respiration, because they have such small body volume and seldom stray far from water (and some early fossil tetrapods probably could do this, too).

Another area of water loss is the mouth and nasal cavity. Every time you breathe, you lose a certain amount of moisture (as you notice when you exhale on a very cold day). Fish constantly open their mouths and swallow water, which then passes through the pharynx and out the gills. Tetrapods have restricted the flow of external fluids by breathing through a passage (the choana) that connects the nasal opening directly to the mouth cavity. (In mammals, it is the connection between your nasal sinuses and the back of your throat that is often blocked when you are congested). In many tetrapods, this restricted nasal passage is lined with an area of folded tissues that further trap moisture before exhalation. Even without these, however, the choana conserves most of the moisture in the body of the animal, while allowing normal breathing.

Sense Organs: The senses needed in an aquatic life are very different from those used on land. In water, the light is distorted by the odd refraction patterns (think of the distorted images in a fish tank), and most bodies of water are too muddy or too deep for light penetration, so sight is not as important. However, sight is a very important sense in the air, and most early tetrapods had large eyes, with well-developed eyelids to protect the eyeball from drying out. Most fish have a series of pits and canals along the side of the head and body called the lateral line system, which senses changes in pressure in the water (especially the pulses of energy waves caused by nearby obstacles or moving animals). This system is obsolete on land, because air is so much less dense than water that the lateral line is not sensitive enough to detect changing air pressure. The earliest tetrapods (which were still aquatic) still have traces of the lateral line system, but later forms lose it completely. Instead, the best way for a land vertebrate to detect changes in air pressure is a large membrane for hearing. Most fish cannot hear directly, although the ostariophysan teleosts pick up sound in their swim bladders using the Weberian ossicles mentioned in Chapter 5. In the earliest tetrapods, however, the bones of the gill covers are attached to the bones of the first gill arch in the throat region (including the hyomandibular bone), which in turn is attached to the braincase and inner ear region. When the gill covers were reduced, they left the hyomandibular behind, supporting a skin membrane that became the eardrum of tetrapods. Living amphibians have a large eardrum, whose vibrations are transmitted by the hyomandibular directly to the inner ear region. In your ear, the hyomandibular cartilage that you had as an embryo became the stirrup bone (stapes) of your inner ear, transmitting sound to the sensory region.

Different stages of evolution of all these features can be seen in the earliest tetrapods, such as *Ichthyostega* (**Figure 6.6**) and *Acanthostega* (**Figure 6.7**), and other specimens known primarily from the Upper Devonian rocks of Spitzbergen and Greenland. Like a fish, *Ichthyostega* still had a large tail fin for underwater propulsion, large gill slits on the side of the head, and the lateral line system. Yet like an amphibian, it clearly had well-developed arms and legs with fingers and toes that would propel it across a hard surface. (Later research has shown that the forelimbs were not strong enough to do much walking, but moved in short hops, dragging its flipper-like hind limbs.) Like modern newts and salamanders, its limbs were mostly used for pushing through the obstacles in the water, not for lifting their bodies above the ground in fast walking. The ribs of *Ichthyostega* had robust flanges on them, supporting their chest cavity for breathing out of water, and preventing collapse of their rib cage while on land—but preventing them from the rib-propelled breathing used by many amphibians. *Ichthyostega* also had a long flat snout with eyes that looked upward, and a short braincase, in contrast to the deep, cylindrical skull of *Eusthenopteron* and many other lobed-finned fish, with a short snout and large braincase, and eyes that looked sideways.

Acanthostega was much more fish-like than *Ichthyostega*, with more fish-like limbs that could never have crawled on land, lacking wrists, elbows,

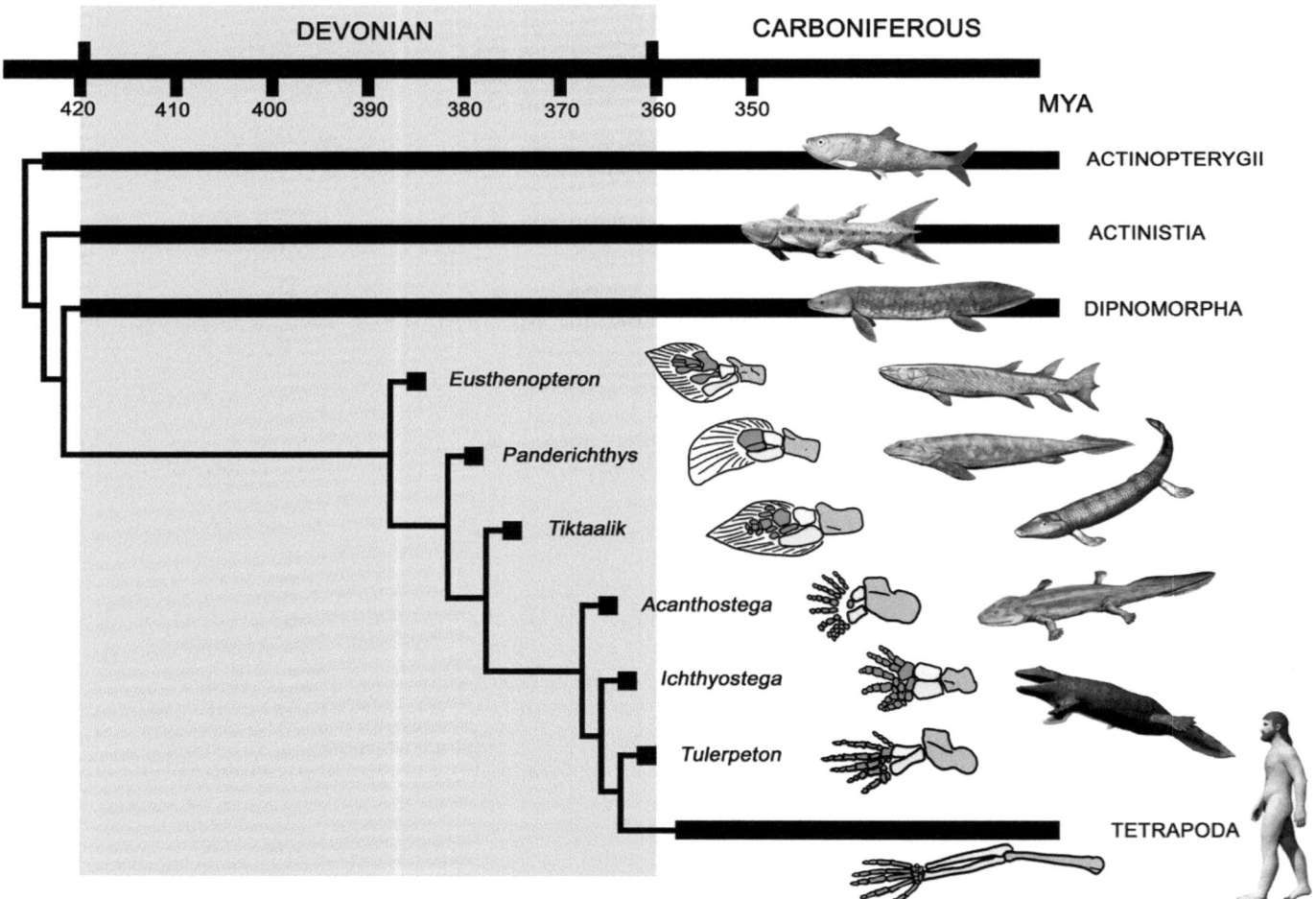

Figure 6.8 Diagram of the transitional series from "rhipidistians" like *Eusthenopteron* to primitive tetrapods like *Ichthyostega* and *Acanthostega* to modern amphibians.

or knees. Although it still had arms and legs rather than fins, they were more for pulling itself through the water and could not propel it across land much.

More and more discoveries continue to be made, so the family tree of the fish to amphibian transition is remarkably complete (**Figure 6.8**). Moving up from primitive lobe-fins like lungfish and coelacanths, we have the very fish-like lobed-finned *Eusthenopteron*, and slightly more advanced fossils called *Panderichthys*, *Elginerpeton*, *Ventastega*, and *Metaxygnathus*. *Tiktaalik* is slightly more amphibian-like but still does not have hands or feet like *Acanthostega*, and finally the most amphibian-like fossil *Ichthyostega*. From there, we have fossils that everyone recognizes as amphibians.

FURTHER READING

Boisvert, C.A. 2005. The pelvic fin and girdle of *Panderichthys* and the origin of tetrapod locomotion. *Nature*. 438 (7071): 1145–1147.

Callier, V.; Clack, J.A.; Ahlberg, P.E. 2009. From fish to landlubber: Fossils suggest earlier land-water transition of tetrapod. *Science* 324 (5925): 364–367.

Clack, J.A. 2002. *Gaining Ground: The Origin and Early Evolution of Tetrapods*. Indiana University Press, Bloomington, IN.

Clack, J.A. 2006. The emergence of early tetrapods. *Palaeogeography, Palaeoclimatology, Palaeoecology*. 232 (2–4): 167–189.

Clack, J.A. 2009. The fin to limb transition: New data, interpretations, and hypotheses from paleontology and developmental biology. *Annual Review of Earth and Planetary Sciences*. 37: 163–179.

Laurin, M. 2010. *How Vertebrates Left the Water*. University of California Press, Berkeley.

Long, J.A. 1995. *The Rise of Fishes*. Johns Hopkins University Press, Baltimore, MD.

Maisey, J.G. 1996. *Discovering Fossil Fishes*. Henry Holt and Company, New York.

Moy-Thomas, J.; Miles, R.S. 1971. *Palaeozoic Fishes*. Saunders, Philadelphia.

Shubin, N. 2008. *Your Inner Fish: A Journey into the 3.5 Billion History of the Human Body*. Pantheon, New York.

Smith, J.L.B. 1956. *Old Fourlegs: The Story of the Coelacanth*. Longman, Green, London.

Thomson, K.S. 1969. The biology of lobe-finned fishes. *Biological Review*. 44: 91–154.

Zimmer, C. 1998. *At the Water's Edge: Macroevolution and the Transformation of Life*. Free Press, New York.

TETRAPOD DIVERSIFY

7

Theories pass. The frog remains.

—Jean Rostand

AMPHIBIANS AND THEIR RELATIVES

Once they had established a foothold on land, early tetrapods soon underwent a spectacular evolutionary radiation into a wide variety of body forms, including huge crocodile-like predators and strange creatures with weird heads shaped like boomerangs, and many other types. These animals were fully able to walk on land, although they were still bound to the water to lay their eggs.

Traditionally, all of these fossils which had four legs (tetrapods), but were not reptiles, have been called "amphibians". But this uses the word "amphibian" as a sort of "evolutionary grade" between fishes and reptiles, and thus it is not a natural group (**Figure 7.1**). The living groups of amphibians (frogs, salamanders, and caecilians) are certainly a natural group, and can be properly called "amphibians". But most of the fossil groups called "amphibians" are a mixed bag of creatures that might be closely related to living amphibians (such as the temnospondyls discussed next) but others are related to reptiles, birds, and mammals (the "anthracosaurs" discussed in Chapter 8). Thus, we will avoid using the word "amphibian" unless it refers to a clear natural group. All of these creatures, including *Ichthyostega*, *Acanthostega*, and more advanced transitional fossils, are clearly members of the Tetrapoda, the natural group that includes all four-legged animals: amphibians, reptiles, birds, mammals, and their fossil relatives. This bit of terminology may seem trivial, but to the scientist, these "wastebasket" groups which were not natural and not clearly defined by specialized anatomical features have long hampered research, so they are now being abandoned for more precise usage.

TEMNOSPONDYLS

The largest and most diverse group of these late Paleozoic-early Mesozoic tetrapods is the temnospondyls. Formerly called "labyrinthodonts", most of them had long narrow bodies and large flat skulls, with relatively short legs. Some of the earliest temnospondyls are represented by *Greererpeton* from the Early Carboniferous coal mine near Greer, West Virginia (**Figure 7.2[A]**). It was a large newt-shaped creature almost 1.5 meters (5 feet) in length, with a long-flattened slender body and long tail with a vertical fin, and tiny legs. Like *Ichthyostega* and other early tetrapods, it still had grooves on the side of its skull for the lateral line system. It also had a thick massive stapes, so it did not have good hearing in air like later tetrapods, but still was adapted for hearing in water. Another early temnospondyl was the Early Carboniferous *Crassigyrinus* from Scotland

DOI: 10.1201/9781003128205-7

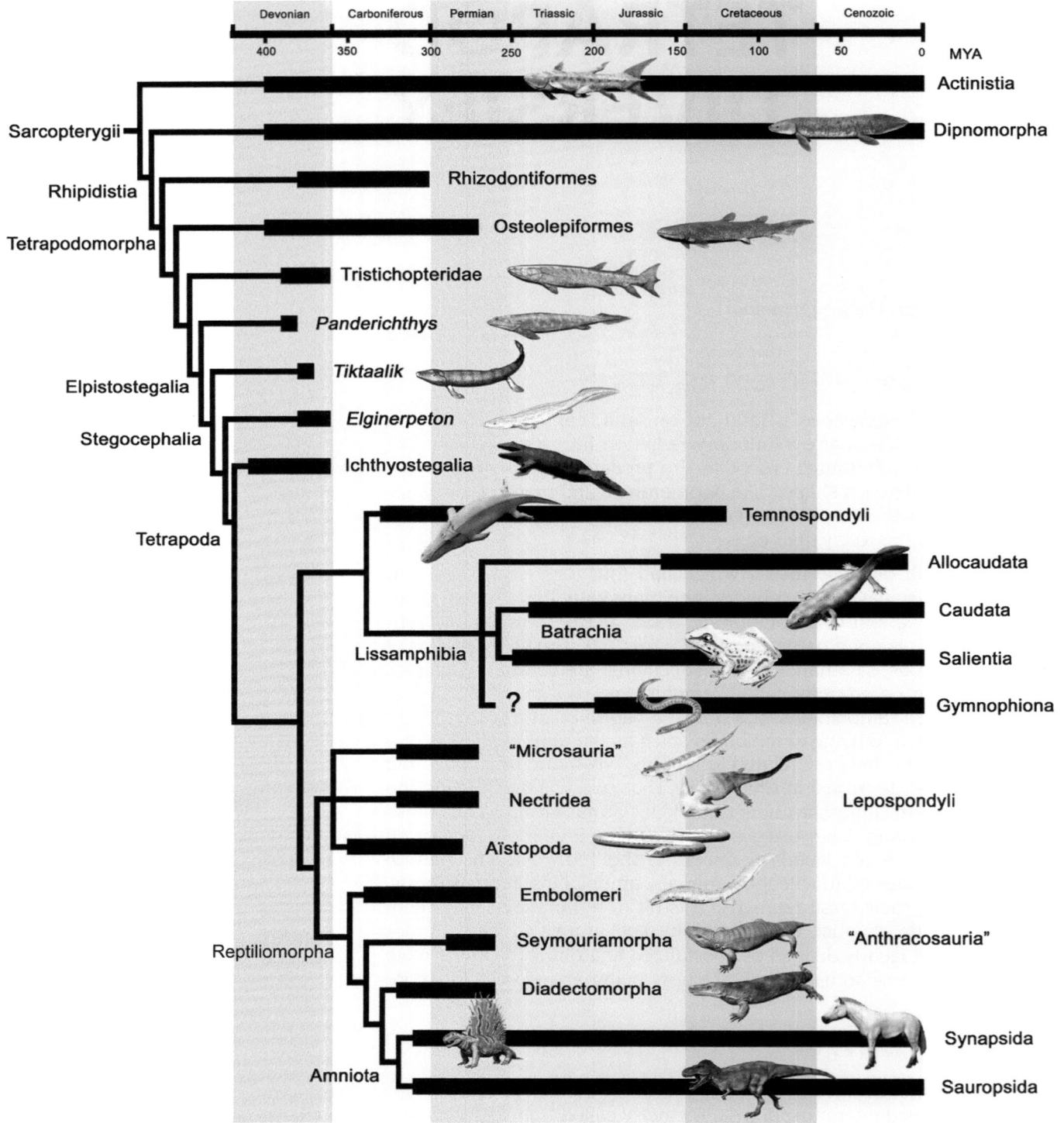

Figure 7.1 A family tree of the tetrapods and their relatives.

(**Figure 7.2[B]**). It was over 2 meters (6.6 feet) long, with a long slender eel-like body and tiny, almost useless limbs, so it was completely aquatic and could not walk on land. It had a relatively large head which enabled it to see in murky water, and a mouth with wide jaws, suggested it preyed on larger animals. Inside the large mouth were two rows of teeth, one on the edge of the mouth with large fangs, and one on the palate.

One of these huge temnospondyls was *Eryops*, a big crocodile-shaped fossil known from several complete skeletons (**Figures 7.2[C]** and **7.3[A]**).

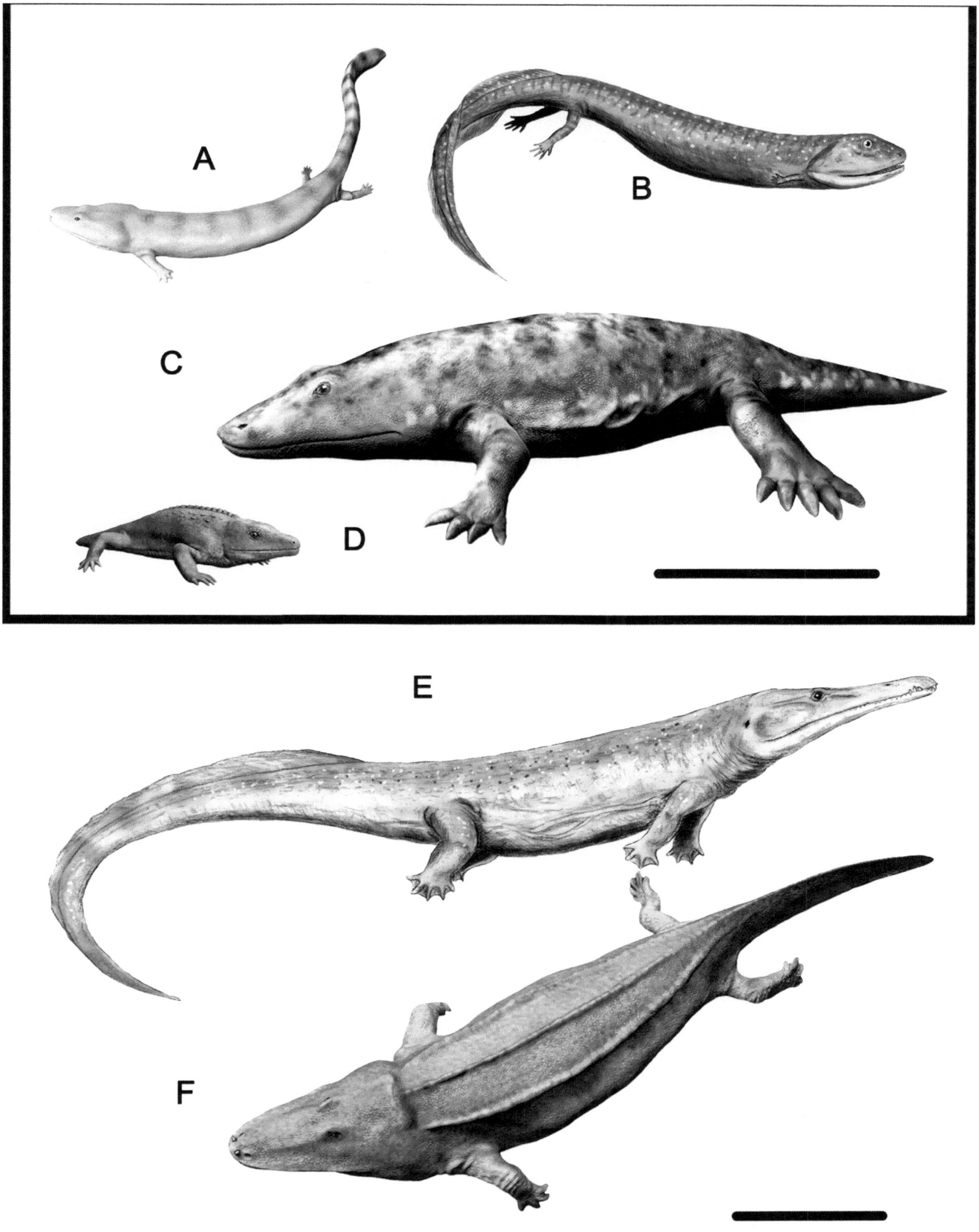

Figure 7.2 Reconstructions of a variety of different temnospondyls: (A) *Greererpeton*, (B) *Crassigyrinus*, (C) *Eryops*, (D) *Cacops*, (E) *Prionosuchus*, (F) *Mastodonsaurus*. (Scale bar = 1 meter.)

It had a large sprawling body over 2 meters (6.5 feet) long, with a robust tail and limbs, and a skull well over 60 cm (2 feet) long in big individuals. This was one of the largest terrestrial animals of the Early Permian, capable of hunting prey both in the water and on the land. The slightly more primitive *Edops* from Early Permian redbeds of Texas had an even longer skull, and would have been even larger than *Eryops* if its skeleton were known. In the Triassic, the temnospondyls were in decline, but huge flat-bodied metoposaurs (**Figure 7.2[B,C]**) like *Mastodonsaurus* (**Figure 7.2[D]**) and *Anaschisma* (formerly *Koskinodon* or *Buettneria*) (**Figure 7.3[B]**) were up to 6 meters (20 feet) long, and often found in logjams of bones, apparently dying in shallow ponds and then fossilized. These temnospondyls were common in the Late Triassic of Germany, Africa, South America, and the Petrified Forest in Arizona. A few taxa, such as armored *Cacops* (**Figures 7.2[D]** and **7.3[E]**) were smaller (only 40 cm long) and less flattened, but these are exceptions to the general trend of temnospondyls being large, aquatic ambush predators.

Some temnospondyls were immense, the largest amphibians to ever occur. One of these came from the Permian of Brazil. Dubbed *Prionosuchus* (**Figure 7.2[E]**), it was up to 10 meters (33 feet) long and weighed at least 360 kg (over 800 pounds)! The skull of some specimens is over 1.6 meters (5.2 feet) long! It was by far the largest fossil temnospondyl ever found. It had a long narrow snout like that of the modern fish-eating crocodilian known as the gavial (or gharial), with hundreds of sharp teeth. These were suitable for snagging aquatic creature that lived in the lagoon deposits and river beds where it was found in the Pedra do Fogo Formation of northern Brazil. Thus, temnospondyls occupied the aquatic crocodilian niche in the Permian, long before true crocodilians appeared in the Late Triassic.

Figure 7.3 Specimens of some of the important early tetrapods. (A) The huge temnospondyl *Eryops* from the Lower Permian rocks of north Texas, with a large modern salamander skeleton in the foreground. [(A,B,D,E) Courtesy Wikimedia Commons. (C) Photo by the author.]

B

C

Figure 7.3 (Continued) (B) The skeleton of the large flat-headed temnospondyl *Anaschisma* (formerly *Buettneria* and *Koskinodon*) from the Triassic of the southwestern U.S. (C) A life-sized reconstruction of *Antarctosuchus*, a huge Triassic metoposaur temnospondyl from Antarctica.

D

E

Figure 7.3 (Continued) (D) *Cacops*, a small dissorophid closely related to the modern amphibians. (E) *Diplocaulus*, a nectridian lepospondyl with the broad pointed "horns" on its skull, giving the head a boomerang shape.

All over the world in the Permian and Triassic, temnospondyls probably lurked in the water like crocodiles, using their huge mouths to gulp
down smaller prey. They were the largest land animals in the Pennsylvanian coal swamps, and major predators of the Permian. Over 170 genera
in 30 families of temnospondyls have been recorded, ranging from the
Early Carboniferous to the Late Cretaceous, although most of their diversity occurred in the late Paleozoic and Triassic.

LEPOSPONDYLS

A second radiation of tetrapods in the late Paleozoic was the lepospondyls, which tended to be smaller and built more like salamanders. One
group, the microsaurs, was extremely lizard-like, with a deep skull, cylindrical body, and relatively tiny limbs (**Figure 7.4[A]**). Another group, the
aistopods (**Figure 7.4[C]**), secondarily evolved a limbless snake-like body,
convergent on many other animals (including not only snakes, but also
apodan amphibians, and amphisbaenid reptiles) that have lost their legs
and become snake-like. The oddest-looking lepospondyls from the Lower
Permian redbeds were the nectrideans like *Diplocaulus*, which had a wide
boomerang-shaped head on a salamander-like body (**Figures 7.3[E]** and
7.4[B]). The skull is flattened with two large triangular "horns" sticking
out from the side and eye sockets that pointed straight up. The function
of these odd "horns" on the skull is still controversial. Some have argued
that it was used as a hydrofoil, allowing them to swim smoothly in an

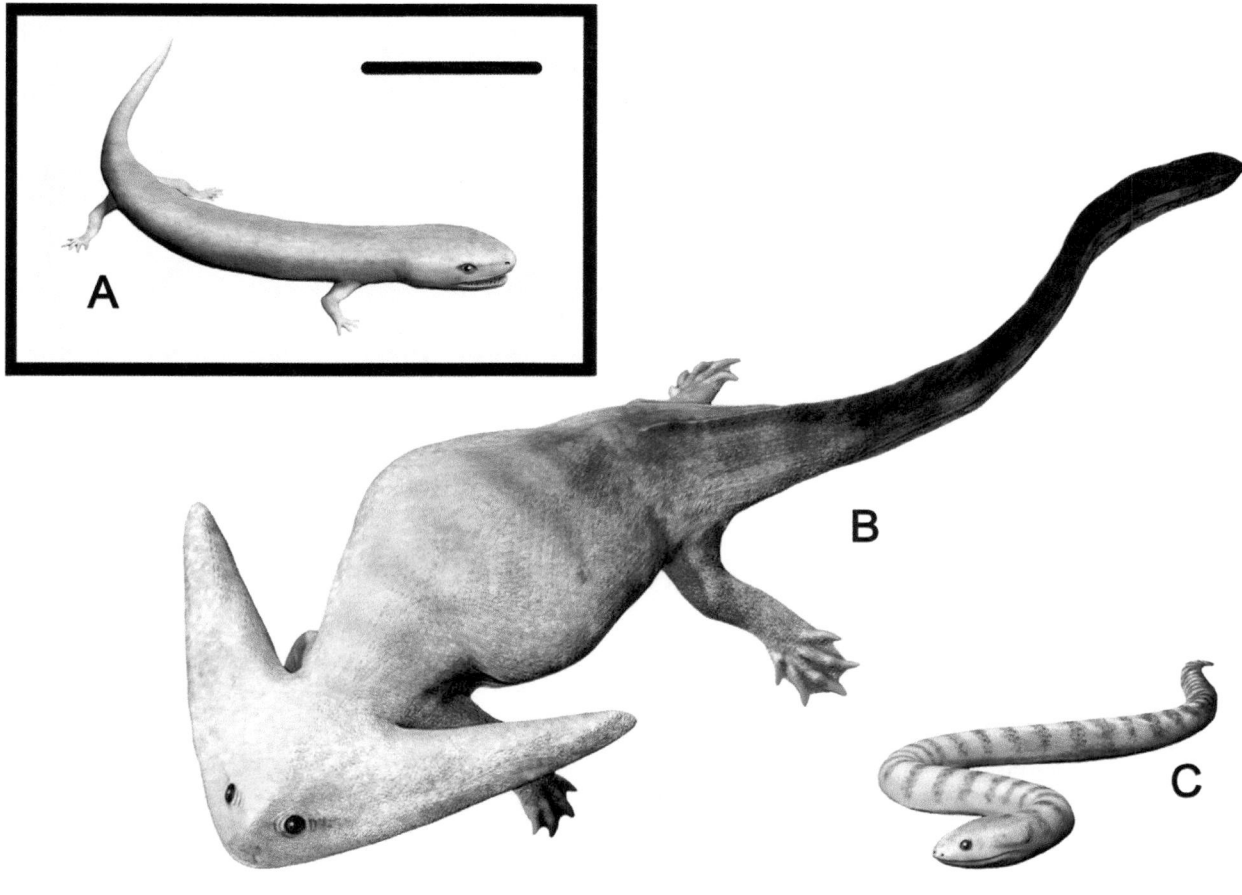

Figure 7.4 A variety of different lepospondyls. (A) *Cardiocephalus*, a microsaur. (Scale bar = 5 cm.) (B) The nectridian
Diplocaulus. (C) The snake-like aistopod *Ophiderpeton*. (Scale bar = 10 cm.)

up-and-down motion with the boomerang head shape providing lift. But their bodies are relatively weakly built and did not have the robust bones to support strong swimming muscles. Others have suggested that it would make it difficult for a predator to eat them head first, since the horns would have made them too wide to swallow, even for the largest Early Permian predators. The upward-pointing eyes suggest that *Diplocaulus* was more of an ambush predator that lay in the bottom of streams and ponds, and then lunged forward and upward to catch its prey with its strong jaws, possibly stunning their prey with a blow from their horns in the process. The most likely hypothesis, however, it is that it is comparable to the main function of horns and antlers of antelope and deer. These creatures use their horns and antlers primarily as a display structure to advertise the strength and dominance of males trying to find mates. The fact that we can trace the growth of these "horns" through their younger stages and there seem to be both robust males and less large-horned females in the collections seems to make this hypothesis most likely.

LISSAMPHIBIANS

Of these extinct groups, where did the living amphibians, or Lissamphibia (frogs, salamanders, and caecilians) come from? The best evidence comes from certain temnospondyls known as dissorophids, with creatures like *Amphibamus*, *Doleserpeton*, *Cacops* (**Figure 7.2[D]**). They have the primitive amphibian body form, and distinctive teeth with their crowns sitting on pedestals, a feature all modern amphibians with teeth have. From these fossils, we have a true transitional fossil showing the transition from salamanders to frogs. Known as *Gerobatrachus hottoni* ("Hotton's ancient salamander"), it was dubbed the "Frogamander" by the media (**Figures 7.5** and **7.6[B]**). Known from a single fossil from the Lower Permian redbeds of north Texas about 11 cm (7.3 inches) long,

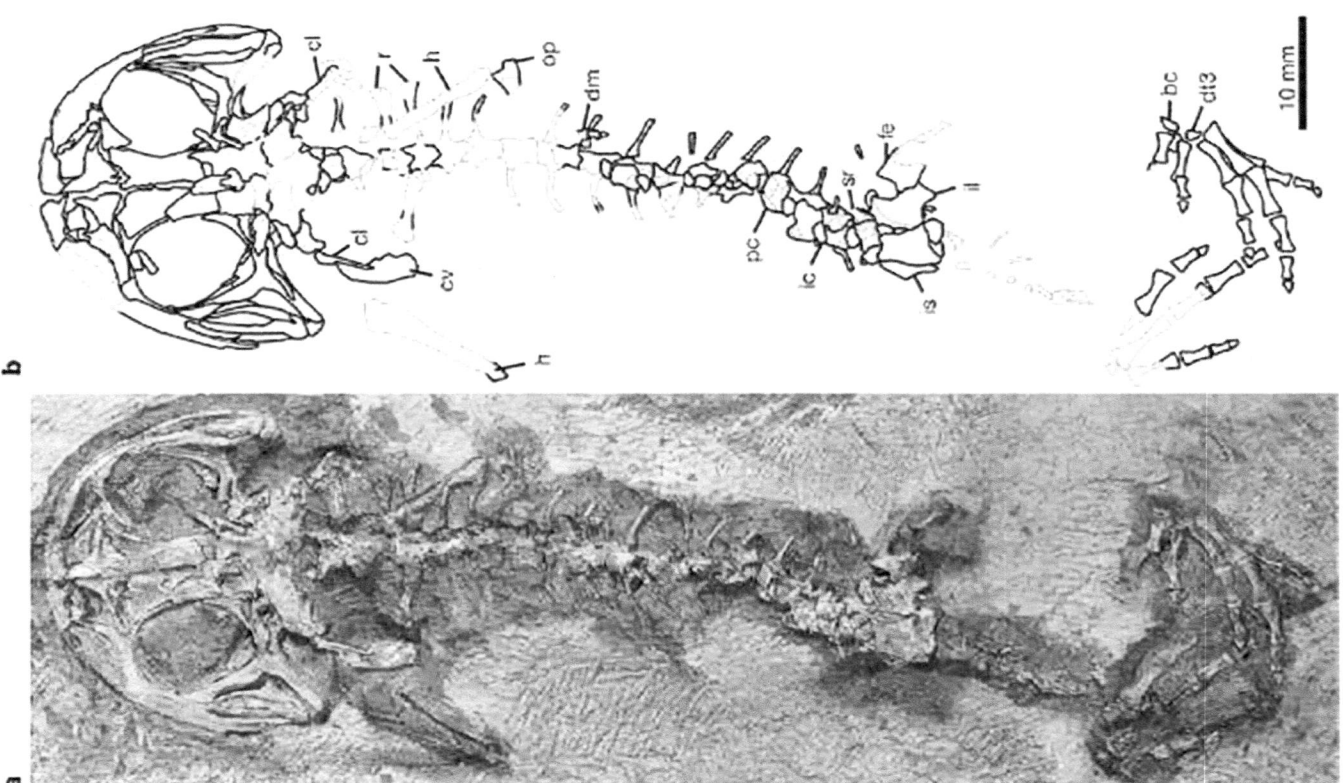

Figure 7.5 *Gerobatrachus hottoni*, the "Frogamander", a Permian fossil with a head like a frog and a body like a salamander. (Photos courtesy J. Anderson.)

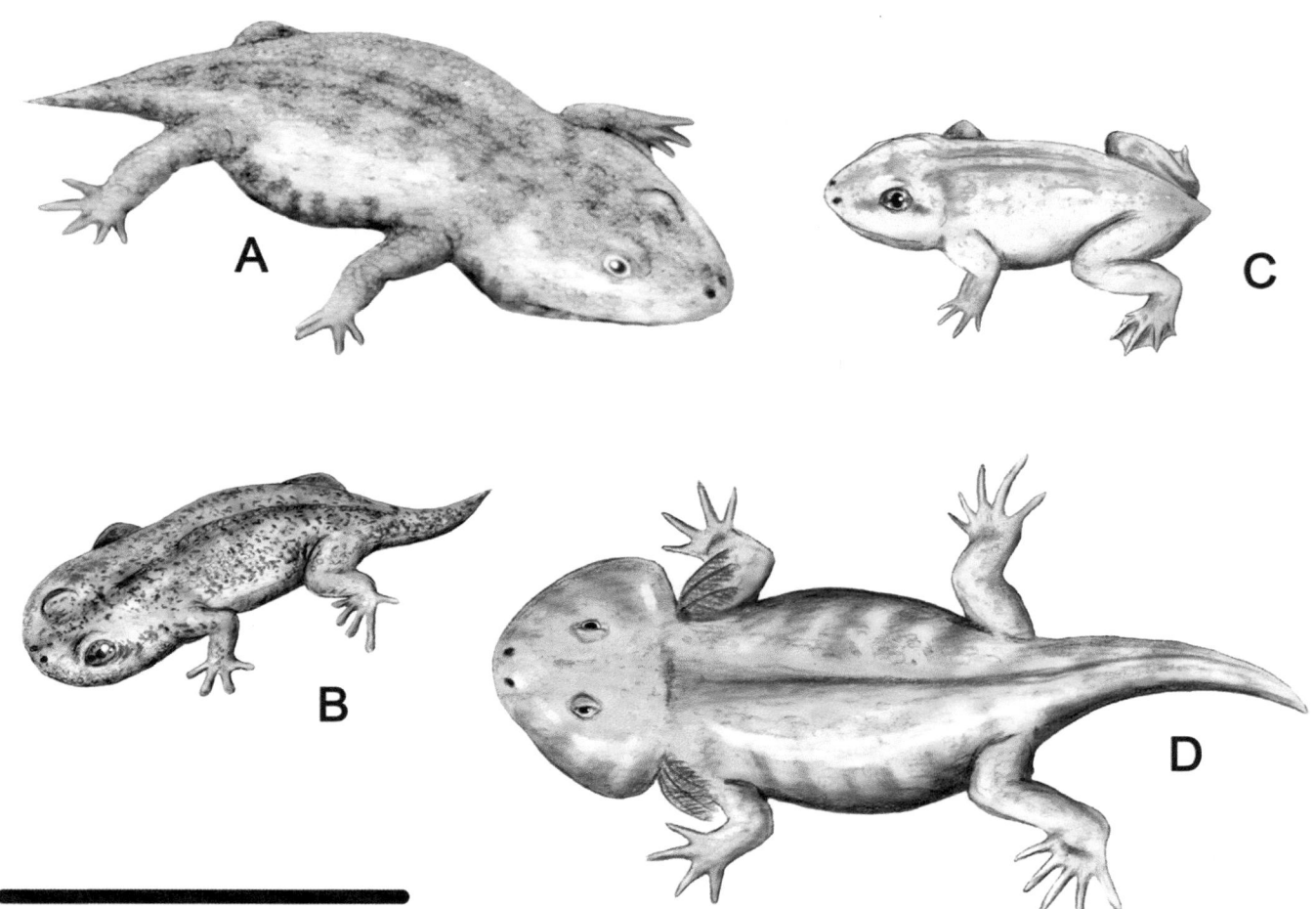

Figure 7.6 Some of the earliest known lissamphibians. (A) *Amphibamus*, (B) *Gerobatrachus*, (C) *Triadobatrachus*, (D) *Karaurus*. (Scale bar = 10 cm.)

the specimen is nearly complete lying on its back with only a part of the hip bones, shoulder bones, and part of the tail missing. The most striking thing about the fossil is that it combines the long-tailed body of a salamander with the broad rounded snout of a frog, so it shows how frogs might have begun to evolve from salamander-like forms. It has a few other froggy features, like a large eardrum, and teeth that sit on tiny pedestals with a distinct base, an anatomical condition found only in the living amphibians and their close fossil relatives. But otherwise it has the primitive salamander-like body, so it is a perfect transition between the two groups.

After the Early Permian *Gerobatrachus*, the next good frog fossil is *Triado-batrachus* from the Early Triassic (240 Ma) of Madagascar (**Figures 7.6[C]** and **7.7**). It has the typical froggy broad snout and long webbed feet, but unlike any living frog, it still has a long trunk region, with 14 vertebrae in its spine, not the 4–9 found in living frogs (**Figure 7.7**). It even retained a short tail that was not lost when its tadpoles grew to adulthood. It had longer hind legs than any salamander, but not the huge muscular legs found in living frogs, so it could swim but it could not jump. By the Early Jurassic (about 200 Ma), there are fossils of the first true frog, *Vieraella*, from Argentina. A tiny creature only 3 cm (2 inches) long, its skull was completely frog-like, the hind limbs were capable of jumping, but it still did not have the short trunk region or extremely modified hip region of modern frogs. By the Cretaceous, frogs looked almost completely

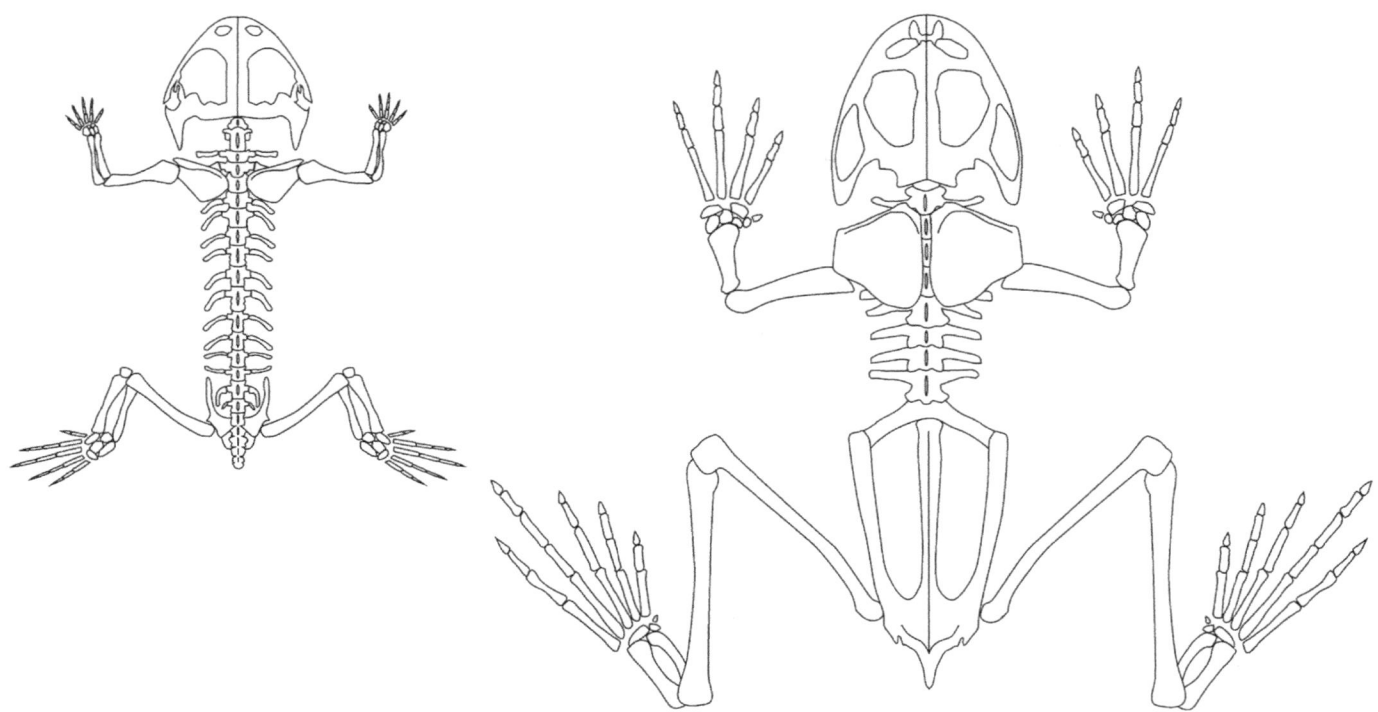

Figure 7.7 Comparison the skeleton of the Triassic frog *Triadobatrachus* (left) with a modern frog (right).
Triadobatrachus has some features of a frog, like a broad snout and longer hind legs, but it still had many trunk vertebrae, small hip bones, a short tail (not lost when it grows out of the tadpole stage), and its hindlegs are not nearly so long and powerful as in modern frogs. (Courtesy Wikimedia Commons.)

modern in their anatomy, and had diversified into many of the groups that are alive today, with dozens of families that include over 4800 living species.

Meanwhile, salamanders first show up in the Jurassic as well, with *Kokartus* from the Middle Jurassic of Kyrgyzstan, *Marmorerpeton* from the Middle Jurassic of England, and *Karaurus* from the Upper Jurassic of Kazakhstan (**Figure 7.6[D]**). By the Cretaceous, we have fossils of some of the modern families of newts and salamanders in many places around the world.

Today, the living amphibians are tremendously diverse, with over 5700 species reported. More than 4800 of these species are frogs and toads, but there are only 655 species of salamanders and newts. In addition, there are about 200 species in a third group of living amphibians, the apodans or caecilians. These are legless amphibians that burrow underground mostly in tropical soils of South America, Africa, and Asia. They have tiny eyes that can sense light and dark, and some have their eyes at the tip of sensory tentacles, but most are blind. To the non-specialist, they look almost like giant earthworms.

Living amphibians range enormously in size, from the tiny New Guinean frog *Paedophryne amanuensis*, which is only 7.7 mm (0.3 inches) long, to the huge Chinese giant salamander (**Figure 7.8[A]**) over a meter long (39 inches). Salamanders and newts retain the simple elongate body form with a long tail and four simple limbs, as found in the most primitive amphibians (such as *Tiktaalik*, *Ichthyostega*, and *Acanthostega*, discussed in the previous chapter).

However, the frogs are the most spectacularly divergent from this ancestral body form of all the living amphibians. As anyone who has dissected a frog in high school biology knows, they are truly unique in their body

A

B

Figure 7.8 Some of the largest lissamphibians known. (A) The living Chinese giant salamander. (B) The gigantic Cretaceous frog from Madagascar, *Beelzebufo*. [(A) Courtesy Wikimedia Commons.]

design. Although adult frogs and toads have no tail, their larvae (tadpoles) hatch with tails that are resorbed into their bodies as they mature. The frog's head is short with a blunt broad snout that allows them to open their mouths wide as they capture food (often using a long sticky tongue). Their very long muscular hind legs allow them to make huge leaps (both to catch prey and escape predators) as well as swim with great power. The trunk of the frog skeleton is also shorter, with tiny stumpy ribs, and very elongated hip bones to support the hind leg muscles (**Figure 7.7**). Since they cannot use their ribs for breathing, they use an inflatable pouch in their throat that can pump air in and out (also used for making a variety of sounds). Frogs range tremendously in size, from the tiny New Guinean frog mentioned previously, to the Goliath frog, which is over 300 mm (12 inches) long, and weighs 3 kg (7 lb). It is so big that it eats birds and small mammals, as well as insects.

If the Goliath frog were not impressive enough, in 1993 a group of scientists working in the Upper Cretaceous rocks of Madagascar found an even bigger frog. After 15 years of fitting all the pieces together (including most of the skull among the 75 fragments), they published a description of it in 2008. They named it *Beelzebufo ampinga*, or the "Devil's toad" (**Figure 7.8[B]**). The name is a composite of Beelzebub, the "Lord of the Flies", another name for the Devil, and *Bufo*, the genus of common toads; *ampinga* means "shield" in Malagasy. It was a ceratophrynine toad, a group known as the "horned toads" of South America, so this family once extended across Gondwana and lived in Madagascar as well. (Not to be confused with the common name "horny toad", which is a horned lizard of the family Phrynosomidae, not an amphibian.) But its most remarkable feature was its size. On the basis of the nearly complete skeleton, *Beelzebufo* was 40 cm (16 in) long and 4 kg (9 lb) in weight—a third again as large as the Goliath frog! It had a very large head and wide mouth, and it is speculated that it could even eat baby dinosaurs, which roamed Madagascar at the time.

Sadly, many living amphibians are disappearing rapidly due to environmental factors, such as destruction of their habitat (especially in rain forests), a number of diseases that are spreading through frog populations and wiping them out, and the increasing acidity of fresh waters due to acid rain, which amphibians cannot tolerate with their porous skins. It would be truly sad if frogs and salamanders, which date back to the Triassic and Jurassic, showed no effects of the extinction that wiped out the dinosaurs, and have diversified in the last 66 million years, were to vanish in a century thanks to humans destroying their habitat.

FURTHER READING

Baird, D. 1965. Paleozoic lepospondyl amphibians. *Integrative and Comparative Biology*. 5 (2): 287–294.

Clack, J.A. 2002. *Gaining Ground: The Origin and Early Evolution of Tetrapods*. Indiana University Press, Bloomington, IN.

Crump, M.L. 2009. Amphibian diversity and life history. *Amphibian Ecology and Conservation: A Handbook of Techniques*: 3–20.

Falcon-Lang, H. J.; Benton, M. J.; Braddy, S. J.; Davies, S. J. 2006. The Pennsylvanian tropical biome reconstructed from the Joggins Formation of Nova Scotia, Canada. *Journal of the Geological Society, London*. 163 (3): 561–567.

Laurin, M.; Canoville, A.; Quilhac, A. 2009. Use of paleontological and molecular data in supertrees for comparative studies: The example of lissamphibian femoral microanatomy. *Journal of Anatomy*. 215 (2): 110–123.

Lombard, R.E.; Bolt, J.R. 1979. Evolution of the tetrapod ear: An analysis and reinterpretation. *Biological Journal of the Linnean Society*. 11 (1): 19–76.

Pearson, M.R.; Benson, R.B.J.; Upchurch, P.; Fröbisch, J.; Kammerer, C.F. 2013. Reconstructing the diversity of early terrestrial herbivorous tetrapods. *Palaeogeography, Palaeoclimatology, Palaeoecology*. 372: 42–49.

Sahney, S.; Benton, M.J.; Falcon-Lang, H.J. 2010. Rainforest collapse triggered Pennsylvanian tetrapod diversification in Euramerica. *Geology*. 38 (12): 1079–1082.

Sahney, S.; Benton, M.J.; Ferry, P.A. 2010. Links between global taxonomic diversity, ecological diversity and the expansion of vertebrates on land. *Biology Letters*. 6 (4): 544–547.

San Mauro, D. 2010. A multilocus timescale for the origin of extant amphibians. *Molecular Phylogenetics and Evolution*. 56 (2): 554–561.

San Mauro, D.; Vences, M.; Alcobendas, M.; Zardoya, R.; Meyer, A. 2005. Initial diversification of living amphibians predated the breakup of Pangaea. *The American Naturalist*. 165 (5): 590–599.

PRIMITIVE REPTILES

8

The most fundamental innovation is the evolution of another internal fluid-filled sac, the amnion, in which the embryo floats. Amniotic fluid has roughly the same composition as seawater, so that in a very real sense, the amnion is the continuation of the original fish or amphibian eggs together with its microenvironment, just as a space suit contains an astronaut and a fluid that mimics the Earth's atmosphere. All of the rest of the amniote egg is add-on technology that is also required for life in an alien environment, and in that sense it corresponds to the rest of the space station with its food storage, fuel supply, gas exchangers, and sanitary disposal systems.

—Richard Cowen, *History of Life*, 1990

LAND EGGS AND THE FIRST AMNIOTES

In the previous chapter, we discussed how a number of groups of tetrapods invaded the land, but they were still tied to the water to reproduce. Even amphibians that live in deserts must find a moist place to lay their eggs. Most do so by laying huge masses of tiny eggs that are vulnerable to being eaten or drying up, but enough survive and hatch to form larvae (such as the tadpole stage of frogs) and eventually adults. Some tree frogs push this type of reproduction to the limit by laying their eggs in damp areas glued to leaves on branches, but ultimately their reproduction is tied to a moist environment. The ability to lay an egg that does not confine the animals to such restrictive conditions would give some groups of tetrapods an advantage in exploiting new habitats.

The animals that developed this innovation (**Figure 8.1**) are called amniotes, after the characteristic amniotic egg found in reptiles, birds, and even egg-laying mammals like the platypus (all other mammals keep the eggs inside their body and let them develop there). Instead of the hundreds of tiny, thin-shelled eggs laid by amphibians, amniotes lay fewer, larger eggs with a shell (either brittle and calcareous, or leathery) that protects the egg and resists water loss while allowing limited exchange of gases through its pores. Inside the egg are numerous specialized systems (**Figure 8.2**). The embryo is surrounded by a membrane called the amnion, which is also filled with amniotic fluid that buffers the embryo against shock, temperature change, and other rapid, stressful fluctuations in the environment. Attached to the gut of the embryo but outside the amnion, is the yolk sac, which provides food for the developing embryo so that it can emerge from the egg relatively self-sufficient, unlike the partially developed tadpole found in the amphibians. A third sac off the hindgut of the embryo is the allantois, which serves for waste storage and also for respiration. As the embryo develops and gets larger, the yolk sac dwindles, while the amnion and allantois get larger as their contents expand.

DOI: 10.1201/9781003128205-8

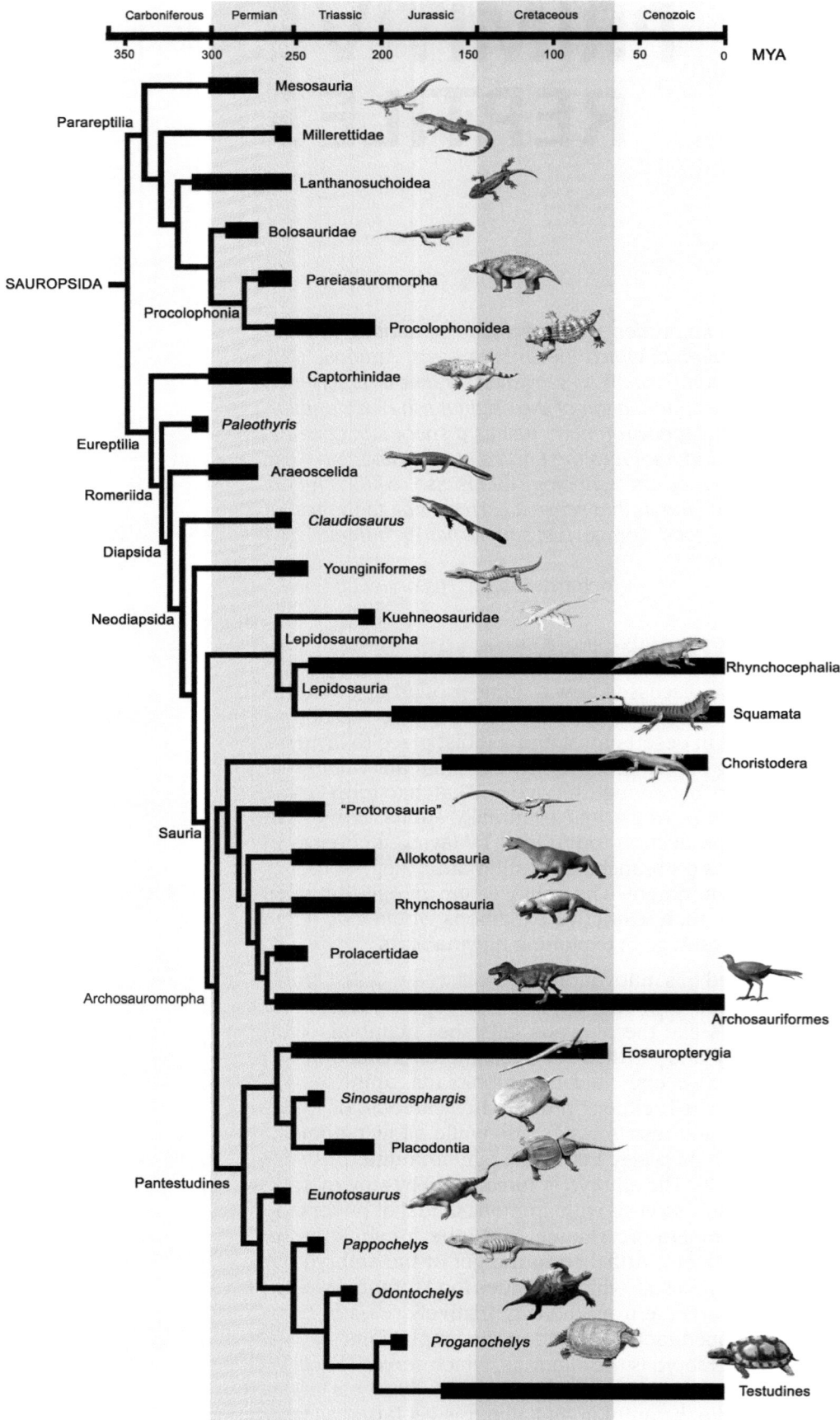

Figure 8.1 Evolutionary radiation of the major groups of amniotes, following the current interpretation based on molecular data that turtles are nested within the Archosauromorpha, and are the closest relatives of marine reptiles.

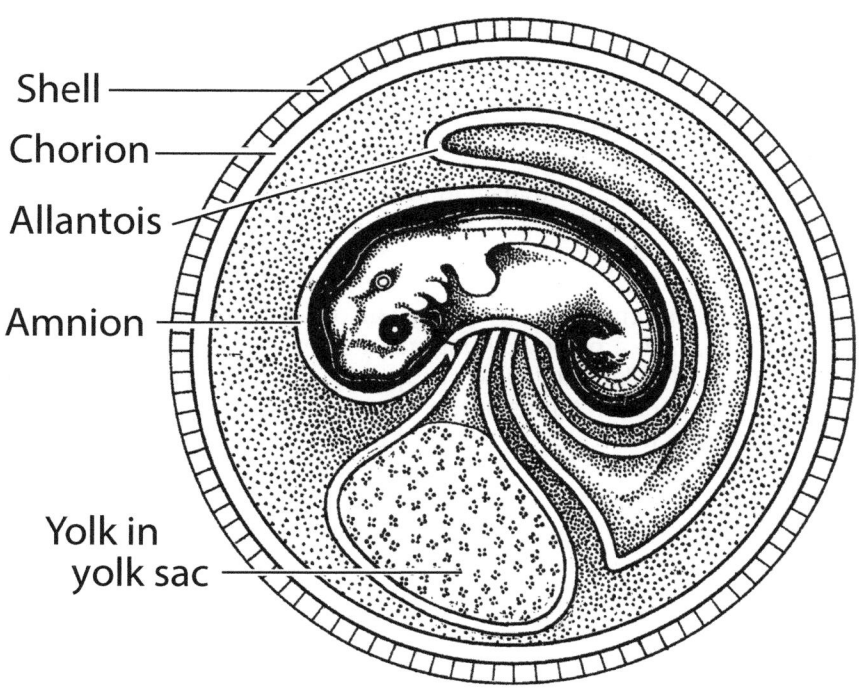

Shell

Chorion

Allantois

Amnion

Yolk in
yolk sac

Figure 8.2 The structure of the amniotic egg. (Redrawn from several sources.)

Amphibian embryos produce waste in the form of urea, which easily disperses in their aquatic habitats, but amniote embryos have a limited water supply, so they secrete their nitrogenous wastes in the form of crystallized uric acid that can be stored without wasting water. Finally, the amnion, yolk sac, and allantois are surrounded by the fluid albumin ("egg white") that fills the rest of the egg cavity, which in turn is completely surrounded by another membrane, the chorion, that lies just beneath the eggshell. You are probably familiar with the chorionic membrane as the thin "skin" just inside the eggshell when you crack and peel a hard-boiled egg, or the thin membrane that resists being punctured when you try to crack a fresh egg.

In addition to protecting the young from drying up and predation, and allowing them to hatch far from water, the amniotic egg has other implications. Each egg is more costly to produce, so fewer can be laid, and each embryo must develop much more before it hatches to ensure that some survive. In addition, the eggs cannot be fertilized by a male who simply swims nearby and sprays the floating egg mass with sperm. The amniotic egg can be produced only by internal fertilization. Males and females must copulate so that the female can carry the sperm to the eggs inside her, where they can begin development. Of course, this is not the first time that internal fertilization appears in the animals. Most land-living arthropods have independently developed it, as have some sharks (which are also capable of giving birth to live young).

Clearly, the amniotic egg is an important innovation, a breakthrough that allowed one group of tetrapods to exploit an entire range of new habitats. Unfortunately, eggs don't fossilize very often, and even more rarely they are found in association with the organism that laid the egg. Although probable amniote skeletal fossils are known from the Early Carboniferous, the oldest known fossil amniotic egg is Permian in age, and some paleontologists question whether it is even an egg.

It is possible that some of the creatures in the wastebasket group called "anthracosaurs" (mentioned in Chapter 7) may have laid an amniotic egg (**Figure 8.3**). If they were not true egg-laying amniotes, they still have many of the skeletal features of amniotes, and so they are the closest relatives. For example, many of these "anthracosaurs" (now commonly called "reptiliomorphs") have lost the amphibian eardrum notch in the back of the skull and instead have a solid notch-less skull roof in the back, with a tubular canal around the stapes for hearing. Most of these "anthracosaurs" have modified the limb girdles for a more upright, efficient form of locomotion; they were not nearly as sprawling as the huge, flat-bodied temnospondyls. They also had high-domed, vertically deeper skulls, with a narrow snout and a short region behind the eyes, in contrast to the flat-skulled temnospondyls and lepospondyls. There are numerous specializations in the wrist and ankle bones that accompany their more active locomotion. Their neck vertebrae become specialized into an atlas (the first neck vertebra, which supports the skull) and axis (the second vertebra, which allows the head to pivot and turn). This allowed them to swivel their heads rapidly to catch prey. Finally, the muscles and bones of the palate region were modified such that they had a much stronger bite force, rather than the relatively weak "snapping"

Figure 8.3 "Anthracosaurs" were probably not amniotes, but they show many skeletal features which are shared with the amniotes, so they are thought to be closer to reptiles than they are to "amphibians". Some examples include: (A) *Seymouria*, (B) *Limnoscelis*, (C) *Diadectes*. (Scale bar = 1 meter.)

motion of temnospondyls. Many even have fangs on the palate to secure their prey.

Various combinations of these characters have been used to recognize early amniotes, although the distinction is not very clear-cut. For example, *Solenodonsaurus* still had the eardrum notch but also had the advanced jaw muscle attachments. *Gephyrostegus* had an eardrum notch and primitive vertebrae and limb girdles, but had an advanced ankle region. *Seymouria* (**Figures 8.3[A]** and **8.4[A]**) was very reptilian in its deep boxy skull, but still had an eardrum notch and primitive ankle bones. *Limnoscelis* (**Figures 8.3[B]** and **8.4[B]**) was much more reptilian, with no eardrum notch, but did not have the advanced ankle bones. *Diadectes* (**Figures 8.3[C]** and **8.4[C]**) was a pig-sized herbivorous animal from the Permian, and had an atlas-axis complex, and strong limbs with an advanced ankle, but still had the eardrum notch. Most paleontologists now regard these "anthracosaurs" or "reptiliomorphs", which are known mostly from the Late Carboniferous or Permian, as extinct side branches of tetrapod evolution that show various combinations of amniote characters long after the amniotes had split off. This is because tiny, lizard-like animals such as *Westlothiana* (**Figure 8.5**) are now known from much older deposits (Lower Carboniferous of Scotland), so the amniote-"reptiliomorph" split must have occurred long before Late Carboniferous-Permian creatures like *Solenodonsaurus*, *Gephyrostegus*, *Limnoscelis*, or *Diadectes* lived.

A

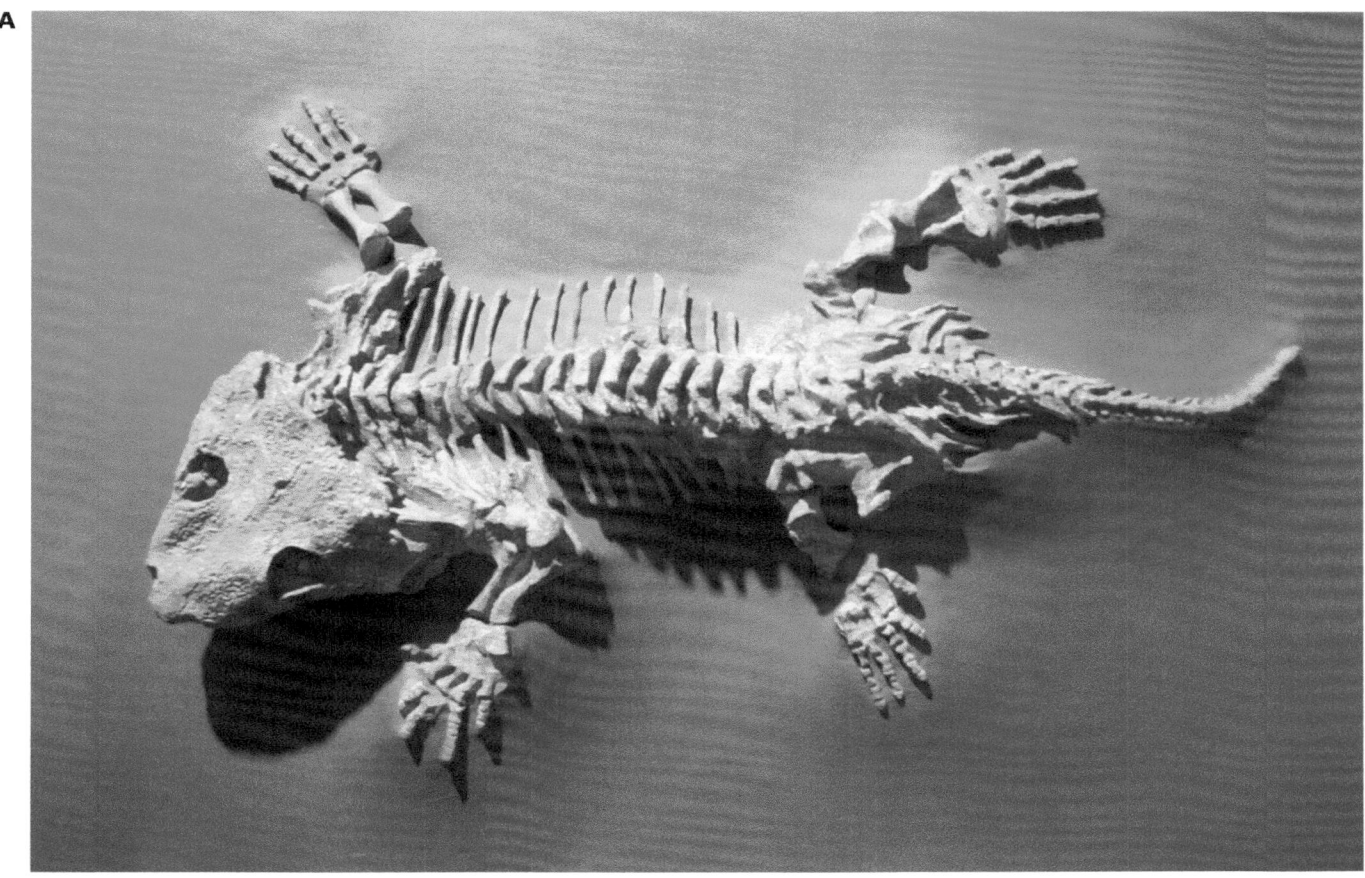

Figure 8.4 Fossils of typical "anthracosaurs". (A) *Seymouria*. (Courtesy Wikimedia Commons.)

B

C

Figure 8.4 (Continued) (B) *Limnoscelis*. (C) *Diadectes*. Behind the *Diadectes* specimen is *Archeria* (formerly *Cricotus*), another "anthracosaur".

Westlothiana from the Early Carboniferous, and *Hylonomus* and *Paleothyris* from the Middle Carboniferous, are now considered as representative of the earliest amniotes (**Figure 8.5**). These animals were built much like slender lizards, with long gracile limbs and toes and a very long trunk and tail. They had relatively small heads in proportion to their bodies, but their deep skulls had very effective jaw muscles, and so they were very effective in catching insects and other small prey. Their relatively large eyes also suggest that they may have been active predators, possibly at night as well as during the day. The best specimens of *Hylonomus* were found inside the hollow trunks of giant club moss trees (lycopods) from

Figure 8.5 Some reconstructions of the earliest amniotes known: (A) *Westlothiana*, (B) *Hylonomus*. Scale bar: 5 cm.

the Middle Carboniferous beds of Joggins, Nova Scotia. Traditionally, it was thought that these animals had been trapped inside half-buried rotting logs, but more recent analysis suggests that they probably lived and hunted in and around these hollow trees.

How the rest of the amniotes are related is a very controversial issue (**Figure 8.1**). For decades, they were grouped based on a very simplistic classification of the holes (called "temporal fenestrae") in the back of the skull roof (**Figure 8.6**). These gaps in the bone serve to lighten the skull, and in many cases, they allow for the bulging and increased attachment area for larger jaw muscles, making a stronger bite. Anapsid amniotes (Greek for "no arches") had a solid skull roof with no holes in it. The only living reptiles without any temporal fenestrae in their skull are the turtles and tortoises and their kin, but nearly all the primitive amniotes lacked any holes in the back of their skulls. Some amniotes had a temporal fenestra low in the skull roof, beneath the postorbital bone behind the eye socket (or "orbit") and the squamosal bone that makes up the back corner of the skull. The condition with a skull bearing a lower temporal fenestra only was called synapsid ("united arch"), and it is typical of all the amniotes that were closely related to mammals, and by extension, mammals themselves.

A third condition had a temporal fenestra high on the skull roof, above the postorbital and squamosal bones. This condition was called euryapsid ("broad arch" in Greek), and it was found mostly in the marine reptile groups known as ichthyosaurs and plesiosaurs. Finally, most reptiles alive today have two holes in their skull roof, one on the top back end of the skull, and one below it on the side, separated

Figure 8.6 Different configurations of the holes in the side of the amniote skull (temporal fenestration). (A) The "anapsid" condition (seen in turtles and many primitive reptiles) has no opening in the back of the skull. (B) The "euryapsid" condition has a single temporal fenestra above the postorbital-squamosal bones. It occurs in marine reptiles like ichthyosaurs and plesiosaurs. (C) The "synapsid" condition has a single temporal fenestra low on the skull roof, with a bar of bone composed of the postorbital bone and the squamosal bone, lying just above the edge of the fenestra. All mammals and their relatives are synapsids. (D) Finally, the "diapsid" skull has two openings, one above and one below the postorbital-squamosal bar. It is found in most reptiles, including snakes and lizards, crocodiles, pterosaurs, dinosaurs, and birds. These skull roof features were the basis for amniote classification for almost a century, but more recent analyses using other anatomical features and molecular similarities shows that they are not that important in classification. (Redrawn from several sources.)

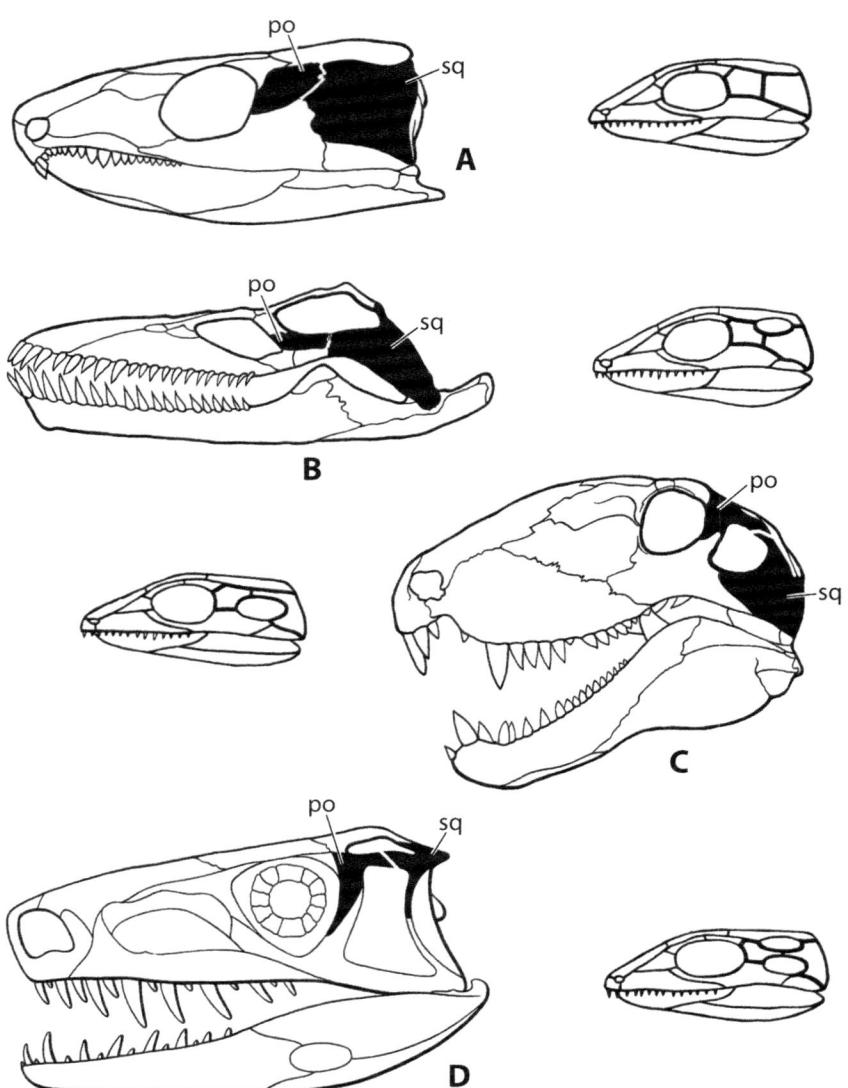

by a bar of bone made of the postorbital and squamosal bones. This condition is known as diapsid ("two arches" in Greek). Diapsid skulls occur in lizards and snakes, crocodilians, and birds and dinosaurs, and many other reptiles. For most of the decades in the twentieth century, the debates that followed were often along the lines of "Did diapsids originate from euryapsids and add a lower temporal fenestra, or did they originate from synapsids and add an upper temporal fenestra?

This simplistic classification scheme goes back to the late 1800s and early 1900s, and dominated paleontology until the 1980s and 1990s, when scientists began to look at other anatomical features besides the number of holes in the side of the skull. Eventually, this simple four-group system was thrown out, when other characteristics of the anatomy showed that it was mostly wrong. For example, anapsids were a wastebasket group with no unique evolutionary specializations—the solid skull roof is something amniotes inherited from more primitive tetrapods, and not an advanced feature that we use to define a natural group. Even more surprising, when molecular evidence began to come in for the living members of these groups, it has been suggested based on molecules that turtles were not close to any of the anapsids, but were actually closely related to more advanced reptiles like lizards, snakes, and crocodilians and other diapsids, even though they had the primitive anapsid skull (**Figure 8.1**). Another surprise was that the rest of the anatomy (other than the skull holes) showed that euryapsids were related to diapsids, and originally had diapsid skulls that lost the lower temporal fenestra. Finally, the idea that synapsids were another group of "mammal-like reptiles" was discredited when it became clear that they branched off from the reptiles at the very beginning of amniote evolution. Both the earliest synapsids (*Asaphestera* and *Protoclepsydrops* from the Early Carboniferous and *Archaeothyris* from the Middle Carboniferous) and the earliest reptiles (*Westlothiana*, *Hylonomus*, and *Paleothyris*) are known from deposits of about the same age, suggesting that synapsids and reptiles are separate, contemporaneous branches, not ancestors and descendants. (The earliest fossil widely regarded as a diapsid, *Petrolacosaurus*, is known from the Late Carboniferous.) Synapsids are *not* "mammal-like reptiles", but a separate clade that split off very early from the branch that led to the groups we recognize as reptiles (turtles, lizards, snakes, crocodiles, and their extinct relatives). Synapsids never had anything to do with reptiles in any sense of the word and should never be called "mammal-like reptiles". This will be discussed further in Chapter 20.

So if you separate the synapsids from the rest of the early amniotes, then the remaining cluster of animals that includes all the living reptiles (turtles, crocodilians, snakes, and lizards) plus the many extinct animals (dinosaurs, pterosaurs, euryapsids like ichthyosaurs and plesiosaurs) forms a natural group that could be called "Reptilia"—as long as you include all their descendants, such as the birds (**Figures 1.5** and **8.1**). Modern classification schemes are based on natural groups defined by unique evolutionary specializations, and include all their descendant groups. Thus, "Reptilia" can be defined as a grouping for the living turtles, snakes plus lizards, crocodilians, and birds, and all their fossil relatives. The reptiles classified this way have many shared evolutionary specializations, including scaly, keratinized skin, color vision, rapid eye focusing, a third eyelid, and nitrogenous wastes secreted as uric acid, not urea. These are all unique features of the living reptiles and birds, but there are some features of the bony skeleton that fossilize as well and can be used to define "Reptilia".

However, some people are accustomed to the older way of classifying animals, and cannot get used to the idea that a group must include all its descendants, so they prefer two parallel and equal groups, "class Reptilia" and "class Aves" (birds). In their view, birds are special, and have all these unique adaptations and a huge evolutionary diversification that justifies them being put in a parallel group equal in rank to "Reptilia". But modern classification does not allow any other criteria, like ecological divergence, a role in classification, so most scientists have long ago accepted the idea that birds are dinosaurs, and are also a group within reptiles.

Some scientists try to avoid all the confusion and misleading implications of the term "reptile" by using a different name. In that case, the amniotes split into parallel groups in the Early Carboniferous: the synapsids (mammals and their extinct relatives) and sauropsids (all the rest of the living and extinct animals called reptiles but also including birds). Whichever term is used, this is the deepest evolutionary split in the amniotes (**Figure 8.1**).

PARAREPTILES

Once we get rid of misleading and outdating classifications including "anapsids", we have a cluster of very primitive sauropsids (or reptiles, if you prefer) with a solid skull roof lacking a temporal fenestra of any kind. These have long been classified in another group, the "Parareptilia" (Greek for "near reptiles"). Now that Anapsida has been abandoned as a natural group, the name Parareptilia seems to work for these earliest sauropsids (**Figure 8.1**). Most of them are united by a number of unique evolutionary specializations and the primitive anapsid skull condition is irrelevant.

The most spectacular of the parareptiles were the huge, sprawling pareiasaurs (pah-RYE-o-saurs) which were often the size of a large pig to a small hippo (**Figure 8.7**). They were fat and stumpy-legged, with a short tail and a barrel-shaped trunk that housed a digestive tract large enough to feed on the relatively indigestible plant life of that time. Pareiasaurs had broad, rounded skulls made of thick bony plates. Their heads were covered in bumps and knobs, and their hides were armored by little bumps and knobs of bone over their entire body. The biggest ones, from Russia and South Africa, were typically about 2 meters (6.5 feet) long and weighed as much as 600 kilograms (1300 pounds). Yet pareiasaurs had only small, leaf-shaped teeth, and they were apparently gentle herbivores, munching on the ferns and conifers that covered the Permian landscape. In fact, they were one of the few herbivorous reptilian groups of any kind in the Permian. The Upper Permian Rio do Rasto Formation in Brazil yields a pareiasaur known as *Provelosaurus*, which was about 2.5 meters (8.5 feet) long.

From the huge ugly hippo-like parieasaurs, at the other extreme were the smaller parareptiles. Closely related to the pareiasaurs were the procolophonids (**Figure 8.8[A]**). They were small forms about 30 cm (1 foot) long, shaped mostly like lizards, except that their skulls were very short and broad, with extremely wide cheekbones and short snout with a set of nipping teeth in front. Some of them, like *Hypsognathus*, had spikes flaring out from around their broad skulls (**Figure 8.7[B]**). Primitive

Figure 8.7 Reconstructions of typical pareiasaurs: (A) *Bradysaurus*, (B) *Pareisaurus*, (C) *Scutosaurus*, (D) *Elginia*. Scale bar: 1 meter. (Courtesy Wikimedia Commons.)

Figure 8.7 (Continued) (E) Fossil skeleton of Bradysaurus.

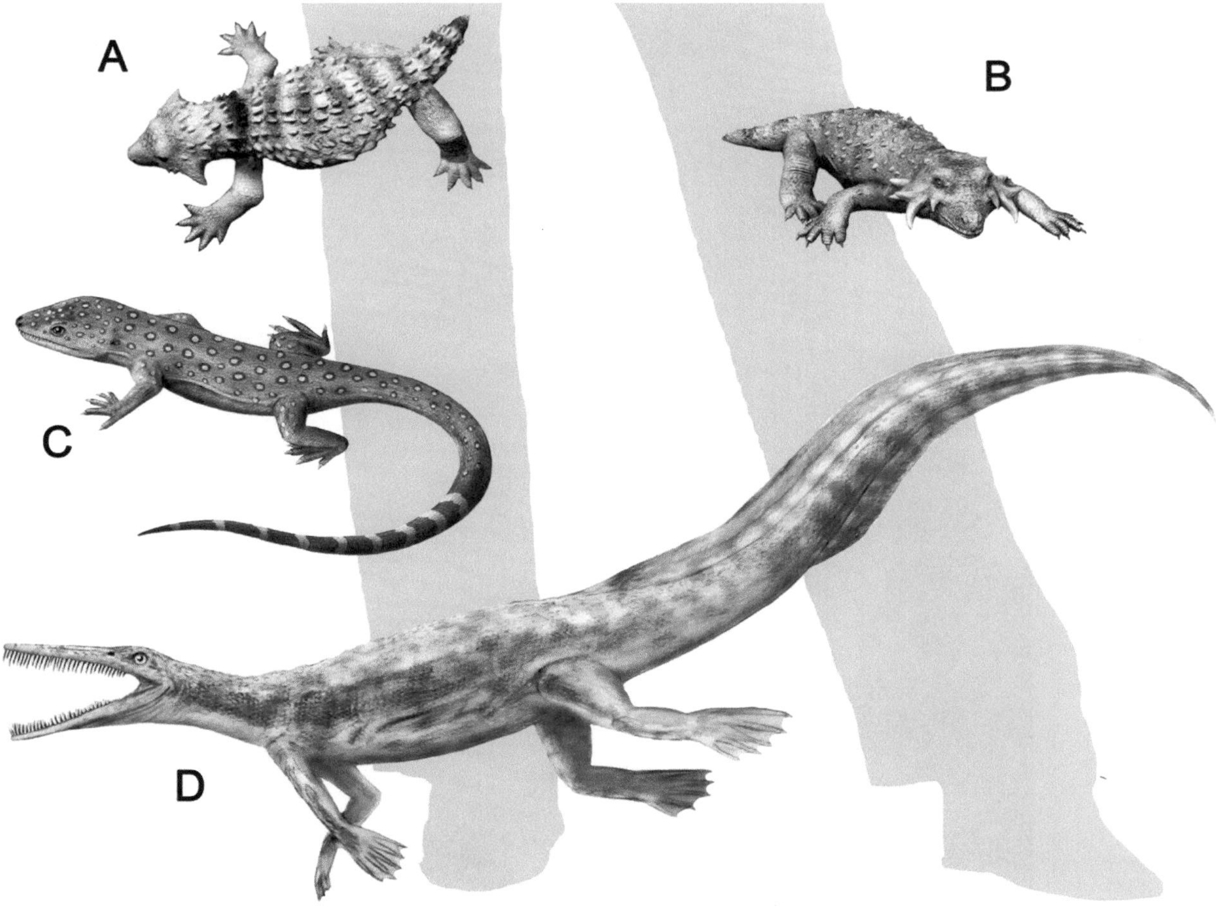

Figure 8.8 Other examples of parareptiles include: (A) *Procolophon*, (B) *Hypsognathus*, (C) *Milleretta*, (D) *Mesosaurus*.

E

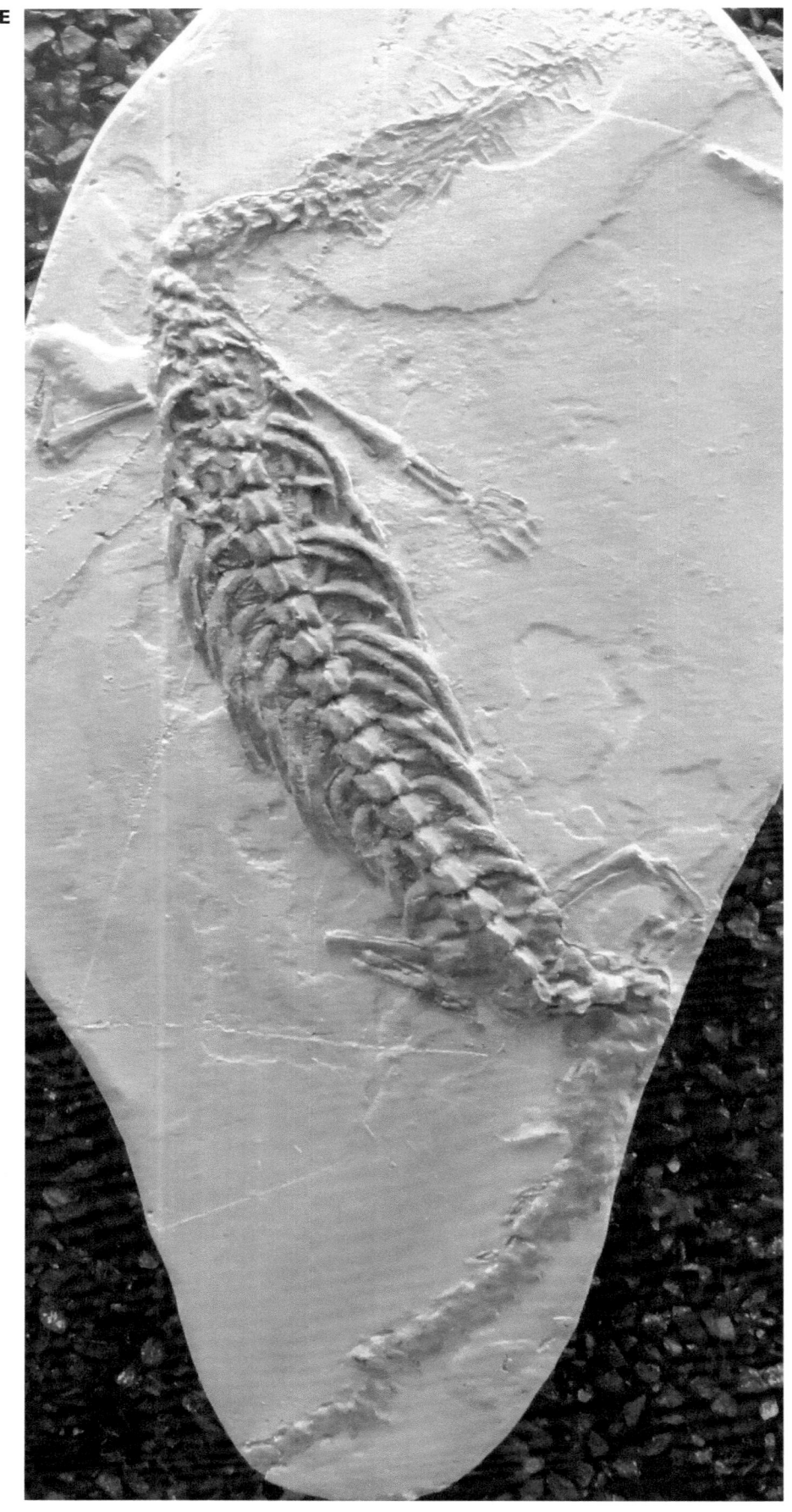

Figure 8.8 (Continued) (E) Fossil specimen of *Mesosaurus*. [(E) Courtesy Wikimedia Commons.]

procolophonids had sharp conical teeth suitable for a diet of insects and small animals, but later in their evolution their teeth became blunt and peg-like, suggesting that they were omnivorous or maybe even herbivorous like pareiasaurs. They were found in many places during the Early Triassic, but vanished by the Late Triassic, as did a number of archaic reptiles.

Another small lizard-like parareptile group is the millerettids (**Figure 8.8[C]**). From the outside view, millerettids like *Milleretta*, *Millerosaurus*, and *Broomia* didn't look much different from a modern lizard, but in their skulls they have a lot of detailed features which place them close to pareiasaurs and procolophonids. Their conical teeth suggest an herbivorous diet. Millerettids are known almost exclusively from the Middle and Late Permian of South Africa, and vanished in the great Permian extinction event, along with about 75% of all land animals.

One of the most significant parareptiles was a little creature known as *Mesosaurus* (**Figure 8.8[D,E]**). Found in the Permian lake beds of Brazil and Uruguay and in South Africa, *Mesosaurus* was a small aquatic form with webbed feet for swimming, a flattened tail that was used for propulsion, and a long snout full of tiny thin needle-like teeth for snagging fish or crustaceans. They also had enlarged ribs made of very dense bone, a common feature of aquatic animals to make them heavier and provide a sort of ballast for them. Most were only about 50–70 cm long (about 2 feet), so they were not big creatures. Their greatest claim to fame, however, is that they show the connection of the Gondwana continents during the Permian, something that was noticed in the very earliest days of continental drift theory in the early twentieth century. Although *Mesosaurus* was aquatic, it was small and always found in lake beds, never in marine rocks. Its small size made it unlikely that it could swim all the way across the width of the modern Atlantic between South America and South Africa. Once plate tectonics came along in the 1960s, the presence of *Mesosaurus* was one of the key fossils to clinch the idea that the Gondwana continents were united during the Permian.

EUREPTILIA

In the branching sequence of amniotes (**Figure 8.1**), the parareptiles seem to be a natural group based on a number of anatomical specializations. All of them were restricted to the Permian and Triassic, during the earliest diversification of amniotes from their Carboniferous ancestors like *Westlothiana* and *Hylonomus*. But there were other early lineages of reptiles evolving in the Permian as well. According to most analyses (**Figure 8.1**), the other major group of sauropsids besides the parareptiles is a group called the Eureptilia, which includes all the reptiles and birds except for the Parareptila.

One of these distinctive primitive early eureptiles are the captorhinids, a group of about a dozen genera typified by *Captorhinus* (**Figures 8.9[A]** and **8.10[A]**). It is known from many complete specimens found everywhere in Lower Permian rocks, ranging from Oklahoma and Texas to Europe, India, Brazil, and Zambia (in other words, almost all of Pangea). Other captorhinids are found in additional parts of Pangea. *Captorhinus* superficially looked like any other small lizard-like reptile (**Figure 8.9[A]**). But it differs in many ways from any lizard we know. It was about 20 cm (1 foot) long, but there are lots of details of the skeleton that establish it as a primitive Eureptile. Its most distinctive features are shown in the skull, which was long and narrow, with the massive thick construction (and a solid skull roof with no temporal fenestrae), and a large eye socket. *Captorhinus* is most easily recognized by the overhanging hooked snout, where the upper "beak" hooks completely over the front of the lower jaw with a series of large conical teeth protruding downward (what would be called incisor teeth in a mammal). Many of the *Captorhinus* species multiple tooth rows, some on the edge of the jaws, and one on the palate as well. *Captorhinus* has a long slender body with very flexible feet, which would have made is a fast runner able to ambush prey, like many lizards do today. Their small body size but agile build suggests that they hunted small vertebrate prey as well as small invertebrates. They lived in an Early Permian world with small predatory synapsids competing with them, and avoiding huge predators that terrorized the Early Permian, such as the giant temnospondyls and huge synapsids like *Dimetrodon*. Another well-known captorhinid was *Labidosaurus*, which was slightly larger than *Captorhinus*, with a massive triangular skull (**Figures 8.9[B]** and **8.10[B]**). For a long time, captorhinids were considered to be ancestral to turtles (based mostly on the overhanging beak), but more recent analyses place them as one of the earliest and most primitive Eureptilia (**Figure 8.1**).

The next branch of the Eureptilia is the split between the primitive captorhinids and the first diapsids (**Figure 8.1**). One of the oldest known diapsids from the Late Carboniferous is *Petrolacosaurus*, a small lizard-like form about 40 cm (16 inches) long. It is the earliest known fossil with the clearly diapsid condition of both upper and lower temporal fenestrae in the back of its skull roof. *Petrolacosaurus* was a very slender and lightly built reptile, with a long neck, long legs and

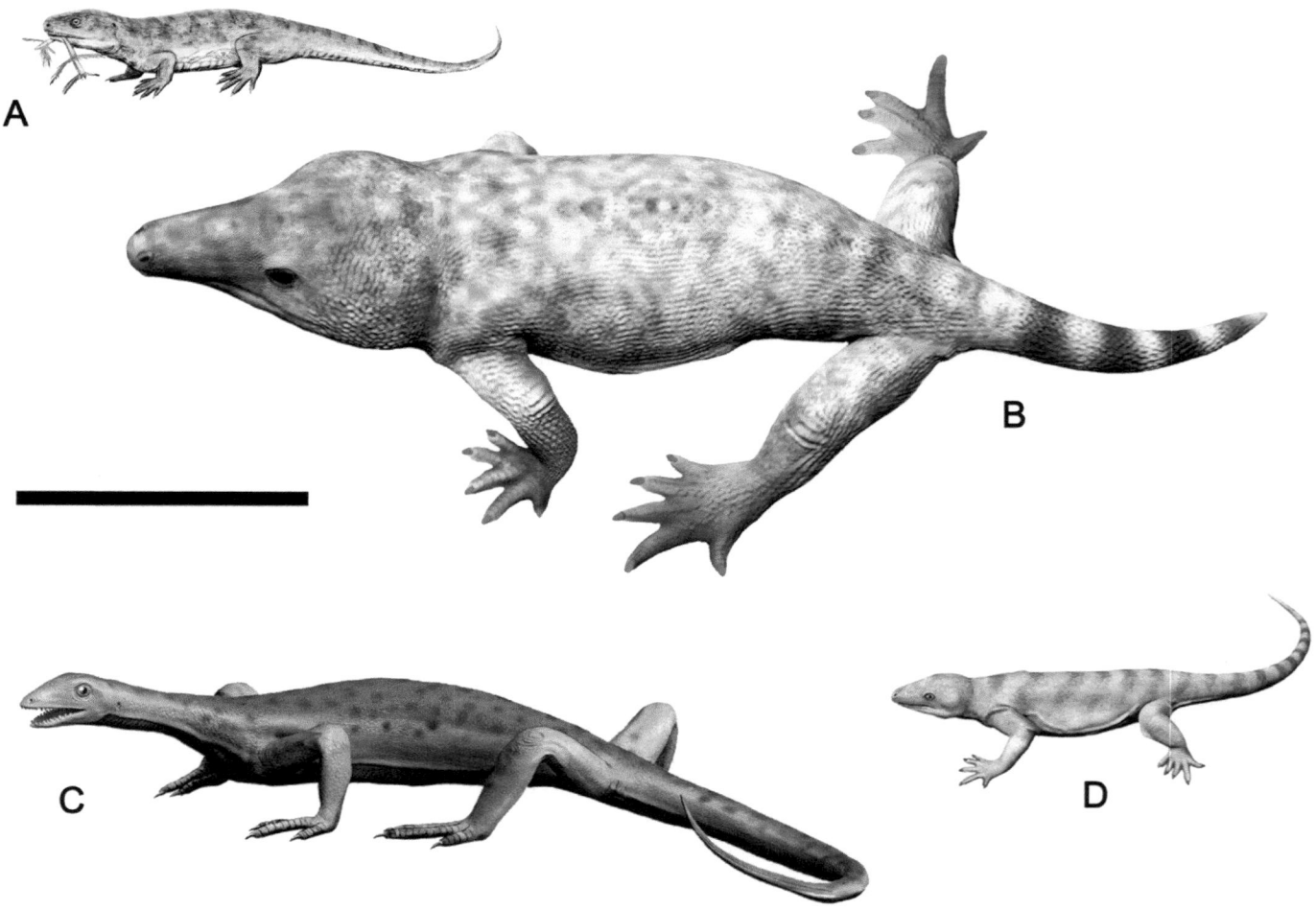

Figure 8.9 Reconstructions of some primitive amniotes: (A) *Captorhinus*, (B) *Labidosaurus*, (C) *Araeoscelis*, (D) *Palaeothyris*. Scale bar is 10 cm.

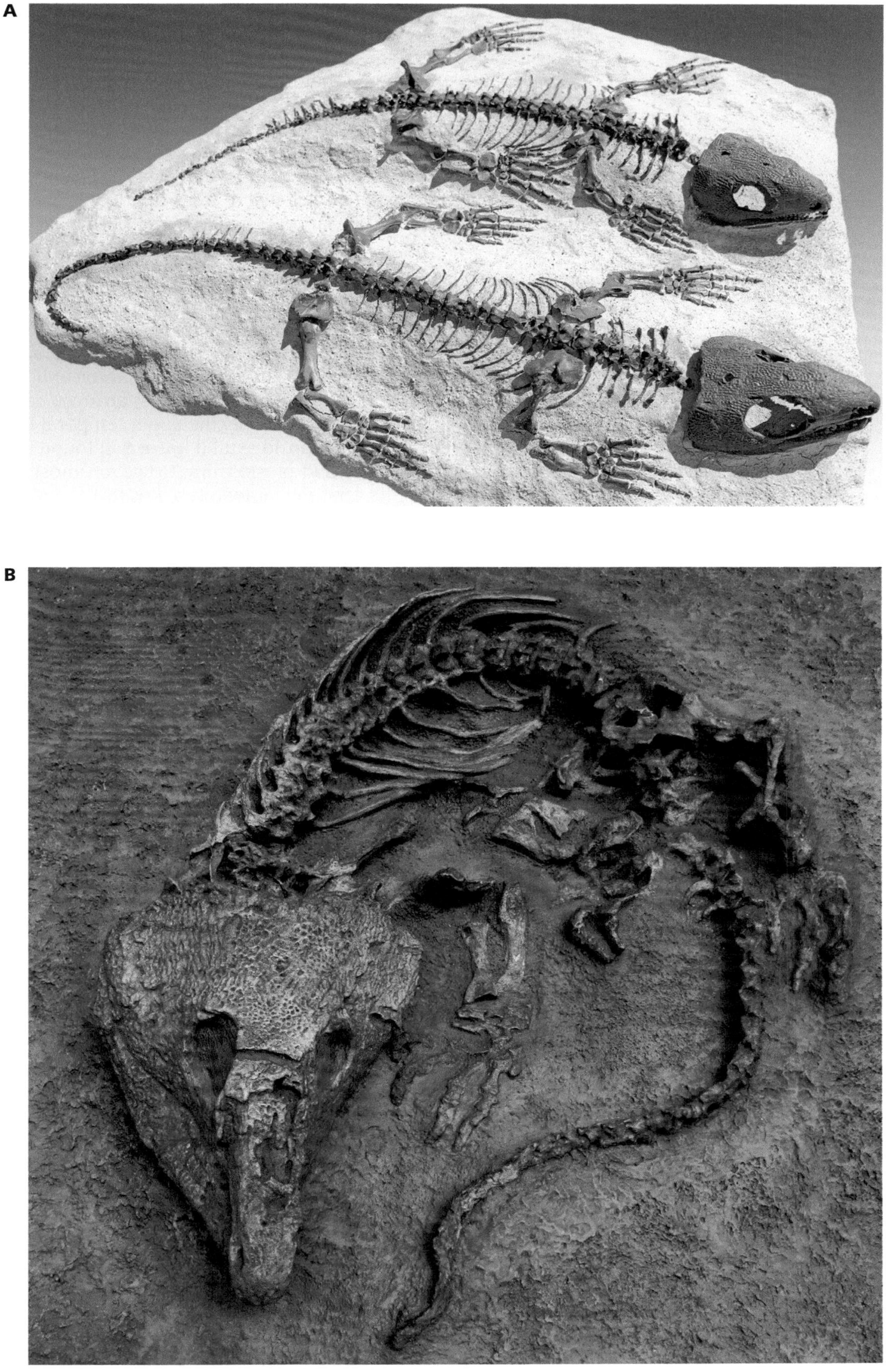

Figure 8.10 Fossils of primitive amniotes. (A) *Captorhinus*, (B) *Labidosaurus*. (Courtesy Wikimedia Commons.)

long slender toes, a long tail, and a strong flexible spine that gave it flexibility for faster running (at a time when most of the reptiles were heavily built and sluggish). Its small size, and numerous conical teeth in the mouth, with a few that are longer fangs in the front of the snout for puncturing and grabbing prey. These features suggest that it preyed most on smaller creatures, especially the abundant hard-shelled insects and other arthropods of the Carboniferous coal swamps. The anatomical features of the skeleton also suggest that it was a decent climber, possibly running up the giant coal swamp trees that would have provided it protection from the many larger predators, and perhaps to grab insects that could also climb.

Just slightly later and more advanced than *Petrolacosaurus* was the primitive reptile *Araeoscelis*, from the Early Permian of Texas (**Figure 8.9[C]**). It was also very lightly built and a bit longer (60 cm, or 2 feet long) compared to typical *Petrolacosaurus*. However, it was a more advanced reptile than *Petrolacosaurus* in that its teeth were larger and blunter, perhaps for cracking thick shells of bigger insects, arthropods, and snails. Its skull is distinctive in that it had only the upper temporal fenestra, so it appears to resemble the euryapsid skull (**Figure 8.6**) found in the marine reptiles like ichthyosaurs and plesiosaurs. However, most recent analyses consider all of its anatomical features, not just the skull roof, and consider it to be a diapsid in the same group with *Petrolacosaurus*. Both genera are now placed in the same group, the Araeoscelida, the most primitive group of diapsids.

TURTLES

As we have already mentioned, turtles and tortoises were long considered among the most primitive reptiles alive, because of their anapsid skull roof. Then a series of discoveries of fossil turtles, and especially evidence of molecular biology, showed the turtles were not primitive at all, but closely related to the diapsids and maybe closer still to some branches with the diapsids. This debate is still going on, so for now we will put them in the advanced diapsids without specifying who their closest relatives were (**Figure 8.1**).

Despite the unique and distinctive turtle body plan, there are transitional fossils that show how turtles evolved from reptiles without shells. The first to be found was a strange fossil known as *Proganochelys*, from the Upper Triassic beds of Germany, and also Greenland and Thailand (**Figures 8.11[D]** and **8.12[C]**). At first glance, it looks just like any other turtle, with a plastron (belly shell) and a carapace (shell on the back). However, a closer look shows that it is a lot more primitive than living turtles, and not a member of any living group. For one thing, the carapace is very different, with many additional plates not seen in any living turtle, especially around the edge of the shell and protecting the legs. In addition, its tail was covered by a spiky bony sheath, with a spiky tail club. Even more primitive is the skull. It looked much more like one of the primitive Permian reptiles, not a turtle with its distinctive arrangement of jaws muscles. Although it has a turtle-like beak, the upper palate still had teeth, the last of the turtles to retain teeth. Most important of all, it could not retract its big head into its shell like all living turtles do, so it had armor and spikes on top of its head to protect it instead. To the less observant person, it is "just a turtle", but it's completely unlike any living turtle in having an unretractable neck and head, and teeth in its mouth.

Then in 2008, an astonishing collection of turtle fossils was announced from the Late Triassic of China (**Figures 8.11[C]** and **8.12[B]**). Known from dozens of complete specimens, the discovery was given the formal

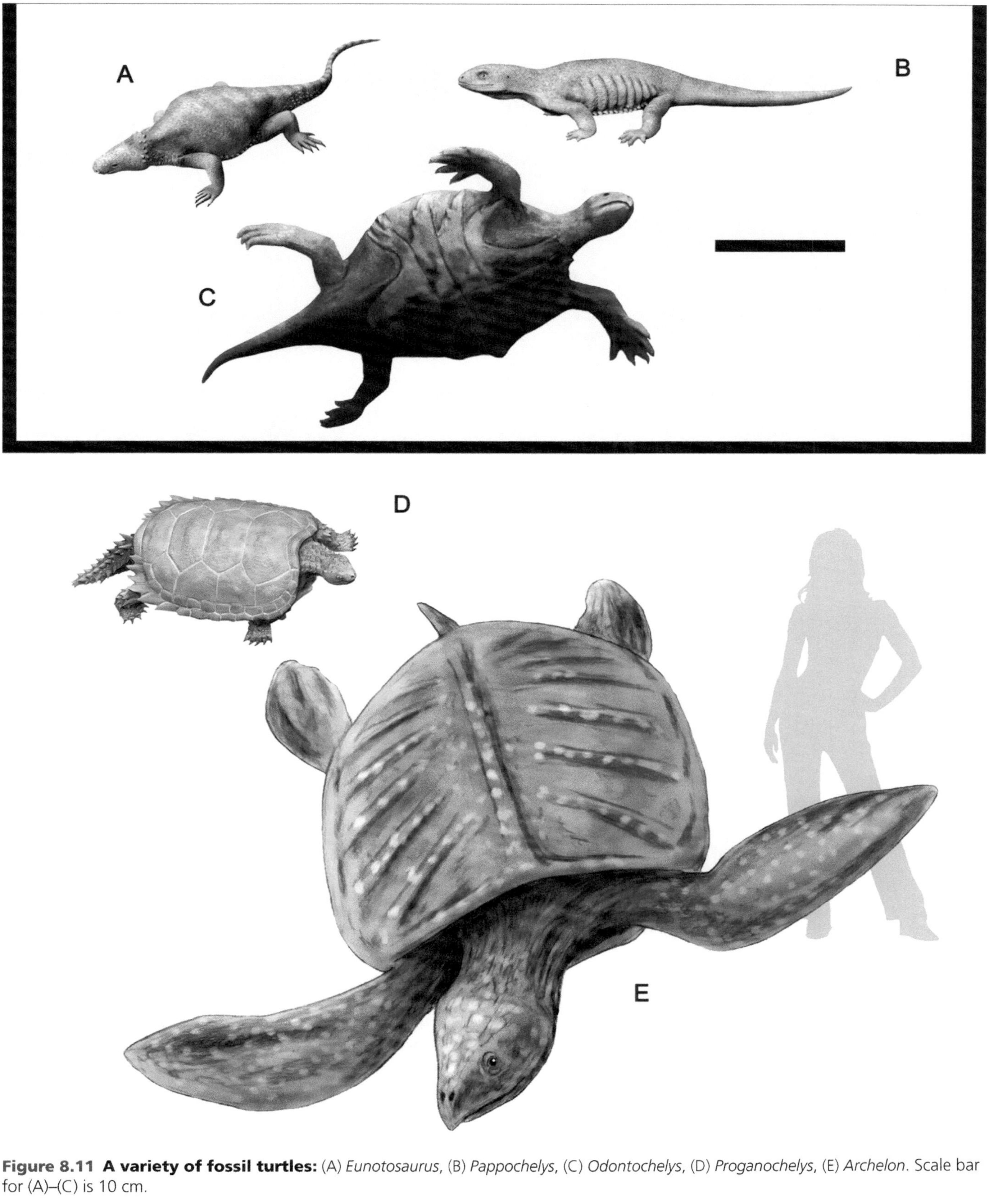

Figure 8.11 A variety of fossil turtles: (A) *Eunotosaurus*, (B) *Pappochelys*, (C) *Odontochelys*, (D) *Proganochelys*, (E) *Archelon*. Scale bar for (A)–(C) is 10 cm.

Figure 8.12 Fossils of some important extinct turtles and their relatives. (A) *Eunotosaurus*, (B) *Odontochelys*. (Courtesy Wikimedia Commons.)

C

D

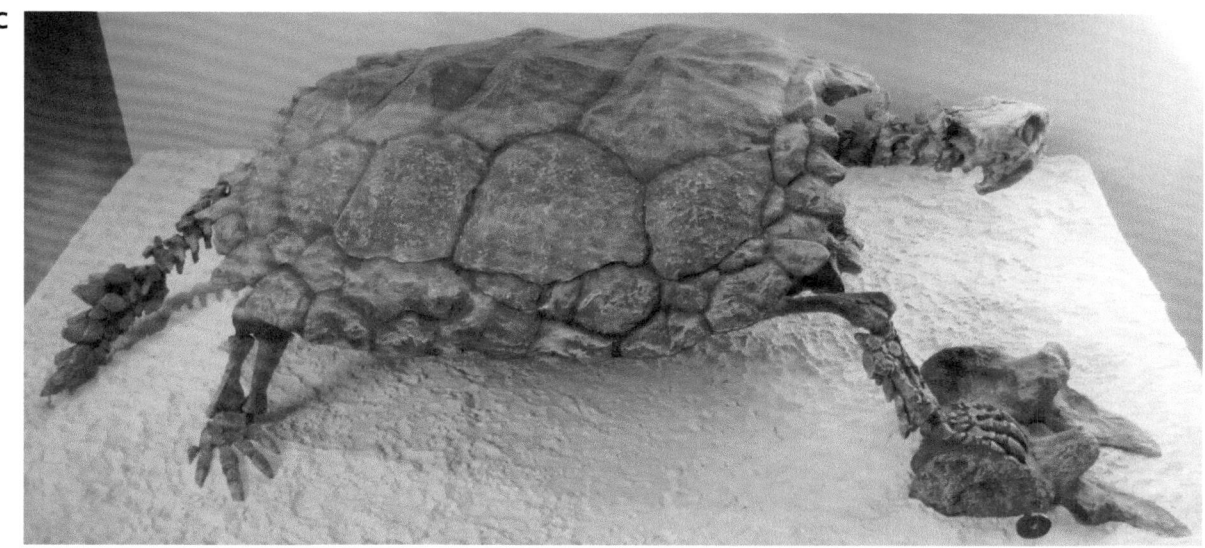

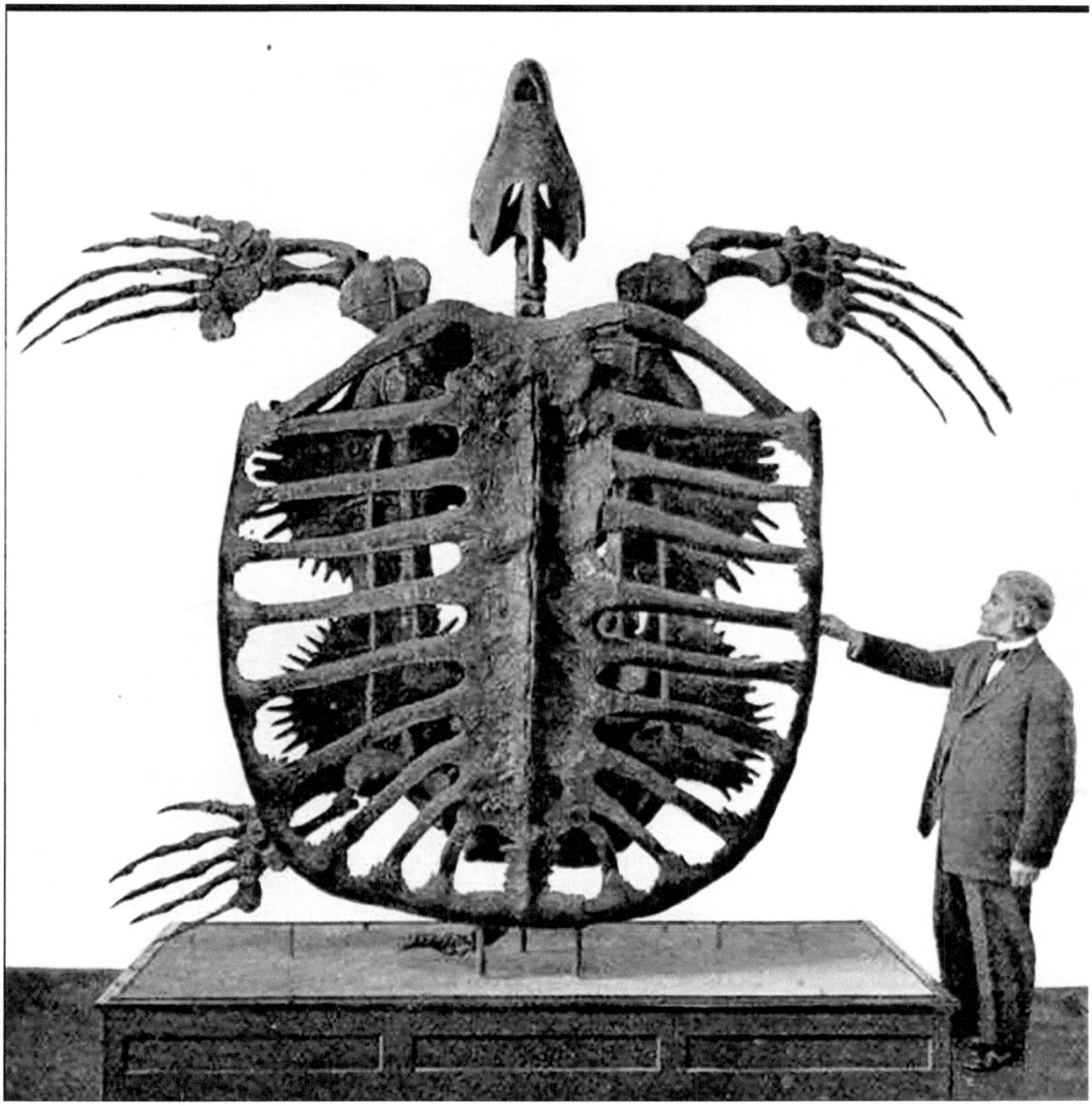

Figure 8.12 (Continued) (C) *Proganochelys*, (D) *Archelon*.

Figure 8.12 (Continued) (E) *Stupendemys*.

scientific name *Odontochelys semitestacea*, or "toothed turtle with half a shell". It solves the riddle of how turtles got their shells, because *Odontochelys* had no shell or carapace on its back (just thick ribs), but it did have a plastron, or belly shield. It is literally a "turtle on the half-shell", a turtle transitional between modern forms with both shells, and its ancestors with no complete shell. Another completely un-turtle-like reptilian feature is a full set of teeth in the mouth, like its ancestors but unlike the toothless beaks of modern turtles.

Odontochelys resolves another longstanding debate as well. For decades, some paleontologists argued that the turtle carapace comes from small plates of bone developed from its skin (osteoderms) which become fused together, while others argued that the carapace was mostly made of expansions of its back ribs. *Odontochelys* shows that the latter argument is correct, since it had broadly expanded back ribs that are beginning to develop and connect into a shell, and there are no osteoderms on top or embedded between the ribs. This was confirmed by embryological studies of turtles, which track the development of the carapace from the developmental changes in the back ribs; no osteoderms are involved.

From the "turtle on a half shell", we can trace the ancestry of turtles even further back to reptiles that have only a few turtle-like features. One of these is *Eunotosaurus*, from the Middle Permian beds of South Africa (**Figures 8.11[A]** and **8.12[A]**). It looked mostly like a large, fat lizard, except for some key features of the skeleton. The most striking of these is the greatly expanded broad flat back ribs that almost connect with each other to form a complete shell on the back. And in 2015, another primitive fossil was announced. Named *Pappochelys* ("grandfather turtle"), it had not only the broad ribs on its back like *Eunotosaurus* but also broad flattened belly ribs ("gastralia") that would eventually fuse into the plastron or belly plate of more advanced turtles. Thus we have a very nice transition from the reptile *Eunotosaurus*, with only the flattened back ribs, to *Pappochelys* with flattened back and belly ribs, to *Odontochelys*, with a belly shield but only flattened ribs on the back, to *Proganochelys*, with the a complete (but primitive) turtle shell, but still retaining some teeth, and unlike modern turtles in having not yet developed the ability to retract its head.

Once turtles with complete shells appeared in the fossil record, it was such a successful body plan that it evolved and diversified, but always with a dome of shell on top (the carapace) and a belly shield (plastron). There are over 356 species of turtles alive today, and the roots of the modern groups can be traced back to the Jurassic. They are separated into two main categories: cryptodires and pleurodires. The cryptodires are the more familiar and most diverse group of turtles on the planet, making up about 260 species of living turtles. Their name *cryptodire* means "hidden joint" and refers to the fact that when they pull their head in their shell, the neck folds upon itself in an S-bend in the vertical plane inside the front of the shell. The much rarer and less diverse pleurodires ("side joint") or side-necked turtles fold their necks sideways like closing a jackknife, and pull their heads in under the overhanging lip of the front of the shell. They not only are rare and endangered, but also are found today only on the Gondwana remnant continents like Australia, South America, and Africa.

Most fossil turtles are relatively small, roughly in the same size range as the living ones, although there are giant tortoises on isolated islands, such as the Galápagos, west of Ecuador, and the Aldabaras, in the Indian Ocean. These creatures independently evolved huge body size from smaller ancestors, since they had few island predators to worry about, and much less competition for food. The largest living turtles are sea turtles, whose immense size is supported by the buoyancy of the water in which they live. Of these, the leatherback sea turtle is the biggest (and the fourth biggest of all the reptiles). Large individuals can be more than 2.2 meters (7 feet) long and weigh up to 700 kilograms (1540 pounds). The leatherback gets its name because most of its bony shell has been reduced, and the skeleton of its back is covered with only a thick tough hide. This loss of bony armor keeps the leatherback from being too dense and sinking too fast, since its skin is thick enough to deter most predators (and full-grown leatherbacks have very few predators).

In the geologic past, however, there were some true monster turtles. The largest was the sea turtle *Archelon* (Greek for "ruling turtle"), which swam in the shallow inland seas of what is now western Kansas, along with such other marine reptiles as plesiosaurs, ichthyosaurs, and mosasaurs (**Figures 8.11[E]** and **8.12[D]**). The largest specimens of *Archelon* are more than 4 meters (13 feet) long and about 5 meters (16 feet) wide from the tip of one flipper to that of the other. It probably weighed more than 2200 kilograms (4850 pounds). Like many sea turtles, it had just an open framework of bone on its back and four jagged plates on its belly. Like the modern leatherback, it probably was covered mostly by thick skin, rather than a bony external shell.

Extinct giant land turtles could not grow quite this large without the support of the buoyancy of water, but nonetheless they dwarfed any modern giant tortoises. One of the largest was *Colossochelys*, which was more than 2.7 meters (9 feet) long and 2.7 meters wide and weighed about 1000 kilograms (2200 pounds) or more. Discovered in Pakistan in the 1840s, its fossils have been found from Europe to India to Indonesia and date from 10 (million years ago) Ma to 10,000 years ago, the end of the last Ice Age. It would have looked like a gigantic version of the Galápagos tortoise. Even bigger was *Carbonemys*, from swamp deposits about 60 million years old in Colombia. It was actually the size of a Smart Car, more than 1.7 meters (5.5 feet) long, and it could have eaten just about any creature it encountered, including crocodilians. It was one of the largest creatures in its world during the Paleocene. Like most South American turtles, *Carbonemys* was a pelomedusoid, a group of side-necked turtles that is common in South America.

The largest of all land turtles was another monster from South America, the appropriately named *Stupendemys*, found in swamp beds of the Urumaco Formation in Venezuela that date to about 5–6 Ma as well as in Brazil (**Figure 8.12[E]**). Like *Carbonemys*, it was a member of the pleurodire group known as pelomedusoids. It was most similar to the living Arrau turtle (*Podocnemis expansa*), except that it was much larger. As the name says, its size was truly stupendous: its shell was more than 3.3 meters (11 feet) long and 1.8 meters (6 feet) wide (**Figure 8.10[E]**). These extreme examples give a small indication of the huge evolutionary diversification of turtles and tortoises.

FURTHER READING

Anquetin, J. 2012. Reassessment of the phylogenetic interrelationships of basal turtles (Testudinata). *Journal of Systematic Palaeontology*. 10 (1): 3–45.

Asher, J.L.; Lucas, S.G.; Klein, H.; Lovelace, D.M. 2018. Triassic turtle tracks and the origin of turtles. *Historical Biology*. 30 (8): 1112–1122.

Benton, M.J. 1985. Classification and phylogeny of diapsid reptiles. *Zoological Journal of the Linnean Society (London)*. 84: 97–164.

Benton, M.J., ed. 1988. *The Phylogeny and Classification of the Tetrapods, vol. 1: Amphibians, Reptiles, Birds*. Oxford Clarendon Press, Oxford.

Carroll, R.L. 1964. The earliest reptiles. *Zoological Journal of the Linnean Society (London)*. 45: 61–83.

Carroll, R.L. 1982. Early evolution of the reptiles. *Annual Reviews of Ecology and Systematics*. 13: 87–109.

Chiari, Y.; Cahais, V.; Galtier, N.; Delsuc, F. 2012. Phylogenomic analyses support the position of turtles as the sister group of birds and crocodiles (Archosauria). *BMC Biology*. 10 (65): 65.

DeBraga, M.; Rieppel, O. 1997. Reptile phylogeny and the interrelationships of turtles. *Zoological Journal of the Linnean Society (London)*. 120 (3): 281–354.

Evans, S.E. 1988. The early history and relationships of the Diapsida, pp. 221–260. In Benton, M. J., ed. *The Phylogeny and Classification of the Tetrapods, vol. 1: Amphibians, Reptiles, Birds*. Clarendon Press, Oxford.

Ford, D.; Benson, R.B. 2020. The phylogeny of early amniotes and the affinities of Parareptilia and Varanopidae. *Nature Ecology & Evolution*. 4 (1): 57–65.

Gaffney, E.S. 1975. A phylogeny and classification of the higher categories of turtles. *Bulletin of the American Museum of Natural History*. 155: 387–436.

Gaffney, E.S.; Hutchinson, H.; Jenkins, F.; Meeker, L. 1987. Modern turtle origins: The oldest known cryptodire. *Science*. 237 (4812): 289–291.

Gaffney, E.S.; Tong, H.; Meylan, P.A. 2006. Evolution of the side-necked turtles: The families Bothremydidae, Euraxemydidae, and Araripemydidae. *Bulletin of the American Museum of Natural History*. 300: 1–698.

Gauthier, J. 1994. The diversification of the amniotes, pp. 129–159. In Prothero, D.R.; Schoch, R.M., eds. *Major Features of Vertebrate Evolution*. Paleontological Society Short Course 7. Paleontological Society, Lawrence, KS.

Gauthier, J.; Kluge, A.G.; Rowe, T. 1988. Amniote phylogeny and the importance of fossils. *Cladistics*. 4: 105–209.

Gauthier, J.; Kluge, A.G.; Rowe, T. 1988. The early evolution of the Amniota, pp. 103–155. In Benton, M. J., ed. *The Phylogeny and Classification of the Tetrapods, vol. 1: Amphibians, Reptiles, Birds*. Clarendon Press, Oxford.

Guillon, J.-M.; Guéry, L.; Hulin, V.; Girondot, M.; Arntzen, J.W. 2012. A large phylogeny of turtles (Testudines) using molecular data. *Contributions to Zoology*. 81 (3): 147–158.

Heaton, M.J.; Reisz, R.R. 1986. Phylogenetic relationships of captorhinomorph reptiles. *Canadian Journal of Earth Sciences*. 23: 402–418.

Iwabe, N.; Hara, Y.; Kumazawa, Y.; Shibamoto, K.; Saito, Y.; Miyata, T.; Katoh, K. 2004. Sister-group relationship of turtles to the bird-crocodilian clade revealed by nuclear DNA-coded proteins. *Molecular Biology and Evolution*. 22 (4): 810–813.

Joyce, W.G. 2007. Phylogenetic relationships of Mesozoic turtles. *Bulletin of the Peabody Museum of Natural History*. 48 (1): 3–102.

Laurin, M. 2004. The evolution of body size, Cope's rule and the origin of amniotes. *Systematic Biology*. 53 (4): 594–622.

Laurin, M.; Reisz, R.R. 1996. A reevaluation of early amniote phylogeny. *Zoological Journal of the Linnean Society*. 113: 165–223.

Lee, M.S.Y. 1997. Pareiasaur phylogeny and the origin of turtles. *Zoological Journal of the Linnean Society*. 120 (3): 197–280.

Lee, M.S.Y. 2013. Turtle origins: Insights from phylogenetic retrofitting and molecular scaffolds. *Journal of Evolutionary Biology*. 26 (12): 2729–2738.

Lee, M.S.Y.; Spencer, P.S. 1997. Crown clades, key characters and taxonomic stability: When is an amniote not an amniote?, pp. 61–84. In Sumida, S. S.; Martin, K. L. M., eds. *Amniote Origins: Completing the Transition to Land*. Academic Press, New York.

Li, C.; Wu, X.-C.; Rieppel, O.; Wang, L.-T.; Zhao, L.-J. 2008. An ancestral turtle from the Late Triassic of southwestern China. *Nature*. 456: 497–450.

Lyson, T.R.; Bever, G.S.; Bhullar, B.-A.S.; Joyce, W.G.; Gauthier, J.A. 2010. Transitional fossils and the origin of turtles. *Biology Letters*. 6 (6): 830–833.

Lyson, T.R.; Sperling, E.A.; Heimberg, A.M.; Gauthier, J.A.; King, B.L.; Peterson, K.J. 2012. MicroRNAs support a turtle + lizard clade. *Biology Letters*. **8** (1): 104–107.

Paton, R.L.; Smithson, T.R.; Clack, J.A. 1999. An amniote-like skeleton from the Early Carboniferous of Scotland. *Nature*. 398 (6727): 508–513.

Reisz, R.R. 1997. The origin and early evolutionary history of amniotes". *TREE*. 2: 218–222.

Rieppel, O. 1995. Studies on skeleton formation in reptiles: Implications for turtle relationships. *Zoology* 98: 298–308.

Rieppel, O. 2017. *Turtles as Hopeful Monsters: Origins and Evolution*. Indiana University Press, Bloomington, IN.

Rieppel, O.; DeBraga, M. 1996. Turtles as diapsid reptiles. *Nature*. 384 (6608): 453–455.

Schoch, R.R.; Sues, H.-D. 2015. A Middle Triassic stem-turtle and the evolution of the turtle body plan. *Nature*. 523 (7562): 584–587.

Schultze, H.-P.; Trueb, L., eds. 1991. *Origins of the Higher Groups of Tetrapods: Controversy and Consensus*. Cornell University Press, Ithaca, NY.

Smithson, T.R.; Carroll, R.L.; Panchen, A.L.; Andrews, S.M. 1994. *Westlothiana lizziae* from the Visean of East Kirkton, West Lothian, Scotland, and the amniote stem. *Transactions of the Royal Society of Edinburgh*. 84: 383–412.

Sues, H.-D. 2019. *The Rise of the Reptiles: 320 Million Years of Evolution*. Johns Hopkins University Press, Baltimore, MD.

Tsuji, L.A.; Müller, J. 2009. Assembling the history of the Parareptilia: Phylogeny, diversification, and a new definition of the clade. *Fossil Record*. 12 (1): 71–81.

Wang, Z.; Pascual-Anaya, J.; Zadissa, A.; et al. 2013. The draft genomes of soft-shell turtle and green sea turtle yield insights into the development and evolution of the turtle-specific body plan. *Nature Genetics*. 45 (6): 701–706.

Zardoya, R.; Meyer, A. 1998. Complete mitochondrial genome suggests diapsid affinities of turtles. *Proceedings of the National Academy of Sciences*. 95 (24): 14226–14231.

BACK TO THE SEA

MARINE REPTILES

9

There were no real sea serpents in the Mesozoic Era, but the plesiosaurs were the next thing to it. The plesiosaurs were reptiles who had gone back to the water because it seemed like a good idea at the time. As they knew little or nothing about swimming, they rowed themselves around in the water with their four paddles, instead of using their tails for propulsion like the brighter marine animals. (Such as the ichthyosaurs, who used their paddles for balancing and steering. The plesiosaurs did everything wrong.) This made them too slow to catch fish, so they kept adding vertebrae to their necks until their necks were longer than all the rest of their body. . . . There was nobody to scare except fish, and that was hardly worthwhile. Their heart was not in their work. As they were made so poorly, plesiosaurs had little fun. They had to go ashore to lay their eggs and that sort of thing. (The ichthyosaurs stayed right in the water and gave birth to living young. It can be done if you know how.)
—Will Cuppy, *How to Become Extinct*

ICHTHYOSAURS AND PLESIOSAURS

As mentioned in Chapter 8, the two major groups of Mesozoic marine reptiles, the ichthyosaurs and plesiosaurs, have a euryapsid skull roof with only an upper temporal fenestra above the postorbital and squamosal bones (**Figure 8.6**), and for a long time they were placed in their own group, the Euryapsida. More recent analyses based on the total anatomical evidence from these fossils shows that ichthyosaurs and plesiosaurs are sub-groups of the Diapsida, but together they are probably not a natural group "Euryapsida" (**Figure 8.1**). These creatures dominated the seas during most of the Mesozoic, and vanished near the end of the Cretaceous. We will discuss their history and evolution in this chapter. In addition to these, there were a few other Mesozoic marine reptiles, such as mosasaurs (descendants of monitor lizards—Chapter 10) and gigantic sea turtles (Chapter 8), as well as marine crocodiles (Chapter 12), but by far the biggest and most common marine reptiles were the ichthyosaurs and plesiosaurs.

It's amazing to think that amniotes, which had adapted for life on dry land, returned to the sea, and paid the price by having to modify their biological systems, especially physiology, respiration, and reproduction. Apparently, the food resources were so great in the oceans that land-dwelling reptiles found their way to reap this bountiful harvest— and they did it independently in at least five different groups. This is amazing in itself, yet the fact that it has happened many times shows how powerful the selection forces for this lifestyle must be. Such a radical change in ecology usually caused much convergence in body form as well, so we can see how ichthyosaurs and whales have independently evolved the streamlined torpedo-like shape that is also found in fish.

DOI: 10.1201/9781003128205-9

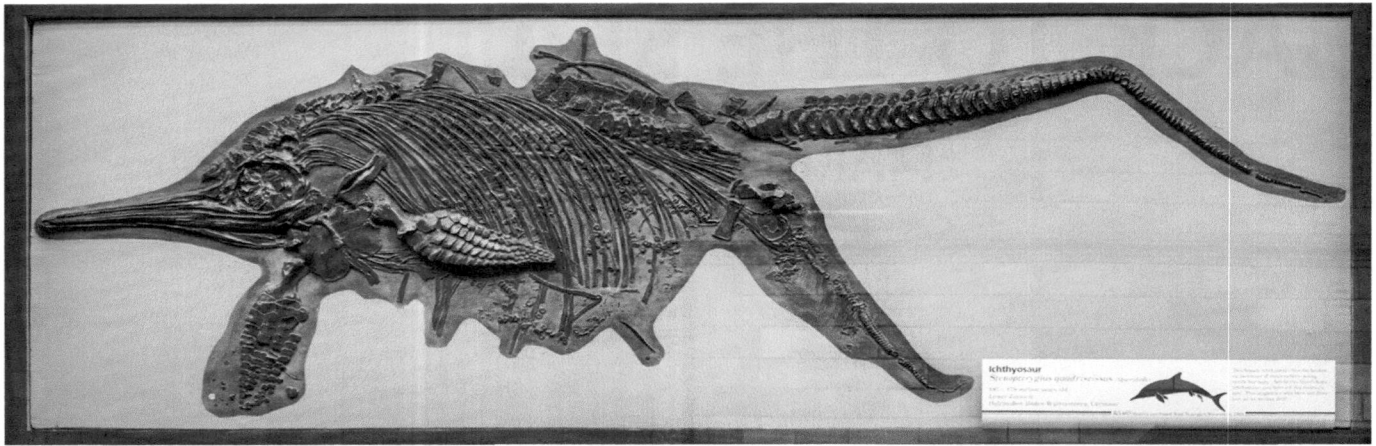

Figure 9.1 Skeleton of the Jurassic ichthyosaur *Stenopterygius* killed and fossilized in the process of giving live birth, with the newborn ichthyosaur just emerging from the birth canal. (Courtesy Wikimedia Commons.)

Returning to the ocean makes certain reproductive and physiological demands, in addition to streamlining the body for swimming and modifying the hands and feet into flippers. For example, marine reptiles had a land egg, yet they must still reproduce somehow. We know that sea turtles and saltwater crocodilians crawl out on land and lay eggs in a nest, and presumably mosasaurs could have done so, too. But ichthyosaurs are so dolphin-like in body form that they could not have wriggled onto a beach and dug a nest with their flippers. We know that whales and dolphins give live birth in the ocean, expelling the young from the uterus and raising it up to take its first breath, after which it can swim on its own.

Apparently, ichthyosaurs could also, since there are several remarkable specimens from the Jurassic Holzmaden shales in Germany that appear to have been in the process if giving birth to a live baby when it died and was fossilized (**Figure 9.1**). And now plesiosaur fossils with embryos inside them have also been discovered, showing that they too gave live birth.

ICHTHYOSAURS

Let us focus on the ichthyosaurs first. They represent the greatest transformation because they are the most highly modified and specialized for marine life. Advanced ichthyosaurs (**Figures 9.1** and **9.2**) have highly fish-like streamlined bodies with long toothy snouts for catching fish and squid, and some of them had huge eyes for seeing in dark murky waters. The bones of later Jurassic ichthyosaurs show signs of decompression sickness, demonstrating that they were deep divers that often suffered the effects of holding their breath for a very long time and of nitrogen being released from their blood as they rose from the deep waters.

The head of ichthyosaurs merged with their body, so they had no visible neck, as in many aquatic animals that are streamlined for full-time swimming. Recent estimates put their fastest speeds at 2 kilometers an hour (1.2 miles an hour), a bit slower than the fastest living dolphins and whales. Their dolphin-like body sported a dorsal fin (analogous to those in dolphins and fish), supported by cartilage but not by bone and visible only on specimens with soft-tissue preservation. But their hands were modified into flippers made of dozens of little closely packed cylinders of bone formed by the multiplication and division of individual finger bones

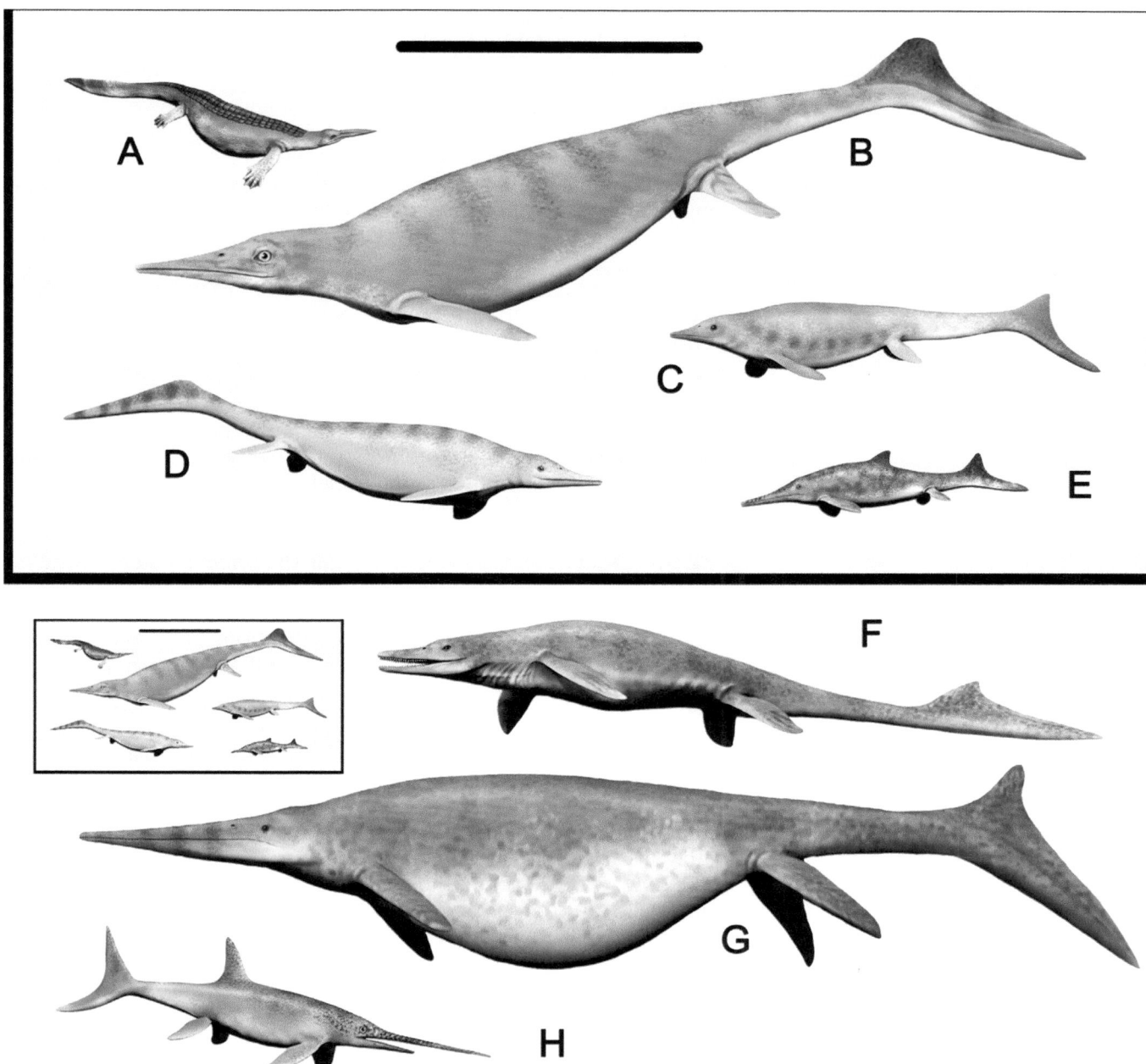

Figure 9.2 Reconstructions of a variety of ichthyosauromorphs: Upper panel: (A) *Nanchangosaurus*, (B) *Utatsusaurus*, (C) *Grippia*, (D) *Chaohusaurus*, (E) *Mixosaurus*. Scale bar: 1 meter. Lower panel: (F) *Cymbospondylus*, (G) *Shonisaurus*, (H) *Eurhinosaurus*. Scale bar: 2 meters.

into many tiny parts. Their hind feet were modified into much smaller paddles (by contrast, the hind limbs are lost altogether in whales and dolphins). These rear paddles in ichthyosaurs were apparently not used much for propulsion during swimming. The rear of the body tapered into a fish-like tail with flukes aligned in the vertical plane, so ichthyosaurs swam with a side-to-side motion of the tail part of the body (as do most fish). Finally, their tails also have a vertically oriented tail fin. In contrast to the tailfin of most fish, the supporting rod of the spinal column of ichthyosaurs flexes downward into the lower lobe of the fin, not upward as in sharks and other primitive fish. (For a long time, early naturalists wondered why the tail vertebrae flexed downward and decided that their fossils must be deformed; only later was the outline of the soft tissue discovered which showed they had upper and lower lobes on the fin.)

Much is known about ichthyosaur paleobiology, since there are numerous well-preserved complete articulated skeletons, often with soft-tissue outlines and stomach contents. Most ichthyosaurs are thought to have fed like dolphins and whales, rapidly chasing after and catching swimming prey (squids and belemnites, ammonites, fish, and the like) with their long toothy beaks, and this is confirmed by preserved stomach contents. Some early ichthyosaurs had blunt crushing teeth for eating mollusks, while others had toothless bills and were thought to have fed by suction (as do many fish). A number of predators were willing to attack them, leaving scars on their faces and bones.

A number of striking fossils demonstrating the origin of the ichthyosaurs are known from the early Mesozoic (**Figure 9.2[A]**). First, there is *Nanchangosaurus* from the Triassic of China. Although it has a slightly streamlined body and a long (but toothless) snout like an ichthyosaur, all of the rest of the features of the skeleton are primitive, including the vertebrae, the limbs that are not modified into flippers but have normal proportions with all the regular wrist and ankle and toe bones, and a long straight tail with no sign of a tail fin. The original authors were not sure where to classify this fossil because it is so primitive, but based on the skull, it seems to be an aquatic lizard on the way to becoming an ichthyosaur. Another group, the Hupehsuchia from the Triassic of China, has been suggested as the closest relatives of the ichthyosaur lineage.

The oldest known fossil that can definitely be called an ichthyosaur is *Utatsusaurus* from the Early Triassic of Japan (**Figure 9.2[B]**). Although it has the general body form of an ichthyosaur, it has a mosaic of primitive features found in its reptilian ancestors. These include a skull with only a short snout and unspecialized teeth, very primitive vertebrae (especially in the neck), hands and feet whose finger and toe bones are not yet completely modified into flippers, and a long straight tail with no evidence of the downward flexion of the spine to support a vertical tailfin. *Grippia* (**Figure 9.2[C]**) from the Early Triassic of Spitsbergen is known primarily from the skull, but it shows a relatively short snout, small eyes, and simple knob-like teeth for crushing mollusks, not the spiky teeth of most fish-eating ichthyosaurs. Yet another Early Triassic form from China, *Chaohusaurus* (**Figure 9.2[D]**) also had a short snout, simple teeth, primitive vertebrae, and robust limbs that are beginning to form a paddle but still have discrete rows of finger bones with the normal count (not the extra bones of an advanced ichthyosaur paddle). *Cymbospondylus* (**Figure 9.2[F]**) from the Middle Triassic of Nevada still retains the primitive hand and foot structure as well and has a relatively short snout with small eyes, and the tail is beginning to show the downward bend that indicates the presence of a small tail fin.

The best known and best preserved of the early ichthyosaurs is *Mixosaurus* from the Middle Triassic of Germany (**Figure 9.2[E]**) and elsewhere in Europe. The body has the classic ichthyosaur shape, with the long snout, large eyes, and dorsal fin. The hands and feet are beginning to form flippers, although they have still not multiplied the finger and toe bones into small cylinders, as in later ichthyosaurs. And the tail shows just a slight downward bend, with some specimens preserving the body outline and showing that it had a small upper lobe on its tail. Thus, it is advanced in many features but still retains the primitive hand and feet and does not yet have the fully bilobed ichthyosaurian tail.

During the Late Triassic, ichthyosaurs got huge. The biggest was *Shonisaurus*, about the size of a large whale, roughly 15 meters (almost 50 feet) in length (**Figure 9.2[G]**). It had a long toothless snout (except when it was young), suggesting to some scientists that it did not swim fast to catch prey. Rather, it inhaled its meals as they swam by or, like baleen whales and the whale sharks, may have fed more on plankton than on large animals. It had a deep, round body and relatively long pectoral and pelvic fins, made entirely of the huge round finger elements that result when finger bones turn into a flipper. There was apparently no dorsal fin, and like many other Triassic ichthyosaurs, it had only a small upper lobe on its tail, with just a slight downward turn of the tip of the tail vertebrae, not the sharp kink seen in ichthyosaurs of the Jurassic.

Finally, during the Jurassic and Cretaceous, ichthyosaurs became really diverse, with a burst of diversification in the Early Jurassic, and over two dozen lineages by the Late Jurassic, and many continuing into the Late Cretaceous (at least 10 lineages). Some were highly specialized with a long sword-like bone on their upper snout (*Eurhinosaurus*) and short lower jaw, apparently for slashing through a school of fish as the modern swordfish does today (**Figure 9.2[H]**).

PLESIOSAURS

If the ichthyosaurs seem very highly specialized to resemble fish or dolphins, the plesiosaurs are specialized in a very different direction. All of the advanced forms had stout bodies with robust shoulder and hip bones and four large well-developed paddles (**Figure 9.3**) with a convergently evolved tendency to transform long finger bones into a flipper built of closely packed cylinders of bone. Some plesiosaurs (especially the elasmosaurs) had long serpentine necks and small heads, while another group (the pliosaurs) had long heads and snouts and much shorter, more robust necks. Instead of speedy dolphin-like swimming employed by ichthyosaurs, plesiosaurs were apparently adapted for slow steady swimming by rowing with their fins (like a sea turtle does) and used their long heads and necks to snap at prey that came within reach.

How could such peculiar creatures evolve? We have an even better series of intermediates for plesiosaurs than we do for ichthyosaurs. They start with a group of small Triassic marine reptiles known as nothosaurs (**Figure 9.3[C]**). These animals had skulls and bodies that were not noticeably different from primitive euryapsids like *Claudiosaurus* (**Figure 9.3[B]**). The biggest difference occurs in the neck, which is much longer and anticipates the long necks of many plesiosaurs. The limbs are not much more specialized for aquatic locomotion than those of its primitive relatives, but they have further reduced a lot of the bone to cartilage, another sign of a largely aquatic lifestyle. However, the shoulder girdle and hip bones are becoming much more robust and plate-like in support of the limbs, a hallmark of later plesiosaurs.

Our final step into full-fledged plesiosaurs is *Pistosaurus* from the Middle Triassic of Germany (**Figure 9.3[D]**). This creature has a relatively primitive head with a slightly longer snout than did nothosaurs (but still retaining the nasal bones, which are lost in plesiosaurs), but its palate is more like that of plesiosaurs. The rest of its body is also quite advanced, with a fairly long neck, deep body, many extra bones along the belly (gastralia), and limbs that are intermediate between the unspecialized nothosaur foot and the highly specialized plesiosaur paddles, which have dozens of extra finger bones. Pistosaurs had weak shoulder girdles and did not have the pelvic support for a strong swimming stroke, but they had true flippers rather than the swimming legs with feet found in nothosaurs. Their tails were shorter than those of nothosaurs, but not short and stumpy like many later plesiosaurs, and their trunk was more

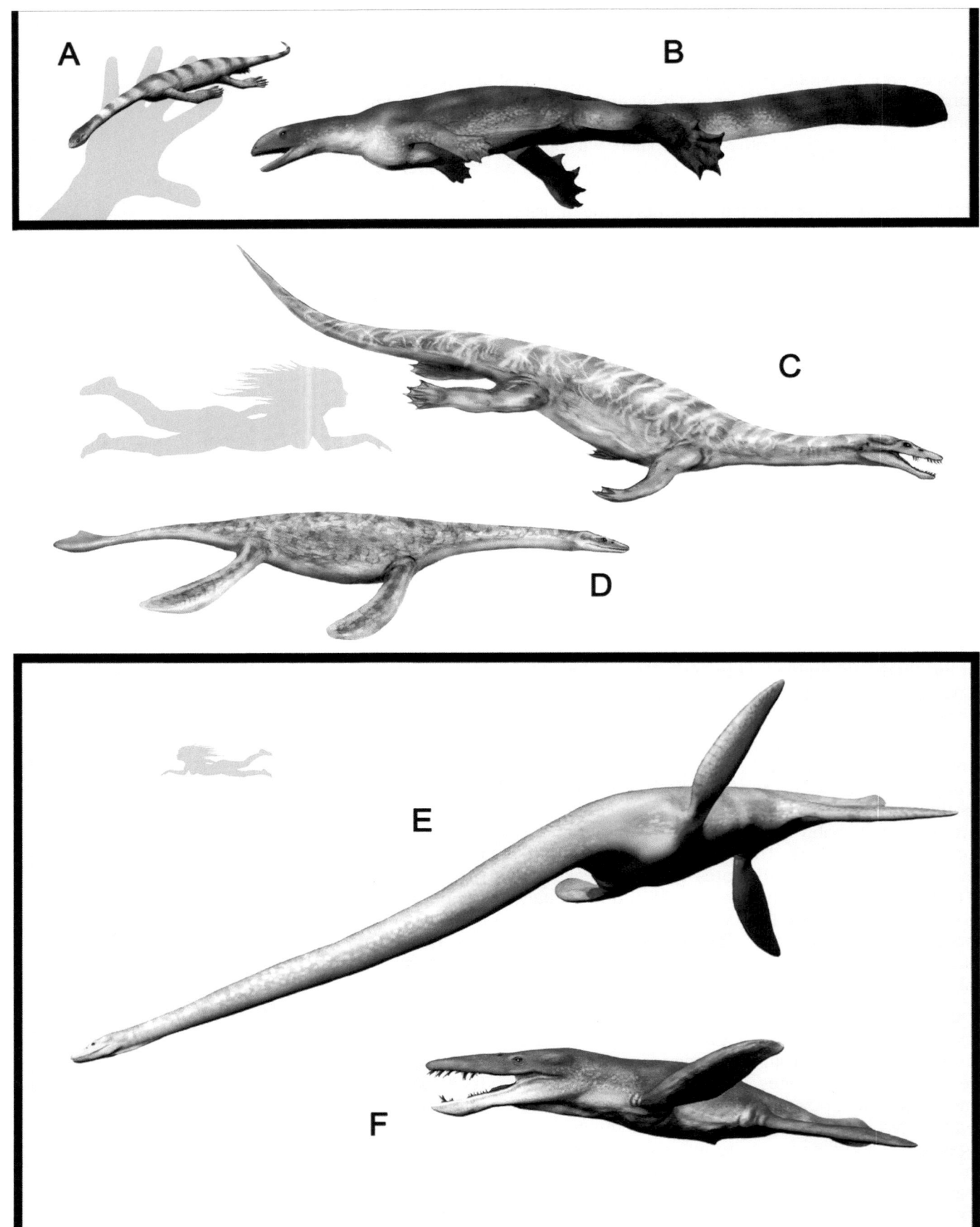

Figure 9.3 Reconstructions of a variety of sauropterygians: (A) *Keichousaurus*, (B) *Claudiosaurus*, (C) *Nothosaurus*, (D) *Pistosaurus*, (E) *Elasmosaurus*, (F) *Liopleurodon*.

robust and incapable of side-to-side wriggling motion which was still possible for nothosaurs.

The biggest of the Jurassic and Cretaceous plesiosaurs were the Pliosauria. Of these, the largest of all was *Kronosaurus* from the Cretaceous beds of Australia. It had a skull almost 3 meters (10 feet) long (**Figure 9.4[A]**), with the front paddles reaching 3.3 meters (11 feet) in length and a total length of about 12.8 meters (42 feet). However, a recent study has suggested that in reconstructing the missing parts, the preparators may have put in too many vertebrae. Its total length may have been closer to 10 meters (33 feet). The specimen at the Harvard Museum of Natural History covers the entire wall of one gallery and takes your breath away when you first see it.

All plesiosaurs had a similar basic build, other than their heads and necks. They were active swimmers that rowed their way across the Jurassic and Cretaceous seas using their huge front and back flippers. Plesiosaurs had a huge shoulder and hip girdle made of several bony plates on their belly for anchoring their powerful swimming muscles. Between the girdles was a mesh of belly ribs (gastralia) that gave their abdomens additional strength and support. In many specimens, smooth polished stones were found where the stomach was inside the rib cage, suggesting that plesiosaurs swallowed stones to provide ballast. Also found in the stomachs of the specimens from Queensland were fossils of their meals, which prove that creatures like *Kronosaurus* ate marine turtles and smaller

A

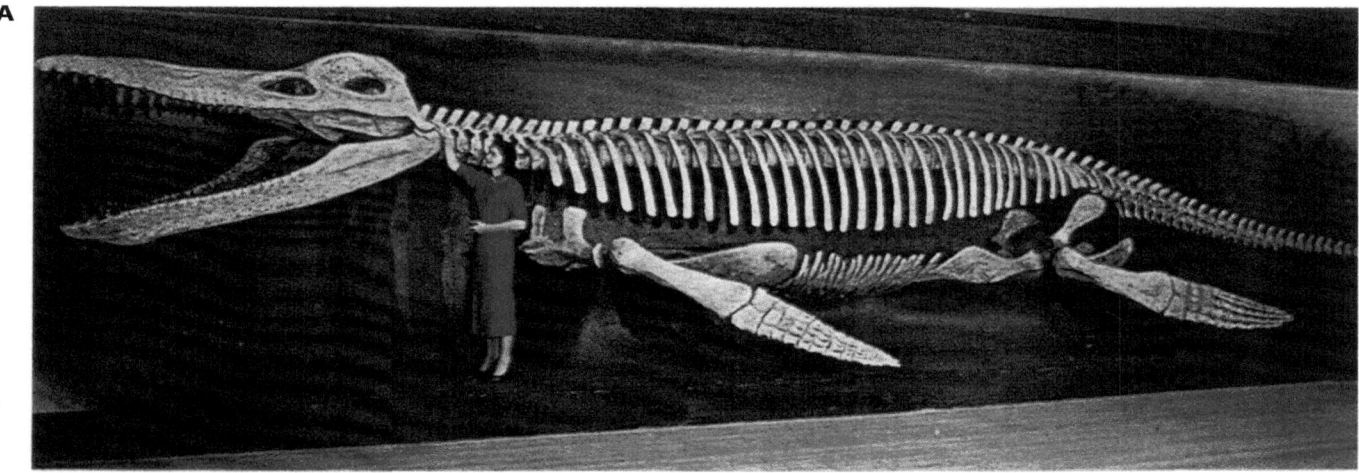

B

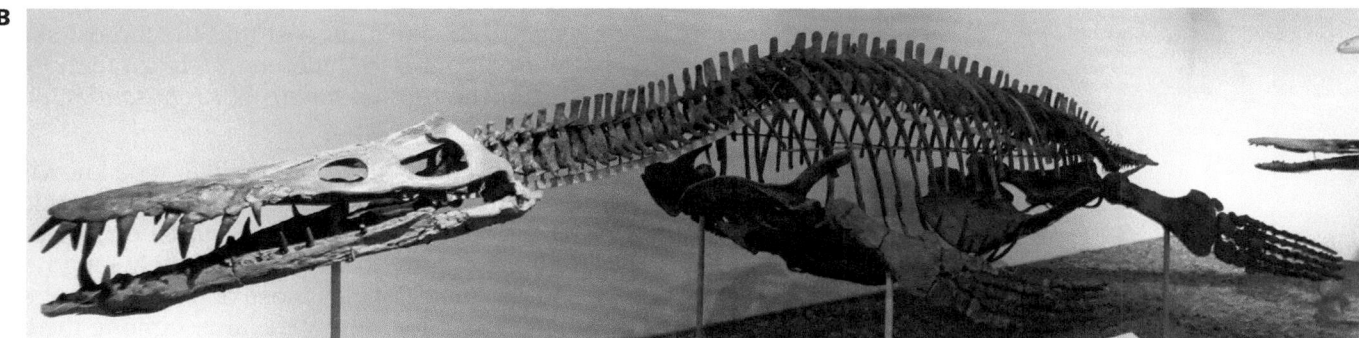

Figure 9.4 Some of the largest plesiosaurs. (A) The skeleton of *Kronosaurus*, the gigantic pliosaur from the Cretaceous of Australia, now on display at the Harvard Museum of Natural History. (B) The skeleton of *Liopleurodon*, from the Jurassic of Europe. (Courtesy Wikimedia Commons.)

C

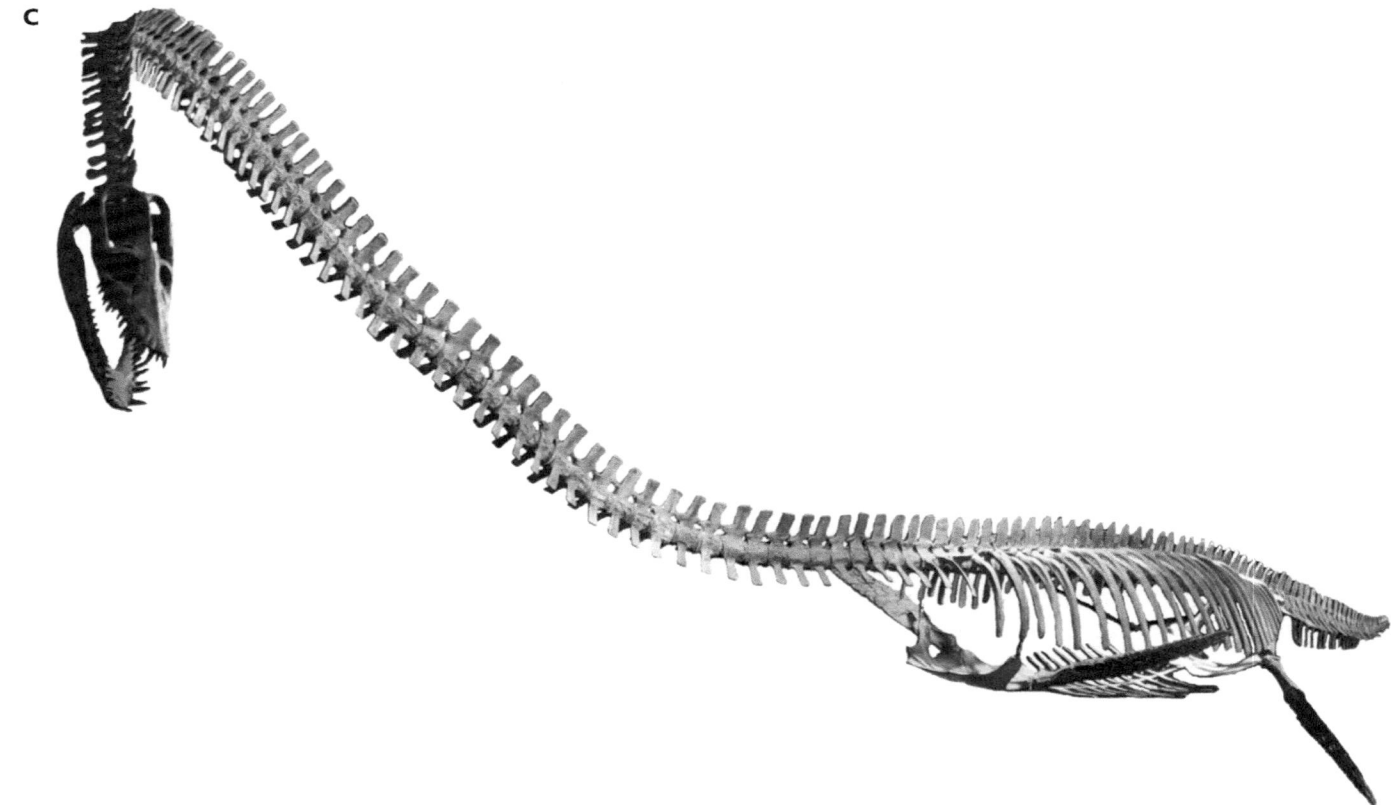

Figure 9.4 (Continued) (C) The skeleton of the long-necked *Elasmosaurus*, from the Kansas Cretaceous chalk beds of the Western Interior Seaway.

plesiosaurs. Fossils of huge ammonites and giant squid lay in the same beds, and they almost certainly were food for such a gigantic predator. The plesiosaur *Eromangasaurus*, also from the same beds, has large bite marks on its skull, suggesting an attack by *Kronosaurus*.

Viewers of the popular television series *Walking with Dinosaurs* may have seen depictions of a large plesiosaur from Europe called *Liopleurodon* (**Figure 9.3[F]**). The creature was animated as a monster more than 25 meters (82 feet) long, preying on dinosaurs and every other form of life during the Jurassic. In this size range, it approaches the size of the largest whales, including the blue whale. In fact, there are no relatively complete specimens of *Liopleurodon* that suggest such a large size. Instead, there the fossils consist of mostly a few large skulls and jaws, as well as other isolated bones. The largest complete skeleton, on display at the Museum für Geologie und Paläontologie in Tübingen, is only 4.5 meters (15 feet) long (**Figure 9.4[B]**). New methods of estimating size from skulls suggest that the largest skulls belong to animals that were about 5 to 7 meters (16 to 23 feet) long, not even close to the size of the revised length of *Kronosaurus*, at 10 meters (33 feet).

The other branch of the plesiosaurs is the more familiar type known as the elasmosaurs (**Figures 9.3[E]** and **9.4[C]**). Instead of the heavy long snout and short neck of pliosaurs such as *Kronosaurus*, elasmosaurs evolved in the opposite direction: tiny head and extremely long neck. These creatures had a body about as long as those of pliosaurs, but certainly not as heavy. Nonetheless, they were very large. Among the biggest was *Elasmosaurus*, which is known from complete specimens up to 14 meters (46 feet) in length and was estimated at 2000 kilograms (4400 pounds) in weight (**Figures 9.3[E]** and **9.4[C]**). In contrast to pliosaurs, elasmosaurs were probably much slower swimmers, but paddled slowly along using all four flippers for propulsion.

Since the discovery of fossils of elasmosaurs, paleontologists often reconstructed them with a long, flexible snake-like neck and a head that could whip around easily in any direction, and most reconstructions still show them that way. More recent analyses of the weight of their neck and head, the limited muscles of their neck, and the constraints on the movement of the neck vertebrae show that the neck was probably not very flexible. These studies suggest that the elasmosaur neck would have been semi-rigid and incapable of bending very far, more like a fishing pole than like a snake neck. It also could not have been lifted out of the water in a swan-like fashion.

If the neck could not flex sharply and allow the elasmosaurs to snap in any direction, how could they catch prey? Paleontologists have suggested alternative methods of feeding that do not require a flexible neck. One proposal is that their long neck allowed them to lurk in deeper waters below the prey without being detected. Then they could poke their head into a school of fish or squid or ammonites and grab a meal before the shock wave of their massive body arrived to alert their prey to their movements. Their huge eyes are also consistent with this idea.

Another suggestion is that elasmosaurs were bottom feeders, using their neck to plow through the mud of the seafloor in order to grab prey. Most elasmosaurs had long conical peg-like teeth that pointed forward, a common adaptation for spearing fish and other aquatic prey. Some elasmosaurs, like *Cryptoclidus* and *Aristonectes*, had hundreds of tiny pencil-like teeth that suggest they could have strained out small food items from either the plankton or the sea bottom. Other scientists are not so sure that elasmosaurs had a semi-rigid neck. They point out that a lot of soft tissue is missing from the fossils (especially the cartilage between the vertebrae), and with so many neck vertebrae, their neck would still have been fairly flexible. The neck was certainly not as flexible as a snake's body, or capable of curling into an *S* shape, but these scientists argue that elasmosaurs could still have curled their neck into a fairly tight arc to reach prey. If so, then the elaborate behaviors suggested by the "rigid-neck" hypothesis are less likely.

The large body size, the flippers directly beneath their body, the lack of attachment of their hind limb bones to their spine, and other features of plesiosaurs make it unlikely that plesiosaurs could have crawled far onto land or dug a hole in which to lay eggs, as do sea turtles. Still, many artists persist in showing plesiosaurs awkwardly splayed across rocks or a beach, with flippers far too short to drag their body across the surface. Finally, these myths were punctured and their purely aquatic reproduction finally was confirmed by the recent description of a plesiosaur fossil with an embryo in its body, showing that they gave birth to live young in the sea.

PLACODONTS

Finally, we should mention a small group of plesiosaur relatives, the placodonts (**Figures 9.5** and **9.6**). These were found only in the Middle to Late Triassic, dominating the seas before ichthyosaurs or plesiosaurs had diversified. Placodonts are best known from Germany and central Europe, but also from North Africa, the Middle East, and China. They had barrel-shaped bodies, a long tail, and relatively short limbs, so they were not fast swimmers. But their short necks, heavy skulls, and blunt pavement-like teeth (**Figure 9.5[B]**) show why they didn't need to swim fast: their prey were molluscs, lying on the seafloor, and unable to swim away from them. Of all the specialized mollusc-eaters with pavement-shaped teeth for crushing shells, placodonts have the best developed teeth for that purpose, with a large blunt domed-shaped surface covered in thick enamel that could easily smash a snail or a clam or possibly even brachiopods. In fact, their named "placodont" means "plate tooth" in reference to the layer of hard enamel covering their teeth. In addition, placodonts had long protruding teeth in the front of the mouth for grabbing and manipulating the shells of their prey.

Typical specimens of *Placodus* (**Figures 9.5[A,B]** and **9.6[B]**) were up to 2 meters (6.6 feet) long, but most specimens were slightly smaller than

Figure 9.5 Fossils of the bizarre creatures known as placodonts. (A) *Placodus* from the Triassic of Europe. (Courtesy Wikimedia Commons.)

B

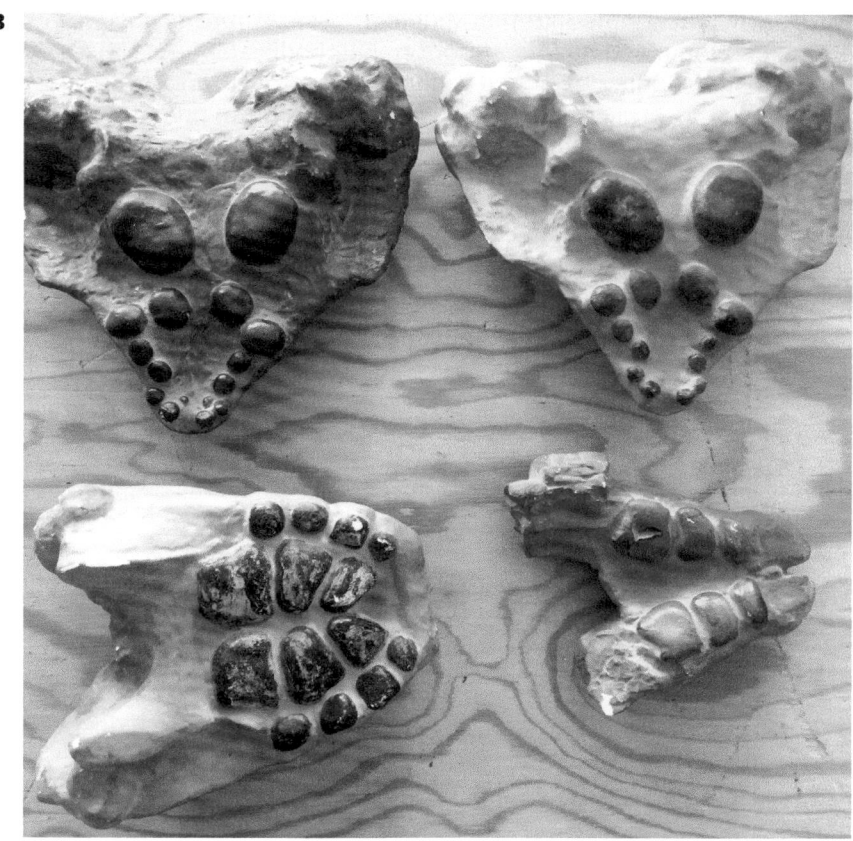

C

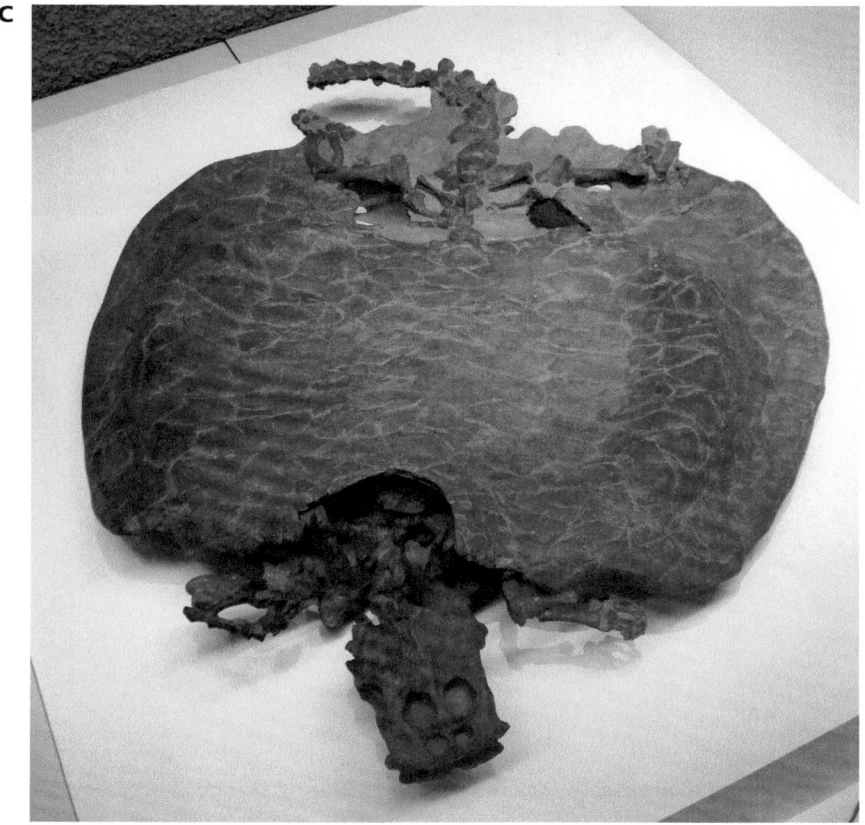

Figure 9.5 (Continued) (B) Upper jaws of *Placodus*, showing the blunt teeth used for crushing shelled prey. (C) The weird turtle-like placodont *Henodus*.

this. Early placodonts like the Middle Triassic *Paraplacodus* were only 1.6 (5 feet) long, and had no armor on their bodies (**Figure 9.6[A]**). Later Triassic placodonts developed different kinds of armor in their backs, presumably to protects themselves from large new predators, like the newly developed predatory marine reptiles. Other experiments in placodonts were *Cyamodus*, which had a carapace on its back that superficially resembled that of a turtle (**Figure 9.6[D]**).

Even weirder was the meter-long *Henodus* (**Figures 9.5[C]** and **9.6[C]**) from the non-marine lagoonal deposits of the Upper Triassic of Germany,

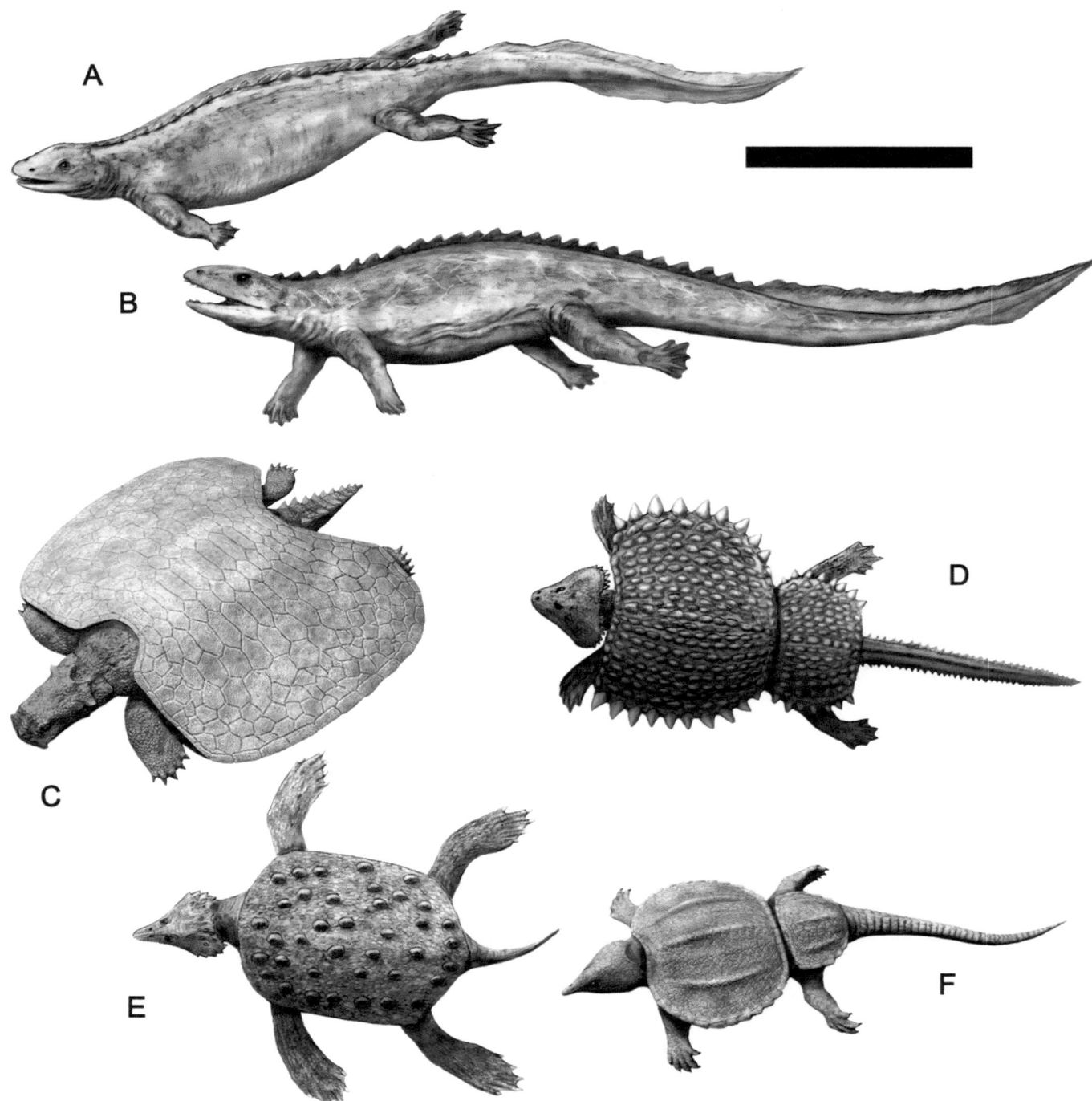

Figure 9.6 Reconstructions of some different types of placodonts: (A) *Paraplacodus*, (B) *Placodus*, (C) *Henodus*, (D) *Cyamodus*, (E) *Placochelys*, (F) *Psephoderma*. Scale bar is 50 cm.

which had a plastron and carapace like that of a turtle—except that it was broader and almost rectangular in top view, compared to the rounded or oval shape of a true turtle shell. In detail, the armor is not homologous with a turtle shell at all, because it has a completely different pattern of dermal bones forming plates, all of which were fused to the spine, and not formed like the turtle shell, which is made of broadened ribs fused to dermal bone armor. The weird flat-snouted skull of *Henodus* had just a single tooth in each side of the upper and lower jaws, a toothless beak in front, and a mouth full of baleen-like denticles, suggesting that they were filter-feeders rather than shell-crushers. The bones of the throat region suggest that they could expand their throat cavity and gulp in a large volume of water and then pass it through the sieve of denticles and trap their prey inside their mouth.

The seas of the Mesozoic were ruled by a wide variety of marine reptiles that had independently made the transition from water to land, and occupied a variety of niches, from filter feeders to shell crushers, to predators specialized for a wide variety of prey. These two groups, ichthyosaurs and plesiosaurs, made up most of this marine reptile diversity. However, the return to the sea also happened with sea turtles (Chapter 8), mosasaurs (descendants of monitor lizards—Chapter 10) and marine crocodiles (Chapter 12), which are discussed elsewhere.

FURTHER READING

Bernard, A.; Lécuye, C.; Vincent, P.; Amiot, R.; Bardet, N.; Buffetaut, E.; Cuny, G.; Fourel, F.; Martineau, F.; Prieur, A. 2010. Regulation of body temperature by some Mesozoic marine reptiles. *Science*. 328 (5984): 1379–1382.

Callaway, J.; Nicholls, E.L. 1997. *Ancient Marine Reptiles*. Academic Press, San Diego, CA.

Ellis, R. 2003. *Sea Dragons-Predators of the Prehistoric Oceans*. University Press of Kansas, Lawrence, KS.

Everhart, M.J. 2005. *Oceans of Kansas: A Natural History of the Western Interior Sea*. Indiana University Press, Bloomington, IN.

Maisch, M.W. 2010. Phylogeny, systematics, and origin of the Ichthyosauria: The state of the art. *Palaeodiversity*. 3: 151–214.

Massare, J.A. 1987. Tooth morphology and prey preference of Mesozoic marine reptiles. *Journal of Vertebrate Paleontology*. 7 (2): 121–137.

Massare, J.A.; Callaway, J. M. (1990). The affinities and ecology of Triassic ichthyosaurs. *Geological Society of America Bulletin*. 102 (4): 409–416.

McGowan, C. 1983. *The Successful Dragons: A Natural History of Extinct Reptiles*. Samuel Stevens & Company, New York.

McGowan, C. 1991. *Dinosaurs, Spitfires, and Sea Dragons*. Harvard University Press, Cambridge.

Motani, R. 1999. Phylogeny of the Ichthyopterygia. *Journal of Vertebrate Paleontology*. 19 (3): 472–495.

Motani, R. 2005. Evolution of fish-shaped reptiles (Reptilia: Ichthyopterygia) in their physical environments and constraints. *Annual Review of Earth and Planetary Sciences*. 33: 395–420.

O'Keefe, F.R. 2001. A cladistic analysis and taxonomic revision of the Plesiosauria (Reptilia: Sauropterygia). *Acta Zoologica Fennica*. 213: 1–63.

Sander, P.M. 2000. Ichthyosauria: Their diversity, distribution, and phylogeny. *Paläontologische Zeitschrift*. 74 (1–2): 1–35.

Taylor, M.A. 1987. A reinterpretation of ichthyosaur swimming and buoyancy. *Palaeontology*. 30: 531–535.

Thorne, P.M.; Ruta, M.; Benton, J. 2011. Resetting the evolution of marine reptiles at the Triassic-Jurassic boundary. *Proceedings of the National Academy of Sciences*. 108 (20): 8339–8344.

Welles, S.P. 1952. A review of the North American Cretaceous elasmosaurs. *University of California Publications in Geological Science*. 29: 46–144, figs. 1–25.

Zammit, M. 2012. Cretaceous ichthyosaurs: Dwindling diversity, or the Empire strikes back? *Geosciences*. 2 (2): 11–24.

THE SCALY ONES

LEPIDOSAURIA—LIZARDS AND SNAKES

10

Snakes are vertebrates and vertebrates are classified as higher animals, whether you like it or not. I mean you can be a higher animal and still be a snake. This seems to be a rather peculiar arrangement, to be sure. If you can think of a better, let's have it. . . . Snakes in a word, are well worth knowing, unless you'd rather know something else. In closing, I have a little message which I wish you'd relay to some of those people who won't read a snake article because it gives them the jumps: there are no snakes in Iceland, Ireland, or New Zealand. And no snake articles.
—Will Cuppy, *How to Become Extinct*

LEPIDOSAURIA

By far the most numerous and diverse group of reptiles alive today are the Lepidosauria, or the lizards and snakes and their kin. There are almost 10,000 species of lizards and snakes on the planet now, the second largest order of vertebrates after the percomorph fish. The other branches of living reptiles have nowhere near this diversity; there are only 27 species of crocodilians, and only about 365 species of turtles and tortoises. Living lepidosaurs include the numerous lizards and snakes (the Squamata) plus the tuatara (**Figure 10.1[A]**), a "living fossil" of a group of Mesozoic reptiles represented by a single species, *Sphenodon punctatus*, on a handful of islands off the shores of New Zealand.

The lepidosaurs are defined by a number of distinctive characteristics. The name Lepidosauria means "scaly lizard" in Greek, and Squamata comes from the Latin for "scale", so scales are an important feature of all lepidosaurs. Their bodies are covered by relatively thin scales covered by layers of keratin, the protein found in hair, fingernails, feathers, and many other dermal structures. Most lepidosaurs shed the outer layer of their scales by flaking them off as they are worn out, although snakes shed their skins in one big continuous sheath. By contrast, turtles and crocodilians do have scales, but these features tend to be much thicker, and most of their bodies are covered by bony plates called scutes. Unlike scutes, most lepidosaur scales overlap in shingled fashion like roof tiles.

Another characteristic is that male squamates have a forked set of reproductive organs called "hemipenes" or "half-penises". These organs are filled with fluid (like many mammalian and bird penises), but they reside inside their bodies most of the time. When engorged with fluid, they turn inside-out like the fingers in a rubber glove. The male lepidosaur then inserts whichever side of the hemipenis is in the right position to find the cloaca, or urogenital opening of the female, and releases his sperm. Another lepidosaur feature is that the cloacal slit on both males and

DOI: 10.1201/9781003128205-10

Figure 10.1 Rhynchocephalians.
(A) The tuatara, *Sphenodon punctatus*, the only living member of the Rhynchocephalia branch of the Lepidosauria. Today, it lives only on some islands off the coast of New Zealand, but during the Jurassic, sphenodontids were the most common lepidosaurs.
(B) The skull of *Clevosaurus*, a primitive sphenodontid from the Triassic of Brazil. It shows the characteristics of the group, including the hooked upper beak, the pineal opening between the eyes, and the large multi-cusped tooth plates.
(C) *Homeosaurus*, a sphenodontid known from numerous complete specimens from the Upper Jurassic Solnhofen Limestone of Bavaria, Germany. (Courtesy Wikimedia Commons.)

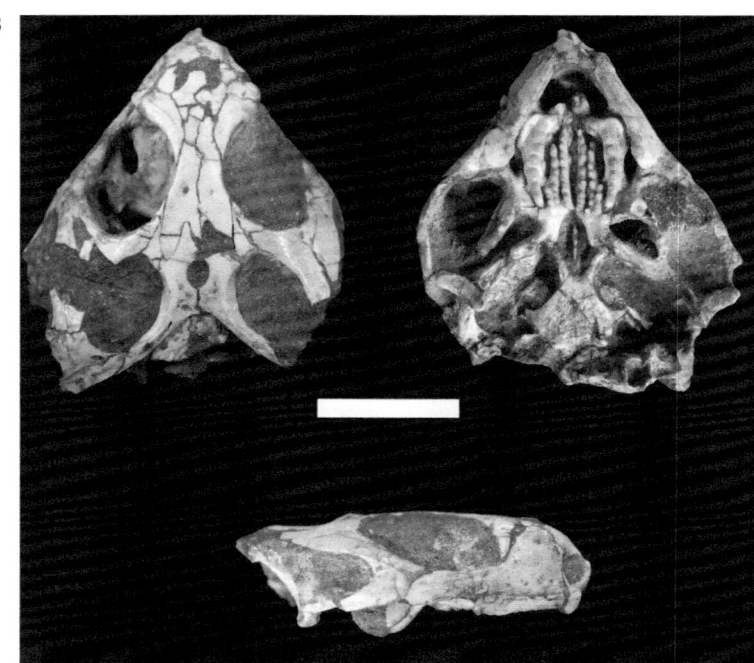

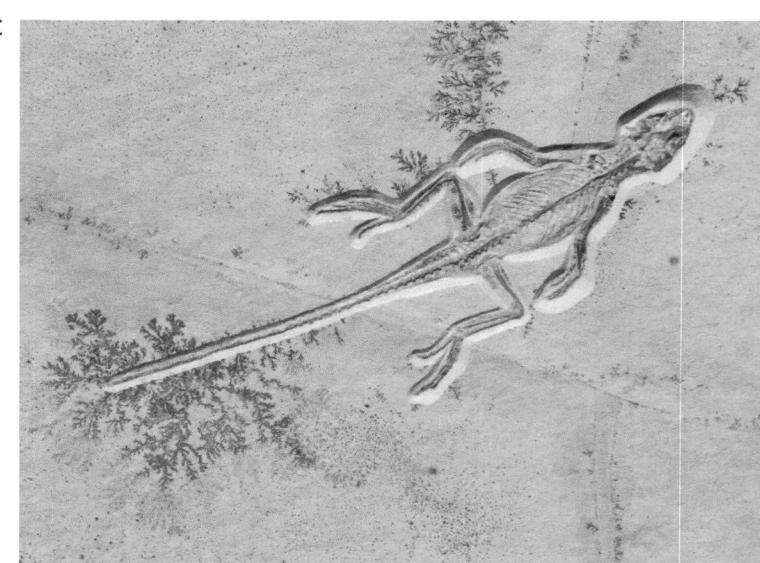

females is transverse across the body axis, while the rest of the reptiles have a single cloacal hole, or a slit parallel to their body axis.

Many lepidosaurs (especially lizards but also some snakes) can break off their tails to distract a predator who focuses on the tail while the rest of the reptile gets away. They have a special set of vertebrae in the spine of the tail which forms a natural break point. Finally, features like a forked tongue is used to taste the air for scents and carries these molecules to the vomerona-sal organ on the roof of the mouth, where they are "tasted". Forked tongues are found only in lepidosaurs, and never in any other group of reptiles. This is one of the glaring errors in the low-budget movies about prehistoric life done in the mid-twentieth century: they would dress up a modern lizard (usually some kind of iguana) with horns or a frill as a "dinosaur" but the dead giveaway that it's *not* a dinosaur is the flicking of their forked tongue.

Most primitive lepidosaurs (including the earliest relatives of the tuatara, plus most snakes and lizards) have teeth which form in rows along a trough on the inside edge of the jaw, a pleurodont dentition, and very different from the teeth in sockets found in archosaurs. Primitive tuatara relatives also have a feature defining squamates in that the bone below the lower temporal fenestra of their diapsid skull is incomplete. Later tuatara relatives fuse this bone back together, but as we shall see later in the chapter, it is a well-developed and defining feature of squamates. In addition, there are number details of the bones in the skull that are unique to the lepidosaurs. One of the most distinctive of these features is a shell-shaped capsule, or "conch", in the skull, which encloses the eardrum and ear apparatus. Finally, numerous molecular analyses have consistently proven that the Lepidosauria are a natural group composed of the squamates plus the tuatara, so the reality of this group is well established.

Yet lepidosaurs tend to be overshadowed in the fossil record and in the public understanding of prehistoric life for a number of reasons. The most obvious is that the vast majority of them were small or even tiny in body size, and thus have fragile bones that tended to break up and leave very few fossils behind. By contrast, most dinosaurs left huge bones behind that are easily fossilized, and turtle shells tend to be common fossils in some settings, and even crocodilian skeletons tend to be rather large and robust. Some lizard fossils can be identified by key bones like the jaw or certain parts of the limb girdles, but small incomplete bones of most lizards simply cannot be identified to genus or species at all. The major-ity of the fossils of lizards are collected using techniques developed for other vertebrate microfossils, such as washing loads of fossil-rich sand and gravel through sieves to recover tiny lizard jaws and bones. Snakes have even worse odds in terms of fossilization, since their fragile skulls made of thin rods of bone, and their skeleton made of delicate ribs and vertebrae, tend to break up and fall apart very easily. There are a handful of extinct lizards and snakes known from excellent complete skeletons, of course, but the vast majority are known from just fragments.

Given their limited ability to fossilize, it is not surprising that their fos-sil record is sparse. Lepidosaurs are the closest relatives of archosaurs (Chapter 12), which arose in the Early Triassic (**Figure 8.1**). Likewise, the oldest lepidosaurs are known from the Early Triassic, but their fossils are rare through most of the Mesozoic, while archosaurs took over the planet and ruled it even after the extinction of the non-bird dinosaurs 66 Ma.

These earliest lepidosaur fossils are isolated jaws not identifiable to a specific group, but consist mostly of primitive jaws. However, there is a skull from the Early Triassic of South Africa known as *Paliguana* which has many of the distinctive lepidosaur features, so most scientists regard it as the first undoubted member of the group. In the past, some Late Per-mian and Early Triassic fossils like *Palaeagama* and *Saurosternon* have been called lepidosaurs, but more recent analyses have only shown that they are primitive diapsids.

And then there are the strange primitive lepidosaurs known as *Kuehneosaurus* and the very similar *Icarosaurus* from the Late Triassic of England and New Jersey, respectively (**Figure 10.2[A]**). Both were small reptiles (about 10 cm, or 4 inches long), with enormously elongated ribs along their bodies, which could be folded back against the body. These bones apparently supported a wing-like membrane on the side of their torso, so they could glide from one high perch to another. This feature appeared more than once in the vertebrates, including in the "flying lizards" (genus *Draco*) which are still common in the jungles of southeast Asia today, as well as an extinct diapsid reptile, *Coelurosauravus*, from the Permian of Madagascar, England, and Germany, and the Early Cretaceous Chinese iguanian lizard *Xianglong* (discussed in the following section). In addition to reptiles, many other vertebrates have found ways of parachuting

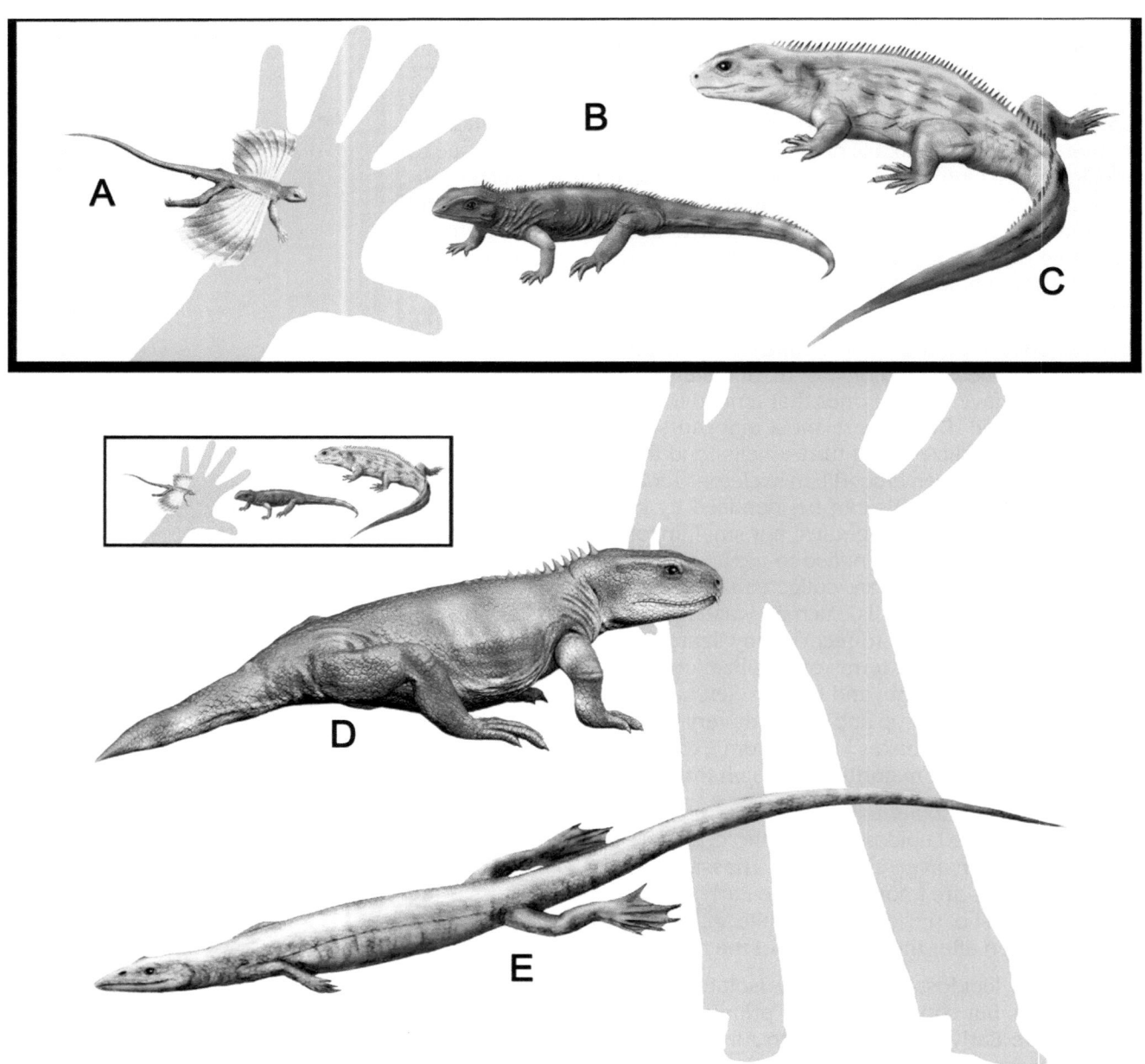

Figure 10.2 Reconstructions of some primitive lepidosaurs: (A) *Icarosaurus*, (B) *Gephyrosaurus*, (C) *Brachyrhinodon*, (D) *Priosphenodon*, (E) *Pleurosaurus*.

or gliding down from trees, including flying squirrels, the marsupials known as sugar gliders, the colugos ("flying lemurs") of southeast Asia, the Malabar flying frog *Rhacophorus*, and a few other examples.

The specialized gliding adaptation in all these animals is complete evolutionary convergence, since the kuehneosaurs were probably very primitive lepidosaurs, while *Draco* is a specialized lizard in the family Agamidae, *Xianglong* is an unrelated lizard, *Coelurosauravus* is a primitive diapsid, and the rest are mammals unrelated to lizards at all. Modern *Draco* "flying lizards" hunt insects and other small prey on tree branches, and can glide 60 meters (200 feet) while losing only 10 meters (33 feet) in elevation, a remarkable feat for an animal only 20 cm (8 inches long). Presumably the kueheosaurs had very similar adaptations, and had comparable abilities to glide long distances.

RHYNCHOCEPHALIA

During the Late Triassic and Jurassic, most of the lepidosaurs were members of one main branch, the Rhynchocephalia ("beak heads" in Greek). At one time this group also included a group of pig-sized herbivorous archosaurs known as rhynchosaurs, but the only thing those two groups had in common was the hooked beak, so they are no longer considered related. Today, only one species of the Rhynchocephalia, the living tuatara *Sphenodon*, still survives on 28 small islands off the coast of New Zealand (**Figure 10.1[A]**). Tuataras are odd-looking greenish-brown reptiles, resembling a thick-bodied lizard. Their size ranges up to 80 cm (31 inches) long, with a short snout with an overhanging upper "beak", a well-developed "third eye" (pineal gland) for detecting light and seasonal change, and a row of spines along their back. They live on a diet of small invertebrates, especially snails, slugs, worms, larvae, insects, and also eat bird's eggs. They have a row of wedge-shaped teeth on the outer part of the jaw for crushing these creatures, along with additional teeth on the palate that aid this process. The jaw mechanism allows their teeth and jaws to slide front-to-back a bit, aiding in their food processing. They were once numerous all over New Zealand, but were wiped out when rats were introduced by humans, and now tuataras only survive on those 28 islands around New Zealand, where they are considered vulnerable to extinction.

But in the Triassic and Jurassic, rhynchocephalians were by far the most common squamates. They were represented by at least six genera by the late Triassic, and about 17 Mesozoic genera altogether. Unlike modern *Sphenodon*, these Mesozoic rhynchocephalians had teeth indicating a diverse diet, with strictly carnivorous forms, as well as herbivores and omnivores. The most primitive of these was *Gephyrosaurus* from the Late Triassic and Early Jurassic of Wales and England, which had long slim legs and may have been able to climb trees (**Figure 10.2**). Its jaws and teeth suggest that it was an insectivore. *Brachyrhinodon* (**Figure 10.2[C]**) was another Late Triassic primitive rhynchocephalian from the famous Elgin Quarries of Elgin in northern Scotland. Like *Clevosaurus* from the Late Triassic and Early Jurassic of Brazil (**Figure 10.1[B]**), Nova Scotia, England, Wales, and South Africa, *Brachyrhinodon* had a short boxy bulldog-like skull with a robust beak, giving it a powerful crushing bite, possibly for eating hard-shelled prey. Other primitive rhynchocephalians include *Homeosaurus*, which is known from complete articulated skeletons from the Upper Jurassic Solnhofen Limestone of southern Germany (**Figure 10.1[C]**). These are the same beds that yield *Archaeopteryx* and *Pterodactylus*. More advanced sphenodontids include *Opisthias*, which is a common fossil among the tiny vertebrates of the Upper Jurassic Morrison Formation that yields so many giant dinosaurs. *Oeneosaurus* from the Late Jurassic of Germany had large crushing tooth plates, probably for eating mollusks or hard-shelled arthropods.

A very different rhynchocephalian was *Pleurosaurus*, also from the Late Jurassic of Germany (**Figure 10.2[E]**). It was over 1.5 meters (5 feet) long, with a long slender body and short limbs, and a powerful tail, and presumably was a swimmer that caught fish in the shallow water setting where it is found. Its nostrils were far back on its head near its eyes, another adaptation for breathing just before diving, as we see in aquatic animals like whales and dolphins. *Priosphenodon* from the Late Cretaceous of Argentina (**Figure 10.2[D]**) is a close relative of the modern tuatara, and one of the youngest fossil sphenodontids known. It was a much larger lizard-like form (over a meter long), with a long snout and teeth adapted for shearing plants. Then the fossil record of rhynchocephalians becomes very sparse in the Cenozoic, culminating in the "living fossil" tuataras of New Zealand.

SQUAMATES

As mentioned earlier, except for the tuatara, all the 10,000 or so living species of lepidosaurs are squamates, the lizard, snakes, and a legless group called the amphibaenids. In addition to the distinctive features of their scales, and how they are shed, that are mentioned earlier, the most important feature of the squamates is that the bone at the bottom of the lower temporal fenestra is lost, so the bone at the back end of the skull (the quadrate), which hinges with the jaw bone, is free to swing in a front-to-back plane (**Figure 10.3**), a condition known as streptostyly. This allows the lower jaw to be extended or opened even wider, and gives squamates the ability to swallow very large prey.

Snakes have taken this to an extreme. They are capable of swallowing a prey item whole that is bigger than their head diameter. To do this, their skulls have reduced most of the bones (except the braincase) into long thin splints hinged together and connected by stretchy tendons and ligaments (**Figure 10.3[E]**). When they start to work their jaws around a larger prey animal, they stretch their entire mouth around it, ratcheting the jaw bones with their rear-facing teeth over the prey one slow step at a time. Eventually they stretch their entire mouth around the food, then slowly work it back into their gullet and stomach, where it will show up as a huge bulge in the body of the snake, often for weeks as it is slowly digested. During this time, snakes are often in torpor and in hiding while the difficult process of digestion of a complete unchewed carcass takes place. The bulge of the prey can be seen moving through their bodies as the digestion proceeds.

Where do squamates come from? All the anatomical and molecular analyses and fossil evidence show they split from a common ancestor with the rhynchocephalians, which go back to at least the Late Triassic. Then in 2018, a fossil from the Middle Triassic of the Austrian Alps called *Megachirella* was restudied and identified as a very primitive squamate, pushing their fossil record back to about the age of the oldest rhynchocephalian. Still, their fossil record is relatively sparse during the Jurassic, since sphenodontids like *Opisthias* and *Homeosaurus* and others mentioned previously greatly outnumber the Jurassic squamates. By the Middle Jurassic, however, there are fossils that belong to living groups of lizards, such as geckos and skinks, and by the Cretaceous there are fossils of iguanas and varanoids (monitor lizards, like the goanna and Komodo dragon). A few groups, like the polyglyphanodontids, are only known from the Mesozoic, and vanished at the end of the Cretaceous. Altogether, there are about 6000 species of fossil squamates known from Cretaceous and Cenozoic.

Today, there are a number of distinct natural groups within the squamates that nearly all scientists recognize. The amphisbaenids, or "worm lizards", consist of about 180 living species of legless worm-shaped squamates that have adapted to a fully subterranean life of burrowing. They used to be put in a separate but equal category with snakes and

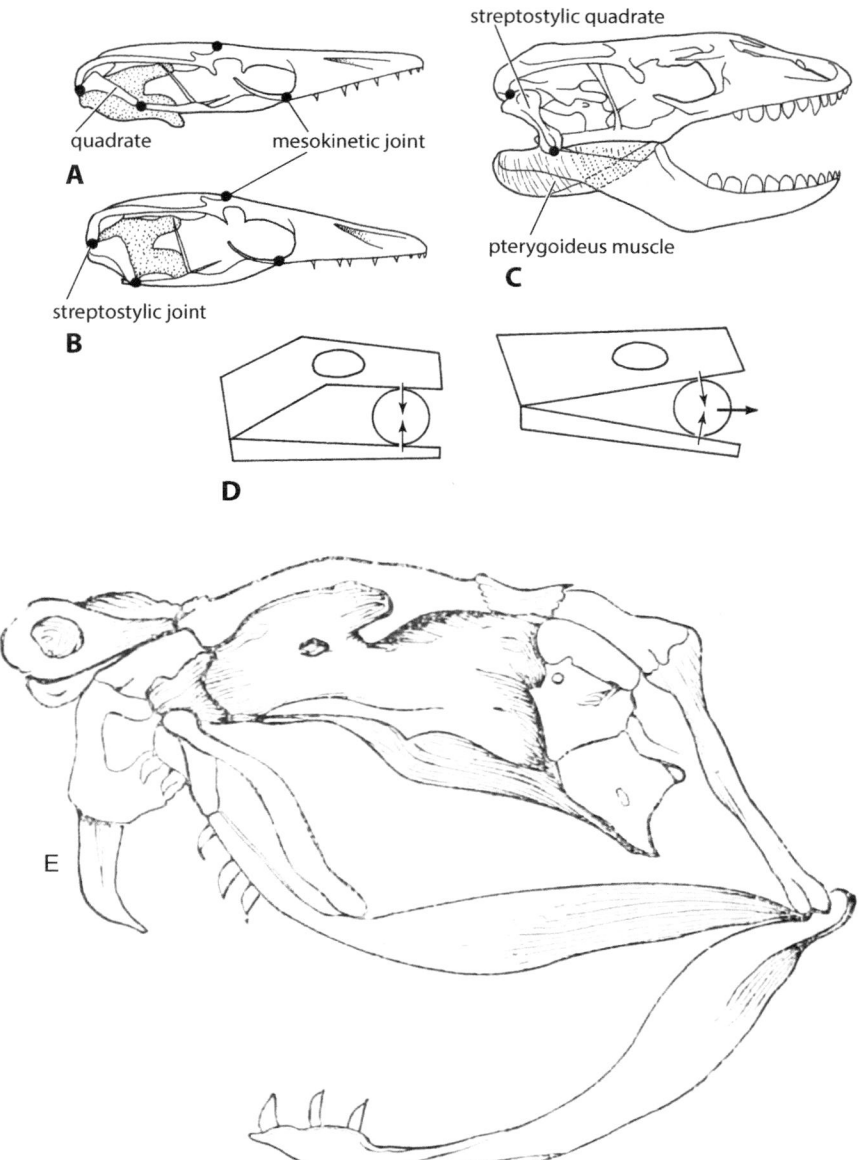

Figure 10.3 Mechanics of the squamate kinetic skull (this is based on a monitor lizard) (A,B) The lower temporal bar is gone in the squamate skull, so the quadrate bone of the back of the skull is a free-swinging hinge connected with the bones of the upper and lower jaw. As it swings forward and backward, it allows the jaw to move freely, especially when expanding the gape to increase the mouth opening and swallow prey, or pulling backward with its recurved teeth to drag the prey item further into its throat, aided by the pterygoideus muscles around the back of the jaw (C). This kind of motion is called streptostyly, and it is unique to squamates. (D) The squamate streptostylic jaw (left) distributes the bite forces differently that then simple "snap-trap" jaws of most reptiles (right). (E) The snake skull is reduced to thin rods and struts of bone, which can stretch around a prey item with long flexible tendons and ligaments. (Redrawn from several sources.)

lizards, but now they are recognized as just a branch within the lizards. The Gekkota include not only the tropical geckos that have ridged footpads allowing them to cling to walls and ceilings, but also the blind lizards (Dibamidae), and the "legless lizards" (Pygopodidae), another group that lost their legs and focused on burrowing. The Iguania include not only the familiar herbivorous iguana lizards, but also the agamids (including the bearded dragons), chameleons, basilisks, collared lizards, horned lizards ("horny toads"), anoles (the New World color-changing lizards that are mistakenly called "chameleons"), and many more obscure groups. The Lacertoidea include the true lizards (Lacertidae, the main lizard family in Eurasia and Africa), the tegus or whiptails, and the spectacled lizards. The Anguimorpha include the anguids (alligator lizards, glass lizards, and slowworms), the anniellids (California legless lizards), the Gila monsters, the knob-scaled lizards, the earless lizards, and the monitors (goannas and Komodo Dragons and other monitors). The Scincomorpha include not only the skinks, but also the spiny-tailed lizards, the night lizards, and the plated lizards. Finally, there are over a dozen families of snakes as well. We will not go into greater detail about all these groups, because that is outside the scope of a book like this, but it is something every herpetologist learns early in their career.

With only a sparse fossil record of these relatively incomplete specimens, most scientists by necessity focused on the abundance of evidence in the soft tissues of the 10,000 living species of squamates to sort out how they are all related. But this has still not given us a clear-cut answer, even after numerous analyses. The conventional viewpoint (going back to the work of Richard Estes in 1988 and earlier), based on anatomy and fossils, considered iguanians and their relatives as the most primitive group of living squamates, and all other squamates were put in the Scleroglossa (meaning "rough tongue" in Greek), including the anguid lizards, the skinks, and the rest of the lizards (plus snakes and amphisbaenids). The iguanians use their tongues to capture prey (as do more primitive lepidosaurs like the tuatara), whereas the Scleroglossa use their jaws, and the tongue is not as important (since it is often a thin forked tongue used to taste the air, not to feed). This arrangement appeared to be confirmed by a molecular study done in 2004 by Lee and others.

But since 2005, analyses of DNA sequences seem to be giving a different answer from the molecular result of Lee and others in 2004. This evidence places geckos as the earliest branch, followed by skinks, lacertids, amphisbaenids, snakes, anguids, and finally the iguanians as the most advanced group. The most recent analysis by Reeder and colleagues in 2015, and even later analyses, seems to be giving this result as well. The issue is still not resolved, but this seems to be the consensus from molecular studies done in the last decade.

For a book about extinct animals, a detailed examination of mostly living squamate groups is not really appropriate. However, there are some spectacular fossil lizards to mention. One of these was the curious Chinese gliding lizard *Xianglong* (**Figure 10.4[A]**) from the Lower Cretaceous lake beds of Liaoning Province which have produced so many spectacular complete fossils with soft tissues. Like the kueneosaurines and the living "flying lizard" *Draco*, this creature also had elongated ribs on its torso that seemed to have formed a gliding membrane so it could swoop from tree to tree. This appears to be yet another convergent and independent development of this gliding membrane supported by ribs, since it is grouped with the iguanian lizards. The living *Draco* is also an iguanine, but not closely related to *Xianglong*.

The most spectacular of all lizards was the giant Pleistocene monitor lizard *Megalania*, from Ice Age beds in Australia (**Figure 10.4[B]**). It is known from fairly incomplete fossils but scaling up from the bones and comparable bones of living monitor lizards, it was probably 4.5 meters (15 feet) long and weighed about 330 kg (730 pounds). Widely published estimates of 7 meters in length (23 feet) are probably in error. Still, this creature was the largest lizard that ever lived, and is thought to be closely related to a number of different kinds of goannas and monitors from southeast Asia and Australia, perhaps even the giant Komodo Dragon (the largest lizard alive today, but just over half the size of *Megalania*). At such large size, it was the largest predator in Australia during the Ice Ages, dwarfing the handful of mammalian predators like the marsupial lion, *Thylacoleo*. It could capture just about every large mammal and bird in the outback in the Pleistocene, from the rhino-sized wombat relatives known as diprotodonts, to the giant kangaroos, and of course smaller prey as well. Its jaws were filled with serrated blade-like teeth, capable of shearing the flesh of almost any animal. Its sprinting speed was estimated at up to 3 meters/sec (11 km/hr), which is fast enough to catch all but the fastest, nimblest prey, and that velocity is comparable to the speed of large crocodiles when they charge.

Figure 10.4 Reconstructions of extinct squamates: (A) The Cretaceous Chinese gliding lizard *Xianglong*, (B) the gigantic monitor lizard *Megalania*.

The *Megalania* vanished from Australia about 40,000 years ago, about the same time that many of the mega-mammals that it preyed on vanished. The early Aboriginal peoples were in Australia about 60,000 years ago, so there is no direct evidence that humans drove it to extinction. Instead, it must have been a frightening part of their daily lives, and something they learned to avoid because of its size, speed, and strength. It probably haunted the nightmares of Aborigines long after it was extinct and only a legend.

SNAKES

Many people have a deathly fear of snakes, and in some cases it is deep and irrational and given its own name, ophidiophobia. When our hominin ancestors roamed Africa, they had good reason to fear any snake, because many of them (cobras, vipers, mambas) are highly venomous. Whatever your personal feelings about snakes, they are clearly one of the most successful and diverse groups of animals on land. Despite their highly specialized predatory lifestyle (they only eat live prey), there are over 2900 species alive today, clustered in 29 separate families and dozens of genera. They occur from the Arctic Circle in Scandinavia to Australia in the south, and on every continent except Antarctica. They occur as high as 16,000 feet (4900 meters) in the Himalayas, and the sea snakes are fully marine predators. Many islands have no snakes (New Zealand, Ireland, Iceland, Hawaii, most of the South Pacific), but not necessarily because St. Patrick or anyone else drove them out. More likely, these islands were cut off from other areas on the mainland that did have snakes, so it was impossible for them to reach these places, even when sea level dropped at the last peak glaciation and most land mammals were able to walk or swim to distant islands. Some of these islands (such as Ireland) were almost completely under ice caps during the Pleistocene, while others (such as Hawaii and many other remote islands) were just too remote for snakes to reach from any continent.

Although some snakes have good eyesight, the majority can only see a blurry vision of their surroundings, and tend to be best at tracking movement; a few are blind. Instead of eyesight, most snakes use their forked tongues to "taste" the smells in the air, using the Jacobsen's organ on the roof of the mouth that tastes the scents brought in by the tongue. In addition, many snakes have heat-sensing pits on their snouts that allow them to detect the presence of warm-blooded animals (both predator threats and their own prey). Snakes have lost their external ears, but instead most of them hear with their lower jaws. That's one of the reasons "snake charming" is bunk, largely aimed at fleecing the tourists. The snake needs its jaw on the ground to feel the vibrations, so when it rears up, it is responding to the movements of the "snake charmer" and cannot hear his flute at all.

Behind the skull are almost 200–400 vertebrae, adding more as they grow. In contrast, you have only 33 vertebrae and never add any, and most animals with long tails have about 50. The attached ribs make up nearly the entire body of the snake. The ribs are covered by a criss-crossing truss-work of muscles that allow the snake to control its movement, as well as propel it along with a variety of sinuous motions. The body mostly consists of a very elongated trunk region (ribcage) and short tail. Inside this long body is a single right lung, with the left lung highly reduced due to the limited space in their narrow bodies. All their other paired organs, such as the kidneys or gonads, are also staggered with one part ahead of the other to fit in their narrow bodies. The most primitive snakes (especially the boas and their relatives) still retain tiny vestiges of their hip-bones and thighbones, which no longer function as limbs but do still serve in courtship and sexual combat. They demonstrate their ancestry in four-legged animals.

Snakes show an enormous range of size for such a restrictive body plan. The smallest are the Barbados threadsnake, only about 10 cm (4 inches) in length, which could curl up easily on a dime. Most snakes are about 1 meter (3.3 feet) long, big enough to subdue their normal prey of rodents and other small mammals and birds (and occasionally other snakes). At the other extreme is the reticulated python and anaconda, two huge boa constrictors. The anaconda is a specialized swimmer who drags its prey underwater as it crushes the air out of it. They can reach 6.6 meters (21 feet) in length, and up to 70 kg (154 lb) in weight. The reticulated python is not as heavy, but can be a bit longer, reaching 7.4 meters (24 feet). Both of these snakes are so large that they can swallow large prey, such as goats, sheep, small cattle, capybaras, and other larger animals. But these are nothing compared to the giants of the past.

The recently discovered *Titanoboa* from the Paleocene (60–58 Ma) deposits of Colombia shatters the records held by living snakes like the anaconda (**Figure 10.5**). Now known from hundreds of vertebrae, and parts of the skull, the size of these bones is so enormous that the entire snake is estimated to have reached about 15 meters (50 feet) and weighed about 1135 kg (2500 lb), as long as a school bus. *Titanoboa* lived in a time just after the giant dinosaurs had just died out 5 million years earlier. In the tropical swamps of Colombia, it lived side-by-side with gigantic crocodilians and turtles as well as other huge reptiles. Their gigantic size was

Figure 10.5 Reconstruction of the gigantic Paleocene snake *Titanoboa*, which was the length of a school bus. It could eat large reptiles like this crocodilian. (Courtesy K. Beck.)

probably due to the lack of large mammalian predators (yet to evolve) or large dinosaurs (extinct). In their absence, the niche for giant predator was occupied by reptiles such as snakes, crocodiles, and turtles.

Titanoboa broke the previous record held by *Gigantophis*, a monster snake in the ancient extinct Gondwana family Madtsoiidae, from the Eocene (40 Ma) beds of Egypt and Algeria. *Gigantophis* reached 10.7 meters (35 feet) in length, still much larger than the largest anaconda or reticulated python. Another huge snake of the family Madtsoiidae was *Wonambi*, an Ice Age snake from Australia. It reached 6 meters (20 feet), one of the largest reptiles and largest predators that Australia has ever seen. Its head, however, was small, so it could not have eaten the rhino-sized wombat relatives called diprotodonts or the gigantic kangaroos of the Ice Age in Australia, but most other game was within reach. It died out about 40–50,000 years ago, along with the bulk of the Australian "megafauna" of gigantic marsupial mammals.

Snakes are a marvel of adaptation and success, and have been so ever since the dinosaurs vanished from the planet. But where did they come from? How do we turn some other reptile into a snake? Where are the transitional fossils that demonstrate this process?

Actually, becoming legless is the simplest part of the whole process. It has already happened in many different groups of four-legged animals, all independently evolved. The examples of leglessness include not only the snakes, but also an entire group of living reptiles called the amphisbaenians mentioned earlier, as well as several different groups of legless lizards, including some skinks, the Australian flap-footed lizards, "slow worms", "glass lizards", and several others (also mentioned earlier). Among amphibians, an entire group (the caecilians or apodans) developed worm-like bodies, plus a group called the sirens that have only stunted front limbs and no hind limbs. In addition, there are at least two extinct groups of amphibians, the aistopods and lysorophids, which became limbless as well (Chapter 7). Nearly every one of these examples are burrowing animals, so the loss of limbs appears to aid in digging through the ground or soft mud. There's a simple reason why losing the limbs is so easy. The development of the limb buds and eventually the limbs is controlled by a specific set of Hox genes and Tbx genes, so all it takes is for those genes to shut off the commands to develop limbs, and the limbs will vanish.

Nonetheless, finding a fossil snake caught in the act of losing its limb would seem to be extremely unlikely. Most snakes don't fossilize at all, since they are built of hundreds of delicate vertebrae and ribs that are usually broken and disassociated, and only a handful of snakes are known from partial or complete articulated skeletons. The vast majority of fossil snakes are known only from a few vertebrae, so the diagnostic characteristics of these creatures must come from little details of the spinal column.

Despite all these obstacles, the geologic record has produced a remarkable set of fossils that document the transition from four-legged lizards to legless snakes. The first stage is represented by a fossil known as *Adriosaurus microbrachis* (**Figure 10.6[A]**) found in 2007 in rocks from Slovenia dating to the middle Cretaceous (about 95 Ma). Its name means "Adriatic lizard with small arms". *Adriosaurus* is an extremely slender, long-bodied marine lizard that had fully functional forelimbs but vestigial, non-functional hind limbs.

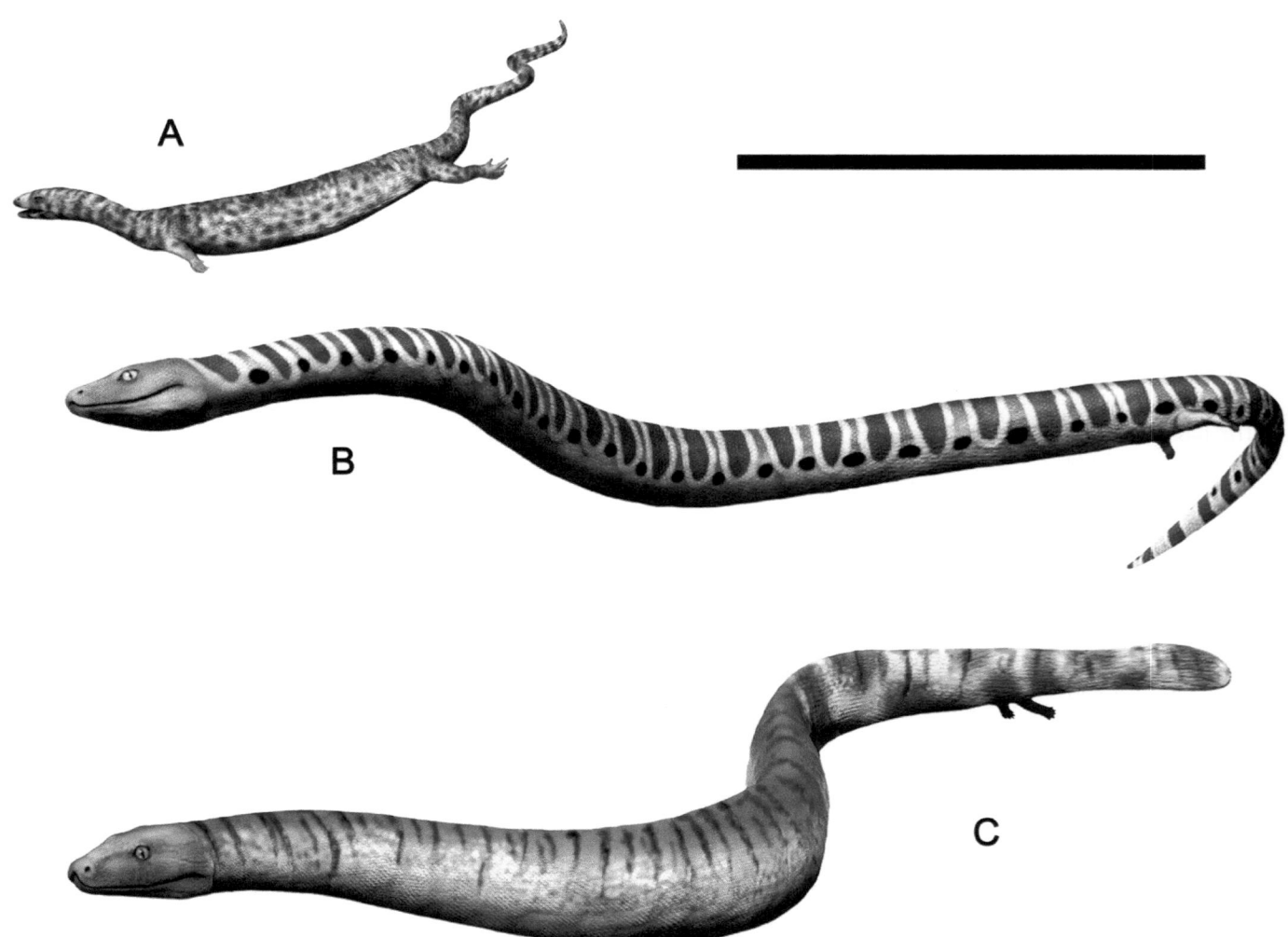

Figure 10.6 Reconstructions of a variety of extinct snakes: (A) *Adriosaurus*, (B) *Haasiophis*, (C) *Eupodophis*. Scale bar is 20 cm.

Next comes a wide variety of fossil snakes that have lost their forelimbs, but still have their tiny functionless hind limbs. For example, the fossil *Najash rionegrina* is a burrowing land snake described in 2006 from the Candeleros Formation in Argentina and dating to about 90 Ma. ("Nahash" is an old Biblical Hebrew name for the Serpent in the Garden of Eden.) *Najash* still has the pelvic bones, the vertebrae that attach to the pelvis, and vestigial hind limbs that still retain the thigh bone, shin bone, and tibia. In 2015, another four-legged snake, *Tetrapodophis*, was described, showing even more steps in the evolution of snakes.

Even more specialized and snake-like are a series of extraordinary fossil snakes from the Upper Cretaceous marine rocks of the Middle East (Lebanon and Israel). The most complete of these fossils is *Haasiophis terrasanctus*. Its name means "Haas's snake from the Holy Land", named after the Austrian paleontologist Georg Haas, who found the locality and was working on the fossil before he died in 1981 (**Figure 10.6[B]**). *Haasiophis* was found in the marine limestones of the Ein Yabrud locality in the Judean Hills of Palestine, near Ramallah on the West Bank, and is about 94 million years old. It is a nearly complete skeleton, missing only the tip of its tail, and is about 88 cm long. The skull and most of the vertebrae look much like the other primitive snakes. But the hind limbs are still present and very tiny, including the thigh bone, both shin bones, and part of the feet. Unlike the hind limbs of *Najash*, the hip bones of *Haasiophis* are tiny and are no longer attached to the spinal column, so they are completely vestigial and useless. *Haasiophis* and many other of these Cretaceous marine snakes apparently had a vertical fin or paddle-shaped tail, much as living sea snakes do.

A slightly larger snake is *Pachyrhachis* from the same Ein Yabrud locality in Israel, described by Georg Haas in 1979. Although its fossils are less complete than those of *Haasiophis*, it also has tiny vestigial hind limbs on its 1 meter (3.3 foot) long body. It also has very thick dense bones in its ribs and vertebrae, which would help it in diving in the Cretaceous seas. A third legged snake from the marine limestones of the Middle East is *Eupodophis descouensi* (**Figures 10.6[C]** and **10.7**), which was found in rocks about 92 million years old from Lebanon (not far from Ein Yabrud). The name *Eupodophis* means "good limbed snake" and its species was named after the French paleontologist Didier Descouens. It was 85 cm (34 inches) long, about the same size as *Haasiophis*, but its limbs are even more reduced and tiny than the other two Cretaceous two-legged snakes, *Haasiophis* and *Pachyrhachis*.

Thus, there were several marine snakes from the Late Cretaceous that only had vestigial hind limbs. As we mentioned earlier, very primitive living snakes like the boas and their relatives still have vestigial hip bones and thigh bones, sometimes with tiny "spurs" projecting from their bodies—mute but powerful testimony that snakes originated from creatures with legs.

But where did snakes originate? The earliest ideas were proposed in the 1880s by the pioneering paleontologist and herpetologist Edward Drinker Cope, who noticed that snakes have many anatomical similarities to the monitor lizards, such as the goannas of Australia and the Komodo dragon (and even more similarities to the Cretaceous marine lizards known as mosasaurs). The anatomical evidence still seems to support this idea, although recent evidence does not put snakes closest to monitor lizards, but to mosasaurs.

The idea that snakes got their leglessness by becoming marine swimmers seems to be supported by the many marine snake fossils from the Cretaceous of the eastern Mediterranean (Israel, Lebanon, Slovenia). Under this scenario, the loss of external ears and the fused transparent eyelids of snakes make sense as marine adaptations, rather than adaptations for burrowing.

Figure 10.7 Numerous fossils of Cretaceous snakes still have tiny vestigial hind limbs. (A) Complete skeleton of *Eupodophis*, with the vestigial leg bones near the center of the photo. (B) Detail of the vestigial hind limbs in *Eupodophis*. (Courtesy Wikimedia Commons.)

A

B

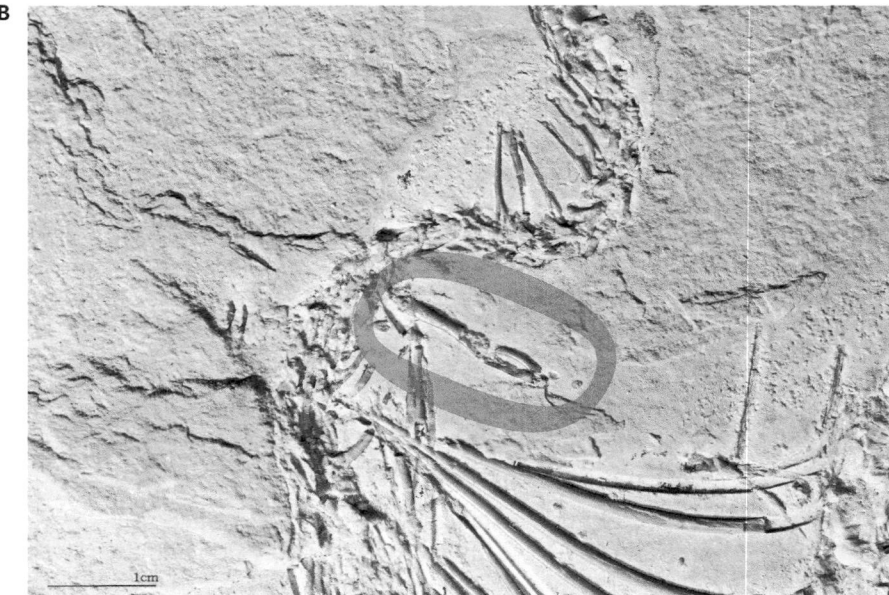

Another school of thought argues that snakes evolved from burrowing lizards, not swimming lizards, like the earless burrowing monitor *Lanthanotus* that lives in Borneo today. To these scientists, the clear eyelids would protect the eyes against the abrasion of grit while burrowing, and the lack of external ears is also good for keeping dirt out of the ear region. The terrestrial adaptations of *Najash* would be consistent with this idea, although it is slightly later in time than the earliest snakes such as *Haasiophis*, *Pachyrhachis*, and *Eupodophis*. The most primitive of all known snakes, however, is *Coniophis*, which had the head of a lizard but a body like a snake, although the fossil is too incomplete to determine what limbs it might have had. Nevertheless, it was terrestrial, not marine. Yet the aquatic lizard *Adriosaurus* is an even more primitive snake relative, and it had four limbs and swam in the ocean.

MOSASAURS

One of the first fossils described by Baron Georges Cuvier, the founder of vertebrate paleontology and comparative anatomy, was an enormous skull found around 1770 in a mine in the Cretaceous marine limestones near Maastricht, Holland. It was confiscated by the French army (for a reward of 600 bottles of wine since the Dutch had hidden it), and brought to Paris in 1794 after they conquered the Low Countries. It was called the "monster of Maastricht", but in 1808 Cuvier realized it was not a crocodile or a whale (as some suggested), but thought it was a giant marine lizard closely related to the monitor lizards. In 1822, William Daniel Conybeare named it *Mosasaurus*, or the "lizard of the Meuse River", which flowed near the site of the discovery. This was the first specimen ever found of the group known as the mosasaurs, and they looked essentially like gigantic Komodo dragons adapted for swimming—and Cuvier was right. Mosasaurs are indeed members of the family Varanidae, or monitor lizards, which includes not only the Komodo dragon but also all the Australian goannas. Although not as highly specialized as plesiosaurs and ichthyosaurs, mosasaurs were completely aquatic, with long bodies, hands and feet fully developed into flippers, and a vertical fin on the tail. But they still have the characteristic skulls of squamates, with the streptostylic quadrate bone that enhanced their gape, and a hinge in the middle of the lower jaw that allowed the tip of the jaw to flex up and down relative to the rest of the head.

Their roots can be traced to reptiles known as the aigialosaurs (**Figure 10.8[A]**) such as *Aigialosaurus* and *Opetiosaurus* from the middle Cretaceous of the Adriatic region. The aigialosaurs were semiaquatic lizards about a meter long, perfectly intermediate between mosasaurs and their varanid ancestors. The aigialosaurs have at least 42 characteristics that make them more advanced than varanids; most of these are concentrated in the skull region and the semiaquatic limbs—but otherwise aigialosaurs retain the primitive varanid skeleton. There are another 33 anatomical transformations between aigialosaurs and the most primitive true mosasaurs, most of which involve developing flippers, extending the body, and developing a vertical fin on the tip of the tail. The most primitive of these was the recently described *Dallasaurus* from the earliest Late Cretaceous of Texas, which was also only a meter or so long, but had definite mosasaur features.

From creatures like these, mosasaurs underwent an explosive radiation in the Late Cretaceous into at least 27 genera grouped into at least 6 subfamilies (**Figures 10.8** and **10.9**). They took over the role as the dominant large predator in Late Cretaceous seas, as ichthyosaurs were vanishing and plesiosaurs were less diverse than they were back in the Jurassic and Early Cretaceous. Mosasaurs were clearly very active swimmers, but with their elongate bodies adapted for undulation as they swam (somewhat like eels), they probably were ambush predators that lunged at prey over a short distance. By comparison, the plesiosaurs were relatively slow paddlers who relied on their long neck or long jaws to reach prey, and the ichthyosaurs, with their bodies shaped more like tunas and dolphins, were much more steady and continuous swimmers. With their huge gape enhanced by their double-jointed lower jaws, mosasaurs could catch and swallow large prey almost as large as their heads, and they probably ate anything they could catch. One specimen of *Tylosaurus* (**Figure 10.9**) had remains of the loon-like bird *Hesperornis*, a bony fish, a shark, and another smaller mosasaur in its stomach contents. In addition, fossil ammonite shells (squid-like creatures related to the chambered

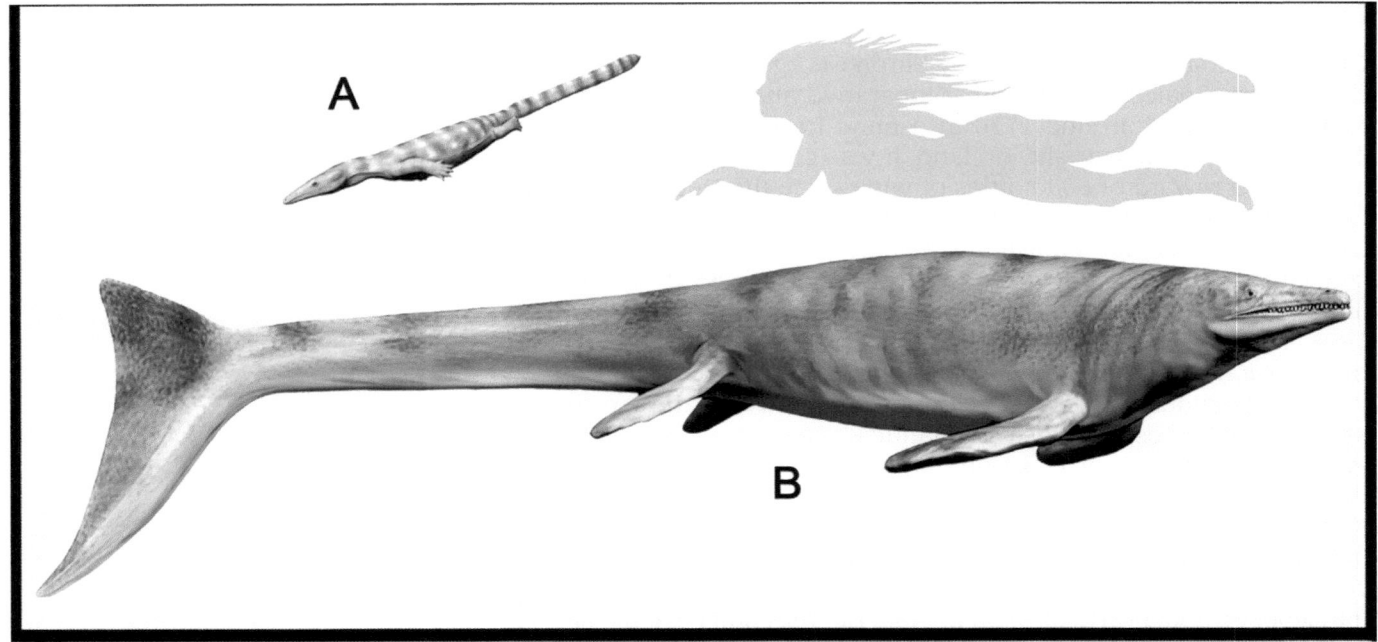

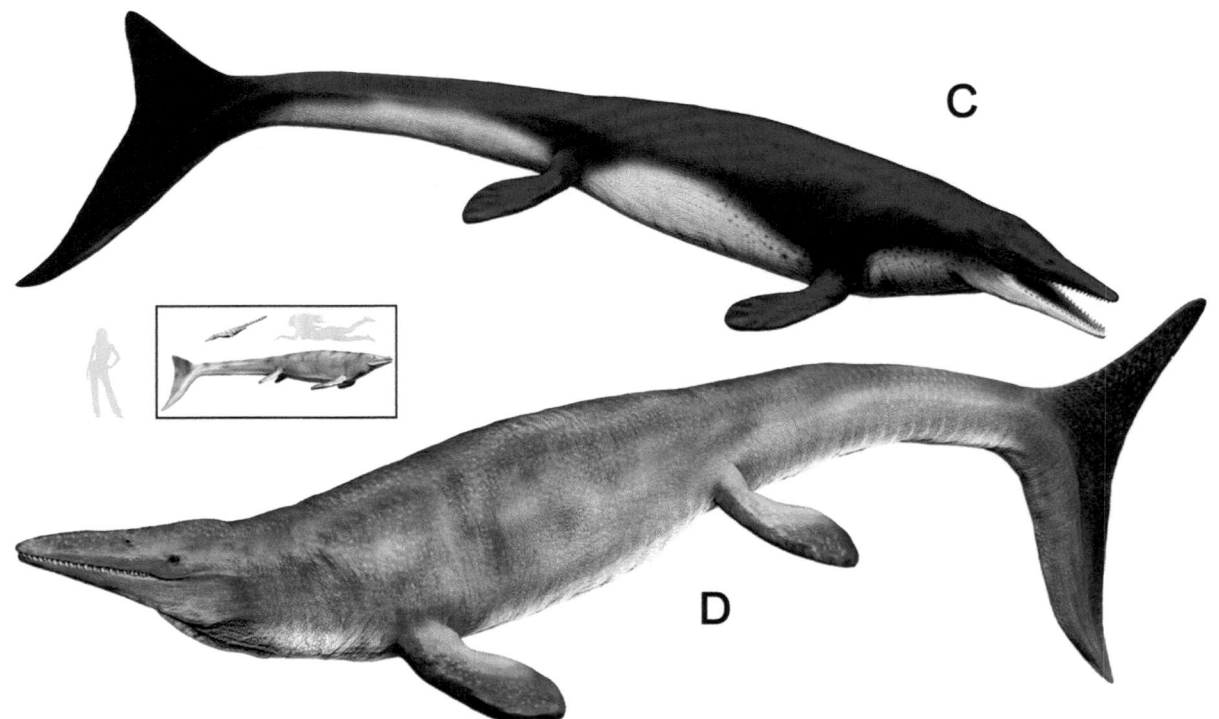

Figure 10.8 Reconstructions of mosasaurs and their relatives: (A) *Aigialosaurus*, (B) *Platecarpus*, (C) *Tylosaurus*, (D) *Mosasaurus*.

nautilus) have been found with a "V"-shaped row of puncture marks that exactly match the bite marks of a mosasaur. Other mosasaurs, like *Globidens*, had blunt rounded tooth crowns, and probably used these pavement teeth to crush mollusk shell. The biggest mosasaurs like *Tylosaurus*, *Hainosaurus*, and *Mosasaurus* (**Figures 10.8** and **10.9**) reached up to 15 meters (50 feet) in length, which made them the biggest predator in the seas.

However, mosasaurs were not as big as the movies portray them. In the first two *Jurassic World* movies, an enormous whale-sized *Mosasaurus*

Figure 10.9 Skeleton of the large mosasaur *Tylosaurus*, from the Cretaceous chalk beds of Kansas. (Courtesy Wikimedia Commons.)

is shown, which is least 36 meters (120 feet) long, more than twice as large as the largest known mosasaur fossils of *Tylosaurus* or *Mosasaurus*. Instead of this unnecessary mistake, the filmmakers could have shown true giants that were nearly this large, like the enormous pliosaur *Kronosaurus* (**Figure 9.4[A]**). There is another glaring error in the film as well. When *Mosasaurus* opens its mouth to gulp down a great white shark (an endangered animal and a huge waste of rare marine life), you see a thick tongue like in some iguanian lizards. But the *Mosasaurus* was a varanoid and closely related to Komodo dragons and snakes, so it would have had a forked tongue; the holes in the palate confirm this. In addition, the coloration is wrong. Recent discoveries show that mosasaurs were countershaded, dark on the top (so they are not visible from above) and light on the bottom (also to help conceal them from below), like many sharks and other fish. Their scales were thought to have a rough texture so they were not very reflective, an important feature to keep them from alerting potential prey before they lunged in ambush.

Mosasaurs were found in marine rocks on every continent, including Antarctica, and dominated the Late Cretaceous seas. They were the last of the great marine reptiles of the Mesozoic seas, and vanished at the end of the Cretaceous along with the rest of the victims of Cretaceous extinctions. Since that extinction event started with the plankton in the oceans, wiped out the ammonites and many other marine invertebrates and probably many of the fish groups, it is no surprise that the top predators, the mosasaurs, also vanished.

FURTHER READING

Benton, M.J. 1985. Classification and phylogeny of diapsid reptiles. *Zoological Journal of the Linnean Society (London)*. 84: 97–164.

Benton, M.J., ed. 1988. *The Phylogeny and Classification of the Tetrapods, vol. 1: Amphibians, Reptiles, Birds.* Oxford Clarendon Press, Oxford.

Estes, R.; Pregill, G. K., eds. 1988. *The Phylogenetic Relationships of the Lizard Families: Essays Commemorating Charles L. Camp.* Stanford University Press, Palo Alto, CA.

Evans, S.E. 1988. The early history and relationships of the Diapsida, pp. 221–260. In Benton, M.J., ed. *The Phylogeny and Classification of the Tetrapods, vol. 1: Amphibians, Reptiles, Birds.* Clarendon Press, Oxford.

Evans, S.E. 2003. At the feet of the dinosaurs: The early history and radiation of lizards". *Biological Reviews.* 78 (4): 513–551.

Hutchinson, M.N.; Skinner, A.; Lee, M.S.Y. 2012. Tikiguania and the antiquity of squamate reptiles (lizards and snakes). *Biology Letters.* 8 (4): 665–669.

Jones, M.E.H.; Anderson, C.L.; Hipsley, C.A.; Müller, J.; Evans, S.E.; Schoch, R.R. 2013. Integration of molecules and new fossils supports a Triassic origin for Lepidosauria (lizards, snakes, and tuatara). *BMC Evolutionary Biology.* 13: 208.

Müller, J. 2004. The relationships among diapsid reptiles and the influence of taxon selection, pp. 379–408. In Arratia, G.; Wilson, M. V. H.; Cloutier, R., eds. *Recent Advances in the Origin and Early Radiation of Vertebrates.* Verlag Dr. Friedrich Pfeil, München, Germany.

Pyron, R.A.; Burbrink, F.T.; Wiens, J.J. 2013. A phylogeny and revised classification of Squamata, including 4,161 species of lizards and snakes. *BMC Evolutionary Biology.* 13: 93.

Reeder, T.W.; Townsend, T.M.; Mulcahy, D.G.; Noonan, B.P.; Wood, P.L.; Sites, J.W.; Wiens, J.J. 2015. Integrated analyses resolve conflicts over squamate reptile phylogeny and reveal unexpected placements for fossil taxa. *PLoS ONE.* 10 (3): e0118199.

Zheng, Y.; Wiens, J.J. 2016. Combining phylogenomic and supermatrix approaches, and a time-calibrated phylogeny for squamate reptiles (lizards and snakes) based on 52 genes and 4162 species. *Molecular Phylogenetics and Evolution.* 94 (Part B): 537–547.

RULING REPTILES
ARCHOSAURS

11

ARCHOSAURIA—This great order of Archosaurs corresponds with the Monimostylica of Müller. . . . The important feature which characterized the order [is] close sutural attachment of the quadrate bone. . . . The order embraces that large series of forms which seem to be equidistant between all the extremes of the reptilian type. It is therefore not a strictly homogeneous group, yet its subdivisions did not appear, with present knowledge, to be sufficiently marked, to render it proper to esteem them of equal value with the other orders here enumerated.
  —Edward Drinker Cope, 1869, *Synopsis of the Extinct Batrachia, Reptilia, and Aves of North America*, p. 30

ARCHOSAURIA

Dinosaurs, pterosaurs, birds, and crocodiles. Most people don't associate these groups of animals in their minds, but indeed they are all closely related, members of a group called the Archosauria, or "ruling reptiles". They have a wide range of evolutionary novelties in their anatomy that diagnose the group. They arose in the Early Triassic, and soon dominated the land realm with a wide spectrum of strange archosaurs long known by the obsolete wastebasket name "thecodonts". By the Late Triassic, three main branches of the archosaurs had replaced their primitive relatives: the dinosaurs, pterosaurs, and crocodylomorphs. These animals then dominated the Jurassic and Cretaceous landscape, ruling the earth with some creatures that reached immense sizes (not only gigantic dinosaurs but also huge crocodilians and pterosaurs as big as a small airplane). Meanwhile by the Late Jurassic one group of carnivorous dinosaurs gave rise to the birds, which also flourished. In fact, all active flying vertebrates except for bats are archosaurs (either pterosaurs or birds). The non-bird dinosaurs and pterosaurs vanished at the end of the Cretaceous, but the crocodilians have been among the dominant aquatic predators for the past 66 million years—and the birds have flourished in the skies for over 150 million years. Together, the birds and crocodilians still count over 10,000 living species, so the group has not vanished just because the Mesozoic dinosaurs and pterosaurs are gone. Along with the Lepidosauria, the Archosauria make up the two main groups of reptiles, and all living reptiles except the turtles and tortoises are members of one of these two groups (**Figure 11.1**). (And some molecular data suggest turtles are in this group as well, but this is controversial.)

So what features diagnose archosaurs (**Figure 11.2**)? In general, they tend to have more upright, less sprawling posture and a more active lifestyle. Dinosaurs and birds have completely upright postures, with their legs directly below their hips, allowing for more efficient locomotion and faster speed. Their limbs are partly or completely under the axis of the body and move in a fore-and-aft plane, and they have a stiffened spine so that they run more efficiently without the side-to-side wiggles of lizards

DOI: 10.1201/9781003128205-11

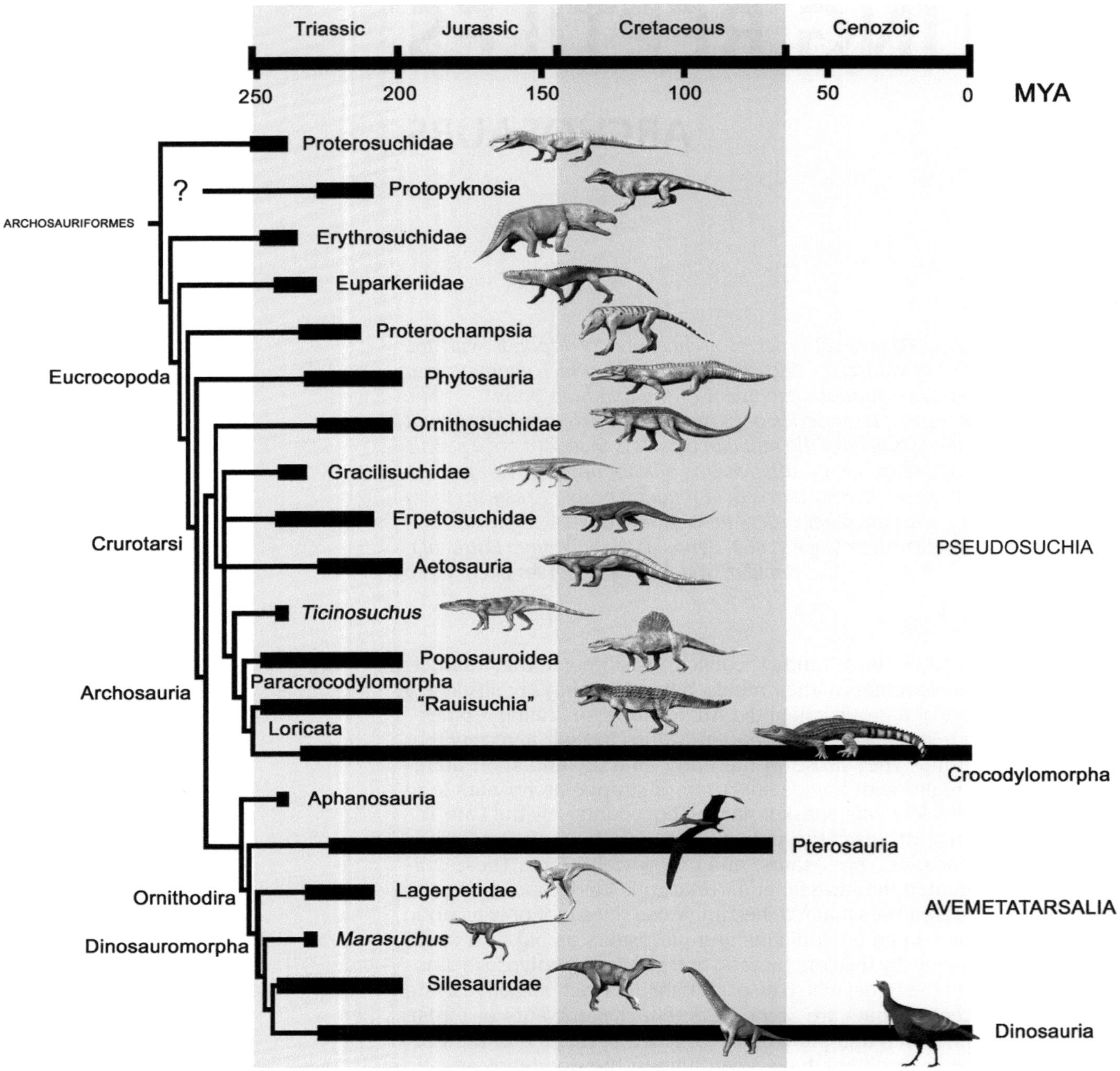

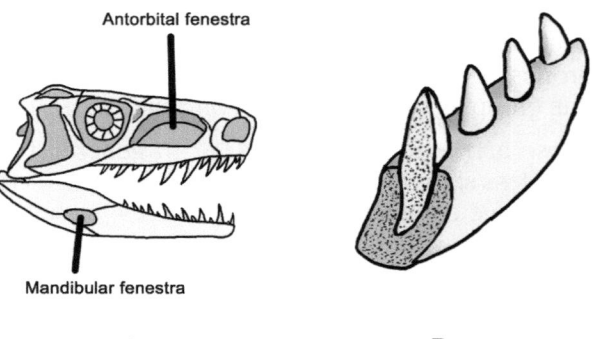

Figure 11.1 Phylogeny of the major groups of archosaurs.

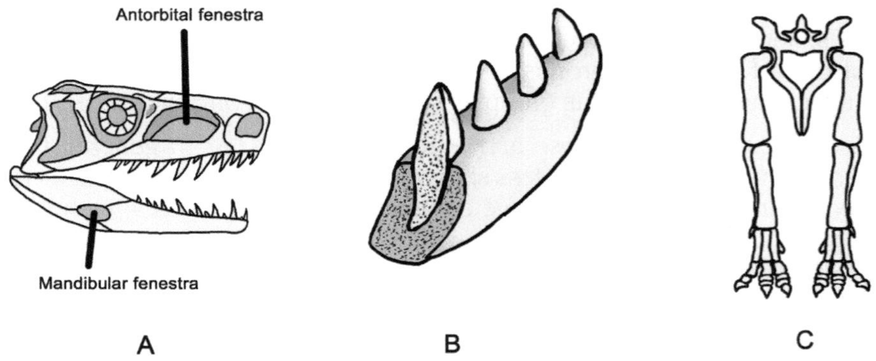

Figure 11.2 Some of the key features that define the archosaurs. (A) They have two distinct holes, or fenestrae, in their skull and jaws. One was an antorbital fenestra in front of the eye socket (orbit) and a mandibular fenestra, or hole in the jaw (mandible). (B) They also have teeth set in sockets ("thecodont teeth"), unlike lepidosaurs, whose teeth are set in a groove on the inside edge of the jaw. (C) Most have upright or semi-upright posture, with their legs directly below their hip bones, not sprawling.

Antorbital fenestra

Mandibular fenestra

A   B   C

or snakes. Most early archosaurs were relatively upright in their posture as well. Although crocodilians spend a lot of time sprawling, when they need to run, they are upright and do not drag their bellies on the ground.

Archosaurs have an additional ridge on the thighbone called the fourth trochanter, which provides an additional anchor point for the muscles that pull the thigh backwards. Thanks to these leg muscles, most archosaurs have moved away from the completely sprawling posture of lizards, who rest on their bellies when not running, to a semi-upright posture seen in crocodilians, or the fully upright posture of dinosaurs. There are other differences in the bones as well, but these are the ones that are easiest to see on the skeleton. There are also many unique specializations in the foot, particularly in the upper rows of ankle bones.

Archosaurs have relatively large lungs that are pumped by a diaphragm muscle, rather than the inefficient rib pumping found in most other amniotes. Archosaurs have completely divided ventricles in the heart, a pulmonary artery with three semilunar valves, and muscular lateral valves to the right of the auriculo-ventricular orifice. They are also unique in having a muscular gizzard filled with rocks or sand for grinding up their food (in the absence of grinding teeth and chewing abilities). If you look at almost any archosaur skull (**Figure 11.2**), you will find an opening before the eye socket called the antorbital ("ante" for "in front", so the name means "in front of the orbit") fenestra, and in the lower jaw, or mandible, there is an additional hole called the mandibular fenestra. In many archosaurs, these extra holes in the skull serve as points for muscle attachment or allow jaw muscles to bulge, or may reduce the bony weight of the skull. In addition, in some of the more primitive archosaurs, the eye opening (orbit) is shaped like an inverted triangle (pointed down).

Archosaur teeth are inserted in sockets (**Figure 11.2**) along the top edge of the jaw (this condition is known as "thecodont teeth"), in contrast to the pleurodont dentition of lepidosaurs, where the teeth insert in a trough on the inside of the jaw. Until recently, all the early members of the Archosauria that were not dinosaurs, pterosaurs, or birds, were lumped into a huge taxonomic wastebasket, the "Thecodontia", based on their thecodont tooth configuration. But in modern classification, all groups must include their descendants, so just as you can't define Reptilia without the birds as their subgroup, likewise you can't create a wastebasket group for all archosaurs that are not members of the advanced groups, crocodilians, dinosaurs, birds, or pterosaurs. Thus, no serious paleontologist uses the obsolete and misleading term "thecodonts" any more, although it still appears in books that copy from older outdated books.

ARCHOSAUROMORPHS

The formal group, known as the Archosauria, includes two main branches (**Figure 11.1**): the crocodilians and their extinct relatives (called the Pseudosuchia, or also called the Crurotarsi), and the pterosaurs, dinosaurs, birds and their relatives (called the Ornithodira or Avemetatarsalia). Early archosauromorphs, however, are extinct primitive relatives of the archosaurs that belong to neither of these main archosaur branches, but are just extinct primitive side branches that do not yet have all the anatomical features that diagnose an archosaur (**Figure 11.1**).

Some of these earliest, most primitive known archosauromorphs include *Protorosaurus*, known from the Late Permian of Europe, and persisting into the Triassic. *Protorosaurus* itself is a unspecialized lizard-like creature (**Figure 11.3[D]** from the Late Permian of Germany. Up to 2 meters (6.6 feet) long, it had a long neck, long tail, and long legs, suggesting that it was a fast-moving predator which could snag almost any smaller prey with its long neck and sharp teeth. Curiously, however, two specimens

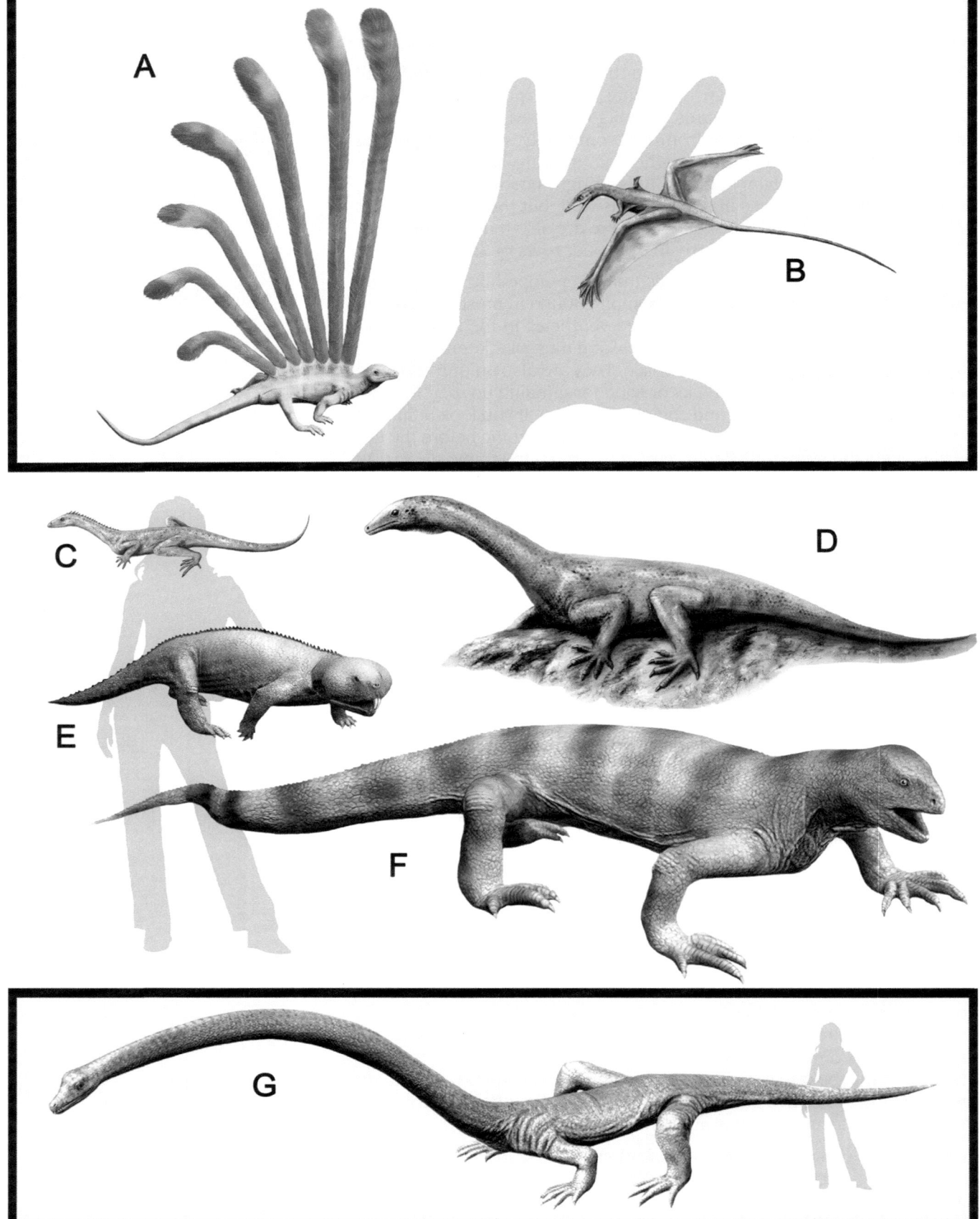

Figure 11.3 Reconstructions of some early archosauromorphs: (A) *Longisquama*, (B) *Sharovipteryx*, (C) *Macrocnemus*, (D) *Protorosaurus*, (E) *Hyperodapedon*, (F) *Trilophosaurus*, (G) *Tanystropheus*.

with gut contents preserved showed that it ate ferns and conifers. A similar creature was the long-bodied *Macrocnemus* from the Triassic of Italy, Switzerland, and China (**Figure 11.3[C]**). Almost a meter long, *Macrocnemus* also had a long neck and long legs, but even a longer tail than *Protorosaurus*. The long limbs and tail suggest a very fast-moving predator, but it is speculated that it could even run bipedally for short distances. Since some specimens are found in marine rocks, *Macrocnemus* may have been aquatic as well. About a dozen other genera of reptiles related to *Protorosaurus* are known from all over Pangea in the Triassic.

The weirdest of these primitive archosaurs was the long-necked *Tanystropheus*, which had a neck almost 4 meters long and the rest of the body less than 2 meters in length (**Figure 11.3[G]**) Its neck was composed of a 12–13 elongate vertebrae, so it was not very flexible or sinuous, but must have flexed more like a fishing rod. The specimens are known mostly from marine rocks in Europe (especially Italy, Germany, and Switzerland), although relatives are known from the Middle East, North America, and China as well. Nobody really knows what these strange animals did with their odd necks, but they apparently didn't have the mobility to snap them sideways to catch mobile prey. Their teeth are simple pegs for catching prey, not for eating immobile plants or other hard food. The latest suggestion is that they were shallow-water predators that used their long necks to stealthily get close to schools of fish or squid or other aquatic prey without generating the shock waves in the water from their body that might alert their prey. As they got near prey, a sudden lunge would give their heads a chance to snag a fish or squid within the school. However, some reconstructions have them walking on land as well.

As if *Tanystropheus* were not weird enough, even stranger-looking is the creature known as *Sharovipteryx* (**Figure 11.3[B]**). Known only from the single type specimen from the Middle Triassic of Kyrgyzstan, it was only about 25 cm (10 inches) long. Its weirdest feature is a triangle-shaped membrane stretched between its hind limbs and tail, so unlike all other gliding animals which suspend a membrane from their hands, or between front and hind limbs (like in flying squirrels, or the marsupials know as sugar gliders, or the living Dermoptera or colugos), or by extending their ribs (as in *Icarosaurus*, *Coelurosauravus*, or the living gliding lizard *Draco volans*), *Sharovipteryx* used only the back half of its body to glide. Various analyses have shown that this is a difficult arrangement for controlled gliding, unless the front limbs also had a separate membrane—and the only specimen has had the matrix around the front limbs prepared away, so it is no longer possible to know this.

Another curious creature from the same Triassic beds in Kyrgyzstan that produced *Sharovipteryx* is *Longisquama* (**Figure 11.3[A]**). Known from a number of smaller specimens about 25 cm (10 inches) long, its most peculiar feature is a fan-like row of long structures that appear to be hockey-stick-shaped spines or fins along its back, regularly spaced apart on the original specimens. Different paleontologists have debated how to reconstruct these structures, and some have suggested that they may have flared out from the side of the animal to support a gliding membrane, not arranged as a fan of 'hockey sticks" along its back.

Rhynchosauria

As weird as creatures like *Tanystropheus*, *Longisquama*, and *Sharovipteryx* were, another odd-looking group of early archosauromorphs were the rhynchosaurs, whose name means "beaked lizards" in Greek (**Figure 11.3[E]**). A diverse group of at least 17 genera found in North America, Brazil and Argentina, South Africa, England, Madagascar, Tanzania, and India, they are the only common herbivorous archosaurs of the Middle and Late Triassic in those places, often making up 40–60% of the specimens. As the common herbivore on land during the Triassic, reconstructions often show them as fodder for all the larger predators that lived at the time. Most rhynchosaurs were pig-sized and pig-like short-legged creatures with a broad triangular skull, deep cheek region, and flattened snout, capped with a hook-like upper and lower beak (hence their name). The broad skull apparently supported powerful jaw muscles, and their beaks acted as powerful scissors for slicing up plant matter. Instead of conventional reptilian peg-like teeth, they had broad tooth plates on their palate for grinding up tough vegetation. They had long barrel-shaped bodies and short legs, and their hind feet also had large claws, presumably for digging up plants and roots and tubers. At one time, they were lumped with the tuatara (see Chapter 10) in the beak-headed group called the "Rhynchocephalia", but more recent analyses show that rhynchosaurs are archosauromorphs, and the similar beaks with the tuatara are convergently evolved.

Allokotosauria

Another early archosauromorph group was the Allokotosauria, typified by the large lizard-like reptile known as *Trilophosaurus* (**Figure 11.3[F]**). *Trilophosaurus* reached up to 2.5 meters (8 feet) long, with a thick neck, long trunk, long sprawling limbs and a very long tail. The skull was heavily built, with a short bulldog-like snout and a turtle-like beak. Most distinctive of all, however, are the teeth, which were not the simple conical or peg-like teeth of most reptiles, but are broad molar-like teeth with three cross-crests on them for shearing up tough plant material. (The name *Trilophosaurus* means "three-crested lizard".) *Trilophosaurus* is known from the Late Triassic of the American Southwest, but the rest of the Allokotosauria include at least five other genera, known from the Middle and Late Triassic of Asia and Africa, as well. *Azendohsaurus*, for example from the Triassic of Morocco and Madagascar, was originally mistaken for an early sauropod dinosaur until better specimens showed that it was an allokotosaur.

One of the weirdest was *Shringasaurus* from the Middle Triassic of India (**Figure 11.4**). It was built like a bulky lizard, with short sprawling limbs, a long trunk and tail, and small boxy head with a long neck. Large individuals reached 3–4 meters (10–13 feet) in length. The mouth is full of leaf-shaped teeth, suggesting a herbivorous diet. Its size and long neck suggest that it fed on the relatively tall plants in its seasonal floodplain environment, and there were no large predators found in these beds that could challenge it once it reached adult size (although there were huge temnospondyl amphibians in the river deposits).

Figure 11.4 Reconstruction of the bizarre horned archosauromorph *Shringasaurus* from the Middle Triassic of India.

Its most distinctive feature, however, was a pair of long curving horns over the eye sockets, reminiscent of the horned dinosaurs, especially *Anchiceratops* with its short forward-curving horns—but this creature was not related to dinosaurs at all. The horns have a rough external texture, suggesting that they were covered in a sheath of keratin (the same compound that makes up horn sheaths in cattle and antelopes). The horns very widely in shape, from small juvenile horns, to a difference between the sexes, with larger more robust horns in the presumed males, and smaller more slender horns in the presumed females. The horns appear too small to be of much use in defense against larger predators, but their differences in size and shape suggest that they were sexually selected, with the males competing for females using their horns to advertise they age and maturity, and also possibly for head-to-head wrestling (as happens in many horned deer and antelopes alive today).

Proterosuchidae

So far, we have seen a great diversity of early archosauromorphs that were either herbivorous, or small-bodied predators. But among the big archosauromorphs were several groups that dominated the large-bodied predatory role before the earliest dinosaurian became huge and predatory. They are known as the Proterosuchidae. The earliest and most primitive of these was *Archosaurus* from the Late Permian of Russia, and *Proterosuchus* from the Early Triassic of South Africa (**Figure 11.5[A]**). Other members of the group were found in China and India. Proterosuchidae were slender animals about 1.5 meters (5 feet) long with body proportions like that of a crocodilian. The skull, however, was far more primitive, and had a distinctive

Figure 11.5 More advanced archosauromorphs of the Triassic: (A) The lizard-like *Proterosuchus*, (B) the huge *Erythrosuchus*, a dominant predator of the Middle Triassic.

hooked upper snout with wicked teeth protruding downward. When *Proterosuchus* appeared in the Early Triassic, they were the first large predators to evolve from the Permian survivors like *Protorosaurus* mentioned earlier.

From the early predators like the Proterosuchidae evolved the dominant predators of the Middle and Late Triassic, the Erythrosuchidae ("bloody crocodiles" in Greek) (**Figure 11.5[B]**). The group of at least seven genera found in China, Russia, and South Africa is typified by *Erythrosuchus* itself. At over 5 meters (16 feet) long, it was one of the largest predators of the entire Triassic, unsurpassed in size until predatory dinosaurs grew larger at the end of the Triassic. It had a massive crocodile-like body with four strong limbs which sprawled out slightly from its sides, much more upright than other Triassic archosaurs, but not as upright as later dinosaurs. However, the most distinctive feature of *Erythrosuchus* is its huge skull over a meter long, which is narrow and deep with powerful jaws and wicked recurved blade-like slashing teeth, completely different from all the other early archosauromorphs—but very similar to the deep powerful skulls found in the predatory dinosaurs. At its large size, it could hunt down and kill nearly every other animal found in the forests of the Middle and Late Triassic wherever it occurred.

MYSTERY REPTILES: CHORISTODERES

This menagerie of strange Permo-Triassic reptiles that were related to more advanced archosaurs has gradually been sorted out over the past 20 years, although there are often cases where their relationships are rethought and their classification reshuffled as new analyses are conducted and better specimens are found. But one group of primitive reptiles has long stood as a paleontological mystery even though there are hundreds of good specimens, including many complete skeletons in at least 12 genera. These are the choristoderes (including a group known as the champsosaurs). Superficially, the largest champsosaurs (**Figures 11.6[B,C]** and **11.7**) look much like crocodiles (especially the narrow-snouted crocodilians known as gavials or gharials), and they are often found in Jurassic and Cretaceous lake and river deposits alongside a variety of crocodilians. Unlike any animal mentioned so far in this chapter, they survived in the end-Cretaceous extinctions and lasted until the Miocene. They were clearly diapsid reptiles, but opinion has swung back and forth as to whether they were part of Lepidosauria, or part of the Archosauria, or more distant relatives of both. The latest analyses, especially the detailed bony structure of their ear regions in the skull, however, have pushed them into the Archosauromorpha, so we will discuss them here. If they are indeed primitive archosauromorphs, then there is a long gap between their closest relatives in the Early Triassic and the first choristodere fossils in the Late Jurassic, almost 60 million years of missing record.

The earliest and most primitive choristodere is known as *Cteniogenys* from the Upper Jurassic Morrison Formation and similar aged beds in England and Portugal. *Cteniogenys* was quite small (only about 25–50 cm, or 10–20 inches long), with long slender jaws with numerous conical teeth. On the basis of its size, it probably fed on insects and smaller prey, and is found mostly in the pond deposits of the Upper Jurassic formations of North America and Europe. It is extremely rare, with only about 60 specimens known out of at least 30,000 vertebrate fossils from the Morrison Formation.

In the Cretaceous, there are extremely well-preserved specimens of *Hyphalosaurus* (**Figures 11.6[A]** and **11.7**) and *Monjurosuchus* from the lake beds of the Yixian Formation in Liaoning Province, China. *Monjurosuchus* was about 40 cm long, with large eyes and a rounded snout (rather than the long narrow fish-catching snout of most choristoderes), a short neck, and reached up to 40 cm (18 inches) long, so in some ways it more closely resembled a salamander rather than a gharial. Specimens

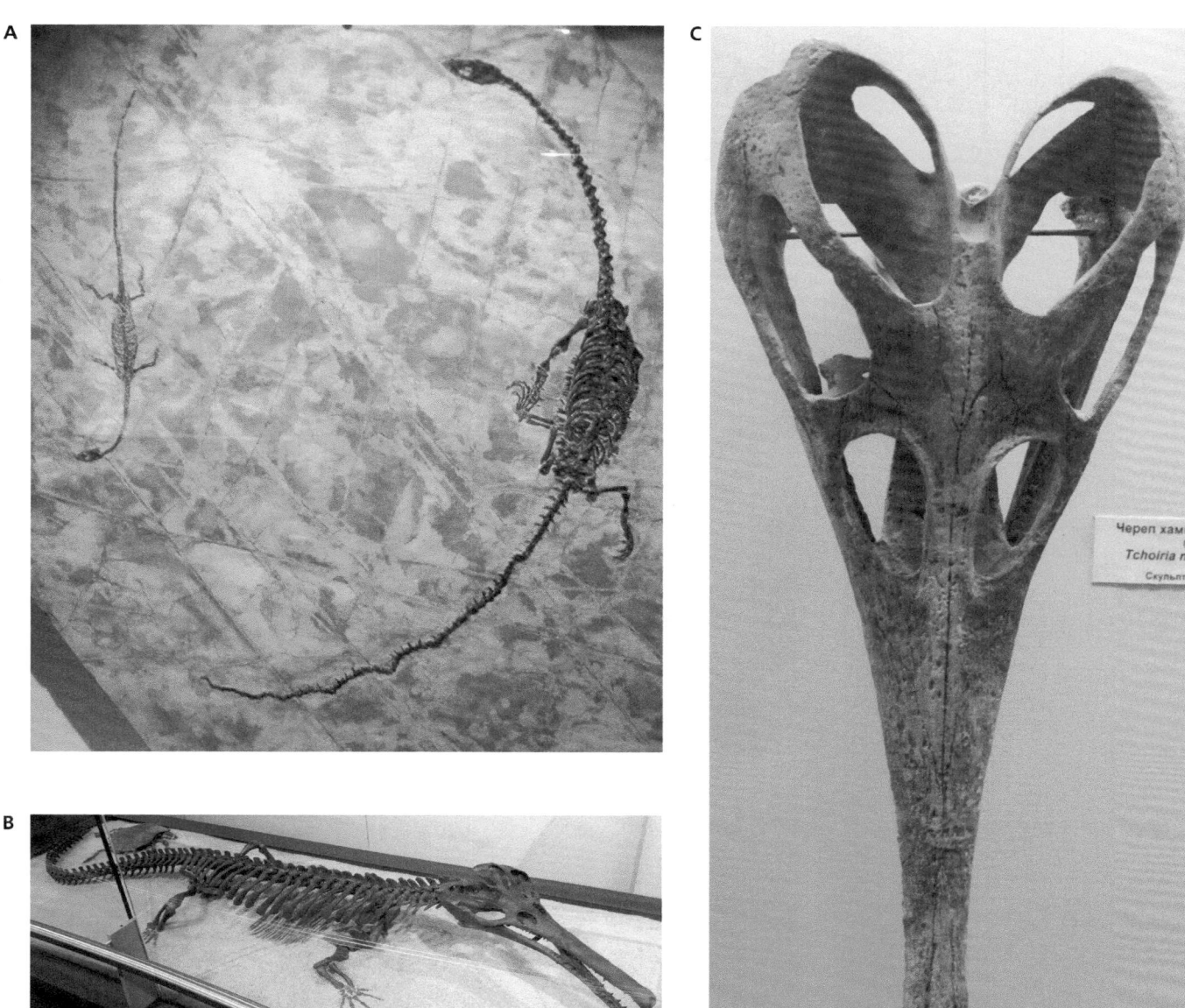

Figure 11.6 Photos of fossils of some choristoderes, including: (A) *Hyphalosaurus*, (B) *Champsosaurus*, (C) skull of *Champsosaurus*. (Courtesy Wikimedia Commons.)

with preserved skin impressions showed that it had webbed feet and soft skin. The most common tetrapod in the entire Yixian Formation, however, is *Hyphalosaurus* (**Figures 11.6[A]** and **11.7[A]**), which was slightly larger (80 cm, or 2.6 feet long), and had a short triangular skull also lacking the long gharial-type fishing snout seen in more advanced champsosaurs. The short, flattened skull and long neck suggest it caught aquatic fish and crustaceans by a rapid sideway strike, like other aquatic predators with flattened skulls do today. *Hyphalosaurus* had a barrel-shaped chest with heavy ribs for ballast (typical of aquatic animals), short limbs with webbed feet, soft skin covered with polygonal scales, and a long, compressed tail for propelling itself through the water. On the basis of the sedimentary geology, it lived in the deeper open waters of the lake, rather than the shallow swampy deposits typical of other Cretaceous formations in China. *Ikechosaurus* from the Late Cretaceous of Mongolia, by contrast, shows the long narrow snout typical of later choristoderes.

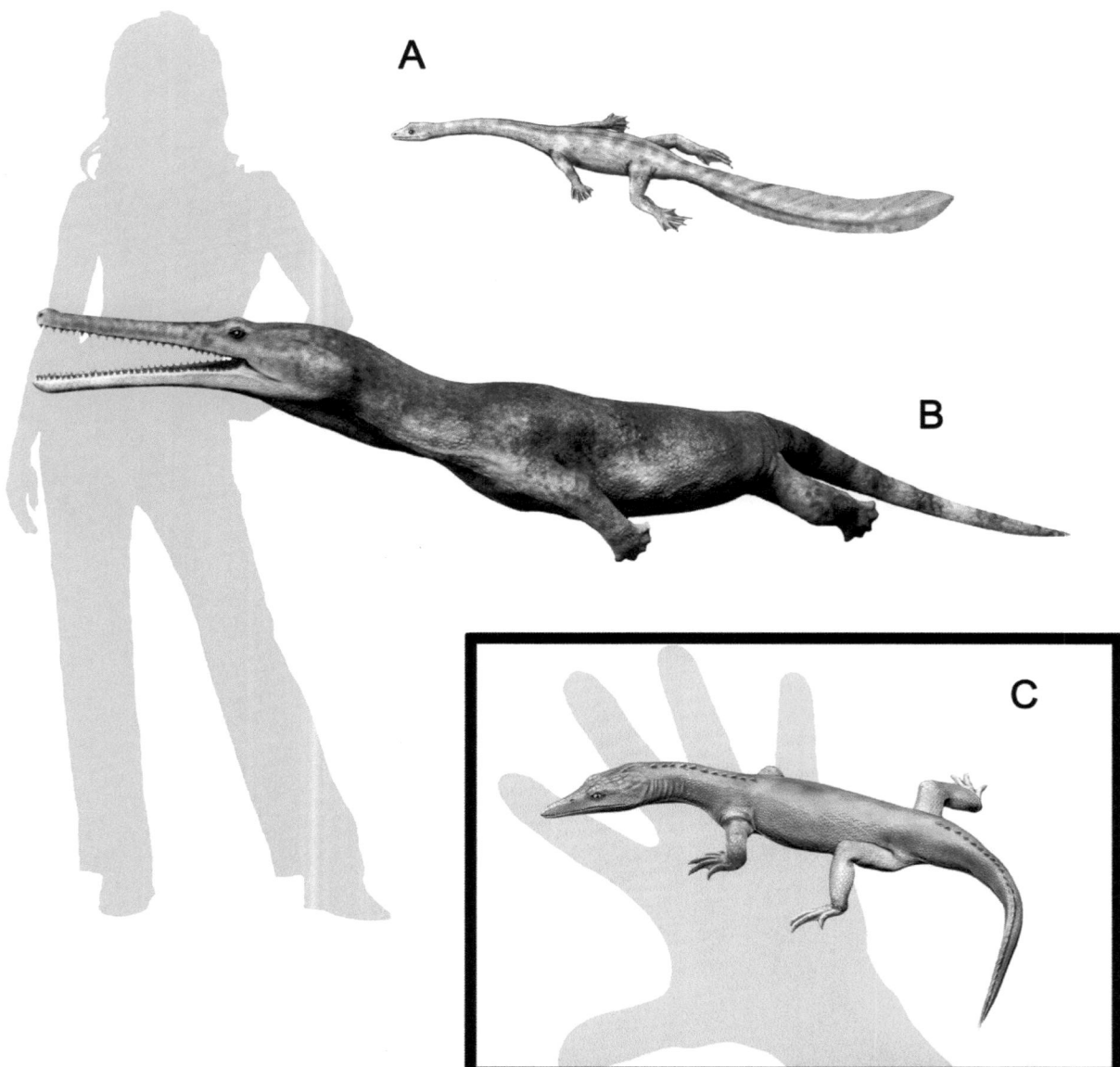

Figure 11.7 Reconstructions of some choristoderes, including: (A) *Hyphalosaurus*, (B) *Champsosaurus*, (C) *Lazarussuchus*.

The best known of the choristoderes is *Champsosaurus* (**Figures 11.6[B,C]** and **11.7[B]**), a common fossil (especially its teeth) in the Upper Cretaceous dinosaur-bearing beds of western North America, as well as Paleocene and Eocene deposits from Wyoming and Montana to New Mexico and Texas. Big specimens were up to 3.5 meters (12 feet) in length, and with bodies very much like that of a gharial or a crocodile, with a long narrow snout full of teeth (**Figure 11.6[B,C]**), a long-flattened body with short legs adapted for swimming, short stout massive ribs (common in aquatic animals) as well as belly ribs (gastralia), and a long tail flattened sideways for propulsion in water. But a closer look at the skull (**Figure 11.6[C]**) reveals that it looks nothing like a crocodilian in detail, with a broad bulbous back end, and a more open structure made of arches of bone in the rear, rather than the dense skull of true crocodilians with very few openings or arches.

Choristodere eyes were located much further forward than in the skull than in the skull of crocodilians, and their nostrils were at the very tip of the snout, not on the top of the snout as in crocodilians. The anatomical evidence of the fusion of the hip bones only in presumed female specimens suggests that they were much more aquatic than crocodilians, with the males never leaving the water, while females did so only to lay eggs in a nest. Champsosaurs also lacked the hard dermal armor of bony oste-oderms found in crocodilians, but instead had relatively soft scaly skins.

By the Cenozoic, there were several species of huge champsosaurs, yet for a long time it was thought that they had vanished during the middle Eocene. Then, collections from the Oligocene and Miocene beds of Europe produced *Lazarussuchus inexpectatus* (**Figure 11.7[C]**). Even though it was the last of the choristoderes, it is relatively primitive in its anatomy, with a much shorter more pointed snout than in seen in champsosaurs, and only about 27 cm (12 inches) long. Most recent analyses show that its closest fossil relatives are known from the Late Jurassic or Early Cretaceous, a gap from about 120 Ma to about 25 Ma, or almost 100 million years of missing fossil record. Hence it was named after Lazarus in the Bible, who was dead and placed in his tomb, but was then miraculously raised from the dead and brought back to life. The species name *inexpectatus* refers to the fact that is it surprising to see a relict of the Cretaceous Period in Oligocene and Miocene beds with no records in between. As we have seen, choristoderes are particularly noted for long gaps in their record—the gap from their closest Triassic relatives and the oldest fossils of choristoderes in the Middle-Late Jurassic, and the gap from the Early Cretaceous relatives of *Lazarussuchus* and the first appearance of these fossils in the late Oligocene. Why this is not so well understood and hotly debated, but the fact that they are restricted mostly to lake beds and only in certain regions may have contributed to their scarcity.

THE CROCODILE BRANCH: PSEUDOSUCHIA

There are two surviving branches of Archosauria (**Figure 11.1**): the branch that includes crocodilians and all their extinct relatives (sometimes called the Pseudosuchia or Crurotarsi), and the branch that includes pterosaurs, dinosaurs, and birds (sometimes called the Ornithodira, or Avemetatarsalia, depending up which scientific classification you follow). For a long time, the Crurotarsi was the preferred name for the crocodile branch, because they were defined on a distinctive crurotarsal ankle configuration (**Figure 11.8**) where the hinge of the angle runs between the first row of two ankle bones, the calcaneum and astragalus. But more recent analyses have suggested that some groups (the phytosaurs, discussed next) are more primitive than the rest of the members of the croc branch, so the croc branch is now called the Pseudosuchia, and phytosaurs are placed either as just outside the Pseudosuchia, or in other cases, as primitive archosauromorphs just out-side the Archosauria (**Figure 11.1**). These classifications continue to change as more and more fossils are found, and additional anatomical features are analyzed, so we will not discuss that debate further here.

Phytosaurs

The phytosaurs (**Figures 11.9** and **11.10**) were very distinctive group of archosaurs. Their name means "plant lizard", which is misleading since they were clearly aquatic predators and are yet another group (like champsosaurs) that looked much like crocodilians before actual crocodylomorphs evolved in the Late Triassic. They were "croc-mimics" that occupied the same ecological niche. Known from about 25 genera, they underwent an explosive evolutionary radiation, and they occurred worldwide almost anywhere in Pangea that Late Triassic terrestrial deposits are found. They are common in such places like the Petrified Forest in Arizona and Triassic beds in New Mexico and Texas, but also Greenland, France, Germany, Poland, Morocco, Zimbabwe, Madagascar, China, Thailand, Brazil, and India. Surprisingly, they are not yet found in South Africa or Argentina (but there is one possible fragment from

Figure 11.8 Diagram of the ankle configuration in archosaurs. (A) The crurotarsal ankle, typical of most of the Crurotarsi, where the hinge of the angle runs between the first row of two ankle bones, the calcaneum and astragalus, and the astragalus fused to the end of the tibia, but the joint runs between the calcaneum and the fibula. (B) The "croc reversed" ankle, where the ankle joint runs between the tibia and the astragalus, and the calcaneum is fused to the fibula. This is found in certain archosaurs, like the ornithosuchians. (C) The mesotarsal joint, where the astragalus and calcaneum fuse to the end of the tibia and fibula, and the hinge runs between them and the second row of ankle bones. This condition is found in all the Avemetarsalia, including the pterosaurs, dinosaurs, birds, and their relatives.

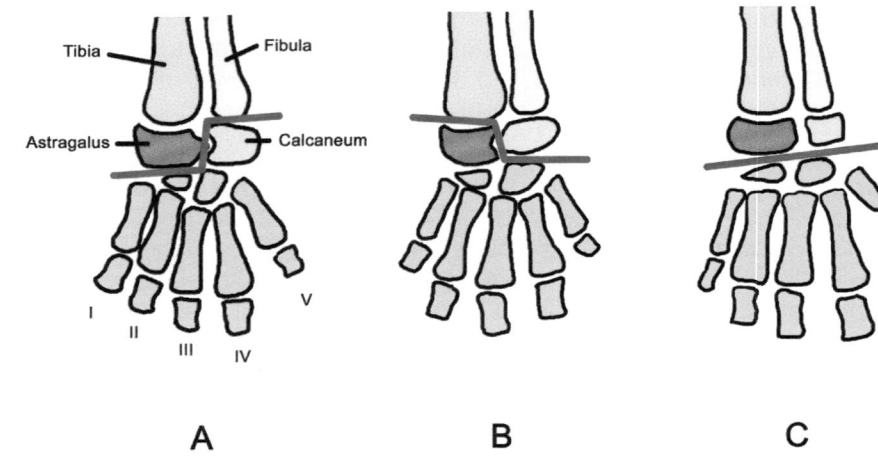

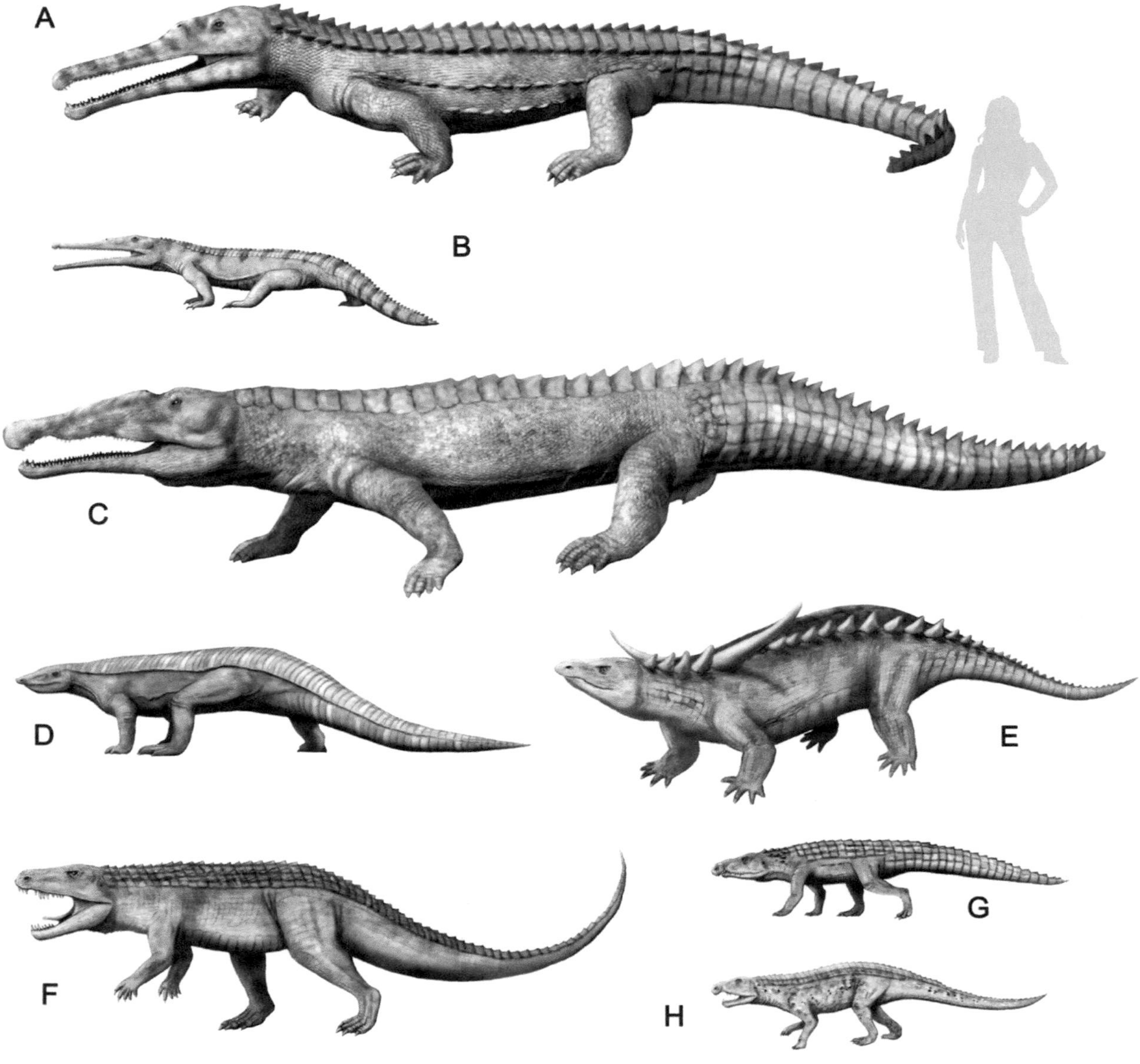

Figure 11.9 Reconstructions of a variety of crurotarsans in the crocodile branch of archosaurs. These include phytosaurs such as: (A) *Rutiodon*, (B) *Parasuchus*, (C) *Smilosuchus*; and aetosaurs such as (D) *Stagonolepis*, (E) *Desmatosuchus*; as well as (F) *Ornithosuchus*, (G) *Riojasuchus*, and (H) *Venaticosuchus*.

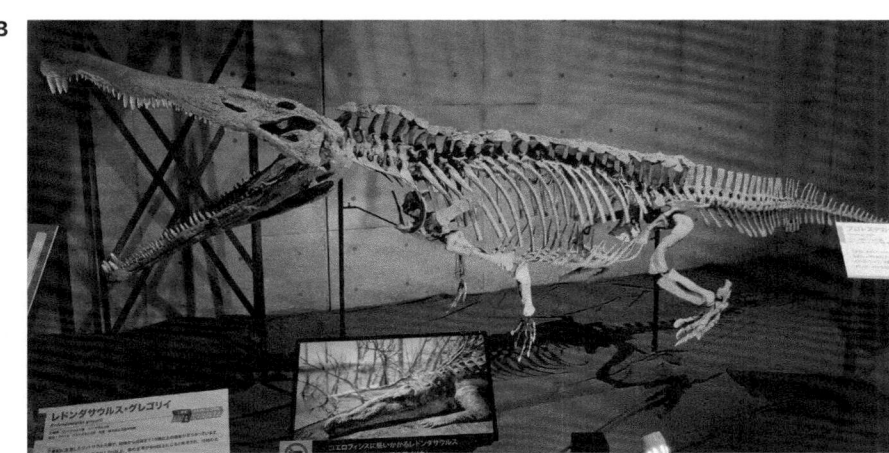

Figure 11.10 Fossils of crurotarsans:
(A) The skeleton of the phytosaur
Rutiodon. (B) A *Redondasaurus* skeleton.
(C) The skeleton of the aetosaur
Desmatosuchus. (Courtesy Wikimedia
Commons.)

Figure 11.10 (Continued) (D) The skeleton of the aetosaur *Typothorax*.

D

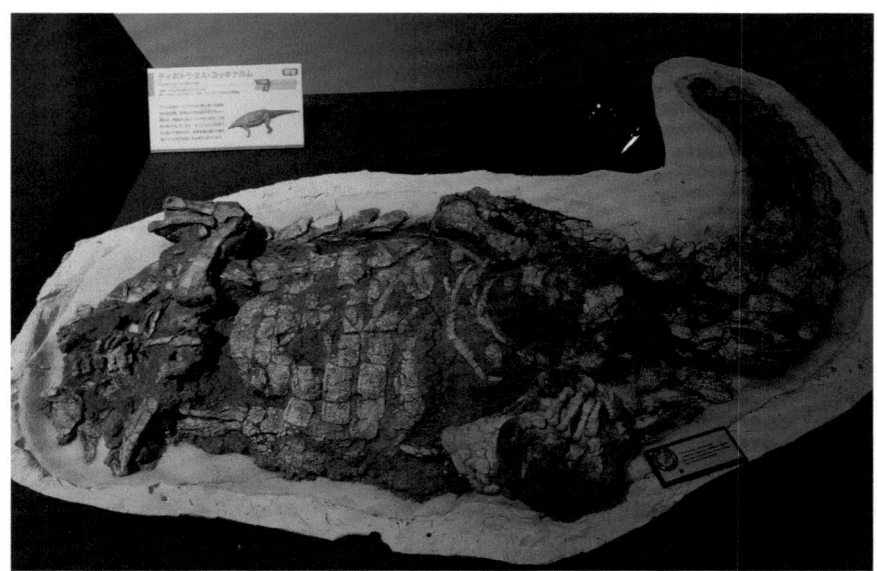

Brazil), both of which have important fossiliferous Late Triassic verte-brate-bearing sequences including some of the earliest dinosaurs.

Most phytosaurs (for example, *Rutiodon* and *Parasuchus*) (**Figures 11.9[A,B]** and **11.10[A]**) had long narrow fish-catching snouts full of elongated peg-like teeth, similar to the gharial (or gavial) that lives in Asia today. Others (like *Machaeroprosopus*, *Redondasaurus*, and *Smi-losuchus*) (**Figures 11.9[C]** and **11.10[B]**) had broader more powerful snouts resembling those of living crocodilians, with different types of teeth: long fangs in front for snaring prey, and slicing teeth in the back of the jaw. These creatures probably ate larger animals that could be ambushed near the water (like living alligators and crocodiles do), as well as aquatic prey. Most were about the size of the range of living crocodilians as well, but the largest phytosaurs were over 12 meters (40 feet) long, bigger than most living or fossil crocodilians except giants like *Sarcosuchus* and *Deinosuchus*. In the details of the skulls, however, the phytosaurs were very different from crocodilians. The most obvious difference is that the nostrils of the phytosaurs are shifted up to the top of their skulls, between and just in front of their eyes, whereas the nostrils of crocodilians are always near the tip of their snouts. They also lacked a secondary palate in their mouth that allowed them to breathe and swal-low at the same time, something all crocodilians can do. Their skulls and bodies are often more heavily armored than those of crocodilians, with bony osteoderm armor all over their backs and sides, and heavily armored belly ribs (gastralia) as well.

After a great diversity of genera in the Late Triassic, phytosaurs were apparently victims of the Triassic-Jurassic extinction event, and there are no confirmed instances of any surviving into the Jurassic. But true crocodylomorphs were evolving during the Late Triassic, and as we shall see later, they took over that niche in the Jurassic and have dominated it ever since.

Aetosaurs

Another even weirder group of archosaurs were the armored forms known as aetosaurs (**Figures 11.9[D,E]** and **11.10[C,D]**). In a world dom-inated by carnivorous archosaurs, aetosaurs were among the few herbi-vores. They were rather unspecialized herbivores, with a beak and small head and jaws lined by small simple leaf-shaped or chisel-like teeth, but they were apparently very successful on their diet of low-growing plants like ferns and cycads and conifers (found in places like the Petrified Forest in Arizona). Their heads were small, covered with armor plates on the top, and many aetosaurs had a toothless pointed snout that curved upward. On the back of all aetosaurs was a thick armored cover made of lots of individual rectangular bony osteoderms packed tightly together in regular rows, so their armor plates on their back had some flexibility. Many aetosaurs had spikes along the sides of the bony armor on their back and in some aetosaurs (like *Desmatosuchus*—**Figures 11.9[E]** and **11.10[C]**) they had very large spikes protruding from their shoulders. Others, like *Typothorax* (**Figure 11.10[D]**) had very broad rounded bodies more like that of a turtle, and emphasized width over length.

Unlike the sprawling posture of earlier archosauromorphs and phytosaurs, aetosaurs were among the earliest quadrupedal archosaurs to have their limbs fully under their heavily armored bodies and standing straight up and down (the "pillar-erect" posture), similar to the condition in the heavily armored ankylosaur dinosaurs. Yet their front limbs were shorter than their hind limbs, so their hips were high while their head and shoulders were much lower to the ground. The primitive early Late Triassic genera like *Aetosaurus* and *Coahomasuchus* were only about a meter in length, but they quickly evolved to larger and larger forms. Most aetosaurs were about 3 meters (10 feet) in length or smaller, but the largest ones like *Desmatosuchus* (**Figures 11.9[E]** and **11.10[C]**) were up to 4 meters (13 feet) long, so they were not only heavily armored, but also fairly large animals compared to the predators of the Late Triassic (except the giant aquatic phytosaurs).

Most of the first described aetosaur fossils were mistaken for the bony osteoderms of phytosaurs, and so their actual appearance and classification was a mess until more complete skeletons were found. Over 27 genera of aetosaurs are known, restricted to the Late Triassic deposits like those of the Chinle Formation in Petrified Forest as well as similar beds in New Mexico, Utah, and Texas, and Triassic beds in Connecticut, New Jersey, and North Carolina. But aetosaurs were also global in their distribution; they are known from Greenland, Scotland, Germany, Italy, Poland, Morocco, Algeria, Madagascar, India, Argentina, and Brazil—only Australia and Antarctica lack aetosaurs (mostly because Upper Triassic outcrops are scarce in those continents). At the end of the Triassic, the aetosaurs vanished completely, just as the phytosaurs did.

Ornithosuchidae

In 1894, one of the first groups of archosaurs to be described was the Upper Triassic Scottish fossil *Ornithosuchus* (**Figure 11.9[F]**) and similar Upper Triassic fossils like *Riojasuchus* and *Venanticosuchus* (**Figure 11.9[G,H]**) from Argentina and *Dynamosuchus* from Brazil. These creatures had relatively strong hind limbs and short front limbs, so they were among the first bipedal archosaurs known (although they were probably bipedal only when running, and quadrupedal most of the time). Large specimens of *Ornithosuchus* were up to 4 meters (13 feet) long, and they had a wicked mouth with sharp recurved teeth, and a downturned overhanging snout that flexed downward at the front, with a gap in the toothrow where the upper jaw flexed downward. Their ankle bones (**Figure 11.8**) were distinctive, with the hinge between the calcaneum and astragalus, but there is a socket in the astragalus with fits with a prominent knob in the calcaneum. This is the opposite condition of the crurotarsal ankle of most crocodilians and other pseudosuchians, so it is known as a "croc-reversed" ankle. There is a concavity on the calcaneum that receives the knob-like bump that extends from the astragalus (**Figure 11.8**). However, nearly all analyses based on characters besides the ankle joint place ornithosuchians within the pseudosuchians.

Poposaurs

In addition to the croc-mimic phytosaurs, the armored aetosaurs, the bipedal ornithosuchids, the Middle and Late Triassic were ruled by a wide range of medium- to large-sized land predators which had long narrow snouts with wicked teeth for killing nearly all the other tetrapods that roamed the Late Triassic landscape. For example, *Ticinosuchus* (**Figure 11.11[A]**) is a Middle Triassic fossil from Switzerland and Italy, and *Mandasuchus* from Tanzania, are close relatives of the croc relatives like the poposauroids and rauisuchians discussed next. They were both about 3 meters (10 feet) long with an elongated body and four long legs, and a fully upright posture, making them fast runners. The entire body of *Ticinosuchus*, even the belly, was covered by thick armored osteoderms. The presence of fish fossils in its stomach suggests that it was partially aquatic and lived on fish much of the time. But *Mandasuchus* was clearly terrestrial, without all the aquatic features of *Ticinosuchus*.

There is not enough room in this chapter to discuss all of them, and their classification is still hotly debated, but a few examples will give a sense of their diversity. The poposauroids (**Figure 11.11[B–H]**) include a wide range of archosaurs that showed many different shapes and lifestyles. The most primitive of the poposauroids is *Qianosuchus* (**Figure 11.11[B]**) from the Middle Triassic of China. Found in a marine limestone, it was over 3 meters (10 feet) long, and its skull was over 35 cm (13 inches) long. *Qianosuchus* had a long narrow snout with many sharp recurved teeth. The nasal opening was large and the eyeball was reinforced with a bony sclerotic ring. They had elongate neck ribs, suggesting that they might have caught aquatic prey by rapidly opening its mouth and expanding its throat cavity and sucking in their food. *Qianosuchus* had a tall but narrow tail, and thickened dense ribs, both clear features of aquatic lifestyle. But they still had well-developed front and hind limbs, so they were capable of land locomotion as well.

In contrast, a completely different poposauroid was *Arizonasaurus* (**Figure 11.11[D]**) from the Middle Triassic Moenkopi Formation of Arizona. Discovered and named in 1947 and originally known just from a few isolated teeth and an upper jaw, a nearly complete skeleton was found in 2002. Its body was long and slender and capable of both bipedal and quadrupedal locomotion. However, it is strikingly different in having tall spines on the middle of its back, which supported a tall "sail" on the back, like the protomammals *Dimetrodon* and *Edaphosaurus*, or the dinosaurs like *Spinosaurus* and *Ouranosaurus*. *Arizonasaurus* was a member of a group called the ctenosauriscids, allied with genera such as *Ctenosauriscus* (**Figure 11.11[C]**), *Lotosaurus* (**Figure 11.11[E]**), *Bromsgrovela*, and *Hypselorhachis*, and apparently many of them had some kind of sail on their backs as well.

Poposaurus itself (**Figure 11.11[F]**) is known from nearly complete skeletons, but with only partial skulls. It was fully bipedal with small arms and long hind legs and tail. It is about 4 meters (13 feet) long but almost half its length is the long tail. *Poposaurus* gets its name from the red Upper Triassic Popo Agie (po-POH-zha) Formation in Wyoming, where it was first found in 1915, but since then it has been found in Arizona, Utah, and Texas. When the scrappy skeleton was first described, it was mistaken for a dinosaur, a phytosaur, a "thecodont", and a rauisuchian. Better specimens found in the 1990s and 2000s helped clear up the understanding of this creature, and since then it is recognized as the original member of the Poposauridae. At that huge size, it was one of the largest bipedal archosaurs of the Late Triassic, surpassed only by some of the early sauropodomorph dinosaurs like *Plateosaurus* (see **Figure 17.1[G]**).

Another odd-looking poposauroid is *Lotosaurus* from the Middle Triassic of Hunan Province, China (**Figure 11.11[H]**). At 2.5 meters (8 feet) long,

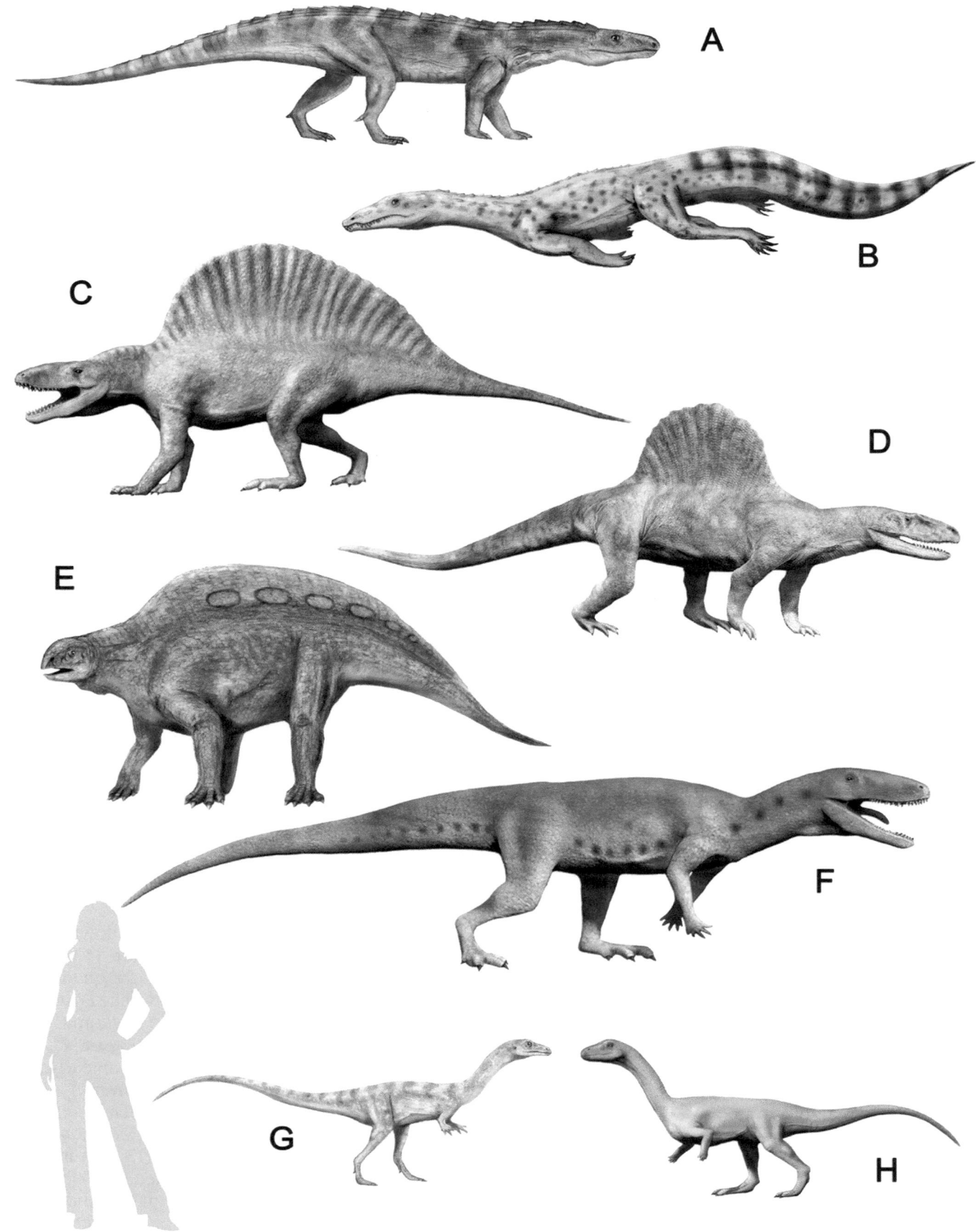

Figure 11.11 Reconstructions of the variety of poposaurs: (A) *Ticinosuchus*, (B) *Qianosuchus*, (C) *Ctenosauriscus*, (D) *Arizonasaurus*, (E) *Lotosaurus*, (F) *Poposaurus*, (G) *Shuvosaurus*, and (H) *Effigia*.

it was a large heavily built quadruped, but instead of the usual predatory teeth, it had a toothless beak for eating vegetation. Like the Ctenosauricidae, it also had a short sail on its back as well. *Lotosaurus* was found in a single bone-bed with dozens of individuals which died side by side. The bones were well preserved and not very scattered, so they were not buried in a catastrophic mudslide, but apparently died near a waterhole and then were buried more or less intact.

Finally, the most divergent members of the Poposauroidea are the Shuvosauridae, including *Shuvosaurus*, *Effigia*, and *Sillosuchus*. *Shuvosaurus* (**Figure 11.11[G]**) was found in the Upper Triassic Dockum Formation of the Texas Panhandle by Shuvo Chatterjee, the son of Dr. Sankar Chatterjee of Texas Tech University in Lubbock, Texas. Its skull was originally mistaken for an ornithomimid dinosaur, because it was similar in shape to the ostrich dinosaurs, but its skeleton was a lightly built and bipedal (**Figure 11.11[G]**) with a long neck and tail, and a toothless beak; it was recognized as a poposaurid with a different name, *Chatterjeea*. However, in the early 2000s American Museum of Natural History then-graduate student Sterling Nesbitt (now at Virginia Tech University) was looking at a number of unprepared field jackets in the storage area of the museum from the famous *Coelophysis* quarry near Ghost Ranch, New Mexico. He opened a number of jackets and found not *Coelophysis* but a new shuvosaurid poposaur which was named *Effigia*, as in "ghost" (**Figure 11.11[H]**). This much more complete specimen showed that not only *Effigia* and *Shuvosaurus* but also *Sillosuchus*, another lightly built bipedal creature which was almost 3 meters (10 feet) long from the Late Triassic of Argentina were all poposauroids. Some bones of *Sillosuchus* suggest a length of almost 10 meters (33 feet), bigger than any other bipedal animal known from the Late Triassic.

Rauisuchians

The closest relatives to crocodilians among the pseudosuchians are a group known as rauisuchids (**Figures 11.12** and **11.13**), which together with the crocodilians, make up a group called the Loricata. With at least 13 genera known from the U.S., China, Russia, Brazil, Brazil, Morocco, Germany, Poland, and India, rauisuchians were one of the largest predators of the Late Triassic all over Pangea, although some of them may date back to the Early Triassic. Superficially, they resembled the "bloody crocs" (erythrosuchids—**Figure 11.5[B]**) of the Middle Triassic, with long slender bodies, a short but narrow, deep skull which strong muscles and sharp teeth for ripping apart large prey animals, but they were actually much more closely related to crocodilians than their large-headed relatives.

Rauisuchids also had a long tail, two rows of osteoderms down its back, and long legs that gave them a fully upright stance (**Figure 11.12**). Some were apparently bipedal, whereas others were clearly quadrupedal. Some of the best known and most complete specimens, like *Postosuchus* (**Figures 11.12[B]** and **11.13[A]**) of the Upper Triassic Dockum Group of Texas were up to 5 meters (16 feet) long. *Teratosaurus* (**Figure 11.12[A]**) from the Late Triassic of Germany was long mistaken for an early theropod dinosaur until it was re-identified as a rauisuchid. *Vivaron*, a genus recently discovered at Ghost Ranch, is very similar to *Teratosaurus* and almost as large. The genus that gave the name of the group, *Rauisuchus* from the Late Triassic of Brazil, was about 4 meters (13 feet) in length. *Polonosuchus* from Poland was even bigger, up to 6 meters (20 feet) long. At least 11 other genera are known in the Triassic, making them a very important predator at the top of the food chain. But like the phytosaurs, aetosaurs, ornithosuchids, and poposauroids, they vanished in the great Triassic-Jurassic extinction event, eventually to be replaced by predatory dinosaurs.

Figure 11.12 Reconstructions of typical rauisuchians: (A) *Teratosaurus*, (B) *Postosuchus*.

A

Figure 11.13 Photo of fossils of rauisuchians such as: (A) *Postosuchus*. (Courtesy Wikimedia Commons.)

B

Figure 11.13 (Continued) (B) *Prestosuchus*.

FURTHER READING

Benton, M.J. 1985. Classification and phylogeny of diapsid reptiles. *Zoological Journal of the Linnean Society (London)*. 84: 97–164.

Benton, M.J., ed. 1988. *The Phylogeny and Classification of the Tetrapods, vol. 1: Amphibians, Reptiles, Birds*. Oxford Clarendon Press, Oxford.

Benton, M.J.; Clark, J. 1988. Archosaur phylogeny and the relationships of the Crocodylia, pp. 295–338. In Benton, M.J., ed. *The Phylogeny and Classification of the Tetrapods, vol. 1: Amphibians, Reptiles, Birds*. Clarendon Press, Oxford.

Brusatte, S.L.; Benton, M.J.; Desojo, J.B.; Langer, M.C. 2010. The higher-level phylogeny of Archosauria (Tetrapoda: Diapsida). *Journal of Systematic Palaeontology*. 8: 1–47.

Evans, S.E. 1988. The early history and relationships of the Diapsida, pp. 221–260. In Benton, M. J., ed. *The Phylogeny and Classification of the Tetrapods, vol. 1: Amphibians, Reptiles, Birds*. Clarendon Press, Oxford.

Gauthier, J. 1994. The diversification of the amniotes, pp. 129–159. In Prothero, D. R.; Schoch, R.M., eds. *Major Features of Vertebrate Evolution*. Paleontological Society Short Course 7. Paleontological Society, Lawrence, KS.

Gower, D.J.; Wilkinson, M. 1996. Is there any consensus on basal archosaur phylogeny?. *Proceedings of the Royal Society B*. 263 (1375): 1399–1406.

Liu, J.; Organ, C.L.; Benton, M.J.; Brandley, M.C.; Aitchison, J.C. 2017. Live birth in an archosauromorph reptile. *Nature Communications*. 8: 14445.

McLoughlin, J.C. 1979. *Archosauria: A New Look at the Old Dinosaur*. Viking Press, New York.

Müller, J. 2004. The relationships among diapsid reptiles and the influence of taxon selection, pp. 379–408. In Arratia, G.; Wilson, M.V.H.; Cloutier, R., eds. *Recent Advances in the Origin and Early Radiation of Vertebrates*. Verlag Dr. Friedrich Pfeil, München, Germany.

Nesbitt, S.J. 2011. The early evolution of archosaurs: Relationships and the origin of major clades. *Bulletin of the American Museum of Natural History*. 352: 1–292.

Sereno, P.C. 1991. Basal archosaurs: Phylogenetic relationships and functional implications. *Journal of Vertebrate Paleontology Memoir*. 2: 1–53.

Sereno, P.C.; Arcucci, A.B. 1990. The monophyly of crurotarsal archosaurs and the origin of bird and crocodile ankle joints. *Neues Jahrbuch für Geologie und Paläontologie, Abhandlungen*. 180: 21–52.

CROCODYLO-MORPHS

12

How doth the little crocodile
Improve his shining tail
And pour the waters of the Nile
On every golden scale!
How cheerfully he seems to grin
How neatly spreads his claws,
And welcomes little fishes in
With gently smiling jaws!
　　　　　—Lewis Carroll, *Alice's Adventures in Wonderland*, 1865

CROCODYLOMORPHS: THE CROCODILES AND THEIR KIN

Crocodilians have been a part of human culture since the days of ancient Egypt, which worshipped the crocodile gods Ammit and Sebek. They appear prominently in the cultures of Africa and southern Asia, where they were an important part of the legends and mythology of the peoples of the tropics, from India to southeast Asia to Australian Aboriginal myths. Herodotus described crocodiles in Egypt, and repeated the myth that the "crocodile bird" or "trochilus" (possibly the Egyptian plover) walked into the opened mouths of crocodiles and picked foot particles from its teeth. There were many stories about alligators and caimans in Mesoamerican cultures as well.

The crocodile was an important part of the bestiaries of the Middle Ages and later, even though they didn't live in Europe and were rarely seen north of Africa and southern Asia. In the *Etymologies* by Isidore of Seville in the 500s, crocodiles allegedly got their name from their saffron color (*croceus* is "saffron" in Latin). He also claimed that crocodiles could be killed by fish with serrated crests sawing into their soft underbelly, and that both male and female crocodiles took turns guarding their nests. Ever since the ninth century *Bibliotheca* by Photios I of Constantinople, crocodiles were reputed to weep over their victims' fate, leading to the myth about "crocodiles tears". When the English traveler Sir John Mandeville visited India in 1400, he wrote:

> In that country [of Prester John] by all Ind [India] be great plenty of cockodrills, that is a manner of a long serpent, as I have said before. And in the night they dwell in the water, and on the day upon the land, in rocks and in caves. And they eat no meat in all the winter, but they lie as in a dream, as do the serpents. These serpents slay men, and they eat them weeping; and when they eat they move the over jaw, and not the nether jaw, and they have no tongue".

In more recent years, crocodilians have become a big part of western children's stories and literature, from the *Just-So Stories* of Rudyard Kipling,

DOI: 10.1201/9781003128205-12

to the poem "How Doth the Little Crocodile" in *Alice in Wonderland*, to the Neverland crocodile who chases Captain Hook in J.M. Barrie's *Peter Pan*, to modern cultural symbols like mascots of athletic teams (such as the University of Florida Gators—where real alligators wander the campus at will), movies like *Crocodile Dundee* and TV shows such as the late Steve Irwin's *Crocodile Hunter*, to the cartoon crocodiles in Walt Disney's 1940 movie *Fantasia* dancing ballet in Ponchielli's "Dance of the Hours", or Wally Gator in the Hanna-Barbera cartoons. Crocodiles and alligators are still a big part of modern culture, even is regions where they never occur naturally.

"BUNNY CROCS"

When we see crocodilians today, we think of these heavy long-bodied short-legged dangerous aquatic predators which can ambush prey on land or in water. There are about 27 species in 9 genera of crocodiles, gharials, alligators, and caimans alive now, but there are over 150 genera of extinct crocodylomorphs and many hundreds of species recognized (**Figure 12.1**). But crocodilians didn't start out as the large aquatic predators that resembled the living members of the group. The earliest crocodylomorphs lived in the Late Triassic among all the other pseudosuchians, but they were delicately built animals with relatively long, slender legs and shorter snouts (**Figure 12.2**). The niche for the large-bodied, short-limbed body ambush predators like living crocodilians was still occupied by phytosaurs in the Triassic. Some of these earliest crocodylomorphs included *Hesperosuchus* from the Chinle Formation in Arizona and New Mexico (**Figure 12.2[A]**). *Hesperosuchus* was a lightly built form only about 1.2–1.5 meters (4–5 feet) long with delicate hollow bones and long delicate limbs and tail, not much different from other pseudosuchians such as *Ornithosuchus* (mentioned in Chapter 11). Its long hind limbs would have allowed them to be fast runners. Their skeletons even reminded some scientists of the limb proportions of rabbits, so some of these early crocodylomorphs have been nicknamed "bunny crocs", although it is unlikely that they were good hoppers or leapers. However, the genus *Lagosuchus* (whose name literally translates to "bunny croc") from the Triassic of Argentina is actually a primitive dinosaur relative. *Hesperosuchus* was found in sediments suggesting it lived near water, and it was fossilized along with abundant fish fossils and teeth of phytosaurs.

Saltoposuchus (whose name means "leaping foot crocodile") from the Upper Triassic of Germany, was similar in size (1.5 meters long), and built for a completely bipedal lifestyle (**Figure 12.2[C]**) A close relative of *Saltoposuchus* was *Terrestrisuchus* (**Figure 12.2[B]**) from the Upper Triassic of Wales. It was shaped much like *Hesperosuchus*, but much smaller, only about 76 cm (30 inches) long, with a much more slender body. There are about a dozen more genera of these "bunny crocs" from the Late Triassic and Early Jurassic, including *Carnufex* and *Dromicosuchus* (Greek for "fast crocodile") from North Carolina, *Redondavenator* from New Mexico, *Erpetosuchus* from Scotland, and *Pseudohesperosuchus* and *Trialestes* from Argentina.

In the Jurassic, some crocodylomorphs had similar body shapes to their Triassic relatives, including *Kayentasuchus* from Arizona, and *Dibothrosuchus* (**Figure 12.2[D]**) and *Junggarsuchus* from China. These and other more advanced early crocodylomorphs may have been more bipedal and tended to be longer and more heavily built. *Litargosuchus*, whose name means "last running crocodile" comes from the Early Jurassic of South Africa, and was indeed one of the last delicately built "bunny crocs", along with *Sphenosuchus* from the same beds. These Jurassic crocodile relatives were beginning to occupy the niche of a larger-bodied quadrupedal predator occupied by modern crocodilians, because the crocodylomorphs had survived the Triassic-Jurassic extinction event that wiped out the phytosaurs, the dominant croc-mimic of the Late Triassic.

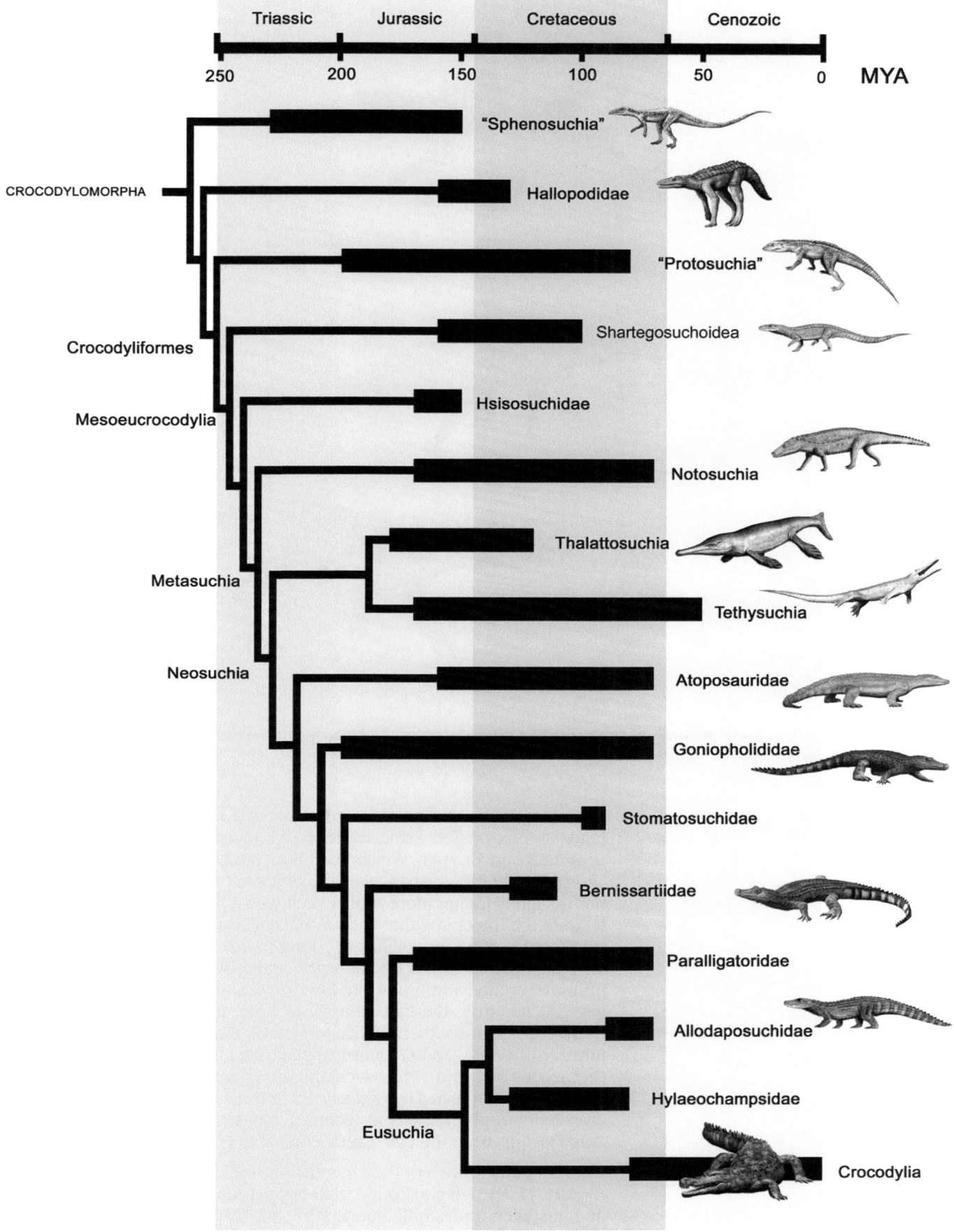

Figure 12.1 Family tree of the crocodylomorphs.

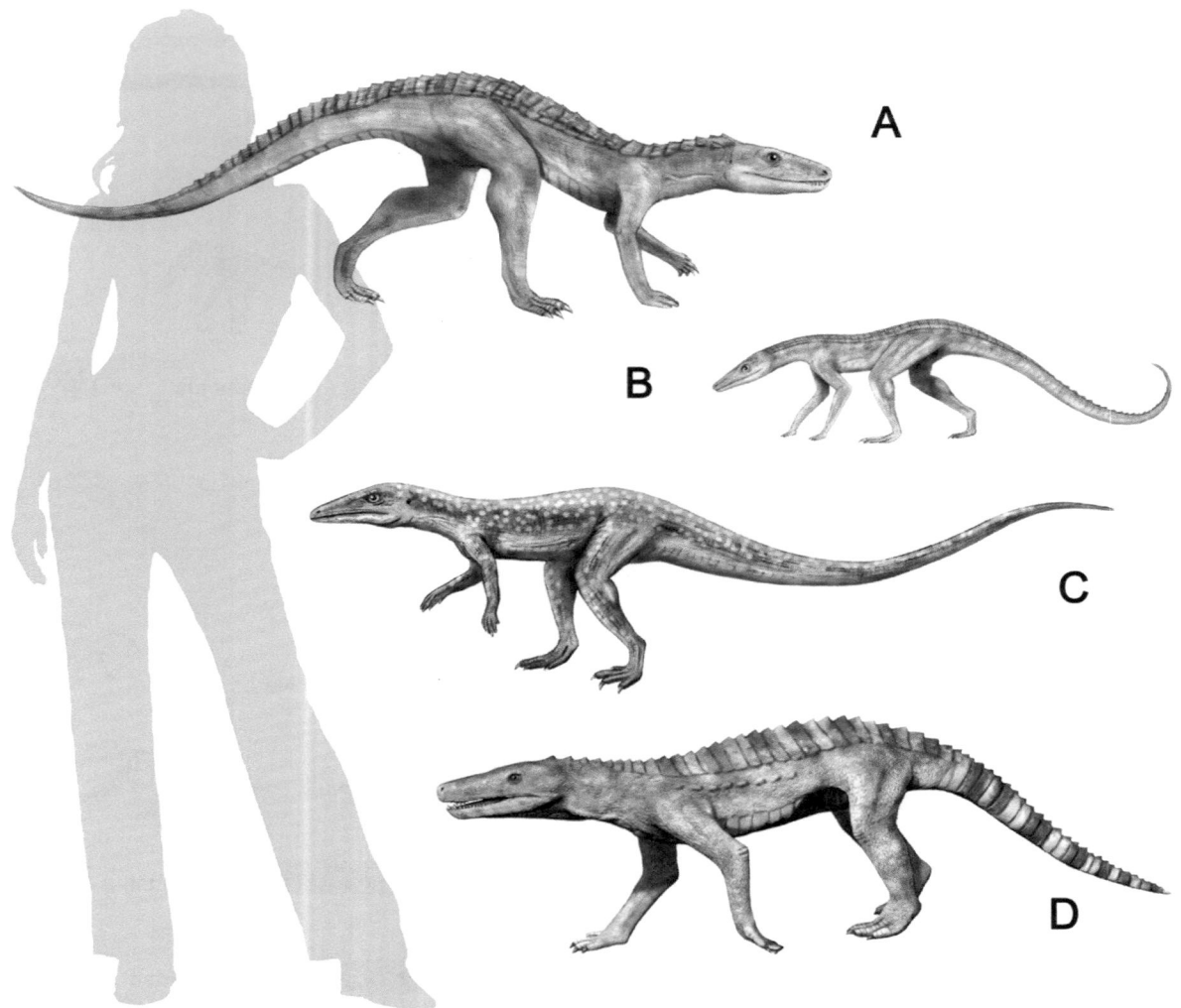

Figure 12.2 Reconstructions of some of the primitive crocodylomorphs known as sphenosuchians: (A) *Hesperosuchus*, (B) *Terrestrisuchus*, (C) *Saltoposuchus*, (D) *Dibothrosuchus*.

Even more common in the Late Triassic and Early Jurassic were the first heavier long-bodied armored crocodylomorphs like *Protosuchus* (**Figures 12.3** and **12.4[A]**), which was about a meter (3.3 feet) long. *Protosuchus* not only had the typical shape of some living crocodilians, but also had a slightly longer more robust skull with a prominent snout, including teeth in the lower jaw that fit into sockets in the upper jaw (a characteristic feature of all crocodilians). Along its back were rows of rectangular plates paired on each side of the spine. At least 11 other genera of primitive crocodylomorphs like *Protosuchus* are known, including *Hemiprotosuchus* from the Late Triassic of Argentina, *Orthosuchus* from the Early Jurassic of southern Africa (Lesotho), *Sichuansuchus* from the Late Jurassic of China, and *Edentosuchus* and *Shantungsuchus* from the Early Cretaceous of China. "Protosuchians" hung around even to the Late Cretaceous as exemplified by *Neuqensuchus* from Argentina, and the strange crocodylomorphs known as gobiosuchines, such as *Gobiosuchus* and *Zaraasuchus*, from the Late Cretaceous of the Gobi Desert in Mongolia.

Another early branch of the crocodylomorphs is exemplified by *Hallopus* (**Figure 12.3[B]**). It was found in the Upper Jurassic Morrison Formation, and mistaken for a small dinosaur by O.C. Marsh (it's only about a meter long). It was originally known only from fossils of its very long hind limbs and slightly shorter front limbs. Later more complete specimens

Figure 12.3 Reconstruction of the larger-bodied primitive crocodylomorphs: (A) *Protosuchus*, (B) *Hallopus*.

showed that it was a crocodylomorph, slightly more primitive than the "protosuchians". Likewise, another Marsh specimen from the Morrison Formation, *Macelognathus* was first known from a lower jaw with a toothless beak, but most recent analyses of better specimens place it with *Hallopus*. Finally, *Alamadasuchus* from the Upper Jurassic of Argentina seems to be closely related to *Hallopus* as well.

Notosuchia: The "Southern Crocodiles"

One of the biggest and more diverse groups of Mesozoic crocodylomorphs was the Notosuchia. Known from at least 25 genera in the Cretaceous, they were found primarily on the Gondwana continents, hence their name which translates to "southern crocodiles" in Greek. They were particularly common in South America (Brazil, Bolivia, Argentina, Uruguay) and Africa (Niger, Madagascar, Malawi, Tanzania), with their only non-Gondwana occurrence in China. Most had long slender bodies with long front and hind legs that were fully upright, so they were relatively fast runners. Unlike living crocodilians, their skulls tended to be short and deep, and many of them had very wide skulls. In addition, many of them had highly specialized teeth, suggesting a wide range of diets.

Beyond these generalities, notosuchians showed an enormous range of variation (**Figure 12.5**). The earliest known notosuchian was from the

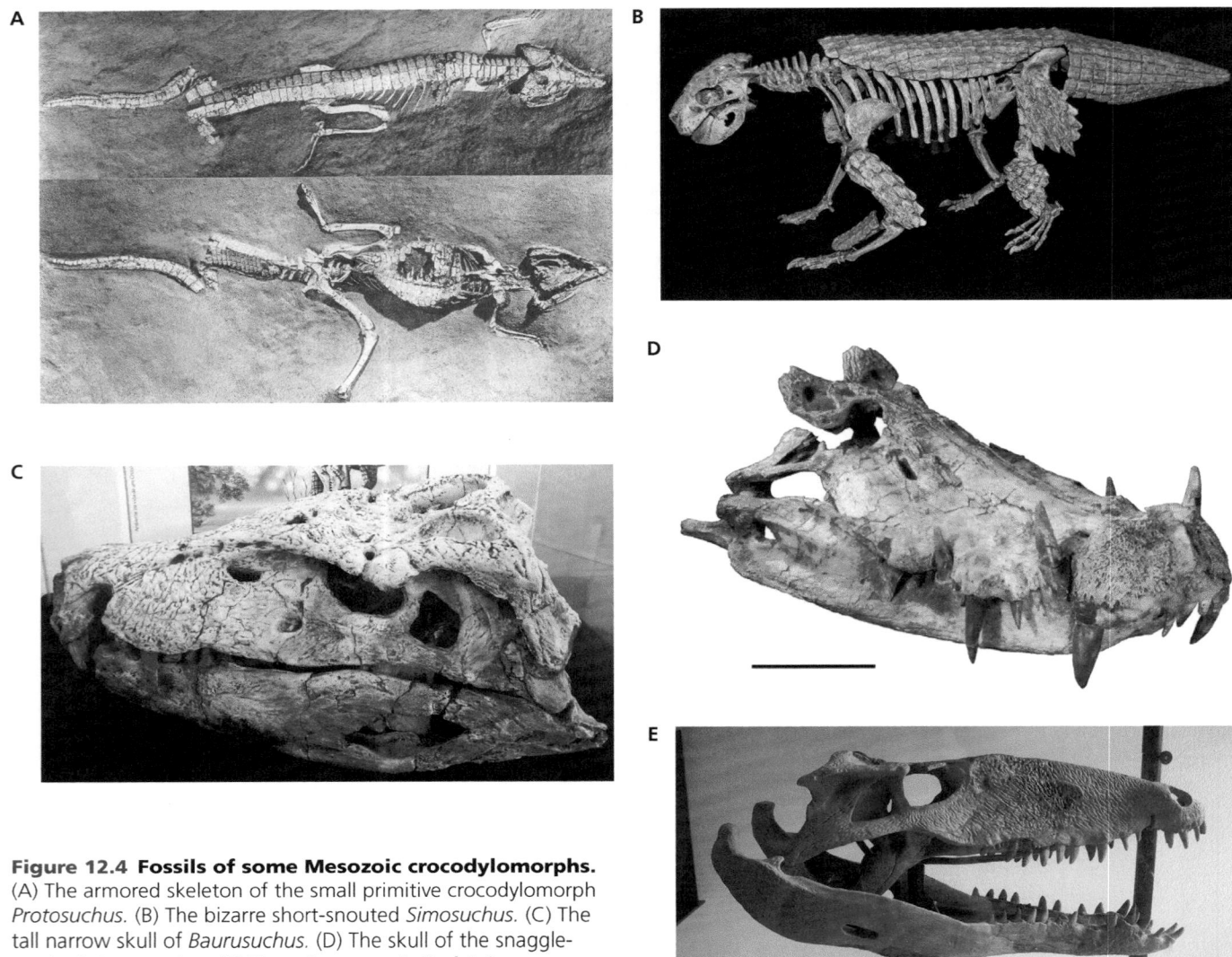

Figure 12.4 Fossils of some Mesozoic crocodylomorphs.
(A) The armored skeleton of the small primitive crocodylomorph
Protosuchus. (B) The bizarre short-snouted *Simosuchus.* (C) The
tall narrow skull of *Baurusuchus.* (D) The skull of the snaggle-
toothed *Kaprosuchus.* (E) The tall narrow skull of *Sebecus.*
(Courtesy Wikimedia Commons.)

Middle Jurassic of Madagascar. Dubbed *Razanandrongobe* (Malagasy for
"large ancestor lizard"), it was one of the largest notosuchians known,
with big sharp peg-like teeth in front, and globular teeth in the back of the
mouth, and a large blunt snout. *Anatosuchus* from the Lower Cretaceous
of Niger was about a meter long, but had a beak like that of a duck (**Figure
12.5[A]**). *Armadillosuchus* from the Upper Cretaceous of Brazil was about
2 meters (6.6 feet) long and had armor around its body arranged in flexible
bands between rigid shields of hexagonal plates, similar to the plates of
an armadillo (**Figure 12.5[B]**). Rather than the simple teeth of most crocodylomorphs, it had pointed incisors, stabbing canines, and conical teeth
with shearing edges in the back part of its mouth. This suggests a diet of
tough plants or maybe a carnivorous diet. In addition, *Armadillosuchus*
had large digging claws on its front feet. *Notosuchus* itself (**Figure 12.5[C]**),
which gave the group its name, was found in the Upper Cretaceous bed
of Argentina. It was described in 1896, one of the very first notosuchians
discovered and named, and the skull seems to have had a pig-like snout.
Adamantinasuchus from the Upper Cretaceous of Brazil was about 60 cm
(2 feet) long, and had a very short tall skull with large eye sockets.

By contrast, *Simosuchus* ("pug-nosed crocodile" in Greek) from the Upper
Cretaceous of Madagascar was a really peculiar creature (**Figures 12.4[B]**
and **12.5[D]**). Although it was only 0.7 meters (2.5 feet) long, it had a very

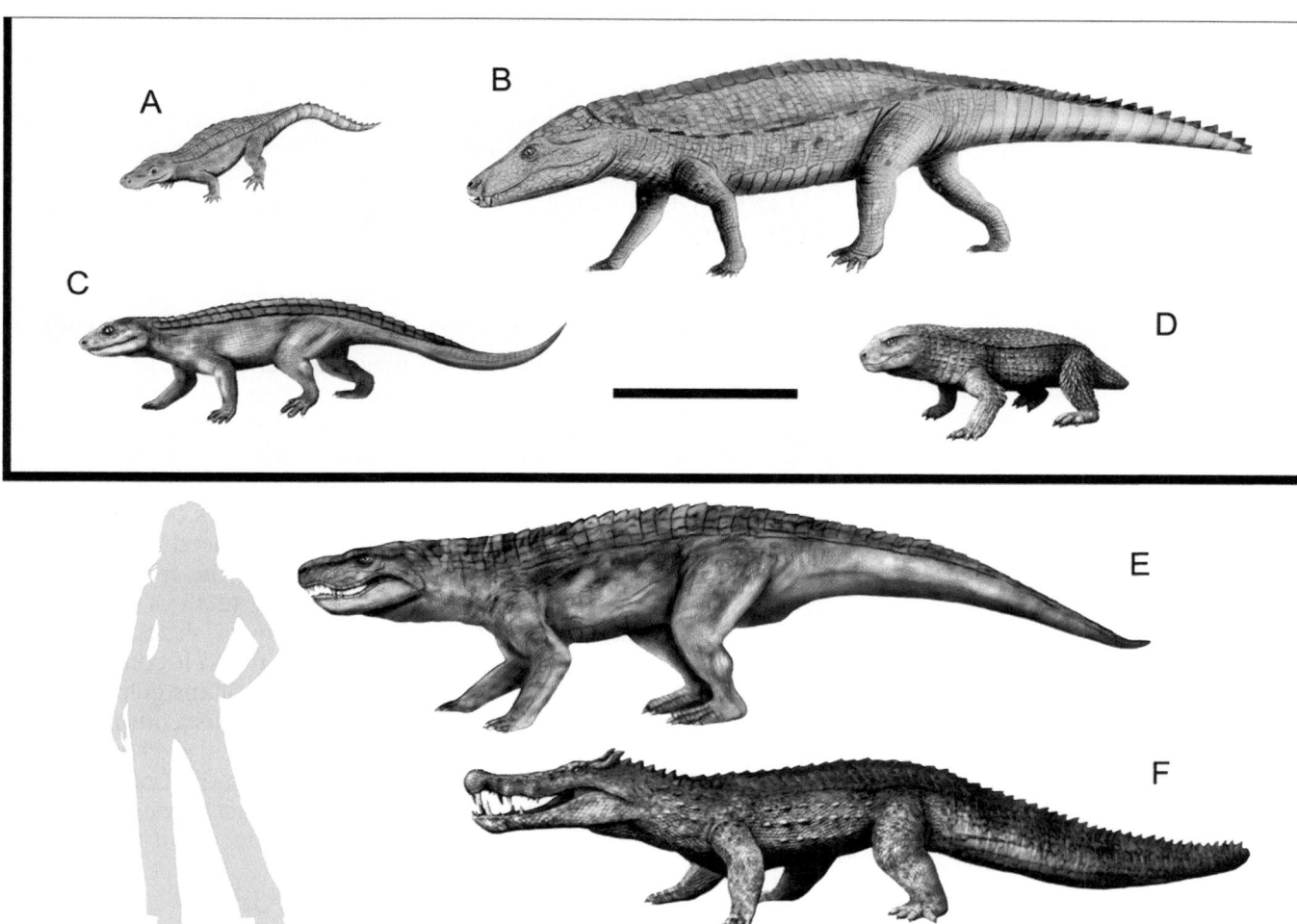

Figure 12.5 Reconstructions of Mesozoic crocodylomorphs such as: (A) *Anatosuchus*, (B) *Armadillosuchus*, (C) *Notosuchus*, (D) *Simosuchus*, (E) *Baurusuchus*, (F) *Kaprosuchus*. Top scale bar is 50 cm.

short snout that gave it a "pug-nosed" appearance. The short small teeth shaped like maple leaves were probably used for an omnivorous diet. *Malawisuchus* from the Lower Cretaceous of Malawi was only about 60 cm (2 feet) long, and had a highly specialized dentition that was quite mammal-like. It also had strong articulations in the head and neck, and it was apparently found inside burrows where it had once lived. Unlike almost any known archosaur, it had hinges in its lower jaw that allowed it to chew in a back-and-forth motion. *Pakasuchus* from the Early Creta-ceous of Tanzania also had a different kind of complex molar-like teeth in its mouth, suggesting an ability to chew up its food as well. The range of interesting lifestyles is amplified by *Yacarerani* from the middle Cretaceous of Bolivia, which had front incisors like those of a rabbit and was found on its probable nest site. *Araripesuchus* was the most widespread of all the notosuchians, known from six different species in the Late Cretaceous of Madagascar, Niger, Brazil, and Argentina. Large specimens were up to 1.8 meters (6 feet) long, with a relatively short snout that bulged out side-ways, very large eye sockets, thin osteoderms over the entire body, and upright limbs and hip joints that suggest a relatively fast runner.

Probably the most terrifying notosuchian was *Baurusuchus* for the Upper Cretaceous of Brazil (**Figures 12.4[C]** and **12.5[E]**). Up to 4 meters (13 feet) long, it was a huge predator that appeared to be completely terres-trial in its habits, judging from the long, strong upright limbs, and low nostrils on its head which would not allow them to submerge as modern crocodilians do. The skull was heavy and boxy and very deep, more like a theropod dinosaur than a crocodilian, suggesting that they had a powerful bite force, and the long wicked recurved teeth also resemble those of a predatory dinosaur. Even more spectacular was *Kaprosuchus* ("boar crocodile" in Greek) from the Late Cretaceous of Niger (**Figures 12.4[D] and 12.5[F]**). It had three sets of long sharp curved caniniform teeth protruding out from its jaws, like those of a boar, which explains their scientific name. Its eye sockets are angled forward, so it had good stereovision and ability to track prey, and its other anatomical features are also consistent with a terrestrial hunter, not an aquatic ambush predator.

Finally, notosuchians survived the mass extinction at the end of the Cretaceous and persisted in South America. The Eocene notosuchian *Sebecus* from Argentina is named after one of the Egyptian crocodile gods, Sebek (**Figure 12.4[E]**). *Sebecus* had a very tall narrow snout that tapered down to a point. The top of the skull was flat from front to back, without the usual raised areas for the eyes on the back of the skull seen in most crocodylomorphs. The teeth are evenly spaced an alternating between the upper and lower jaw, so they allow the mouth to close tightly with interlocking teeth. Unlike the conical teeth of most crocodilians, *Sebecus* had narrow blade-shaped teeth for slicing flesh, with precise shear between the upper and lower teeth, so it could chop up its prey a bit before swallowing; most living crocodilians gulp their prey down whole. *Sebecus* and its relatives included *Baurusuchus* mentioned previously, and about 16 other genera known almost exclusively from South America, although *Bergisuchus* was found in the Eocene Messel lake beds of Germany. Sebecosuchians were the last of the Notosuchia, while managed to survive until the middle Miocene, about 11 Ma.

THALATTOSUCHIA AND DYROSAURIDAE: BACK TO THE OCEAN

As partially or completely aquatic animals, crocodylomorphs have always been good swimmers and capable of doing most tasks in the water. Although most live in fresh water, the saltwater crocodile of Australia spends a lot of time in the sea as well. It is no surprise, then, that shortly after more specialized crocodylomorphs evolved in the Jurassic, they returned to the sea and became completely marine animals. Their bodies became even more adapted for full-time swimming, with the feet developed into paddles, and some of them even developed tail fins. This group of crocodylomorphs are known as the Thalattosuchia ("sea crocodiles" in Greek), and are known from at least 20 genera in the Jurassic and Early Cretaceous, arranged in two families: the Teleosauridae and the Metriorhychidae (**Figure 12.1**).

The oldest known genus is *Pelagosaurus* found in marine sediments from the Lower Jurassic of England, Germany, Switzerland, and France (**Figure 12.6[A]**). At about 3 meters (10 feet) long, it had the narrow fish-catching snout found on modern gharials, very streamlined body, and forward-facing eyes, so it was a pursuit predator, not an ambush predator like most crocodylomorphs. Comparisons with other crocodilians suggest that it fed mostly on smaller prey such as fish and squid, and did not have the capability of taking larger prey. For a long time, there were arguments as to whether *Pelagosaurus* was a teleosaur or a metriorhychid, but the current consensus suggests that it not a member of either highly specialized group, but the primitive relative of both.

From an ancestor like *Pelagosaurus*, the Teleosauridae first appeared in the Early Jurassic and radiated into at least 13 genera (**Figure 12.1**), all from Europe (England, France, Germany, all the way to Poland and

Figure 12.6 Reconstructions of marine crocodilians Thalattosuchia from Mesozoic marine rocks, including: (A) *Pelagosaurus*, (B) *Steneosaurus*, (C) *Dakosaurus*, (D) *Metriorhynchus*, (E) *Neptunidraco*.

Russia) except for one genus found in Africa (Ethiopia, Madagascar, Morocco). In addition, there are fragmentary specimens not assigned to a valid genus from Oregon, Argentina, India, China, and Thailand, so teleosaurids had a global distribution in the Jurassic. They all vanished at the end of the Jurassic except for *Machimosaurus*, the only genus known from Africa, which was the biggest of all, reaching 7.2 meters (24 feet) in length. They are best known from the genus *Steneosaurus* (first named in 1824) (**Figure 12.6[B]**), which became a giant "wastebasket" taxon for primitive teleosaurids and probably should be split into multiple genera, including one species which is now *Macrospondylus bollensis* (**Figure 12.7[A]**). *Steneosaurus*, *Teleosaurus*, and many of the European specimens are sometimes known from nearly complete articulated skeletons found in marine rocks, such as the Middle Jurassic Holzmaden Shale of Germany. Like *Pelagosaurus*, most teleosaurids had a long narrow snout for catching fish and squid, a streamlined body with webbed hands and feet, and a long narrow tail for propelling themselves through the water.

The metriorhynchids were equally diverse, with 15 genera (**Figure 12.1**) mostly from the Jurassic of Europe, but a few were very widespread. The genera *Dakosaurus* (**Figure 12.6[C]**) and *Cricosaurus* lived not only in Europe (from Spain to Russia) but also in Mexico and Argentina. The genus *Purranisaurus* is restricted to Argentina and Chile and not found in Europe. Most of the metriorhynchids died out at the end of the Jurassic, but a few of them (*Geosaurus*, *Dakosaurus*) survived into the Early Cretaceous.

Figure 12.7 Photos of marine thalattosuchian fossils, including: (A) An articulated skeleton of *Macrospondylus*. (B) The skeleton of *Metriorhynchus*. (C) The partial skull of the huge *Dakosaurus*. (Courtesy Wikimedia Commons.)

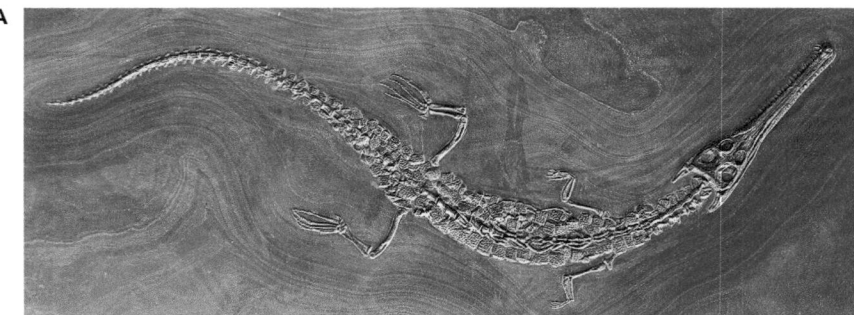

Metriorhynchids were also much more adapted to a fully marine life than were the teleosaurids (**Figures 12.6[D]** and **12.7[B]**). Their hands and feet were fully paddle-like and they have lost their armor of osteoderms, making them much more streamlined in water. Most had a tail fluke that stuck up from the spine of their tail, which tended to flex downward. This can be seen well in genera like *Metriorhynchus* itself (**Figure 12.6[D]**), or in *Neptunidraco* (**Figure 12.6[E]**). Since their small paddle-like hands and feet and short limbs were useless for crawling on land to lay eggs, it is thought that they gave live birth in water, as has already been demonstrated in other marine reptiles like plesiosaurs and ichthyosaurs (Chapter 10).

Most metriorhychids have the long narrow fish-catching snouts similar to that of gavials and also found in teleosaurs. *Dakosaurus*, on the other hand, had a robust triangular tapered snout that superficially looked like the snout of a mosasaur (**Figures 12.6[C]** and **12.7[C]**). Reaching up to 4.5 meters (14 feet) in length, it was one of the largest marine crocodiles, and its narrow teeth with serrated edges plus it robust jaws suggest it was not exclusively feeding on fish, but probably took larger marine animals as prey. In one Upper Jurassic unit in Germany, four different marine crocodiles are found, and they are the top predators since no other marine reptiles occur there. It has been suggested that they partitioned their niche

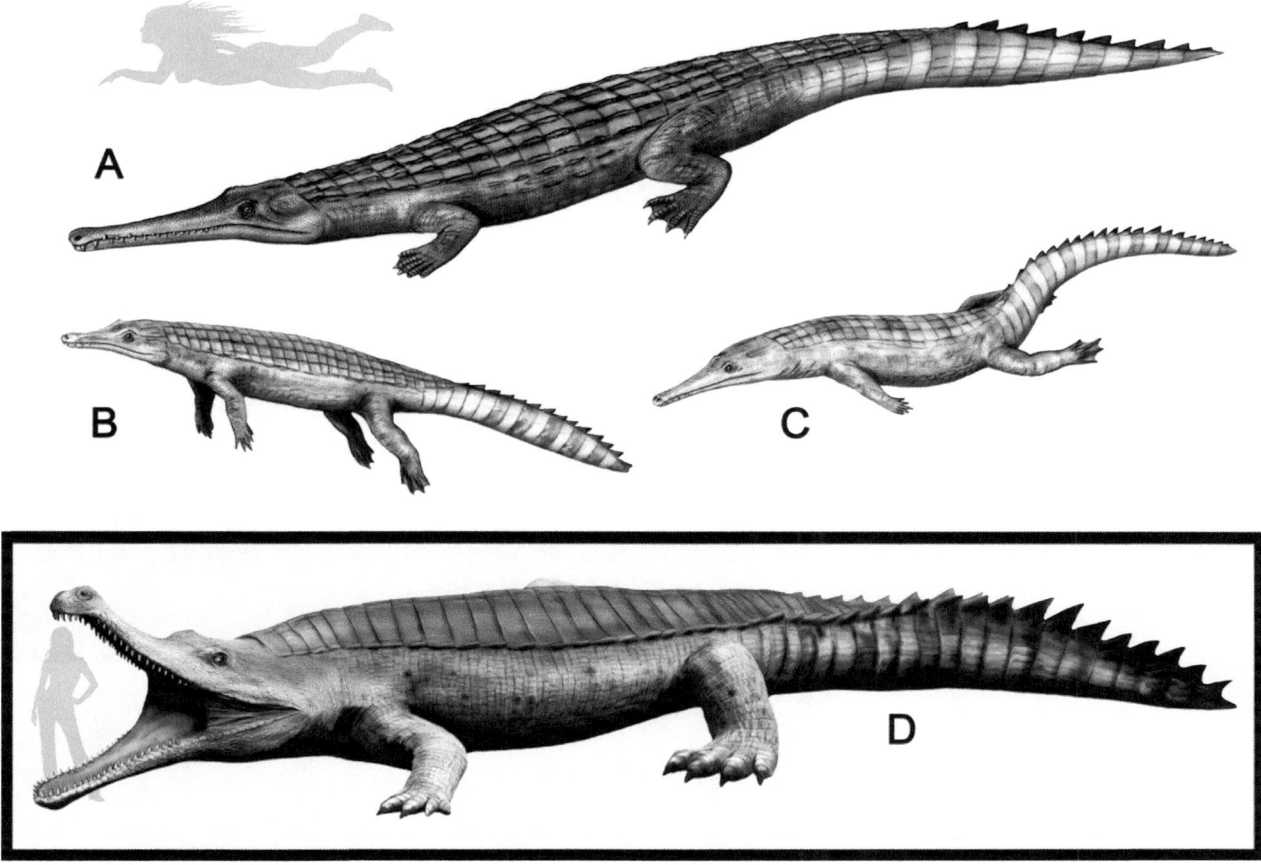

Figure 12.8 Reconstructions of some of the Tethysuchia: (A) *Dyrosaurus*, (B) *Chenanisuchus*, (C) *Guarinisuchus*, (D) *Sarcosuchus*.

so each was feeding on different kinds of prey. *Dakosaurus maximus* and *Geosaurus giganteus* are the largest, and probably fed on any other smaller animals that lived in that seaway. *Cricosaurus suevicus* and *Rhacheosaurus gracilis* were much smaller metriorhynchids, and along with *Steneosaurus*, a teleosaurid, they would have fed only on smaller fish and squid.

The nearest relatives of the Thalattosuchia are a group of mostly Late Cretaceous to Eocene crocodilians known as Dyrosauridae (**Figure 12.8[A]**). At least 17 genera are known, with more than two dozen species recognized so far. Originally, they were thought to be an African group (Nigeria, Angola, Mali, Tunisia, Algeria, Morocco, Niger), but now they have been found in Sweden and the United States to Africa to Central and South America (Mexico, Brazil, and Colombia). They were both freshwater and marine, and apparently their radiation in the Late Cretaceous took over the niche once occupied by the Thalattosuchia in the Early Cretaceous. Dyrosaurs were apparently competing primarily with mosasaurs, which had very different habitats and prey preferences. Unlike all other marine reptiles, which vanished by the end of the Cretaceous, the dyrosaurids *Chenanisuchus* (**Figure 12.8[B]**) and *Hyposaurus* survived the great end-Cretaceous extinction event, and led to another big radiation of these crocodilians in the Paleocene and Eocene with Paleocene forms like *Guarinisuchus* (**Figure 12.8[C]**) (from Brazil) and *Dyrosaurus* from the Eocene of Tunisia (**Figure 12.8[A]**).

Dyrosaurids superficially resembled the Thalattosuchia in having a crocodylomorph body with a long narrow snout, but they differed in important details. Their long narrow snouts were proportionately longer than in most Thalattosuchia, reaching about 70% of the total length of the skull. The nostrils were not on the tip of the snout, but the back end of the snout, allowing their body to be mostly submerged while they hunted or

Figure 12.9 Photos of some extreme crocodilians: (A) Paul Sereno posing with the skull of the African Cretaceous *Sarcosuchus*. (B) Ashley Hall with the South American Miocene giant caiman *Purussaurus*. (C) The skull of the gigantic crocodilian *Deinosuchus*, from the Cretaceous of Texas, compared to a modern crocodile skull (white skull inside its mouth). [(A) Courtesy P. Sereno. (B) Photo by the author. (C) Courtesy Wikimedia Commons.]

A

B

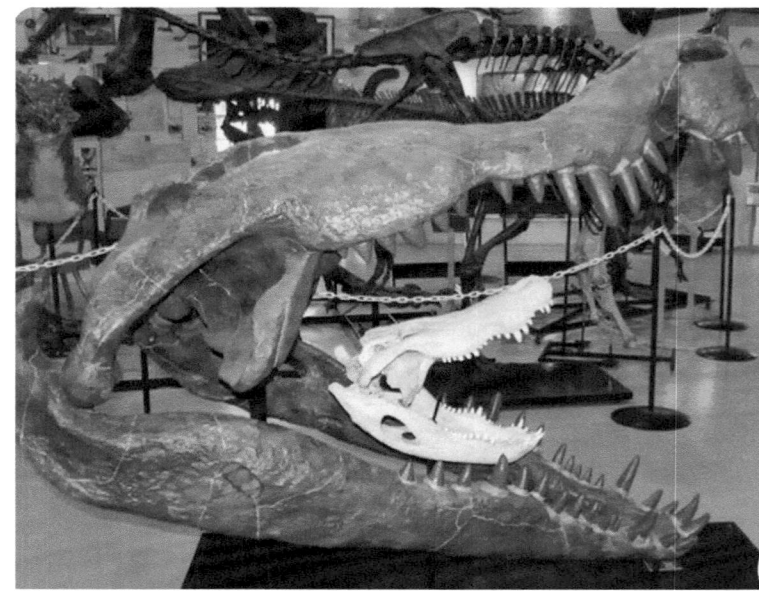

C

swam. In the jaws, there are pits or sockets to accept the points of the teeth from the opposite jaws, so they had very precise interlocking occlusion of the upper and lower teeth when the mouth was closed. They also reached up to 6 meters (20 feet) in length, larger than any Thalattosuchia.

A close relative of the Dyrosauridae is a group known as the Pholidosauridae. One of the largest of the pholidosaurids was the gigantic crocodylomorph *Sarcosuchus* from the Early Cretaceous of the Sahara Desert (**Figures 12.8[D]** and **12.9[A]**). At about 9.5 meters (32 feet) in length, and weighing up to 4.3 tonnes, it was an immense creature, probably preying on smaller dinosaurs and other large terrestrial and aquatic reptiles. Yet despite its size, it still has the long narrow snout of the dyrosaurs, but the snout was still more robust than typical dyrosaurs, with huge robust teeth that did not interlock, so it fed on large prey like a modern crocodilian, not on fish. In addition, it was apparently a freshwater form, not a marine crocodile, and it was found in beds that include not only large fish and even coelacanths, but also a variety of dinosaurs including iguanodonts and sauropods.

NEOSUCHIA

The thalattosuchians, dyrosaurs, and crocodylomorphs related to and including all the living species form a group known as the Neosuchia. These include over 100 additional genera, with extinct families such as the Jurassic-Cretaceous marine forms from Europe known as Atoposauridae (7 genera), and the Jurassic-Cretaceous freshwater neosuchians known as the Goniopholidae (7 genera), known from all over the world, and another 15 extinct genera that are closely related to the living Crocodylia, which include 7 primitive genera not closely related to the living species, plus 19 genera of gharials or gavials (only two are surviving), 18 species of alligators (only three survive), 11 species of caimans (only three survive), and 21 genera of Crocodylidae (only 2 survive). There is no room in a book like this to discuss all these hundred or so genera, let alone convey the subtle distinctions between them. Instead, let us look at some of the more striking examples of Crocodylia that have evolved over the past 150 million years.

Most impressive are some the gigantic crocodilians that have evolved at various times and places. During the late Miocene in Brazil, Colombia, and Peru, for example, there were some enormous caimans. The biggest of these was *Purussaurus*, a monster caiman (**Figure 12.9[B]**) that reached 12.5 meters (41 feet) in length, and weighed about 8.4 metric tonnes (9.25 short tons). It was larger than any other predator (reptile or mammal) at its time, so it would have eaten even the largest mammals. Its bite force would have been about 7 metric tonnes, stronger than any crocodylomorph ever, and stronger than most dinosaurs.

Almost as big as *Purussaurus* was another monster caiman from the same beds known as *Mourasuchus*. Instead of the powerful bulldog bite, however, this caiman has a broad snout shaped more like a duck's bill, weak lower jaws, and rows of small conical teeth that would not have been good for a strong bite and grappling with large prey. In addition, it had a hugely expandable throat sac. Taken together, these features suggest that *Mourasuchus* fed more like a pelican or a baleen whale, using its snout to disturb the water and the bottom muds, then taking a huge gulp of water and food into its throat, and finally forcing the water back out through its teeth and swallowing the food. It was apparently a crocodilian trying to feed like a baleen whale.

Besides these two monstrous caimans, the late Miocene swamps of South America also hosted a gigantic fish-eating gharial called *Gryphosuchus*. In this land of giants, it was also enormous, reaching at least 10 meters (33 feet) in length, only slightly shorter than *Purussaurus*, and weighing about 1745 kg. These three were the largest crocodilians ever to live on the planet since the death of the dinosaurs. *Gryphosaurus*, *Purussaurus*, and *Mourasuchus* lived in the late Miocene swamps that also supported another huge crocodile (*Charactosuchus*) that was convergent on the narrow-snouted gharials.

But the biggest of all the crocodilians was the legendary *Deinosuchus* (Greek for "terrible crocodile") from the Late Cretaceous of North America (**Figure 12.9[C]**). Formerly known as *Phobosuchus* ("fear crocodile" in Greek), it has been found in localities from the East Coast to Texas to Colorado to Montana, as well as northern Mexico. It has mostly been found in deposits that were parts of deltas or estuaries along what is now the East Coast and Gulf Coast of North America, or along the edge of Cretaceous Interior Seaway, which covered the Great Plains of North America from Hudson's Bay to the Gulf of Mexico. At one time, it was thought to be related to the crocodylids, but better specimens have shown that it was an alligatorid, and indeed the broad snout with the bulbous tip make it look more like an alligator than any other type of crocodilian.

The largest specimens of this genus suggest a body length over 12 meters (40 feet), with a skull about 1.3 meters (4.3 feet) long, by far the largest crocodylomorph known. Its bite force has been estimated at 18,000 to 102,000 Newtons, far greater than any crocodilian known and probably even more powerful than a *Tyrannosaurus rex*. At such size, it could have preyed on many of the dinosaurs known at that time, especially the common duckbill dinosaurs. There are tail vertebrae of a duckbill from the Big Bend of Texas that have *Deinosuchus* tooth marks. It has also been suggested that it fed on turtles, since it has robust flatter teeth in the back of its jaws that would be good for shell crushing. Indeed, there are a number of specimens of the Cretaceous side-necked turtle *Bothremys* that show tooth marks consistent with *Deinosuchus*.

The archosaur branch that includes crocodilians and all their extinct relatives have a long and distinguished history. But the other branch has been even more successful, since it led to the pterosaurs, dinosaurs, and birds. That is the subject of the next six chapters.

FURTHER READING

Benton, M.J. 1985. Classification and phylogeny of diapsid reptiles. *Zoological Journal of the Linnean Society (London)*. 84: 97–164.

Benton, M.J., ed. 1988. *The Phylogeny and Classification of the Tetrapods, vol. 1: Amphibians, Reptiles, Birds*. Oxford Clarendon Press, Oxford.

Benton, M.J.; Clark, J. 1988. Archosaur phylogeny and the relationships of the Crocodylia, pp. 295–338. In Benton, M.J., ed. *The Phylogeny and Classification of the Tetrapods, vol. 1: Amphibians, Reptiles, Birds*. Clarendon Press, Oxford.

Brochu, C.A. 2003. Phylogenetic approaches toward crocodilian history. *Annual Review of Earth and Planetary Sciences*. 3: 357–397.

Brusatte, S.L.; Benton, M.J.; Desojo, J.B.; Langer, M.C. 2010. The higher-level phylogeny of Archosauria (Tetrapoda: Diapsida). *Journal of Systematic Palaeontology*. 8: 1–47.

Buffetaut, E. 1992. Evolution, pp. 26–41. In Ross, C.A., ed. *Crocodiles and Alligators*. Blitz, New York.

Holliday, C.M.; Gardner, N.M. 2012. A new eusuchian crocodyliform with novel cranial integument and its significance for the origin and evolution of Crocodylia. *PLoS ONE*. 7 (1): e30471.

Mateus, O.; Puértolas-Pascual, E.; Callapez, P.M. 2018. A new eusuchian crocodylomorph from the Cenomanian (Late Cretaceous) of Portugal reveals novel implications on the origin of Crocodylia. *Zoological Journal of the Linnean Society*. 186 (2): 501–528.

McLoughlin, J.C. 1979. *Archosauria: A New Look at the Old Dinosaur*. Viking Press, New York.

Ross, C.A., ed. 1992. *Crocodiles and Alligators*. Blitz, New York.

Schwimmer, D.R. 2002. *King of the Crocodilians: The Paleobiology of* Deinosuchus. Indiana University Press, Bloomington, IN.

Stubbs, T.L.; Pierce, S.E.; Rayfield, E.J.; Anderson, P.S.L. 2013. Morphological and biomechanical disparity of crocodile-line archosaurs following the end-Triassic extinction. *Proceedings of the Royal Society B*. 280: 1940.

Sues, H.-D. 1992. The place of crocodilians in the living world, pp. 14–25. In Ross, C.A., ed. *Crocodiles and Alligators*. Blitz, New York.

PTEROSAURS

13

Life was very difficult for the average reptile in the Mesozoic era, what with the dinosaurs and the humidity, so some of them took to the air to get away from it all. The pterodactyls grew leathery wings attached to their outer digits and hind legs, which enabled them to fly in a clumsy sort of way. They tumbled through the air more or less as bats do today, and they were never quite sure where they were going to light. They were even worse off on land, as they were constantly tripping over their wings, involved as these were with the wrong parts of their body . . . It was clever of the pterodactyls think of flying but that's all you can say for them. They were doomed from the start because they had no feathers, and no wishbone, or furcula, as flying vertebrates should have. Pretty soon the Archaeopteryx, *a genuine bird, came along, and the pterodactyls faded away. They didn't belong in the picture and public opinion was against them. The* Archaeopteryx *was not much of a bird, but at least it had feathers. As for the pterodactyls, the best thing to do is just to forget them.*
—Will Cuppy, *How to Become Extinct*, 1941

ORNITHODIRA/AVEMETATARSALIA

As discussed in Chapters 11 and 12, there are two main branches of the archosaurs (**Figure 11.1**). One was the Pseudosuchia (also called Crurotarsi), the crocodilians and their kin reviewed already. The second group is the one that includes pterosaurs, dinosaurs, and birds, and their extinct relatives (**Figures 11.1** and **13.1**). For some time, this group was called the Ornithosuchia, until analysis of the original *Ornithosuchus* specimens showed that they were closer to crocodilians than they are to dinosaurs. Different names have been applied to this group, but Ornithodira ("bird joint") or Avemetatarsalia ("bird ankle") have been used the most in recent analyses.

Most of the members of this group are united by a very distinctive feature in their ankles, known as the mesotarsal joint. Instead of the usual hinge between the shin bone (tibia) and the first row of ankle bones (tarsals), the foot of pterosaurs, dinosaurs, and birds hinges between the first and second row of ankle bones (**Figure 11.8C**). As a consequence, the first row of ankle bones in these archosaurs is usually fused to the end of the tibia, and no longer functions as a separate series of ankle elements. The next time you eat a chicken or turkey drumstick, notice the caps of cartilage on the "handle" end of the bone. These are the remnants of the first row of ankle bones, now fused to the tibia ("drumstick") and no longer performing as separate ankle bones. Instead of a normal tibia, this bone is called a tibiotarsus. The rest of the ankle bones are fused to the foot bones (metatarsals), so the leg bone of most birds (the one with naked scales, not feathers) is called a tarsometatarsus.

Another feature of this group is the presence of a feather-like epidermal covering for insulation. Not only do birds have them, but also since the 1990s we have learned that some sort of feathers occurred in most groups of dinosaurs (although they may have been secondarily lost in

DOI: 10.1201/9781003128205-13

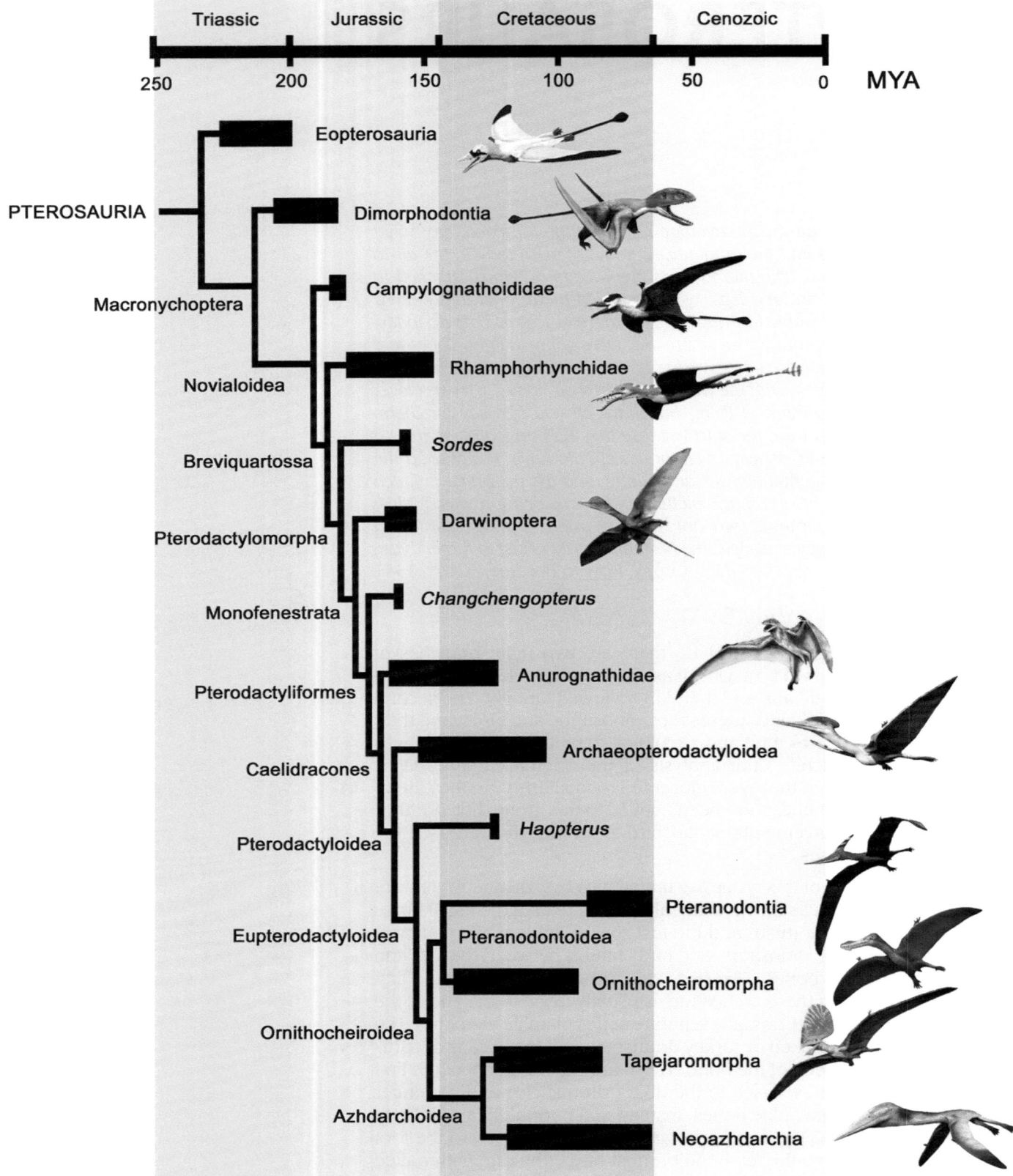

Figure 13.1 Family tree of the major groups of pterosaurs.

Figure 13.2 Restoration of Scleromochlus. Scale bar is 2 cm.

some). In 2018, it was confirmed that pterosaurs had a fine fuzz of what appear to be feathers on their bodies as well.

In addition to pterosaurs, and dinosaurs plus birds, a number of very primitive fossil relatives of the Ornithodira are known. One of the most primitive is the small (18 cm, or 5 inches long) Late Triassic fossil *Scleromochlus* from Scotland (**Figure 13.2**). Even though it is only known from the external molds of the now-vanished skeleton, it was very lightly built, with long delicate bones, and hind legs that were much longer than the front legs, and the foot bones are also long and clustered into a tight bundle. Some paleontologists have argued that *Scleromochlus* is actually closer to pterosaurs, although the evidence is unclear.

In 2017, Sterling Nesbitt and colleagues added a number of other very primitive Middle Triassic fossils to the mix, placing them in a group called "Aphanosauria". These include *Teleocrater* from the Tanzania, *Dongosuchus* from Russia, *Yarasuchus* from India, and *Spondylosoma* from Brazil. They have some derived features of the Ornithodira, but some do not yet have the specialized mesotarsal joint in the ankle.

FLYING REPTILES: THE PTEROSAURIA

Pterosaur Anatomy

Pterosaurs are one of three groups of vertebrates that independently evolved fully powered flight, along with bats and birds. This is distinct from gliding, which has happened in dozens of groups, including flying fish, parachuting frogs, several other groups of gliding reptiles, and at least three cases of gliding mammals. Unlike other flying animals, pterosaurs supported their wing membranes with a single elongate finger bone (the fourth, or "ring" finger). This bony support was supplemented with parallel rows of stiffened fibrous rods called actinofibrils that held the wing membrane semi-rigid (**Figure 13.3**). The wings were made of three different layers of actinofibrils and muscle layers, which were organized in crisscrossing pattern at angles to the adjacent layers. This may have given them considerable rigidity but could also deform under the control of the muscles, allowing some flexible control over the shape of the wing and gave them greater maneuverability in flight. By contrast, bats use all five fingers to support their wings, and birds have fused all their fingers together, supporting their wing with feather shafts instead.

The main membrane of the wing from the arm and fourth finger is known as the brachiopatagium ("arm membrane"). It stretched to the back of the body, and in some pterosaurs it apparently attached to the hind limb,

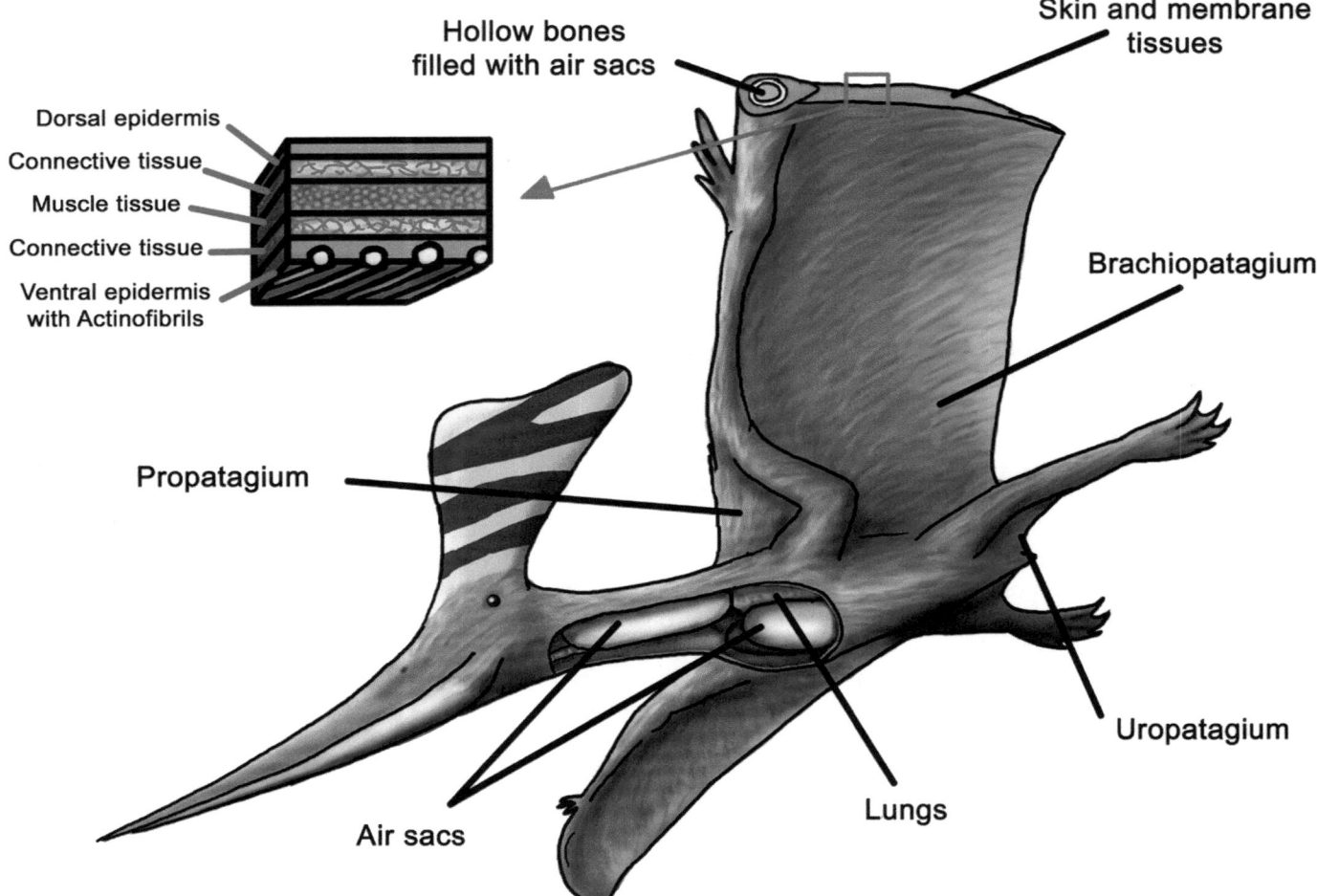

Dorsal epidermis
Connective tissue
Muscle tissue
Connective tissue
Ventral epidermis
with Actinofibrils

Hollow bones
filled with air sacs

Skin and membrane
tissues

Brachiopatagium

Propatagium

Uropatagium

Air sacs

Lungs

Figure 13.3 The detailed anatomy of pterosaurs, showing the unique structure of the wings, the hollow air sacs in their bodies, and their many adaptations to flight.

while in some pterosaurs it apparently did not attach to the legs, but ended just short of them. Some fossils appear to have had a membrane between the hind legs and incorporating the tail, called the uropatagium ("tail membrane"), but the evidence is not clear on many fossils. In addition, all pterosaurs appear to have had a small membrane along the leading edge of the wing in front of the arm, between the wrist and shoulder, called the propatagium. Protruding inward from their wrist was a long thin bone unique to pterosaurs called the pteroid bone, which supported the leading edge of this membrane.

Pterosaurs had many other specializations for flight as well, many of which evolved in parallel in other flying vertebrates, such as birds and bats. Their bones were very light and hollow to minimize their weight. The walls of some of the bones were often only a millimeter or so in thickness, like thin cardboard. The interiors of their hollow bones in many species connected to their respiratory system, aiding in air exchange and cooling the blood (a feature also found in birds). Since the 1870s, this has been interpreted as evidence of high metabolic levels associated with "warm-bloodedness", which makes sense in animals that exert themselves in flight over long periods. Thin hollow bones also making the body less dense (also a feature also found in birds). The breastbone was broadly expanded for the attachment of flight muscles, and a unique structure called the cristospine jutted out in front of the breastbone for the attachment of different muscles. There was only a small bony keel down the middle, unlike the large keel found in birds for much larger flight muscles. However, the large size of the breastbone suggests that there might have been a keel made of cartilage. (Bats also lack a keel on their breastbones, and a have a smaller breastbone than in pterosaurs and birds.) In addition, there was considerable fusion of many of the bones into a single solid element to make them rigid and stronger in flight. In most pterodacyloids, four or five of the chest vertebrae, including the arches on top of the vertebrate, were fused into a structure called a notarium, which had a socket on each site that connected to the shoulder blade. The entire shoulder girdle was fused, and articulated with the front of the breastbone. This allowed the shoulder girdle to "rock" slightly against the notarium and breastbone, absorbing some of the stresses of flight.

The pelvis was fused with the ribs in the hip region of the spine into a single bone called the synsacrum (the same thing occurred in a different way in birds). Pterosaur brains were relatively large, which no doubt helped to coordinate their complicated flight behavior.

The hindlimbs of pterosaurs were actually quite robust and strong, if you compare them to the size of their body. They only appear small from a distance because the wings, neck, and head are so enlarged. However, they were not strong enough to grab large prey or lift a human off the ground, as seen in movies like *Jurassic World*. They also did not have an opposable big toe, so they could not grasp a human or a prey item, or even perch on a branch like a bird can. Pterosaur hindlimbs are very similar to those of birds. The thigh was held in a horizontal position close to the body, and the knees and ankles were the main hinge joints. The feet pointed forward during walking, but could be directed outward, as in birds and dinosaurs. Pterosaurs walked on the soles of their feet (and never on the tips of their toes as in all dinosaurs). In this respect they seem to have been much like crocodiles, who also walk on the soles of their feet. In crocodiles, the whole foot seems to contact the ground at the same time, not in the "heel-toe" contact as humans walk, and this was likely true also of pterosaurs. The fifth toe, or "pinky", was short and stubby in primitive pterosaurs, but in more advanced forms, it got to be a very long hooked structure.

There has been a long debate about how pterosaurs walked. They clearly evolved from bipedal animals like primitive archosaurs. There is no evidence from trackways in the first pterosaurs. It is not clear when they adopted a quadrupedal pose, but it was likely associated with the elongation of the wrist bones in pterodacyloids, which allowed the fingers to touch the ground. The first undoubted quadrupedal tracks of pterosaurs were discovered in the Upper Jurassic beds of France. They were clearly pterosaur tracks, because the hand prints were so far outside the midline of the body. These prints show no evidence of the arms pulling backward during walking. Instead, their hands and wings probably functioned as "walking sticks" while the hindlimbs provided all the propulsion.

Early researchers thought pterosaurs were cold-blooded animals which could only glide. But since the 1980s, scientists have shown that the joints of their wings, including the shoulder socket, the elbow, and the wrist and finger joints naturally articulated to execute the flight stroke, which is a very particular motion which creates a vortex wake behind the animal and pulls it forward. There is no longer any doubt that pterosaurs were excellent fliers, not just gliders. The biggest pterosaurs (as also happens in the largest birds) probably did little flapping, because it takes too much energy. Instead, they soared on thermal air currents and winds, like condors and vultures and eagles do.

A number of specimens have been preserved which show some sort of fuzzy covering, which since 2009 have been called pycnofibers. They appear to be hair-like in some specimens, with central hollow shafts.

Then in 2018, extraordinarily preserved specimens from the Jurassic of Inner Mongolia were reported which showed these structures even better, and they appeared to have frayed ends and resembled certain types of body feathers and down feathers.

Pterosaur Evolution

The first pterosaur found in 1784 was the robin-sized *Pterodactylus* and later in 1839 *Rhamphorhynchus* was discovered (**Figures 13.4[A,B]** and **13.6[A,D]**), both from the Upper Jurassic Solnhofen Limestone of Bavaria, the same rock unit that later yielded *Archaeopteryx*. These small delicate fossils were a puzzle at first, and some early scholars suggested that they swam with their long fingers and hands. But in 1800, Jean Hermann of Manheim argued that the long delicate finger bones supported a wing membrane, and in 1809, Baron Georges Cuvier confirmed this, and gave them the name *Ptéro-dactyle* ("wing finger" in Greek).

One of the earliest pterosaurs to be found outside Germany was *Dimorphodon* (**Figure 13.5[A]**), discovered by the legendary pioneering collector Mary Anning in 1828 in the Lower Jurassic marine rocks near Lyme Regis in England. It was also one of the most unusual pterosaurs in that its skull is very large, with high arches above the eyes and snout, like the condition in other basal archosaurs. It had not yet evolved the long narrow snout that is typical of all other known pterosaurs. It was originally assigned to *Pterodactylus*, but after complete skulls were found in 1859, Richard Owen gave it the name *Dimorphodon* (Greek for "two-shaped teeth"), because not only it had the simple peg-like teeth of most archosaurs in the back of the jaws, but in front of the upper and lower jaws it had four or five long fang-like teeth as well. Since then, seven other genera of archaic pterosaurs have been found, all from the Late Triassic. These include *Preondactylus* (**Figure**

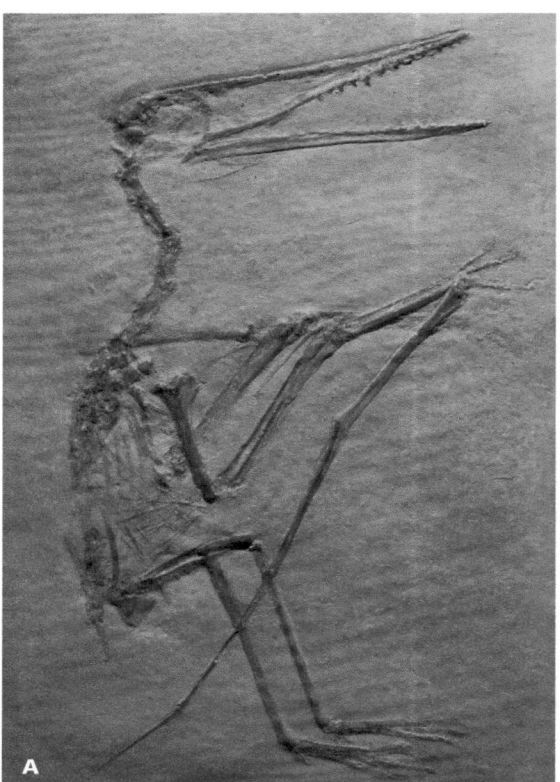

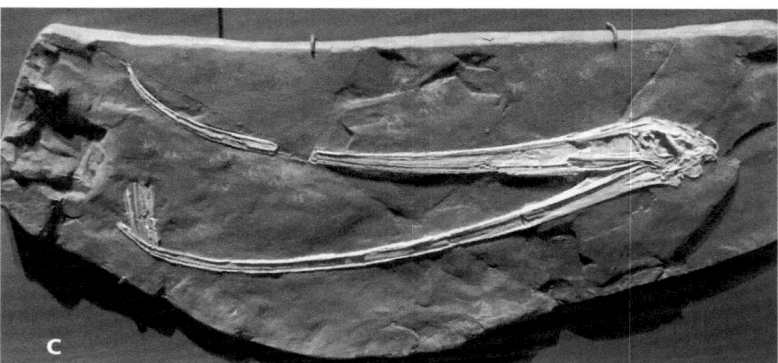

Figure 13.4 Some of the better-known pterosaur fossils. (A) *Pterodactylus*, (B) *Rhamphorhynchus*, (C) *Pterodaustro*. [(C,E,G) By the author; the rest courtesy Wikimedia Commons.]

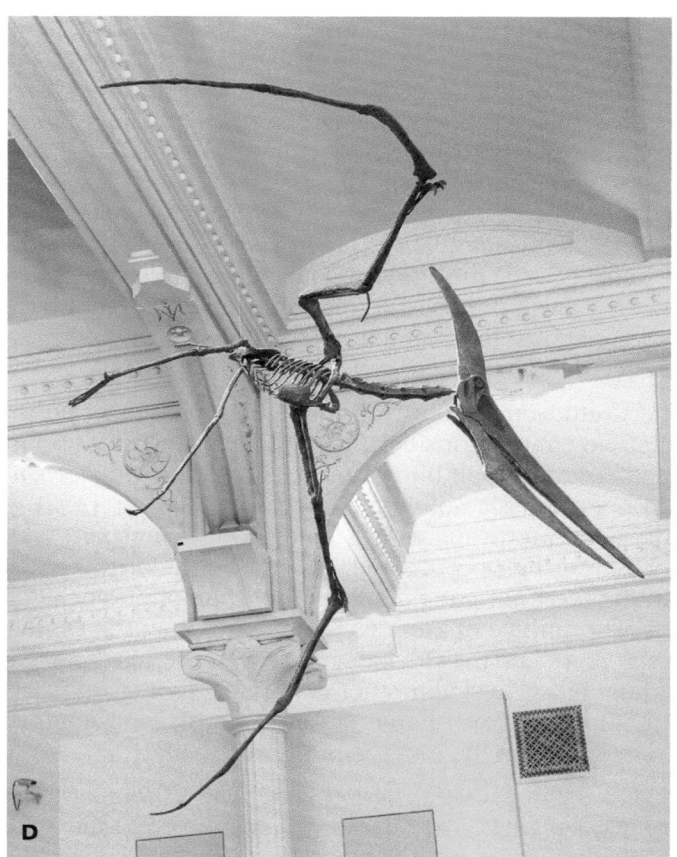

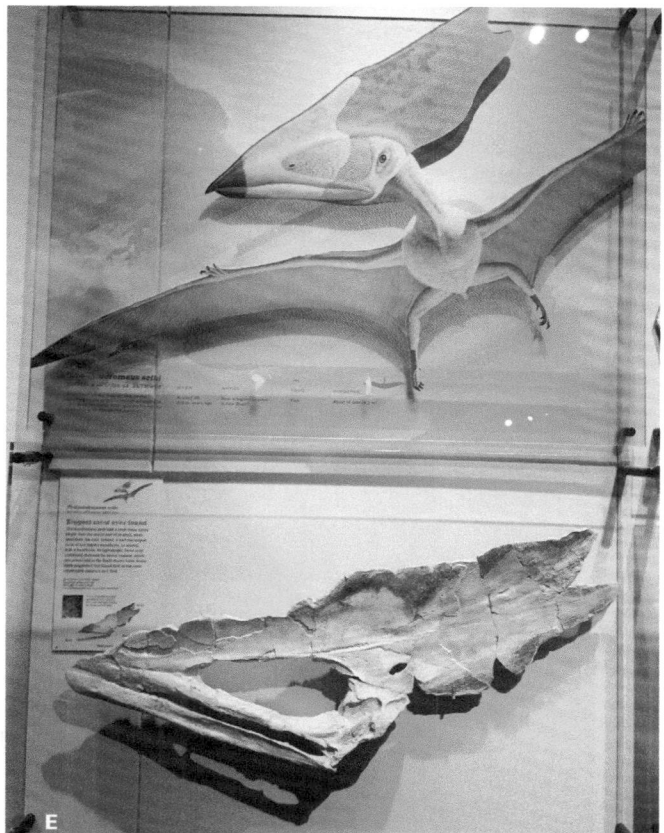

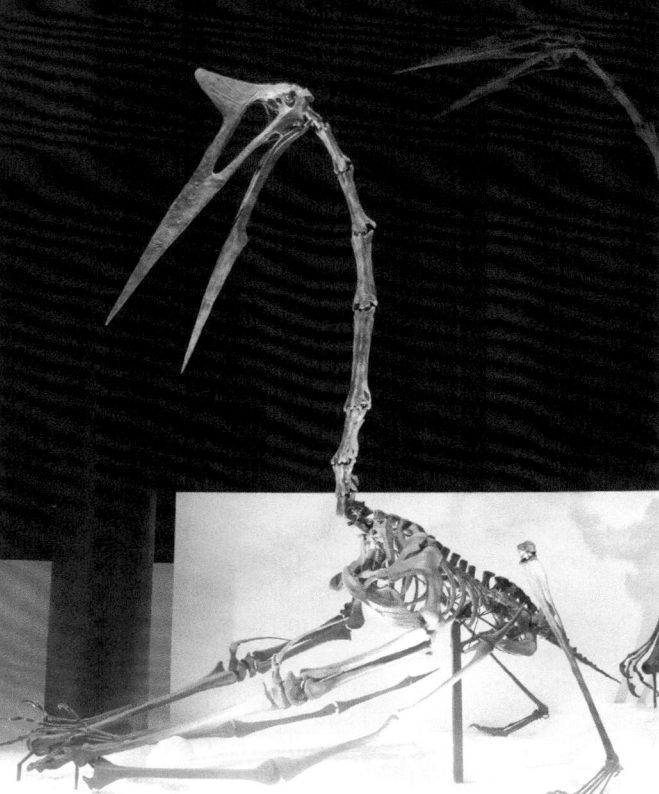

Figure 13.4 (Continued) (D) *Pteranodon*, (E) *Thalassodromeus*, (F) *Quetzalcoatlus*, (G) *Dzugaripterus*.

13.5[B]), *Eudimorphodon* (**Figure 13.5[C]**), *Carniadactylus* (**Figure 13.5[D]**), and *Peteinosaurus* (**Figure 13.5[E]**) from Italy, *Austriadactylus* from Austria (**Figure 13.5[F]**), *Raeticodactylus* and *Caviramus* from Switzerland, and *Arcticodactylus* from Greenland. Some of these early forms had long rows of teeth that had two or three cusps on them, apparently useful for seizing and crushing small fish and other animals. Although the first fossils we recognize as pterosaurs (by their elongate forelimbs) are not know until the Late Triassic, the earliest dinosaur relatives are known by the early Middle Triassic and so the split between the dinosaur and pterosaur lineages must have happened by then (even if those earliest pterosaurs might not have evolved flight by then). Pterosaurs eventually became widespread across the European and Greenland part of Pangea.

In the Jurassic and Cretaceous, there were dozens of genera of pterosaurs, so many that we cannot consider them all here. The most primitive ones include the many small bird-sized forms like *Rhamphorhynchus* found in the Upper Jurassic Solnhofen Limestone (**Figure 13.4[B]**), a distinctive raven-sized pterosaur with a meter-long wingspan, with the oval-shaped vane on the tip of its tail, and the long narrow snout with forward-pointing teeth for catching fish. The Rhamphorhynchidae were among the most primitive of the pterosaurs other than the Dimorphodontidae (**Figure 13.1**), and they include not only the genus that gave its name to the group, but also seven other genera, including *Dorygnathus* from the Early Jurassic of Germany (**Figure 13.6[B]**), and *Angustinaripterus* and *Sercipterus* from the Jurassic of China.

A slightly more advanced group was the Agnurognathidae, a group of about five known genera (**Figure 13.1**). They had a very short (or no) tail, with teeth suitable for catching insects and very large eye sockets suggesting they were night fliers. *Jeholopterus* from the Middle Jurassic of Inner Mongolia is typical of the group (**Figure 13.6[C]**). The fossils of this genus are extremely well preserved, with the pyncnofibers closely resembling feathers; the fossil also show the three-layered structure of the wing membrane. *Jeholopterus* had a short tail, robust wing bones, and a wing membrane that may have reached to the ankle. A slightly more advanced genus was *Kryptodrakon* from the Upper Jurassic of China. With a wingspan of 1.47 meters (4.8 feet), it was a bit larger than the Dimorphodontidae, Rhamphorhynchidae, and Anurognathidae. It had a long pointed toothless beak resembling that of *Pteranodon*, and a small crest on the back of its head.

One major branch of the Pterodactyloidea was the Archaeopterydactyloidea (**Figure 13.1**), a group that includes the original *Pterodactylus* (**Figure 13.4[A]**), *Germanodactylus*, and 13 genera of Ctenochasmatidae. The most bizarre of these was *Pterodaustro* from the Upper Cretaceous of Argentina (**Figure 13.6[E]**), which had a long, curved beak with hundreds of tiny strainer teeth, like the baleen of whales. Presumably it used this beak to strain out small crustaceans and other tiny prey from the water, analogous to the feeding mechanism of living flamingos.

The other major branch of the Pterodactyloidea was the Eupterodactyloidea, which is split into two major branches, the Pteranodontoidea and the Azhdarchoidea (**Figure 13.1**). The Pteranodontoidea are typified by the famous genus *Pteranodon*, which had a 7-m (25-foot) wingspan, and a long crest on the back of its head (**Figures 13.4[D]** and **13.6[G]**). It is best known from the Cretaceous marine rocks of Kansas, so it presumably soared over the sea, plucking fish and squid from the surface waters. Closely related to *Pteranodon* was the family Nyctosauridae, which include five genera, also from the Late Cretaceous marine deposits of Kansas. *Nyctosaurus* (**Figure 13.6[F]**) and its kin were apparently highly specialized soarers who spent their entire lives at sea and possibly seldom came to land, based on the shape of their wings and other skeletal

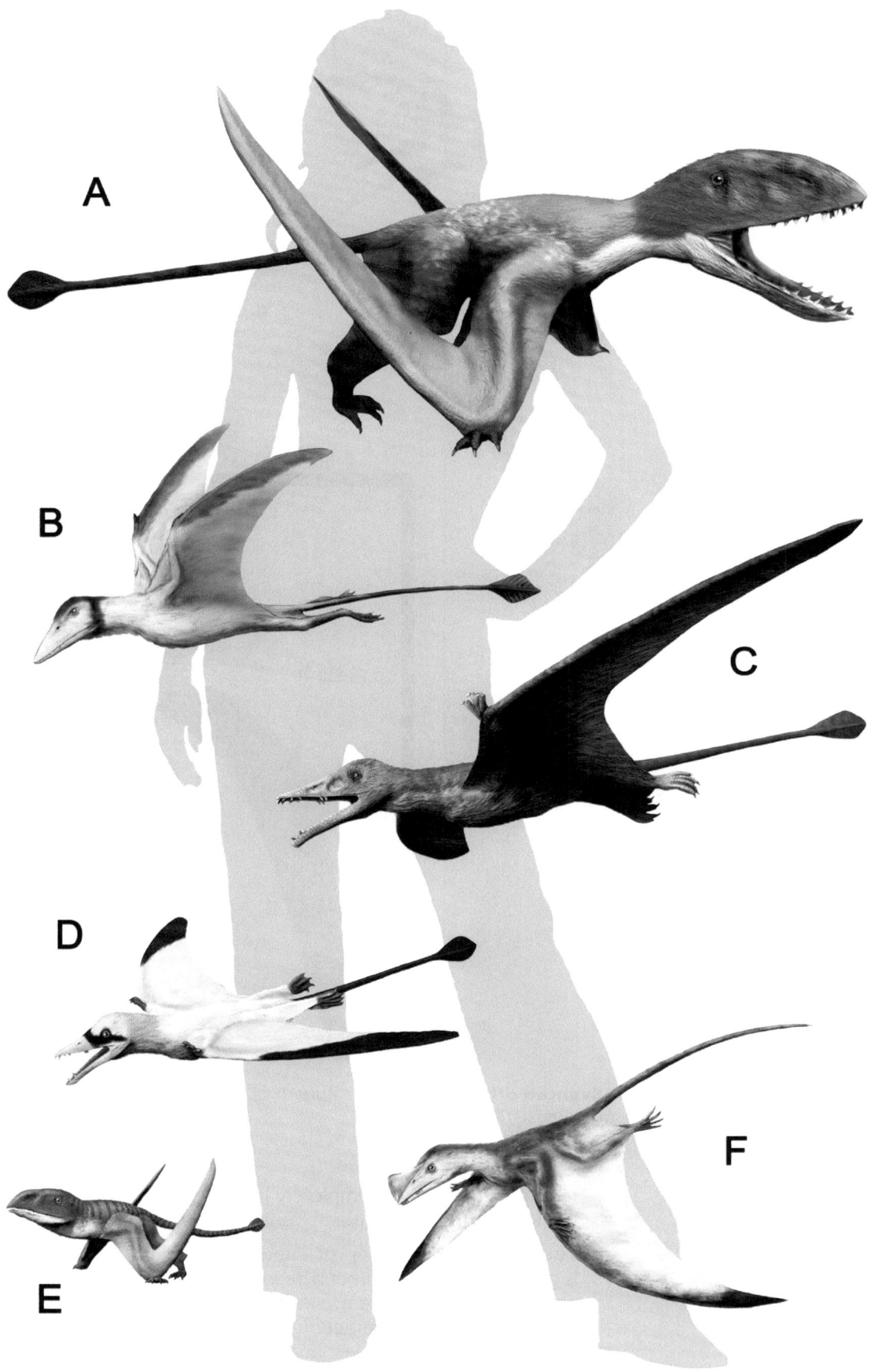

Figure 13.5 Reconstructions of primitive pterosaurs in the Eopterosauria and Dimorphodontia: (A) *Dimorphodon*, (B) *Preondactylus*, (C) *Eudimorphodon*, (D) *Carniadactylus*, (E) *Peteinosaurus*, (F) *Austriadactylus*.

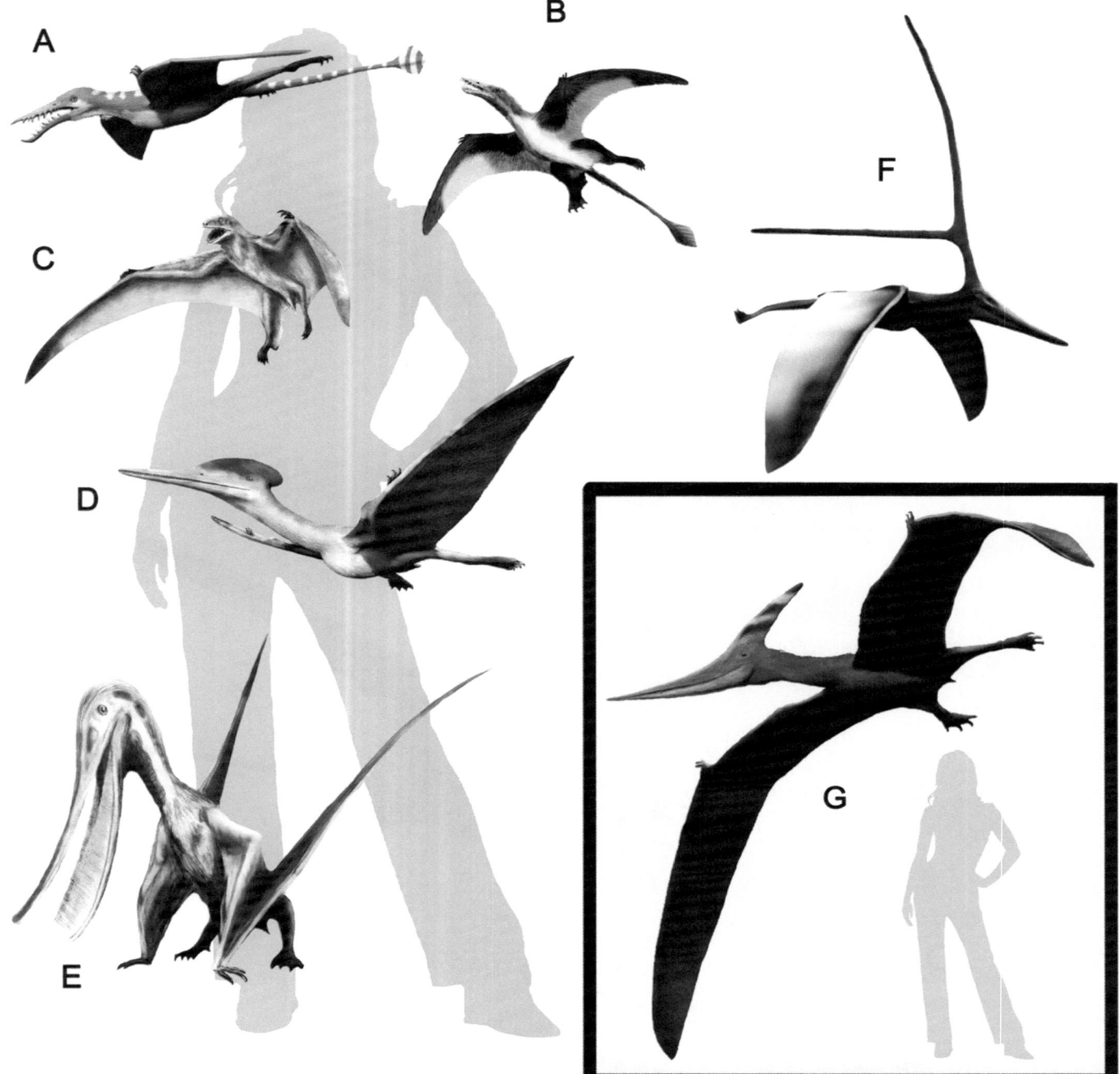

Figure 13.6 Reconstructions of more advanced pterosaurs: (A) *Rhamphorhynchus*, (B) *Dorygnathus*, (C) *Jeholopterus*, (D) *Pterodactylus*, (E) *Pterodaustro*, (F) *Nyctosaurus*, (G) *Pteranodon*.

features. They have completely lost all the fingers in their hand except the long fourth digit that supports the wing membrane.

The second large group of eupterodactyloids was the Azhdarchoidea, a group that has only been recognized and fully understood in recent years. Nineteen genera are currently known from the family Azhdarchidae. Most are Late Cretaceous forms ranging from small pterosaurs the size of a cat to the largest animals ever to fly. They have several somewhat exaggerated pterosaurian features, including extremely long necks and legs, and relatively large heads with toothless jaws that tapered to a point, like those of an egret or stork. On the basis of the shapes of their necks and heads, they were thought to have fed more like storks (**Figure 13.4[F]**), spending most of the time walking on the ground and using their long necks and spear-like beaks to feed on small prey on the ground or in shallow water, like egrets. Typical of these forms is *Hatzegopteryx* of Rumania, *Aerotitan* from Argentina, and many others (**Figure 13.7[A,C]**). The largest of all, however, was the giant Texas pterosaur, *Quetzalcoatlus*, with an 11- to 12-m wingspan (**Figures 13.4[F]** and **13.7[B]**); it was as large as a small airplane.

Figure 13.7 Reconstructions of some azhdarchoid pterosaurs: (A) *Hatzegopteryx*, (B) *Quetzalcoatlus*, (C) *Aerotitan*, (D) *Tupandactylus*, (E) *Tapejara*, (F) *Dsungaripterus*.

The other branches of the Azhdarchoidea include a wide spectrum of pterosaurs, often with bizarre crests. *Thalassodromeus*, *Tupandactylus*, and *Tapejara* from the Early Cretaceous of Brazil had strange tall curved crests on the top of their heads (**Figures 13.4[E]** and **13.7[D,E]**) and *Europejara* from the Early Cretaceous had an even taller, stranger crest. *Dzungaripterus* from the Early Cretaceous of China had a long narrow crest along the midline of the skull, and a weird toothed beak that curved upward at the tip (**Figures 13.4[G]** and **13.7[F]**), and there are 7 other genera in the family Dzugaripteridae, all with equally strange heads.

This is but a very small taste of the diversity of over 100 genera of pterosaurs now known, but it suggests how weird they got. Most of these have been discovered only in the past 20 or 30 years. If you consider how delicate their bones are, and how rarely they fossilize, this suggests that they were an extremely successful group of flying animals. If they had not vanished during the end-Cretaceous event, they might still be flying overhead and competing with the birds today.

FURTHER READING

Alexander, D.E.; Vogel, S. 2004. *Nature's Flyers: Birds, Insects, and the Biomechanics of Flight*. Johns Hopkins University Press, Baltimore, MD.

Alexander, R.M. 1989. *Dynamics of Dinosaurs and Other Extinct Giants*. Columbia University Press, New York.

Andres, B.; Clark, J.M.; Xing, X. 2010. A new rhamphorhynchid pterosaur from the Upper Jurassic of Xinjiang, China, and the phylogenetic relationships of basal pterosaurs. *Journal of Vertebrate Paleontology*. 30 (1): 163–187.

Bennett, S.C. 1994. Taxonomy and systematics of the Late Cretaceous pterosaur *Pteranodon* (Pterosauria, Pterodactyloidea). *Occasional Papers of the Natural History Museum of the University of Kansas*. 169: 1–70.

Bennett, S.C. 1996. The phylogenetic position of the Pterosauria within the Archosauromorpha. *Zoological Journal of the Linnean Society*. 118 (3): 261–308.

Bennett, S.C. 2000. Pterosaur flight: The role of actinofibrils in wing function. *Historical Biology*. 14 (4): 255–284.

Bennett, S.C. 2007. Articulation and function of the pteroid bone of pterosaurs. *Journal of Vertebrate Paleontology*. 27 (4): 881–891.

Benton, M.J. 1999. *Scleromochlus taylori* and the origin of dinosaurs and pterosaurs. *Philosophical Transactions of the Royal Society B: Biological Sciences*. 354 (1388): 1423–1446.

Benton, M.J.; Xu, X.; Orr, P.J.; Kaye, T.G.; Pittman, M.; Kearns, S.L.; McNamara, M.E.; Jiang, B.; Yang, Z. 2019. Pterosaur integumentary structures with complex feather-like branching. *Nature Ecology & Evolution*. 3 (1): 24–30.

Claessens, L.P.; O'Connor, P.M.; Unwin, D.M. 2009. Respiratory evolution facilitated the origin of pterosaur flight and aerial gigantism. *PLoS ONE*. 4 (2): e4497.

Dyke, G.J.; McGowan, C.; Nudds, R.L.; Smith, D. 2009. The shape of pterosaur evolution: Evidence from the fossil record. *Journal of Evolutionary Biology*. 22 (4): 890–898.

Dyke, G.J.; Nudds, R.L.; Rayner, J.M. 2006. Limb disparity and wing shape in pterosaurs. *Journal of Evolutionary Biology*. 19 (4): 1339–1342.

Elgin, R.A.; Hone, D.W.; Frey, E. 2011. The extent of the pterosaur flight membrane. *Acta Palaeontologica Polonica*. 56 (1): 99–111.

Frey, E.; Tischlinger, H.; Buchy, M.-C.; Martill, D.M. 2003. New specimens of Pterosauria (Reptilia) with soft parts with implications for pterosaurian anatomy and locomotion. *Geological Society, London, Special Publications*. 217 (1): 233–266.

Hone, D.W.E.; Benton, M.J. 2007. An evaluation of the phylogenetic relationships of the pterosaurs to the archosauromorph reptiles. *Journal of Systematic Palaeontology*. 5 (4): 465–469.

Hone, D.W.E.; Witton, M.P.; Martill, D.M., eds. 2018. New perspectives on pterosaur palaeobiology. *Geological Society of London Special Publication*. 455: 1–238.

Kellner, A.W.A.; Wang, X.; Tischlinger, H.; Campos, D.; Hone, D.W.E.; Meng, X. 2009. The soft tissue of *Jeholopterus* (Pterosauria, Anurognathidae, Batrachognathinae) and the structure of the pterosaur wing membrane. *Proceedings of the Royal Society B*. 277 (1679): 321–329.

Lawson, D.A. 1975. Pterosaur from the latest Cretaceous of West Texas: Discovery of the largest flying creature. *Science*. 187 (4180): 947–948.

McGowan, C. 1983. *The Successful Dragons: A Natural History of Extinct Reptiles*. Samuel Stevens, Toronto.

McGowan, C. 1991. *Dinosaurs, Spitfires, and Sea Dragons*. Harvard University Press, Cambridge.

Padian, K. 1979. The wings of pterosaurs: A new look. *Discovery*. 14: 20–29.

Padian, K. 1983. A functional analysis of flying and walking in pterosaurs. *Paleobiology*. 9 (3): 218–239.

Unwin, D.M. 2005. *The Pterosaurs from Deep Time*. Pi Press, London.

Unwin, D.M. 2010. *Darwinopterus* and its implications for pterosaur phylogeny. *Acta Geoscientica Sinica*. 31 (1): 68–69.

Wellnhofer, P. 1991. *The Illustrated Encyclopedia of Pterosaurs*. Crescent Books, New York.

Witmer, L.M.; Chatterjee, S.; Franzosa, J.; Rowe, T. 2003. Neuroanatomy of flying reptiles and implications for flight, posture and behaviour. *Nature*. 425 (6961): 950–953.

Witton, M.P. 2013. *Pterosaurs: Natural History, Evolution, Anatomy*. Princeton University Press, Princeton, NJ.

Witton, M.P.; Martill, D.M.; Loveridge, R.F. 2010. Clipping the wings of giant pterosaurs: Comments on wingspan estimations and diversity. *Acta Geoscientica Sinica*. 31: 79–81.

THE ORIGIN OF DINOSAURS

14

[Dinosaurs] *includes two distinct reptilian orders. Consequently, the word dinosaur is now a convenient vernacular name but not a systematic one.*

—Edwin H. Colbert, 1955, *Evolution of the Vertebrates*

WHAT IS A DINOSAUR?

When talking to the public, often the first thing a paleontologist has to deal with is the public's misconceptions about dinosaurs. Many people think that *any* large extinct animal is a dinosaur. Merchants sell bags of plastic dinosaur toys mixed with non-dinosaurs (such as the saber-toothed cats and mammoths), which reinforces this misconception. Likewise, those same plastic "dinosaur toy" sets often include the fin-backed protomammal *Dimetrodon*, since it is also large and prehistoric—even though it is part of the mammal lineage. In movies like *Jurassic World* and TV shows like *Walking with Dinosaurs*, and in many other media, we see images of marine reptiles (mosasaurs, long-necked plesiosaurs, dolphin-like ichthyosaurs). Most people assume that since they are large reptiles of the Age of Dinosaurs, they must be dinosaurs too.

Another common misconception is that all dinosaurs are large—but a great many were small, including a lot that were the size of small birds. A little more excusable is the common misconception that pterosaurs are dinosaurs. The public ignorance is so annoying to paleontologists that Dr. Mark Norell of the American Museum of Natural History in New York wrote a cleverly illustrated children's book called *I Am NOT a Dinosaur!* Every page shows a non-dinosaur that the public thinks is a dinosaur, reinforcing the same point over and over again. (On the other hand, if you show an image of a bird, most of the public will *not* call it a dinosaur—but it is.)

If none of these creatures is a dinosaur, what do paleontologists mean by the term? When the very first known dinosaurs were only *Megalosaurus* and *Iguanodon* and a few others, Richard Owen named and diagnosed it as a group of huge extinct reptiles with a number of distinctive features. As the number of new dinosaur discoveries (and tiny dinosaurs) increased rapidly in the late 1800s and early 1900s, that definition was modified. By the time of Edward Drinker Cope's later work in the 1880s, and especially when Samuel Wendell Williston published *Osteology of the Reptiles* in 1925, the dinosaurs had been clearly separated from other groups of reptiles, such as the marine reptiles, pterosaurs, and others such as the protomammals (still called "mammal-like reptiles" back then). In 1878, Othniel Charles Marsh recognized four groups of dinosaurs: Sauropods, theropods, ornithopods, and stegosaurs, groups that are still valid today (**Figure 14.1**). But few of these authors gave an exact anatomical diagnosis of what constitutes a dinosaur.

DOI: 10.1201/9781003128205-14

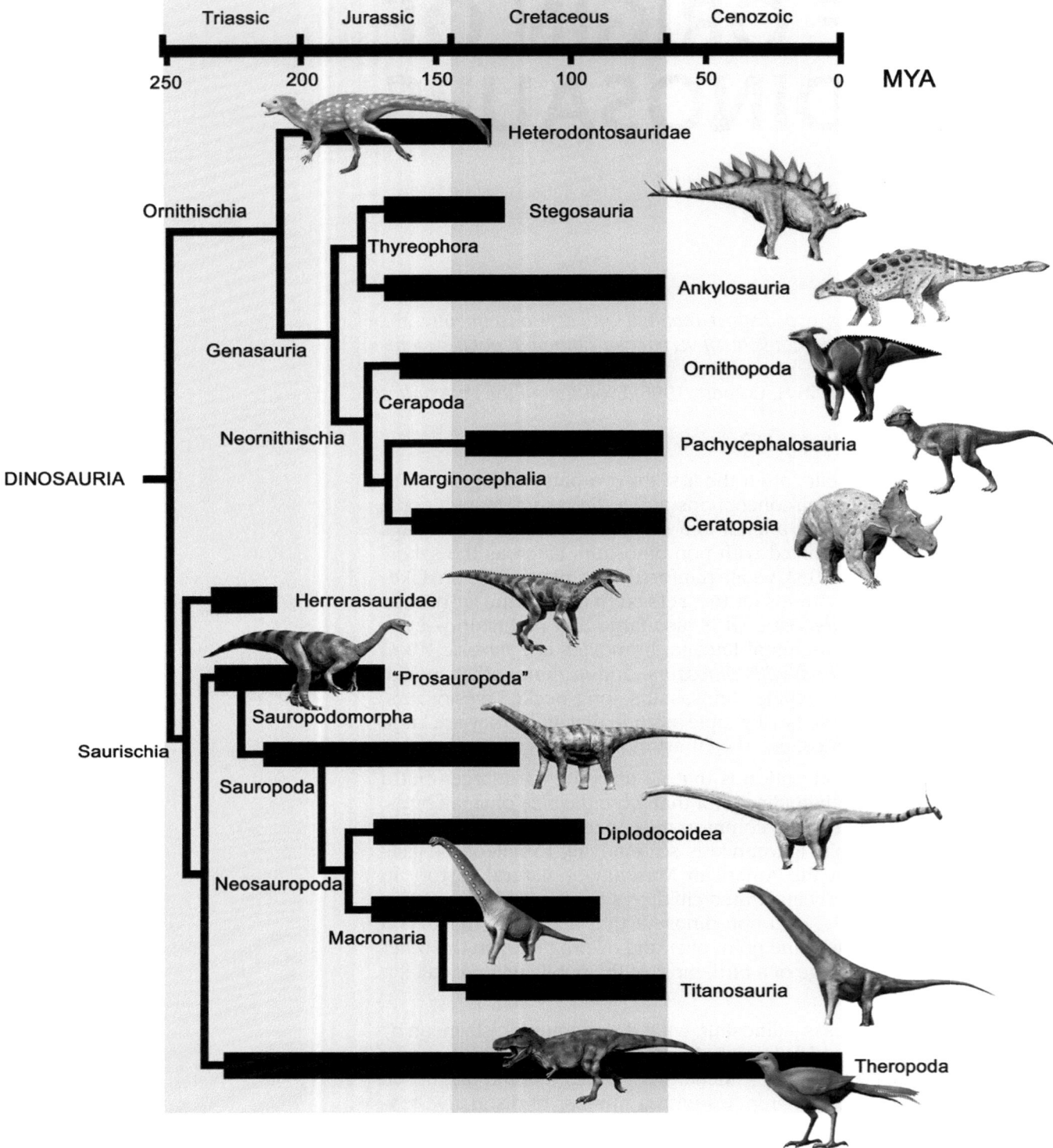

Figure 14.1 Family tree of the major groups of dinosaurs.

In 1888, British paleontologist Harry Govier Seeley recognized two groups of dinosaurs: the "lizard-hipped" dinosaurs or Saurischia (which he used to cluster together theropods and sauropods), and the "bird-hipped" dinosaurs, or Ornithischia (which include most of the herbivorous dinosaurs except sauropods). These ideas gained widespread acceptance for the next 130 years, so most paleontologists agreed that to be a dinosaur, a fossil had to be a member of either one of these two groups.

However, by 1974, Robert Bakker and Peter Galton showed that dinosaurs have a number of unique anatomical features that establish that they are a single natural group, and did not each arise from different "thecodonts" independently (see Chapter 11).

As more and more fossils were found, the differences between the two groups (based on hip structure) seemed to be consistent, and the individual groups (Sauropoda, Theropoda, and so on) continued to work well. The Saurischia were the "lizard-hipped" dinosaurs, with the pubic bone of the hip region pointing forward (**Figure 14.2**). The Ornithischia, or "bird-hipped" dinosaurs", had at least part of the pubic bone shifted backwards, parallel to the rear bone of the hip region, the ischium (**Figure 14.2**).

However major suborders diagnosed by Seeley did not answer the question of how the groups within Saurischia and Ornithischia are interrelated. Even in the early 1970s, paleontologists were not sure that Saurischia and Ornithischia could be combined into the Dinosauria. In his 1955 textbook *Evolution of the Vertebrates*, Edwin Colbert wrote (see epigraph at the beginning of the chapter) that "the term includes two distinct reptilian orders. Consequently, the word dinosaur is now a convenient vernacular name but not a systematic one". This sad misunderstanding might have been typical of thinking in 1955 and 1969, when the first two editions of Colbert's book were published, but unfortunately the text remained the same even in the last edition in 2001, when this idea had been resoundingly debunked.

This was because in the 1970s and 1980s, biologists and paleontologists began to use a new method of classification which searched for unique evolutionary novelties that defined natural groups, and got away from "wastebasket" groups that were unnatural assemblages of unrelated animals. In 1985, Jacques Gauthier showed that one the clearest unique evolutionary specializations of all dinosaurs is that there is a hole right through the hip socket ("acetabulum" in anatomical terms), rather than a closed socket in the bone that holds the head of the thigh bone as it moves (**Figure 14.3**). No other animals have holes right through their hip sockets, so it a unique feature that defines Dinosauria.

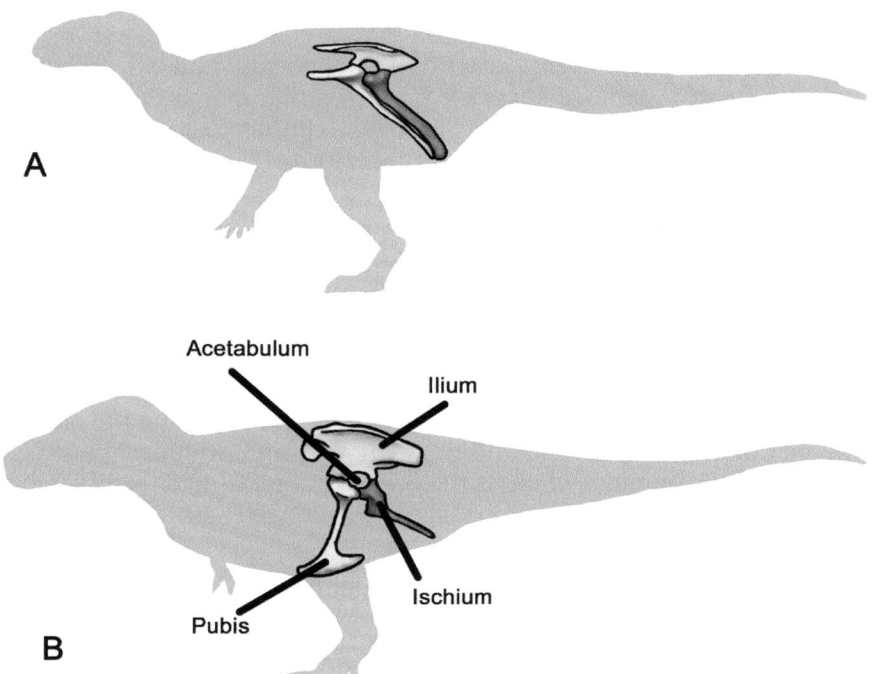

Figure 14.2 Ornithischian vs. saurischian hips. In saurischians (B), the three hip bones point in three different directions, with all the pubic bone pointed forward. In ornithischians (A), at least part or all of the pubic bone points backward.

Figure 14.3 Unique anatomical features that define the dinosaurs. (A) Enlarged deltopectoral crest on the humerus (upper arm bone). (B) Open hip socket (perforate acetabulum) with fusion of three sacral vertebrate to the upper hip (ilium). (C) A shelf on the iliac blade just behind the hip socket. (D) The lower end of the shinbone (tibia) is sub-rectangular in shape, and expanded side-to-side. (E) At the end of the tibia, the astragalus not only is fused (an ornithodiran or avemetatarsalian feature) but also has a distinctive flange of bone ("ascending process") that rises up from the astragalus to the front face of the tibia.

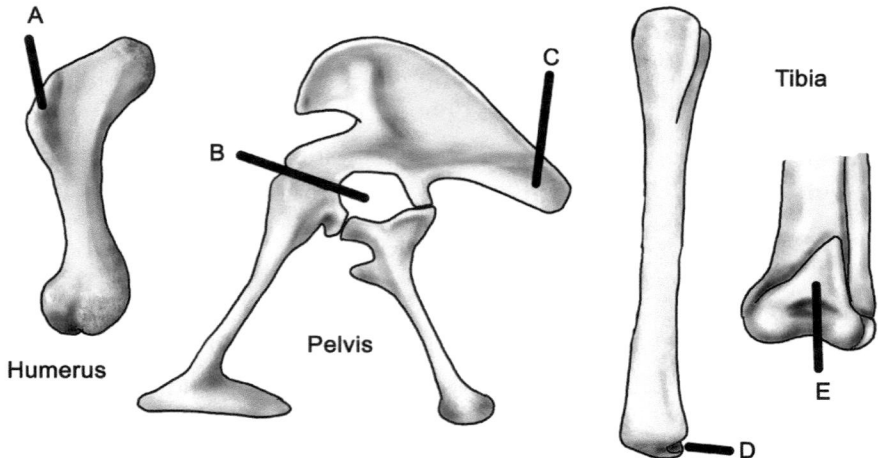

In 2017, British paleontologists Matthew Baron, David Norman, and Paul Barrett published a detailed analysis of dinosaur interrelationships. They hypothesized that when you look at the earliest theropods, sauropods, and ornithischians, there are no unique anatomical specializations that can be used to diagnose the Saurischia. In addition, other evidence suggests that the theropods and ornithischians might be even more closely related, rather than theropods plus sauropods together as "Saurischia". This idea is still controversial, and no other scientists have been able to confirm the hypothesis of Baron, Norman, and Barrett. Such arguments are highly technical and beyond the scope of this book, so we will focus on well-defined groups of genera and families instead.

Even easier to see, all dinosaurs had their limbs held straight beneath their bodies in an upright posture, just like mammals. Nearly all dinosaurs have just three or fewer fully developed fingers in their hands, with the ring finger and pinky highly reduced or missing (**Figure 14.3**). Dinosaurs all walk on the tips of their fingers and toes, with only three toes remaining in their feet (the fourth toe and little toe is reduced or lost). Finally, there are just three vertebrae fused to the upper part of the hip bones, connecting the spine to the hind legs and forming a sacrum. Other animals have fewer or more vertebrae in their hips.

Once the concept of what diagnoses a dinosaur and an archosaur became clearer, then we can look for fossils that match this definition. Indeed, the search has been going for over a century to find actual fossils that most closely resemble the likely ancestors of dinosaurs.

First, all the major groups of dinosaurs, such as the early relatives of sauropods (once called prosauropods) and the earliest theropods and ornithischians are found in Upper Triassic beds in many places around the world. So the likely place to look for these fossils would be Upper Triassic or even upper Middle Triassic beds. Unfortunately, beds of this age are not exposed in most places and most of those outcrops do not yield good terrestrial fossils. Nevertheless, a number of fossils have been suggested as close relatives of dinosaurs over the years. On the basis of the fact that early theropods (like *Coelophysis*) and ornithischians were small bipedal animals, whereas the early sauropods (like the prosauropods) were at least partially bipedal as well, the likeliest ancestor of the dinosaurs was a small bipedal predator.

One of the early candidates was the small bipedal archosaur *Saltopus* (**Figure 14.4[A]**), whose name means "hopping foot". It was only about 80–100 cm long (counting the long tail) and weighed about 1 kg (2.2 pounds). Described by Friedrich von Huene in 1910, it was based on fossils from Upper Triassic Lossiemouth Quarries in the Elgin Formation of northern Scotland. It was frequently featured on the big murals that showed Triassic archosaurs and early dinosaurs by paleoartists such as Rudolph Zallinger, and also in many of the children's dinosaur books before the 1970s as well. More recent studies, however, has shown it to be a close relative of dinosaurs, and not within the Dinosauria.

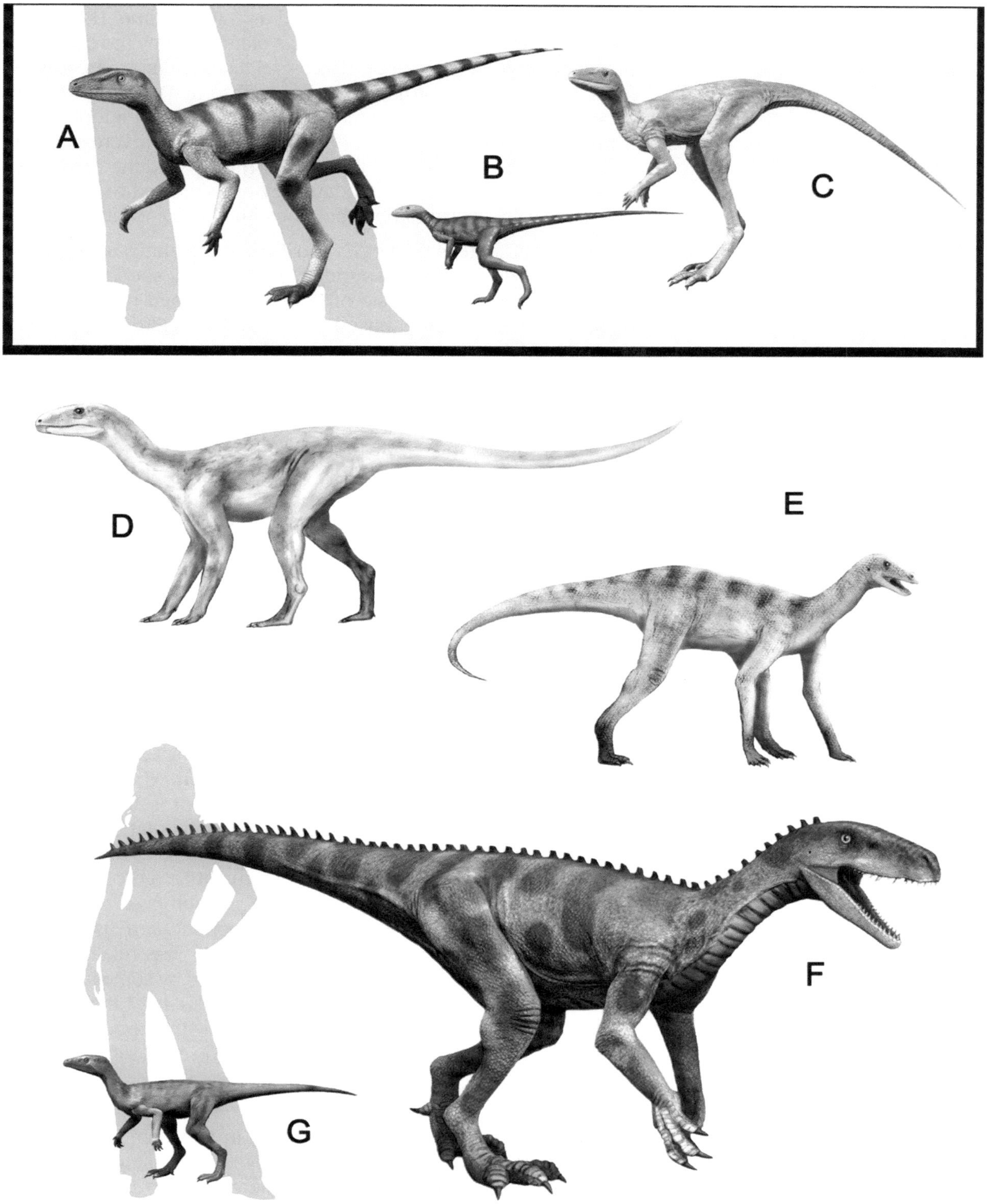

Figure 14.4 Reconstructions of some primitive relatives of dinosaurs: (A) *Saltopus*, (B) *Marasuchus*, (C) *Lagerpeton*, (D) *Silesaurus*, (E) *Asilisaurus*, (F) *Herrerasaurus*, (G) *Eoraptor*.

In 1971, Alfred Sherwood Romer described the little fossil *Lagosuchus* ("bunny crocodile") from the Upper Triassic Chañares Formation of Argentina (**Figure 14.4[B]**). On the basis of the extremely incomplete remains (just a hindleg, pelvis, shoulder blade, many vertebrae, and part of the skull) it was a lightly built archosaur with very long hind legs, suggesting that it could hop like a rabbit as well as run on both four legs and two legs. Later researchers argued that the name *Lagosuchus* was invalid, and the dinosaurian features of the specimen belonged to a different taxon called *Marasuchus*. More recently, *Lagosuchus* has been revived as a valid name. However, in 1975 Jose Bonaparte showed that all aspects of its anatomy are dinosaurian, and now it is considered one of the nearest relatives of dinosaurs. A similar fossil from the same rock formation was described by Romer in the same 1971 paper as *Lagerpeton* ("bunny reptile") (**Figure 14.4[C]**). It too is only known from scraps of the hips, hindlimb, and back and tail vertebrae, but many paleontologists pointed at both the "bunny croc" and the "bunny reptile" as likely dinosaurian relatives. Today, they are classified as near relatives of dinosaurs (Dinosauromorpha), but not within the Dinosauria.

Another candidate is *Silesaurus* and its nearest relatives (**Figure 14.4[D]**), making up the family Silesauridae. The original specimens were found from the early Late Triassic of Poland. Described by Polish paleontologist Jerzy Dzik in 2003 based on some 20 skeletons (so it one of the most completely known of the early dinosaur relatives), it was the size of a Great Dane, about 2.3 meters (7.5 feet) long because of its long tail, but not as high at the shoulders as a Great Dane. However, it was not primarily bipedal like the other candidates, but quadrupedal, and also had teeth suggesting it was herbivore, possibly with a beak on its toothless front part of the lower jaw. Most recent analyses place *Silesaurus* and its relatives (*Sacisaurus* from Brazil, *Eucoelophysis* from New Mexico, *Asilisaurus* from Tanzania (**Figure 14.4[E]**), and *Lewisuchus* from the Chañares Formation of Argentina) as the nearest kin of the dinosaurs (Dinosauromorpha) but not within Dinosauria itself. So the past 40 years have produced many different fossils that are very close to being true dinosaurs, but do not have all the key features that define Dinosauria.

In the year 1959, an Argentinian goatherd by the name of Victorino Herrera was following his flocks when he spotted fossil bones eroding out of the path. He brought in paleontologist Osvaldo Reig, who collected the specimen and described in 1963 it as *Herrerasaurus ischigualastensis* (**Figures 14.5** and **14.4[F]**) in honor of the man who found it; its species name comes from the lower Upper Triassic Ischigualasto Formation, where it was found. The original specimen had such a weird mixture of prosauropod and theropod features, however, that for decades no one was quite sure what kind of creature it was. Reig thought it might be a primitive allosaur or megalosaur, but in 1964 Alick Walker thought it might be a prosauropod. In 1985, Alan Charig noted it has similarity to both prosauropods and theropods, but Romer's 1966 textbook *Vertebrate Paleontology* put *Herrerasaurus* among the prosauropods. Edwin Colbert suggested that it was related to theropods, which was supported by Bonaparte in 1970, and by some later authors. Yet Don Brinkman and Hans-Dieter Sues argued in 1987 that it has features that are present in all early dinosaurs, including theropods, sauropods, and ornithischians; this was confirmed by Fernando Novas in 1992.

The confusion about *Herrerasaurus* was largely due to the incomplete nature of the original specimens. But in 1988, Paul Sereno and his University of Chicago crew were working in the Ischigualasto Formation with legendary Argentinian dinosaur paleontologist Jose Bonaparte, and found the first complete skull and skeleton of *Herrerasaurus*, and finally a lot of key features that were missing from the debate were understood (**Figure 14.5**). Most authors since then have considered *Herrerasaurus*

Figure 14.5 Photo of skeletons of the larger *Herrerasaurus*, and the much smaller primitive dinosaur *Eoraptor*. (Courtesy Wikimedia Commons.)

to be one of the earliest saurischian dinosaurs, with some paleontologists favoring closer relationships to the sauropods, and others to the theropods.

For one of the most primitive dinosaurs known, *Herrerasaurus* was clearly bipedal, but not the tiny creature that many paleontologists had expected. Adults were up to 4 meters (16 feet) long including their very long neck and tail, and weighed about 350 kg (770 pounds). However, some adult specimens were only half this size, so there was enormous variability in their adult body sizes, possibly due to the differences between males and females.

Unlike many of the close relatives of dinosaurs we have just discussed, *Herrerasaurus* was completely bipedal with small front limbs and long, powerful running hind limbs. Like many running animals, the thighbones are relatively short and the toes elongated, with some loss of the side toes. There are stiffening features in the tail, showing that it was held out straight behind it to improve balance during running. Its hip bones did not have a large hole through the hips, but instead a bony hip socket with only a small opening (the beginning of the open hip socket seen in all other dinosaurs). Other features of the hip seem to be more like the condition in theropods.

The skull of *Herrerasaurus* was long and narrow, with long recurved teeth with serrated edges, suitable for slashing prey. It had an odd flexible joint in the lower jaw that let the animal slide its jaw back and forth to give it a grasping bite, pulling the prey back after initially biting down. However, the skull lacks nearly all the specializations that are found in nearly all later dinosaurs, another reason that it is considered a primitive relative of all the dinosaurian groups, but probably closest to sauropods.

Herrerasaurus was a large predator by Triassic standards, and probably could prey on most of the smaller animals found in the Ischigualasto Formation. A few skulls of *Herrerasaurus* show marks consistent with the bites of another *Herrerasaurus*, so there were definitely fights among the animals in the group. However, it was not the largest predator in its time. One of the skulls shows bite marks from an animal with teeth very different from *Herrerasaurus*, so it was probably bitten by the huge crocodilian relative *Saurosuchus* (Chapter 12).

For all its confusing features, *Herrerasaurus* seems to have a lot of specializations and does not closely resemble the likely ancestor of all the dinosaurs. Three years after working with Jose Bonaparte when they found the first complete specimens of *Herrerasaurus*, Paul Sereno and his crew from the University of Chicago were back in the "Valley of the Moon" in the Ischigualasto beds in the austral summer of 1991. Ricardo Martinez, a University of San Juan paleontologist working with Sereno, found some tiny bones sticking out of the rocks. After it was removed and prepared, they named it *Eoraptor lunensis*, "dawn raptor of the moon", in reference to the Valley of the Moon (**Figures 14.5** and **14.4[G]**). It made the cover of *National Geog*raphic and all the news media as the oldest and most primitive known dinosaur. It certainly is very primitive, and also a tiny bipedal animal as well. It was only about 1 meter (3.3 feet) long, about the size of a turkey, and probably weighed about 10 kg (22 pounds). The long bones all have hollow shafts, so *Eoraptor* was very lightly built, compared to the much heavier build of other close relatives of dinosaurs. The skull has relatively large eye sockets and a short snout, so it had great vision but was not built with the vicious jaws and teeth of *Herrerasaurus*. However, it lacked the sliding jaw joint seen in *Herrerasaurus* and many other theropod dinosaurs, which makes it different in this aspect. Unlike *Herrerasaurus* and later theropods, only its upper teeth curved backwards, another archaic feature. Its lower teeth were simple leaf-shaped structures never seen in any theropod, or in most other dinosaurs, either.

Like *Herrerasaurus* and many of the other bipedal archosaurs close to dinosaurs, it had short thighbones and long toes, specializations for rapid running. It had large claws on its three main toes, but apparently the fourth and fifth toes were tiny or lost, as in many dinosaurs. The spool-like centra in the spine were hollow, like many close relatives of dinosaurs. However, it was more similar to later dinosaurs like *Herrerasaurus* in having three vertebrae in the sacral region attached to hip, whereas *Herrerasaurus* has only two as in most other archosaurs.

The place of *Eoraptor* in the dinosaur family tree is controversial. It is clearly more primitive than *Herrerasaurus* and many other very primitive dinosaurs, but it still has features found in theropods. When Sereno and colleagues first described it in 1993, and again in 1995, they pointed out that it was one of the most primitive dinosaurs known, but they assigned it to the Theropoda, like they did *Herrerasaurus*. In their words, it is closest to "the hypothetical dinosaurian condition than any other dinosaurian subgroup". Phil Currie in 1997 thought it was closer to the common ancestor of all the dinosaurs, rather than a primitive theropod. But in 2011, Ricardo Martinez, Paul Sereno, and coauthors described another small early dinosaur from Argentina, *Eodromeus*, and argued that *Eoraptor* was more closely related to sauropods. This was disputed by Michael Benton, yet confirmed by a study by Alpadetti and coauthors in 2011. Then in 2011, Hans-Dieter Sues, Sterling Nesbitt, David Berman, and Amy Henrici argued that *Eoraptor* is a theropod and not a primitive relative of all the dinosaurs. Two years later, Sereno and coauthors reanalyzed the complete skeleton of *Eoraptor*, and returned to the idea that it was related to sauropods. Finally in 2017, the controversial study by Baron, Norman, and Barrett (which broke up the "Saurischia") placed *Eoraptor* at the very base of the theropods.

In short, the confusing mix of features in *Eoraptor* means that it is very close to the ancestral condition of dinosaurs and does not clearly fall within the theropods or sauropods. Paleontologists do not expect to find a fossil that is perfectly ancestral to any group, but *Eoraptor* comes as close as we have to an approximation of how dinosaurs started out.

FURTHER READING

Alexander, R.M. 1989. *Dynamics of Dinosaurs and Other Extinct Giants*. Columbia University Press, New York.

Fastovsky, D.E.; Weishampel, D.B. 2021. *Dinosaurs: A Concise Natural History* (4th ed.). Cambridge University Press, Cambridge.

Gauthier, J. 1986. Saurischian monophyly and the origin of birds. *Memoirs of the California Academy of Sciences*. 8: 1–55.

Gauthier, J. 1994. The diversification of the amniotes, pp. 129–159. In Prothero, D.R.; Schoch, R.M., eds. *Major Features of Vertebrate Evolution*. Paleontological Society Short Course 7. Paleontological Society, Lawrence, KS.

Lucas, S.G. 2005. *Dinosaurs, the Textbook* (5th ed.). W. C. Brown, Dubuque, Iowa.

McGowan, C. 1983. *The Successful Dragons: A Natural History of Extinct Reptiles*. Samuel Stevens, Toronto.

McGowan, C. 1991. *Dinosaurs, Spitfires, and Sea Dragons*. Harvard University Press, Cambridge.

Nesbitt, S.J.; Sues, H.-D. 2021. The osteology of the early-diverging dinosaur *Daemonosaurus chauliodus* (Archosauria: Dinosauria) from the *Coelophysis* Quarry (Triassic: Rhaetian) of New Mexico and its relationships to other early dinosaurs, *Zoological Journal of the Linnean Society*. 191 (1): 150–179.

Norman, D.B. 1985. *The Illustrated Encyclopedia of the Dinosaurs*. Crescent Books, New York.

Sereno, P.C. 1997. The origin and evolution of dinosaurs. *Annual Reviews of Earth and Planetary Sciences*. 25: 435–490.

Sereno, P.C. 1999. The evolution of dinosaurs. *Science*. 284 (5423): 2137–2147.

Sereno, P.C.; Forster, C.A.; Rogers, R.R.; Monetta, A.M. 1993. Primitive dinosaur skeleton from Argentina and the early evolution of Dinosauria. *Nature Research*. 361 (6407): 64–66.

Weishampel, D. B.; Dodson, P.; Osmolska, H., eds. 2007. *The Dinosauria* (2nd ed.). University of California Press, Berkeley.

ORNITHISCHIANS I

ORIGINS AND THE THYREOPHORA

15

The pubes also present two types. First there are the genera in which the bones are directed anteriorly and meet by a median symphysis and have no posterior extension except for the proximal symphysis with the ischium. This type is represented by Cetiosaurus, Ornithopsis, Megalosaurus, *and many genera figured by Professor Marsh. The second form of pubis has one limb which is directed backward parallel to the ischium, and another limb directed forward. It is typically seen in* Omosaurus *and* Iguanodon. *There are many variations in stoutness and details of form of the bones, but so far as I am aware these two plans comprise all the Dinosaurian genera.*

—Harry Govier Seeley, 1887, *On the Classification of Animals Commonly Named Dinosauria.*

THE ORNITHISCHIANS

There is still a controversy over "Saurischia" and whether sauropods and theropods are closely related (see Chapter 14). But the more that paleontologists study the Ornithischia and its members, the stronger the evidence becomes for it being a natural group. In addition to their unique hip configuration (**Figure 14.2**), ornithischians many other unique specializations. Most striking is an extra bone in the tip of their lower jaws that forms a beak, called the predentary (in front of the dentary bone that makes up the tooth-bearing part of the jaw). In fact, the presence of this bone is so distinctive and consistent that Marsh called the group the "Predentata" in 1894. Fortunately, Seeley's name Ornithischia already has priority, or we might be confusing a group of dinosaurs for a grouping of anteaters, sloths, and armadillos that were long called the "Edentata".

Beyond the hip structure and predentary bone, a long list of features confirms the reality of the grouping of ornithischians (**Figure 15.1**). The bones at the tip of the upper jaw and snout are usually toothless, and probably had a horny beak that occluded against the beak on the toothless predentary bone in the lower jaw. In the "eyebrow" area of the eye socket, ornithischians develop a bone called the palpebral, not found in any other dinosaur. The jaw joint was below the line of the tooth row, which was helpful for the leverage of the jaw needed to chew up plants. Nearly all ornithischians have simple "leaf-shaped" teeth, suitable for cropping vegetation. Most ornithischians also have their tooth rows inset deep in the skull, creating a region where fleshy cheeks might have covered the sides of their mouths. This would help in keeping food in their mouths as they chewed. Finally, there are ossified tendons all through the backbone, hips, and tail, and in advanced ornithischians, five of the hip vertebrae were fused to the pelvis (primitively, only 3 were fused in most dinosaurs).

Paradoxically, the name "Ornithischia" means "bird-hipped" dinosaurs—yet birds are actually descended from the Saurischia. The earliest relatives

DOI: 10.1201/9781003128205-15

Figure 15.1 Unique anatomical specializations found only in the ornithischians. These include a predentary bone on the front of the lower jaw (covered with a beak), a palpebral bone in the eyebrow, a reduced antorbital fenestra in the face, a jaw joint which hinges well below the plane of the tooth row, and (not visible in the side view) the teeth are inset in a trough, so there were probably cheeks covering them and holding in the food as they chewed.

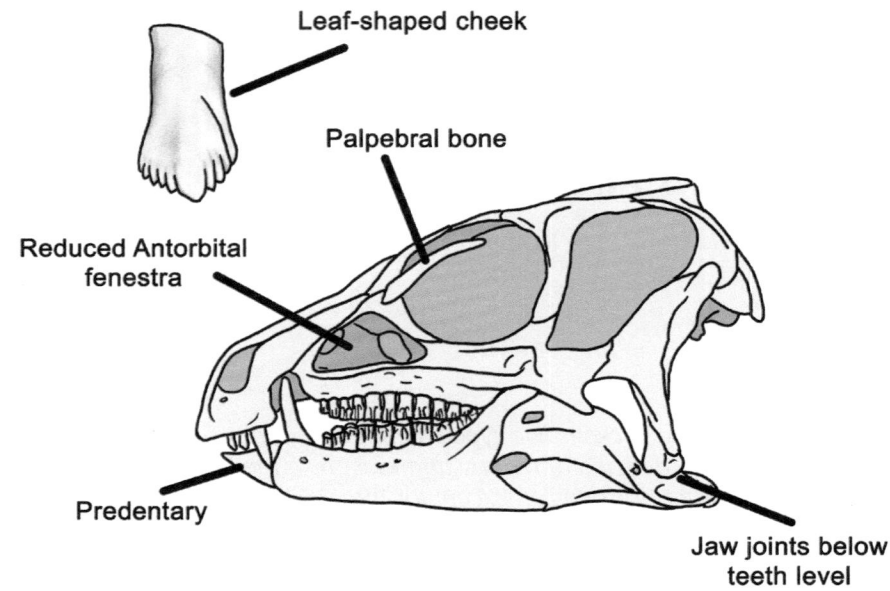

of birds have a forward-pointing pubis like other saurischians, but later in bird evolution the pubis rotated backward, the same condition found in all modern birds. However, bird did not develop this rotation of the pubic bone in the exact same way that Ornithischia did it. In addition, recent discoveries have found that weird herbivorous theropod saurischians like the therizinosaurs and the deinocheirids (Chapter 18) also rotated their pelvis backwards, but not exactly in the ornithischian manner either. Thus, the backward-pointing pubic bone evolved at least three or four times. The ornithischians and therizinosaurs apparently rotated the pubic bone back to allow for a large gut and digestive tract. In birds, the position of the pubic bone is related to the way their skeletons are adapted for flight.

EARLY ORNITHISCHIANS

Once the group was recognized, paleontologists had sorted them into lots of distinct branches of ornithischians: iguanodonts, stegosaurs, ankylosaurs, hadrosaurs, and ceratopsians. For a long time, however, there was no fossil record of how they were related. Some paleontologists were hoping to find fossils that showed how all these diverse groups evolved from a common primitive ornithischian ancestor. Paleontologists were trying to find primitive dinosaurs with ornithischian features but none of the specialized armor, spikes, horns, or jaw features. Naturally, they looked at the relatively generalized ornithischians known as "ornithopods" first. But ornithopods, like the second and third dinosaur ever described, *Iguanodon* and *Hypsilophodon*, had their own weird specializations, so they did not represent the ancestral condition.

The problem with so many of the supposedly ancestral ornithischians was that they were found to be too specialized once we obtained better skeletal material, and nearly all were too late in time. Triassic beds with the preservation potential for the earliest ornithischians are not that common around the world, but they produced nothing that resembled a primitive ornithischian. Other Triassic dinosaurs, like the prosauropods such as *Plateosaurus* and primitive theropods like *Coelophysis* and even primitive saurischians like *Herrerasaurus* and *Eoraptor* were found, but surprisingly few fossils that could be called ornithischian.

Among the early candidates was the English Cretaceous fossil *Hypsilophodon* (**Figure 15.2[B]**), and the Early Jurassic South African fossil formerly called *Fabrosaurus*, but now called *Lesothosaurus* (**Figure 15.2[A]**).

Figure 15.2 Reconstructions of some of the primitive Ornithischia: (A) *Lesothosaurus*, (B) *Hypsilophodon*, (C) *Heterodontosaurus*, (D) *Tianyulong*.

Lesothosaurus is only known from partial skeletal material, but it is clearly a relatively small and very primitive ornithischian. The skeleton is lightly built and completely unspecialized, with delicate arms and legs, and slender tail. It reached a total length of about 2 meters (6.6 feet), counting the long tail. However, the skull was not completely primitive, but had the leaf-shaped teeth, the predentary bone, and the evidence of a horny beak on the upper and lower jaws seen in all ornithischians. It also has the palpebral bone in the eyebrow but does not have the recessed cheek tooth row seen in more advanced ornithischians, suggesting that it didn't have cheeks.

So far, all the possible candidate fossils were too young, too specialized, or too fragmentary to tell us much about the origin of ornithischians. But there is one fossil that does not suffer from these handicaps. In 1962, scientists reported an important discovery: a very primitive Early Jurassic ornithischian (**Figure 15.2[C]**). It had the predentary bone and toothless lower beak in front, the palpebral bone in the eyebrow, and columnar chisel-like plant-eating cheek teeth that were slightly inset, suggesting a set of cheeks. However, it had its own specializations in the teeth, including a set of fang-like canine teeth in the snout just behind the toothless beak, which had a socket in the upper jaw to sheath the tusks, plus incisor like teeth in front of the tusk. These three different types of teeth are extremely unusual for any dinosaur, so it was named *Heterodontosaurus* ("different toothed lizard"). In 1974, a complete articulated skeleton (**Figure 15.3**) was found, and it clearly showed that *Heterodontosaurus* had long arms with five-fingered hands with curved claws, and even longer leg bones. The hind foot had four toes, primitive for all

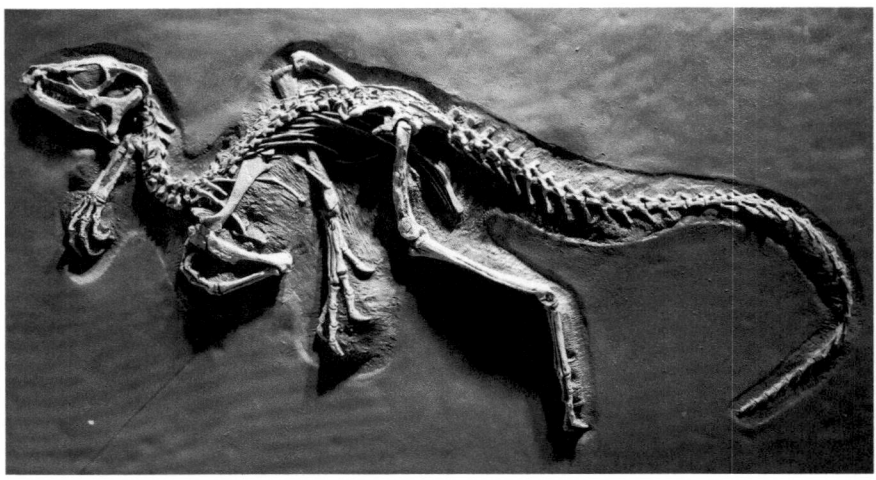

dinosaurs, and not the three seen in more advanced ornithischians. The thigh bone was relatively short, but the shin bone was long, as were the ankle and toe bones. This showed *Heterodontosaurus* was a bipedal fast runner, which ran with its body horizontal, balanced by its long tail. The complete skeleton and other specimens show that *Heterodontosaurus* was about 1.18 meters (3.9 feet) to 1.7 meters (5.7 feet) long, and weighed about 1.8 kg (4 pounds) to 10 kg (22 pounds). The specimen was crucial in many ways. Not only was it the best and most complete archaic ornithischian fossil ever found, but also it showed what kind of anatomy was the starting point for the evolution of all other groups of ornithischians, without the problems that plagued incomplete specimens like *Lesothosaurus*, *Fabrosaurus*, and other primitive forms.

Even though the genus *Heterodontosaurus* is known from just a few specimens, and primitive ornithischians are rare in Jurassic beds worldwide, their diversity has increased with more and more fossils being discovered around the world. Numerous other genera of heterodontosaurines are found in the Early Jurassic of southern Africa, including *Lycorhinus*, *Pegomastax*, *Abrictosaurus*, plus *Manidens* from Chubut Province in Argentina (**Figure 15.2**). In addition to these members of the subfamily Heterodontosaurinae, there are additional more primitive members of the family Heterodontosauridae including *Echinodon* from Lower Cretaceous beds of England (first named and studied by Richard Owen in 1861), *Fruitadens* from the Upper Jurassic Morrison Formation in the Fruita area near Grand Junction, Colorado, and *Geranosaurus* from the Early Jurassic of South Africa.

The most interesting new heterodontosaur is *Tianyulong* from China, which is preserved in fine-grained lake shales so soft tissues are intact (**Figure 15.2[D]**). This fossil shows that some heterodontosaurs had a covering of long filamentous fibers similar to bristles, possibly very primitive feathers, especially along their back and sides. They were arranged almost like the spines of a porcupine along the back. Paul Sereno actually described *Tianyulong* as a "nimble two-legged porcupine". This specimen, plus a number of others like a Chinese specimen of *Psittacosaurus*, all demonstrate that feathers were primitively present across the entire Ornithischia, and most dinosaurs in both the Ornithischia and Saurischia were probably feathered in some way. Added to the other primitive ornithischians such as *Lesothosaurus*, plus the primitive relative of ankylosaurs and stegosaurs known as *Emausaurus*, the outlines of the diversification of the major groups of ornithischians by the Early Jurassic is becoming better and better understood.

Finally, there is an even more primitive fossil from the lower Upper Triassic Ischigualasto Formation of Argentina, source of *Herrerasaurus* and *Eoraptor*.

Dubbed *Pisanosaurus*, it is missing its tail and the pelvis is too broken to see if the pubic bone pointed backwards, but several features of the skull suggest that *Pisanosaurus* might be the most primitive ornithischian ever found. It is probably a dinosaur, because the hip socket is open. The lower jaw has a predentary bone, and the inset row of herbivorous teeth, and a very low jaw joint, all ornithischian features. The top of the skull is missing so we cannot tell if it had a palpebral bone in its eyebrows. However, the missing features leave it open to lots of confusion and different interpretations, so it has been called a heterodontosaurid, a hypsilophodontid, a fabrosaurid, or as the earliest known ornithischian. Some would put it just outside the Dinosauria and within the silesaurids (see Chapter 14). We will probably never be able to resolve this with the specimens we now have.

THYREOPHORANS: STEGOSAURS AND ANKYLOSAURS

The two major armored groups of dinosaurs have both been known for a long time, with their earliest fossils being discovered in the 1830s and 1840s, and the first fairly complete skeletons discovered in the 1870s and 1880s. At one time, the early ankylosaurs were mistaken for stegosaurs, so they were assigned to the same group. Baron Franz Nopsca first coined the name Thyreophora (Greek for "shield bearers") in 1915 for this group, yet the idea that the two groups were related didn't catch on then. It wasn't until the 1990s that paleontologists came back to the idea that stegosaurs and ankylosaurs were closely related and a natural group, defined by such specializations as having body armor plates lined up in rows along the body, relatively small brains, and a quadrupedal posture with hind limbs much longer than their forelimbs.

The most primitive fossil that can confidently be called a thyreophoran is *Scutellosaurus* from the Early Jurassic of northern Arizona (**Figure 15.4[A]**). It was a small bipedal form, about 1.2 meters (4 feet) long, with a very long tail and relatively short forelimbs. The long stiff tail is consistent with a mostly bipedal posture, since it acts as a counterbalance of the front half of the body, yet in most other aspects, *Scutellosaurus* looks much like all the other primitive early bipedal dinosaurs. However, it does have one key feature: parallel rows of osteoderms along the back, with as many as five rows on each side. It also had a double row of osteoderms along the spine from the neck to the tail. That is why it was named *Scutellosaurus* ("little shielded lizard" in Greek). And this is a key feature of the Thyreophora, which is why *Scutellosaurus* shows how giant armored dinosaurs like *Stegosaurus* and *Ankylosaurus* evolved from small bipedal dinosaurs like those of the Late Triassic.

Another key fossil is *Emausaurus* from the Early Jurassic of Germany. At 2.5 meters (8.2 feet) in length, it was larger than *Scutellosaurus*, and probably quadrupedal. It is a very incomplete fossil, but the skull shows the characteristic features of ornithischians, including simple leaf-like teeth, predentary bone in the jaw, and most importantly rows of osteoderms across the body where it is preserved.

The common ancestor of both main branches of thyreophorans seems to be indicated by a remarkable fossil, *Scelidosaurus* (**Figures 15.4[B]** and **15.5[A]**). One of the first relatively complete dinosaurs to be found (certain the earliest in Great Britain), and also one of the first large thyreophorans to evolve, it was first named by legendary anatomist and paleontologist Richard Owen in 1859 based on fragmentary fossils found much earlier. Recently, a nearly complete articulated skeleton was found in England (**Figure 15.5[A]**) which showed how all the formerly fragmentary pieces fit together. On the basis of this complete material, it was nearly 3.8 meters (12.5 feet) long, with long forelimbs and even longer hind limbs, so it was

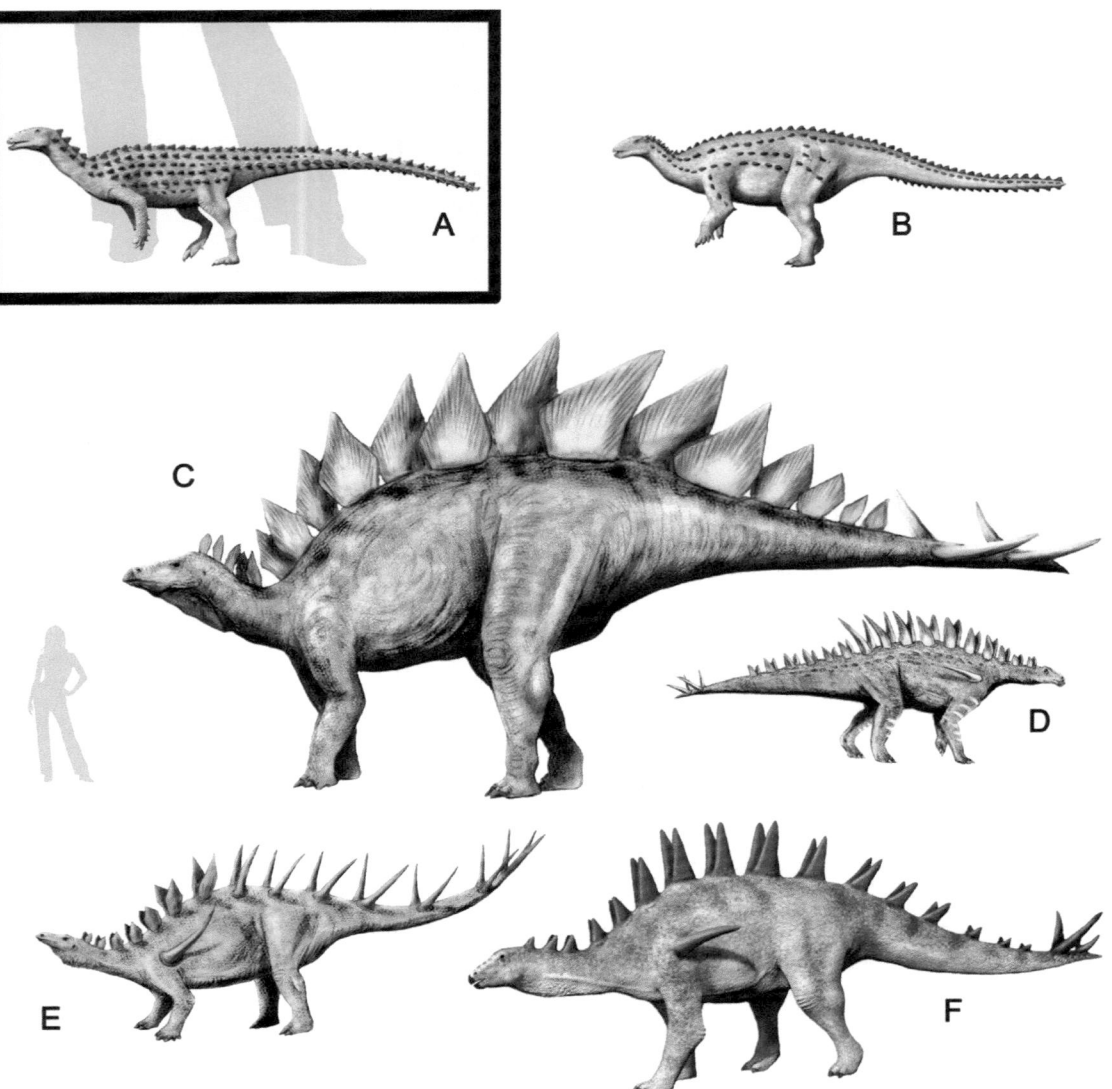

Figure 15.4 Reconstructions of some of the Stegosauria and primitive thyreophorans: (A) *Scutellosaurus*, (B) *Scelidosaurus*, (C) *Stegosaurus*, (D) *Huayangosaurus*, (E) *Kentrosaurus*, (F) *Tuojiangosaurus*.

probably quadrupedal given its great length and the strength of the fore-limbs, although some have suggested it could also be bipedal. The head was small, with a short snout and a horny beak in front, and triangular teeth for eating low-growing vegetation. Most importantly, it has rows of hundreds of small osteoderms embedded in the skin, mostly in rows paralleling the spine, with especially large osteoderms on its neck. The armor extended all the way to the pointed tail, but there were no large plates or spikes or a tail club found in some later thyreophorans.

For decades, *Scelidosaurus* has been classified as a primitive stegosaur or an ankylosaur, or even placed within the ornithopods. But with the nearly complete skeleton and modern methods of analysis, the current consensus is that it's a close relative of both branches of the thyreophorans, showing how both groups could have evolved from a more primitive common ancestor.

ROOFED LIZARDS: THE STEGOSAURS

Stegosaurus is one of the iconic dinosaurs that every kid knows (**Figures 15.4[C]** and **15.5[B]**). Yet when the first fossils were found and described by O.C. Marsh in 1877, they were badly misinterpreted. At first it was

A

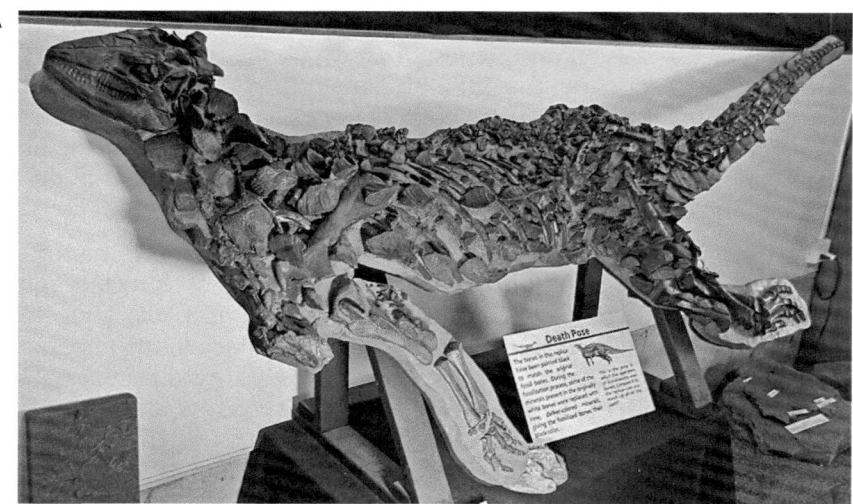

Figure 15.5 Nearly complete skeletons of (A) *Scelidosaurus,* **(B)** *Stegosaurus,* **(C)** *Kentrosaurus.*
(Courtesy Wikimedia Commons.)

B

C

reconstructed as a bipedal creature with its plates lying flat on its back. This is why it got the name *Stegosaurus*, which means "roofed lizard" in Greek. Later reconstructions had the plates standing up in a single row along the back, then lined up in pairs of plates along the back. But when complete articulate specimens were found, they showed that the plates stuck up from the back in two rows, with each plate alternating and overlapping the others in side view.

Even though we have become accustomed to it, with more than a century of familiarity, *Stegosaurus* is still a very weirdly constructed animal. The head was very low to the ground, although the neck was long, so if it reared back, it could reach medium-height plants. Back in the Jurassic and Early Cretaceous, this food would have been mostly ferns and cycads ("sego palms"), along with mosses, horsetails, and short conifers, since there were no grasses or abundant flowering plants until much later in the Cretaceous. The long narrow skull had a pointed snout without teeth, probably covered by a horny beak like that of a turtle. The cheek teeth were small, flat, and triangular, with wear facets showing some evidence of grinding their food. The most complete specimens show that the throat region was protected by a "chain mail" of tiny bony plates called gular osteoderms (**Figure 15.5[B]**).

Famously, *Stegosaurus* had a small brain about 80 g (20.8 oz), about the size of the brain of a dog, which is tiny considering their huge body mass of 4.5 tonnes (5 tons). Scaling brain to body mass, *Stegosaurus* has one of the smallest brains proportional to its size of any ornithischian known. One of Marsh's fossil skulls had a well-preserved brain cavity, allowing him to make a cast of the cavity and describe the brain features in the 1880s. This led to the famous myth that its brain was so small that *Stegosaurus* needed a second brain in its hips just to function. In reality, the "second brain" was just a slightly enlarged ganglion of the nervous system, which would have controlled the muscles in the back of the body; it was not a true "brain". It's also likely that most of the space in the hips housed a glycogen body (also found in bird and sauropods). This feature is typical in living birds and supplements the supply of glycogen (a sugar) to the nervous system. *Stegosaurus* did not need much intelligence to continually munch away at ferns and low-growing vegetation. Apparently, its spiked tail and other defenses and huge size were sufficient that stegosaurs were very successful for millions of years, and spread worldwide.

The body of stegosaurs was weirdly proportioned, with short forelimbs and long hind limbs. This forced their spine into a big arch that flexed upward over the hips but sloped down steeply to the head and tail (**Figure 15.4[C]**). Each hand and foot had three short toes, each of which bore a hoof. In most stegosaurs, the hands had only two finger bones in each finger, and two toe bones in each toe.

What the plates were used for had long been debated. Originally, Marsh and other early paleontologists thought they were protective, although the plates didn't do much of job of shielding their sides or flanks from attack by a theropod like *Allosaurus* (both are found together in several Morrison localities). Some people thought the plates were not adaptive at all. For example, Frederic Loomis argued that the plates adorning the backs of stegosaurs were maladaptive traits that sapped their vigor and signaled their impending extinction.

More recently, a consensus has formed that they were probably for sexual dimorphism and advertising their age and status. Most scientists think that males and females of *Stegosaurus* both appear to have the same sized plates, so it's not clear that their plates were a sexually dimorphic feature in that genus. But a study published in 2015 claimed that the plates were different in males and females, with wider plates in males and taller plates in females.

In the 1970s, Jim Farlow did a series of slices through the plates, and found they had large internal cavities and big canals for a lot of blood vessels. Coupled with the other ideas brewing during the warm-blooded dinosaur debate, this suggested that the plates were for picking up or shedding excess body heat. However, no other group of dinosaurs seemed to need these structures for regulating body temperature. Most other stegosaurs simply have conical spikes or deeply embedded armor plates, so the function of heat regulation would be unique to *Stegosaurus*, and not found in any of its close relatives.

In addition, the surface roughness of the bony plates suggests that they were almost certainly covered with keratinous horny sheaths to increase their size—but this would have also reduced any heat transport through the outside of the plate to the blood beneath. The surface of the plates, however, were covered by bony grooves for blood vessels, so any horny sheath would have covered and protected these. The horny sheath would have not only reinforced defensive function, but also improved their use as display structures. Like most arguments over the function of unusual structures in extinct animals, we may never know the truth. In addition, nature is complex and there is often no simple "right" answer, but it's likely that these structures performed more than one function.

The other distinctive feature of *Stegosaurus* is its spiky tail. Some paleontologists argued these were just for display, although most have regarded them as defensive weapons. Many of the early reconstructions showed *Stegosaurus* with six to eight spikes, but a more careful analysis shows that *Stegosaurus* had only four. (Other stegosaurs have more.) Any model or reconstruction of *Stegosaurus* with more than four is in error. Contrary to many reconstructions, the four tail spikes did not point straight upwards, but stuck out upwards and sideways away from the tail axis, making them much more effective as a weapon with a side-to-side striking motion. Their tails were not held rigid like most dinosaur tails, so they could swing it around. However, the rows of plates on the upper part of the tail restricted movement to some degree. The most important evidence about the tail as a weapon showed that the spikes had a very high incidence of damage (9.8% of specimens examined), suggesting they were used in defense to strike hard objects. In addition, an *Allosaurus* tail vertebra is known that had puncture marks that fit the tail spikes of *Stegosaurus* perfectly.

The tail spikes had no formal anatomical name until cartoonist Gary Larson published a "Far Side" cartoon showing cave men watching a slide slow. The lecturer points to the tail of a *Stegosaurus*, and says, "Now this end is called the thagomizer . . . after the late Thag Simmons". *Far Side* cartoons were always hugely popular with scientists because they often talked about scientific topics or were based on scientific in-jokes. The term "thagomizer" entered the scientific lexicon when Ken Carpenter used it in a lecture at the 1993 Society of Vertebrate Paleontology meeting. Since then, it has been picked up in numerous dinosaur books, used in the displays at Dinosaur National Monument, and in the BBC series *Planet Dinosaur*. Although there is no formal procedure for making popular nicknames into official anatomical terms, "thagomizer" is widely used in paleontology, usually with a smile and a chuckle the first time it is mentioned.

Of course, Larson knowingly committed a scientific boo-boo when he showed "cave men" living with dinosaurs, but Larson was fully aware of this, since it was necessary for the joke. Larson has written that "there should be cartoon confessionals where we could go and say things like, 'Father, I have sinned—I have drawn dinosaurs and hominids in the same cartoon.'" A similar anachronism that is usually overlooked is the common pairing of *Stegosaurus* with *Tyrannosaurus rex*, found in many books and cartoons, and in the animatronic dinosaurs of "Primeval World" in

Disneyland. In reality, *Stegosaurus* vanished about 140 Ma (Late Jurassic), yet *T. rex* did not appear until 68 Ma (latest Cretaceous). It's staggering to think about it, but *T. rex* is closer in time to humans than it is to *Stegosaurus*.

As the years have gone by, more and more different kinds of stegosaurs were found, nearly all with different configurations of armor, plates, and spikes. One of the first stegosaurs to be discovered was the spiky African genus *Kentrosaurus aethiopicus* (**Figures 15.4[E]** and **15.5[C]**) from the Upper Jurassic Tendaguru bone beds in what is now Tanzania. Its name means "sharp point lizard" in Greek. *Kentrosaurus* is known from hundreds of bones found in multiple quarries between 1910 and 1912 (although many were lost during the bombing of German museums in World War II). It was about 4.5 meters (15 feet) long and weighed about 1 tonne (1.1 tons), considerably smaller than some *Stegosaurus*, which reached up to 9 meters (30 feet) in length, and 5.3–7 tonnes (6–7.5 tons) in weight. In most respects, *Kentrosaurus* is much like *Stegosaurus*, with a small but long and narrow head, toothless beak, short front limbs and long hind limbs, and a relatively long tail. Unlike *Stegosaurus*, however, it had small plates only on the front half of its backbone, and most of the rest of the spine was covered with paired spikes that clearly served a defensive function.

So far, we have discussed the common American, European, and African stegosaurs. Their range was extended to China when the amazing fossils that had been unknown to westerners during the political turmoil of the mid-twentieth century finally began to be available for study. The first of these was *Chialingosaurus*, from the Middle Jurassic of China, found during the war years in the 1930s and 1940s, but finally named in 1959 by Yang Zhonjian (also written C-.C. Young), the "Father of Chinese Paleontology". *Chialingosaurus* is based on a partial skeleton, and some do not consider it to be a valid genus for that reason. However, it apparently had small plates in pairs along its neck and backbone along the shoulders, and paired spikes down the rest of its back and tail, like *Kentrosaurus*.

In 1973, Dong Zhiming named *Wuerhosaurus* from the Early Cretaceous of China and Mongolia, one of the very last stegosaurs known. It also consists of a fragmentary skeleton, plus parts of a few more individuals. Its body was much fatter and broader than other stegosaurs, based on the broad pelvis. At one time, it was argued that it had very rounded plates in rows on its back but this has been dismissed as an artifact of breakage of the few plates found. Dong and others described *Huayangosaurus* in 1982, based on a partial skeleton and some other specimens from Middle Jurassic beds of China (**Figure 15.4[D]**). Unlike other stegosaurs, the plates down its back are tall narrow triangles rather than broad polygons. It also had a Thagomizer of four spikes at the tip of its tail.

Its close relative is the Upper Jurassic stegosaur *Chungkingosaurus*, named and described by Dong and others in 1983. *Chungkingosaurus* had an arrangement of narrow tall plates on its back and spikes on its tail similar to that of *Huayangosaurus*. Another Late Jurassic stegosaur with similar armor is *Tuojiangosaurus* (**Figure 15.4[F]**), described by Dong and colleagues in 1977. It may be closely related to *Paranthodon* from Africa. There is also *Gigantospinosaurus* from the Late Jurassic of Sichuan, which had huge spikes protruding from it shoulders, and *Jiangjunosaurus* based on a fragmentary skeleton from the Late Jurassic of Inner Mongolia. That makes at least six Middle Jurassic to Early Cretaceous stegosaurs from China and Mongolia, giving it the highest stegosaur diversity in the world, and almost two dozen genera known from Eurasia, Africa, and North America.

But what about the rest of the Pangea continents: Australia, India, Antarctica, Madagascar and South America? So far none of them have produced unquestioned stegosaurs, although the fossil record in Australia, Madagascar, India, and, of course, Antarctica, is relatively poor during their heyday in the Middle and Late Jurassic. *Dravidosaurus* from the Late Cretaceous of India turned out not to be a stegosaur. In one interpretation, it was based on a weathered set of plesiosaur hip bones and hind limbs. Later authors, however, ruled out the plesiosaur interpretation, but concluded the specimens are too incomplete to tell what they really are. There are trackways in Australia that are claimed to be stegosaurian, but so far, no bones.

However, in 2017 Leonardo Salgado and colleagues described a skull fragment and partial skeleton from the Early Jurassic of Patagonia. The specimen even had gut contents showing that it ate cycads. Named *Isaberrysaura mollensis*, it is definitely an advanced ornithischian, and that is all that Salgado and coauthors would commit to. However, another analysis of it suggested that it might be a very primitive bipedal relative of the stegosaurs, since there are stegosaur-like characteristics in what is known of the skull.

Thus, we have an amazing record of stegosaurs from most of the Jurassic and Early Cretaceous. Aside from the possibility that *Isaberrysaura* may be an Early Jurassic stegosaur, stegosaurs are definitely known from the Middle Jurassic, when they evolved from scelidosaurs, their common ancestor with the ankylosaurs. The earliest known undoubted stegosaur is *Huayangosaurus* from the early Middle Jurassic of China, followed by late Middle Jurassic stegosaurs such as *Chungkingosaurus*, *Chialingosaurus*, *Tuojiangosaurus*, and *Gigantospinosaurus* from China, and *Lexovisaurus* and *Loricatosaurus* from England and France. In the Late Jurassic, stegosaurs were in their heyday in abundance and size, if not diversity, with *Kentrosaurus* in Africa, *Dacentrurus* and *Miragaia* in Europe, *Jiangjunosaurus* in China, and *Stegosaurus* and *Hesperosaurus* in North America.

By the Early Cretaceous, stegosaurs experienced their last phase of evolution, with *Wuerhosaurus* in China, *Paranthodon* in Africa, and *Craterosaurus* from England, plus some undescribed fragments from Russia. Paleontologists have long speculated what might have caused the decline and extinction of stegosaurs. Certainly, the vegetation was changing, with the decline of cycads (possibly their main food source) paralleling the decline in stegosaurs. In addition, by the Early Cretaceous there was a tremendous bloom of flowering plants, including many types of water plants and primitive trees like magnolias. Many paleontologists have suggested that the rapidly reproducing flowering plants may have stimulated the evolution of duckbilled dinosaurs with their complex "dental batteries" of hundreds of prismatic teeth fused together. They were clearly more specialized and efficient plant eaters than the almost toothless stegosaurs, and possibly co-evolved with flowering plants to dominate the Cretaceous landscape. Meanwhile, the stegosaurs were Jurassic relicts, and apparently did not do well facing new competition from herbivores, changing plants in their diet, and possibly new predators as well. For whatever reason, stegosaurs vanished by the end of the Early Cretaceous.

TURTLE-SHELL DINOSAURS: THE ANKYLOSAURS

Another iconic dinosaur is *Ankylosaurus*, the famous dinosaur with the turtle-like armored shell on its back and the huge tail club (**Figure 15.6[A]**). It is known from many books, toys, and even appearances in the movie *Jurassic World*. It is actually one genus out of dozens of ankylosaurs known, widespread around the remnants of Pangea throughout the Jurassic and Cretaceous. Broken parts of the armor of ankylosaurs were known in the 1800s, but it wasn't until 1910 that the first partial skeletons of *Ankylosaurus* were found in the Upper Cretaceous beds of Montana and then Alberta, and the shape and size of the animal were first realized. Large specimens were about 10 meters (33 feet) long, and weighed about 8 tonnes. The huge skull of *Ankylosaurus* was almost a meter wide, with a thick layer of armored osteoderms on the skull

Figure 15.6 Reconstructions of some of the Ankylosauridae: (A) *Ankylosaurus*, (B) *Euoplocephalus*, (C) *Minotaurasaurus*.

roof and snout and protecting the tiny eyes in small eye sockets. On the upper back corners of the skull are strange horns shaped like short blunt pyramids. Yet the teeth were tiny and somewhat leaf-shaped, and wear on their faces, and not on their crowns, suggesting that they ate low-growing plants (probably ferns, cycads, and some primitive flowering plants in the Late Cretaceous). In some specimens, the tongue bones are even preserved, and the jaw was capable of some chewing motion, so they were able to chop up their food a bit more before they swallowed it and it was processed in their huge stomach. The position of the nostrils in the ankylosaurs suggests that they may have also dug with their noses, possibly to eat roots and tubers, although their teeth do not show the wear associated with such a gritty diet. *Ankylosaurus* had large sinuses and nasal chambers, possibly for water and heat balance, or also for sound amplification, as well as an excellent sense of smell.

In *Ankylosaurus*, there were even rings of armor around the neck to protect it. It is most famous for its semi-solid shell of fused osteoderms on the back, composed of a "pavement" of small polygonal osteoderms, alternating with longitudinal and transverse rows of big osteoderms. Ankylosaur legs were relatively short and equal in length, so it did not have the high hips of stegosaurs and was built very low to the ground. The vertebrae of the back and hips were strongly fused together, and partially fused to the shell above them. The base of the tail vertebrae was highly flexible, so it could easily swing its tail club. At the end of the tail in some ankylosaurs was a huge bony club made of fused osteoderms, over 60 cm (2 feet) long, and the last five vertebrae of the tail are rigid and form a straight "handle" on the club that made its impact greater on its predators, such as *Tyrannosaurus* or *Gorgosaurus* or other larger theropods. All in all, *Ankylosaurus* had formidable defenses. It was capable of rapid maneuvering to keep its predators, like *T. rex*, at bay, and bring its body around to break the shins of the predator with its bone-cracking club on the tail.

There are two main branches of ankylosaurs: the Nodosauridae and the Anklylosauridae. The Ankylosauridae was a large and diverse family, with as many as 16 genera confined to the Late Cretaceous. Some, like *Euoplocephalus*, also had a tail club, and even had bony eyelids protecting its eyes from attack (**Figures 15.6[B]** and **15.8[A]**). Most had no tail club at all, but a variety of different kinds of armor on their heads and bodies. *Minotaurasaurus* (**Figure 15.6[C]**) had huge long horns on the corners of its skull, looking a bit like the bull-head of the legendary Minotaur. One of the most interesting is *Crichtonpelta*, a middle Cretaceous ankylosaurid from China that was named in honor of Michael Crichton, the author of the *Jurassic Park* novels. Another nearly complete skeleton from the Hell Creek Formation was described by Victoria Arbor and David Evans in 2017, and given the name *Zuul*, after the evil demi-god and Gatekeeper in the 1984 movie *Ghostbusters*, because in the movie the character Zuul had a head shaped like this dinosaur.

Ankylosaurus itself is one of the more extreme examples of the family, with a relatively solid shell of armor, horns on its head, a tail club, and a smooth boundary along the edge of its shell (not spikes, which are often erroneously shown in illustrations based on the nodosaur *Edmontonia*—**Figure 15.8[B]**). Prior to its discovery, a number of ankylosaurs had been found, but they were all fragmentary specimens that gave little or no reliable indication of what they would have looked like in life.

The other main group of ankylosaurs are called nodosaurs. They had no tail club, and most of them were partially covered in numerous smaller unfused osteoderms making up the flexible armor in their skin of their backs and sides (**Figure 15.8[C]**), rather than the solid bony shell of *Ankylosaurus*, *Crichtonpelta*, *Pinacosaurus*, or *Euoplocephalus* (**Figure 15.8[A]**). Given how fragmentary most of the early specimens were, it is not surprising that no one was able to reconstruct them accurately until nearly complete articulated specimens were finally found. Today, there are at least two dozen genera of nodosaurs known, found on almost every continent including Antarctica.

In the past few decades, the pace of nodosaur discoveries has accelerated, so that there are many new genera from China and North America. The oldest known nodosaurs are *Gargoyleosaurus* and *Mymoorapelta* from the Upper Jurassic Morrison Formation. Nodosaurs become more diverse in the Early Cretaceous with *Hylaeosaurus* from England (**Figure 15.7[A]**), *Polacanthus* from Austria (**Figure 15.7[B]**), *Gastonia*

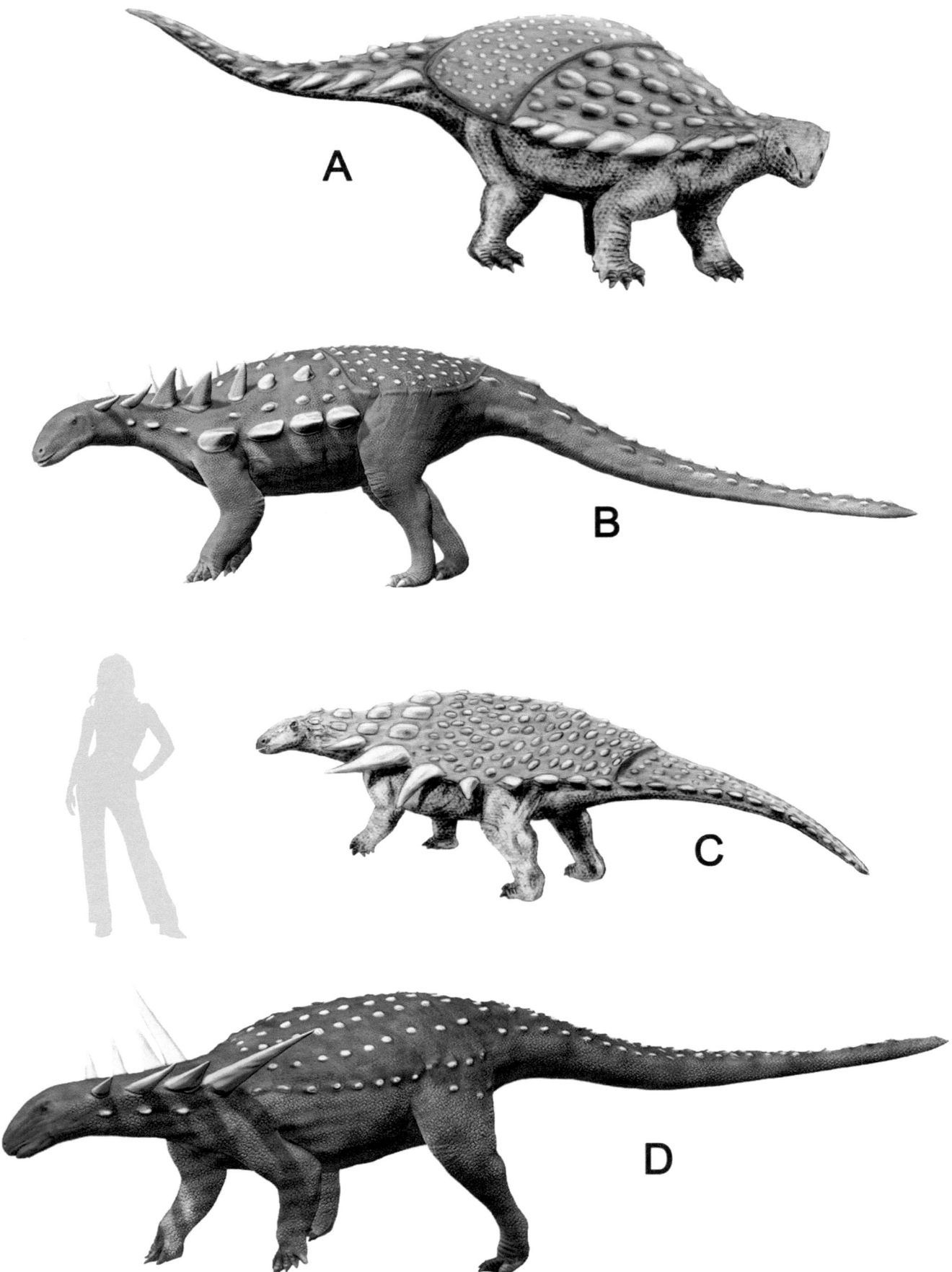

Figure 15.7 Reconstructions of some Nodosauridae: (A) *Hylaeosaurus*, (B) *Polacanthus*, (C) *Acanthopholis*, (D) *Sauropelta*.

A

Figure 15.8 Skeletons of some of the better-known ankylosaurs.
(A) *Euoplocephalus*, an ankylosaurid.
(B) The front end of *Edmontonia*, a nodosaur skeleton found in this position sticking out of a cliff; the back had eroded away. (C) The body armor of a nodosaur, with many small bony elements forming a flexible shield around their backs and sides and tail. (Courtesy Wikimedia Commons.)

B

C

from Utah, and *Hoplitosaurus* from South Dakota. By the late Early Cretaceous, nodosaurs began to diversify explosively with many new genera like *Nodosaurus, Stegopelta, Animantarx, Peloroplites, Tatankacephalus,* and *Sauropelta* (**Figure 15.7[D]**), from the Rocky Mountains of the U.S., *Pawpawsaurus* and *Texasetes* from Texas, *Silvisaurus* from Kansas, teeth referred to *Priconodon* and *Propanoplosaurus* from the Maryland coast, *Europelta* from Spain, *Anoplosaurus* and *Acanthopholis* from England (**Figure 15.7[C]**), *Hungarosaurus* from Hungary and *Dongyangopelta* and *Zhejiangosaurus* from China. In the latest Cretaceous, there are not only the Alberta nodosaurs like *Edmontonia* (**Figure 15.8[B]**) and *Panoplosaurus,* and *Glyptodontopelta* from the very end of the Cretaceous of New Mexico, but also *Struthiosaurus* from eastern Europe, and a handful of teeth from James Ross Island on the Antarctic Peninsula named *Antarctopelta*.

The Heterodontosauridae and Thyreophora are each major branches of ornithischians. In the next chapter we will look at the other branches: the iguanodonts and duckbills, pachycephalosaurs, and the horned dinosaurs.

FURTHER READING

Alexander, R.M. 1989. *Dynamics of Dinosaurs and Other Extinct Giants.* Columbia University Press, New York.

Baron, M.G. 2018. *Pisanosaurus mertii* and the Triassic ornithischian crisis: Could phylogeny offer a solution? *Historical Biology.* 31 (8): 967–981.

Butler, R.J.; Barrett, P.M.; Kenrick, P.; Penn, M.G. 2009. Diversity patterns amongst herbivorous dinosaurs and plants during the Cretaceous: Implications for hypotheses of dinosaur/angiosperm co-evolution. *Journal of Evolutionary Biology.* 22: 446–459.

Butler, R.J.; Galton, P.M.; Porro, L.B.; Chiappe, L.M.; Henderson, D.M.; Erickson, G.M. 2010. Lower limits of ornithischian dinosaur body size inferred from a new Upper Jurassic heterodontosaurid from North America. *Proceedings of the Royal Society of London B: Biological Sciences.* 277 (1680): 375–381.

Butler, R.J.; Upchurch, P.; Norman, D.B. 2008. The phylogeny of ornithischian dinosaurs. *Journal of Systematic Palaeontology.* 6 (1): 1–40.

Carpenter, K., ed. 2001. *The Armored Dinosaurs.* Indiana University Press, Bloomington, IN.

Fastovsky, D.E.; Weishampel, D.B. 2021. *Dinosaurs: A Concise Natural History* (4th ed.). Cambridge University Press, Cambridge.

Ferigolo, J.; Langer, M. C. 2007. A Late Triassic dinosauriform from south Brazil and the origin of the ornithischian predentary bone. *Historical Biology.* 19: 23–33.

Galton, P.; Upchurch, P. 2004. 16: Stegosauria, pp. 361–400. In Weishampel, D.B.; Dodson, P.; Osmólska, H., eds. *Dinosauria* (2nd ed.). University of California Press, Berkeley.

Hayashi, S.; Carpenter, K.; Scheyer, T.M.; Watabe, M.; Suzuki, D. 2010. Function and evolution of ankylosaur dermal armor. *Acta Palaeontologica Polonica.* 55 (2): 213–228.

Lucas, S.G. 2005. *Dinosaurs, the Textbook* (5th ed.). W. C. Brown, Dubuque, Iowa.

Maidment, S.C.R.; Wei, G.; Norman, D.B. 2006. Re-description of the postcranial skeleton of the middle Jurassic stegosaur *Huayangosaurus taibaii. Journal of Vertebrate Paleontology.* 26 (4): 944–956.

Mateus, O.; Maidment, S.C.R.; Christiansen, N.A. 2009. A new long-necked 'sauropod-mimic' stegosaur and the evolution of the plated dinosaurs. *Proceedings of the Royal Society B: Biological Sciences.* 276 (1663): 1815–1821.

McGowan, C. 1983. *The Successful Dragons: A Natural History of Extinct Reptiles.* Samuel Stevens, Toronto.

McGowan, C. 1991. *Dinosaurs, Spitfires, and Sea Dragons.* Harvard University Press, Cambridge.

Norman, D.B. 1985. *The Illustrated Encyclopedia of the Dinosaurs.* Crescent Books, New York.

Norman, D.B. 2021. *Scelidosaurus harrisonii* (Dinosauria: Ornithischia) from the Early Jurassic of Dorset, England: Biology and phylogenetic relationships. *Zoological Journal of the Linnean Society.* 191 (1): 1–86.

Sereno, P. 1986. Phylogeny of the bird-hipped dinosaurs (Order Ornithischia). *National Geographic Research.* (2): 234–256.

Sereno, P.C. 1998. A rationale for phylogenetic definitions, with application to the higher-level taxonomy of dinosauria. *Neues Jahrbuch für Geologie und Paläontologie, Abhandlungen.* 210: 41–83.

Sereno, P.C. 1999. The evolution of dinosaurs. *Science.* 284 (5423): 2137–2147.

Thompson, R.S.; Parish, J.C.; Maidment, S.C.R.; Barrett, P.M. 2012. Phylogeny of the ankylosaurian dinosaurs (Ornithischia: Thyreophora). *Journal of Systematic Palaeontology.* 10 (2): 301–312.

Weishampel, D. B.; Dodson, P.; Osmolska, H., eds. (2007). *The Dinosauria* (2nd ed.). University of California Press, Berkeley.

Zheng, X.-T.; You, H.-L.; Xu, X.; Dong, Z.-M. 2009. An Early Cretaceous heterodontosaurid dinosaur with filamentous integumentary structures. *Nature.* 458 (7236): 333–336.

ORNITHISCHIANS II

HADROSAURS AND MARGINOCEPHALIANS

16

It is hard to walk out into the Hell Creek Formation and not stumble upon a *Triceratops* weathering out of a hillside.

—John Scannella

NEORNITHISCHIA

Primitive ornithischians arose in the Late Triassic, and they quickly diverged into heterodontosaurs, stegosaurs, and ankylosaurs (**Figure 14.1**). The other main branch of the ornithischian is called the Neornithischia, and it includes the duckbills and other ornithopods, pachycephalosaurs, and horned dinosaurs. Although they do not look much like each other, they have some distinctive features that show they are a natural group. The most striking ones occur in the teeth, which are asymmetric in that they have a thicker layer of enamel on the lower teeth. This allows them to shear against the upper teeth, and as the thin enamel on the outer layer wears away faster, the thicker inside layer sticks up and forms a self-sharpening shearing surface against the upper teeth that helps chop up vegetation. This feature is especially well developed in the duckbills and in the ceratopsians, which independently evolved two different versions of a "dental battery"—an array of closely packed teeth that together form a much larger grinding surface. When these features developed, duckbills and horned dinosaurs became superbly adapted for mowing down huge amounts of vegetation. By contrast, primitive ornithischians usually had simple leaf-shaped teeth that did not have very precise occlusion, so they could not handle large amounts of vegetation or process it with great efficiency.

Like all the early ancestors of other dinosaur groups, the earliest neornithischians were lightly built bipedal forms with a long straight tail, usually less than a meter long in total body length (**Figure 16.1**). Almost two dozen genera of these tiny primitive dinosaurs are now known, found all over remnants of Pangea in the Jurassic and Cretaceous. These include not only *Hypsilophodon* from the Early Cretaceous of England and Romania (**Figure 16.1[A]**), but also *Nanosaurus* from the Upper Jurassic Morrison Formation of the Rocky Mountains of Wyoming (**Figure 16.1[B]**), *Orodromeus* (**Figure 16.1[D]**) and *Zephyrosaurus* from the Early Cretaceous of Montana and *Thescelosaurus* from the Late Cretaceous of Montana, *Kulindadromeus* from the Jurassic of Russia (**Figure 16.1[E]**), *Agilisaurus, Yandusaurus* and *Hexinlusaurus* from the Middle Jurassic and *Jeholosaurus* from the Early Cretaceous of China, *Koreanosaurus* from the Cretaceous of Korea, and even one from Australia (*Leaellynasaura*) (**Figure 16.1[C]**). All of these dinosaurs were small lightly built bipeds, with very large eyes, a long straight tail that balanced the front of their body

DOI: 10.1201/9781003128205-16

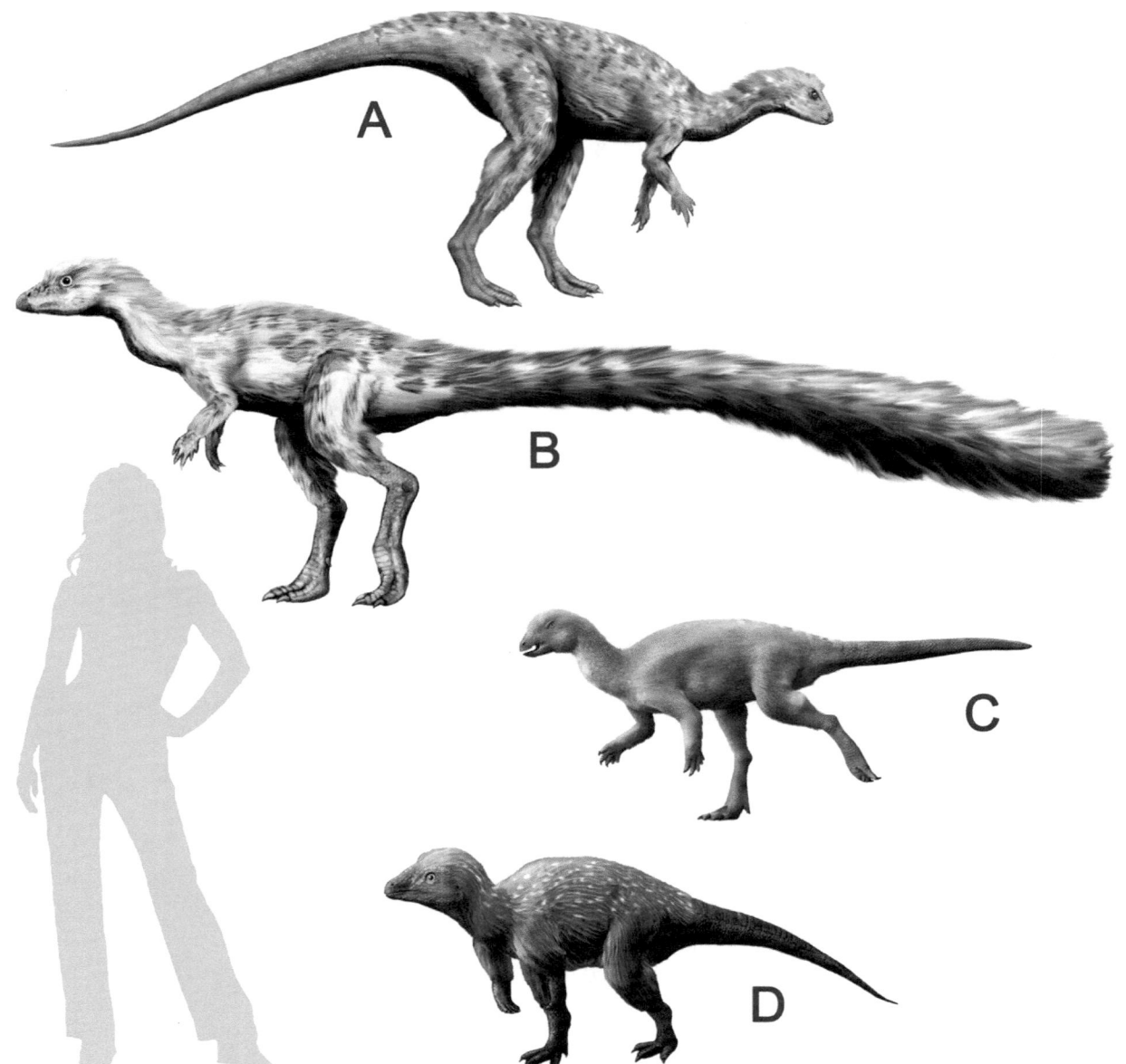

Figure 16.1 Reconstructions of some primitive ornithischians, including: (A) *Hypsilophodon*, (B) *Leaellynasaura*, (C) *Orodromeus*, (D) *Kulindadromeus*.

over their hips, and relatively small front limbs (although some could apparently drop down on all fours). They had a short snout with a sharp, almost turtle-like beak, apparently for feeding on low-growing plants. Some, like *Kulindadromeus*, was preserved with its feathery covering intact, so presumably all of these active warm-blooded small dinosaurs were feathered as well.

ORNITHOPODA: THE "BIRD-FOOTED" DINOSAURS

The first neornithischians to be described were among the first dinosaurs ever discovered. These include *Iguanodon*, only the second dinosaur to be named and described, found in the early 1820s in southeastern England by Gideon Mantell and named in 1825. Another even more primitive member of the group was *Hypsilophodon*, first found in 1849. These dinosaurs were soon joined by a great variety of duckbills found in the United States in the late 1800s, so by 1881 O.C. Marsh had coined the name Ornithopoda for duckbills, iguanodonts, hypsilophodonts, and their more primitive relatives. As dozens more genera of ornithopods have been discovered in the past century, the old definition of Ornithopoda ("bird foot" in Greek, based on their three-toed feet) became insufficient. Some of the primitive ones still have not reduced their side digits, so their feet had more than three toes. But dinosaur paleontologists have found other features that all ornithopods share. Several features are distinctive to them, such as the loss of the primitive archosaurian mandibular fenestra in the lower jaw, the shape and growth pattern of the horny beak over their toothless bony snout, and, most distinctive of all, not only the pubic bone points backward and lies parallel to the ischium (**Figure 14.2**), but also the pubis is usually longer than the ischium, and no longer has a forward extension seen in more primitive ornithischians.

Iguanodon (**Figure 16.2[A]**) was originally the most famous and best-studied of all the ornithopods, especially since it was only the second dinosaur ever found. Early reconstructions in the 1840s and 1850s made it look like a giant iguana lizard with a horn on its nose, since only a handful of broken limb bones were known, plus the teeth that looked an iguana's teeth, but much larger. Then in 1878, dozens of complete articulated skeletons of *Iguanodon* were found in a coal mine in Bernissart, Belgium, and the actual shape of the dinosaur was revealed. *Iguanodon* was mostly bipedal, with a long skull and beaked snout, and the "horn" on the nose was actually a claw on their thumb. But the reconstructions in the late 1800s based on the Bernissart specimens still followed the old, slow, sprawling reptilian model of dinosaur, so the *Iguanodon* are mounted in the "kangaroo" pose, leaning back on their tails. Finally, in the 1970s and 1980s, paleontologists restudied and redescribed the known *Iguanodon* skeletons from Bernissart, and completely rethought their pose and activity levels. The original specimens had a trusswork of criss-crossing tendons in the tail. These held the tail out rigid and straight from the hip as a counterbalance for the body, balanced on the hind legs. But to make the tail curve into the kangaroo pose, the scientists who mounted the Bernissart skeletons had to actually break the tail bones and tendons and ignore this clear evidence of their straight tails. The hands of *Iguanodon* are very different from originally thought as well. Not only is there a thumb spike sticking out perpendicular to the wrist, but also the middle three fingers are thick and robust and resemble three thick parallel pillars, and allowed *Iguanodon* to lean on its hands and feet in a quadrupedal pose. The pinky finger of *Iguanodon* is even stranger. It is very long and capable of curling and gripping things, something never seen on any other dinosaur.

Iguanodonts are known from Europe (including not only *Iguanodon*, but also *Cumnoria*, *Mantellisaurus*, *Hypselospinus*, *Kukufeldia*, and *Barilium* from the same beds in England, and *Proa* and *Magnamanus* from the Early Cretaceous of Spain, as well as others from Portugal), but they occurred all over the world, especially in the Cretaceous (**Figure 16.2**). There are over two dozen genera, and they range from the small primitive North American forms like *Dryosaurus* (**Figure 16.2[B]**), *Uteodon*, and *Camptosaurus* (**Figure 16.2[C]**) (all from the Upper Jurassic Morrison Formation) to *Cedrorlestes*, *Planicoxa*, *Osmakasaurus*, *Dakotadon*, *Iguanacolossus*, *Hippodraco*, and *Theiophytalia* from the Early Cretaceous of North America. One of the best known is the famous iguanodont *Tenontosaurus* (**Figure 16.2[D]**) from Montana (found in the same beds with the raptor known as *Deinonychus*, and often pictured battling with it). In addition, there are *Elrhazosaurus* and *Lurdusaurus* from the Lower Cretaceous of Niger, and the bizarre fin-backed iguanodont *Ouranosaurus* from Niger (**Figure 16.2[F]**), found in some of the same beds that yielded

Figure 16.2 Reconstruction of some neornithischians, including: (A) *Iguanodon*, (B) *Dryosaurus*, (C) *Camptosaurus*, (D) *Tenontosaurus*, (E) *Muttaburrasaurus*, (F) *Ouranosaurus*.

the sail-backed predator *Spinosaurus* (famous from *Jurassic World III*). Iguanodonts occur in many other places, including the Early Cretaceous of China (*Lanzuosaurus*, *Bayannurosaurus*, *Bolong*, and *Penelopognathus*), *Altirhinus* from Mongolia, *Ratchasimasaurus* from Thailand, *Fukuisaurus* from Japan, and *Muttaburrasaurus* from the Cretaceous of Australia (**Figure 16.2[E]**). In short, iguanodonts were almost everywhere except South America and Antarctica in the Early Cretaceous, then most of their diversity vanished, and only a few survived into the Late Cretaceous.

THE HADROSAURS

Duck-billed dinosaurs or hadrosaurs are almost as familiar to the public as large sauropods, *Triceratops*, and *T. rex*. The very first dinosaur discovered in North America was *Hadrosaurus* from Cretaceous beds in New Jersey, found and described by Joseph Leidy in 1858. In the 1870s and 1880s, however, numerous duckbills were discovered and reported from the Rocky Mountain region, especially from the Upper Cretaceous bone beds of Wyoming, Montana, and Alberta, where some were found as complete articulated skeletons. Today, there are at least 50 genera named, and some others that are still not described, so we cannot discuss every genus in a book like this.

Some anatomical features are consistent about almost all duckbills (**Figure 16.3**). While some genera have a narrow snout, in many hadrosaurs, the snout tends to be long and flattened a bit like a duck's bill, because the bones of the snout are very large and elongate. Well-preserved specimens of some genera, like *Edmontosaurus*, show that the flat bony "duck" bill of the upper jaw was capped by a keratinous sheath that hung downward over the front of the lower jaw, giving an "overbite" effect that more efficiently cropped vegetation. Even more impressive was the array of teeth in the back of the jaw. Instead of a handful of individual teeth, hadrosaurs had hundreds of long prism-shaped teeth that were densely packed together into a single dental battery, which formed a broad grinding surface along the top. A typical battery might have several hundred small tooth prisms, so the claim that duckbills had about 1000 teeth in their mouth can be true. The pair of dental batteries in each lower jaw would grind against a pair of dental batteries in the upper jaw, producing a very large efficient grinding mill that made hadrosaurs the most efficient chewing herbivores the planet had ever seen up to that point. The upper jaws were hinged against the rest of the upper skull bones with stretchy tendons and ligaments holding the skull together, so the entire skull could stretch and flex a bit during chewing.

What duckbills ate with this amazing chewing machine is still disputed. Early workers looked at the duck-like bill and imagined them swimming in swamps eating water plants, but all of that has been debunked. Duckbills first underwent an evolutionary radiation in the Early Cretaceous about the same time that flowering plants (angiosperms) were experiencing their own explosive evolutionary diversification. Thus, there were

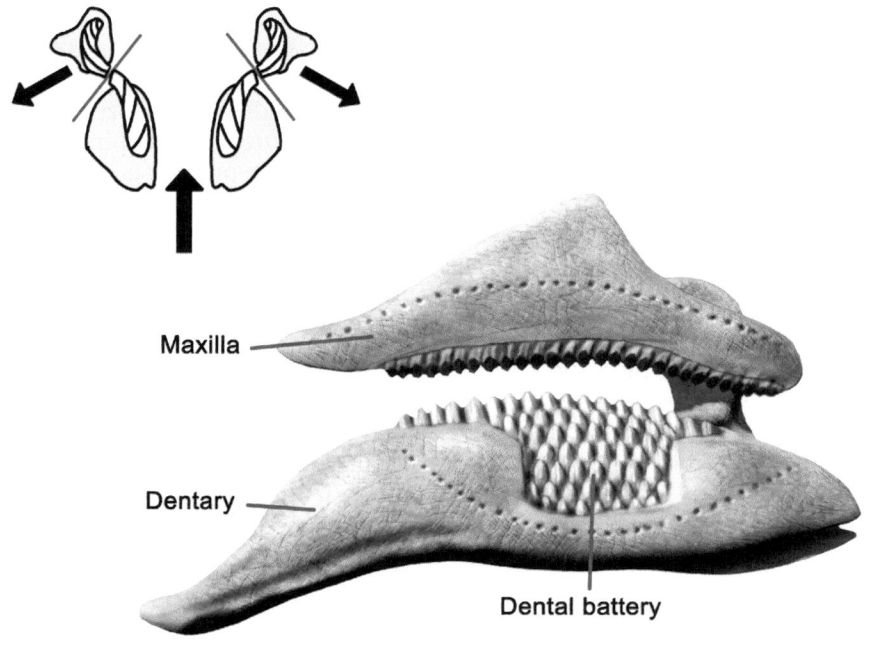

Maxilla

Dentary

Dental battery

Figure 16.3 The skull and lower jaws of a duckbilled dinosaur has thousands of small prism-shaped teeth packed tightly together to form a dental battery, with a flat grinding surface on the top of each battery. (Top left) As the jaws close, the batteries grind against each other, and the upper jaws flex outward as the lower jaws slide between them. This gives duckbills the ability to grind up and consume enormous amounts of vegetation.

abundant different kinds of flowering plants to eat, and stomach contents and coprolites suggest that hadrosaurs ate a variety of leaves and twigs on bushes and trees, and bark (and possibly invertebrates that lived in rotting wood or floating vegetation). Some paleontologists argue that the relentless attach of voracious duckbills might have spurred the evolution of flowering plants, which have a much more efficient mode of reproduction than gymnosperms do. Flowering plants can regrow quickly from dinosaur damage, especially compared to the slow-growing, slow-reproducing gymnosperms, like cycads, ginkgoes, and conifers that dominated the Triassic and Jurassic. Just think of how quickly the grass in a lawn grows back after it is mowed! Flowering plants also can spread by asexual vegetative reproduction, where cuttings of the plant can produce a new plant without sexual reproduction by flowers. Although the evidence is circumstantial, so although it's a bit too much to say that "dinosaurs invented flowers" or "dinosaurs were the first flower children" (as some paleontologists have said), it's likely that flowering plants did diversify faster because their adaptations gave them an advantage over slower-growing, less resilient plants when super-herbivores like duckbills evolved.

Hadrosaurs were primarily bipedal, with their tails held out straight behind them like a balancing pole, and their bodies held out horizontally over their hind limbs. Like iguanodonts, hadrosaurs had a trusswork of tendons making their tails fairly straight and rigid. These tendons often were ossified, or replaced by bone, so they could only flex or bend to a limited degree. In fact, little cylinders of bone from hadrosaur tendons are among the most common fossils one finds in hadrosaur-bearing Upper Cretaceous beds. Although they were primarily quadrupedal, most hadrosaurs had robust hind limbs with hooves, so they could also walk bipedally when necessary.

Thanks to the discoveries of dinosaur nests in central Montana, we know a lot about the reproduction of hadrosaurs. The most famous example are the nests of *Maiasaura* ("good mother lizard"), which showed that the parents guarded their nests and took care of the immature hatchlings and fed them in the nest until they were large enough to leave the nest.

The family Hadrosauridae is now classified into two subfamilies: the Saurolophinae (such as *Saurolophus*, *Prosaurolophus*, *Gryposaurus*, *Kritosaurus*, *Edmontosaurus*, *Maiasaura*, and half a dozen other genera), which have solid crests or lack crests, and the Lambeosaurinae, which have hollow crests, plus a number of primitive duckbills that are related to these subfamilies (**Figure 16.4**).

The argument over what the crests in the Lambeosaurinae were used for has been long and complicated. For a while, their crests were interpreted as snorkels or as air tanks in their heads as they dived for water plants. Others suggested that the crest might be an air trap for keeping water out of the lungs, or that it was attached to a mobile proboscis and aided in feeding, or a weapon for combat with other members of the species. Less outrageous were the ideas that crest housed olfactory tissue for improving sense of smell, or possible salt glands.

It is telling that none of their crests are built like snorkels, since the nasal opening is not at the top, where it would allow them to keep their head below water, but near the bottom of the snout. Nor is the volume of air space inside their nasal cavities enough to give them any advantage in holding their breath under water. In fact, if they had dived deeper than about 3 meters (10 feet), the water pressure on their bodies would have been too great to allow them to inflate their lungs. More to the point, there is good evidence that hadrosaurs were mostly land dwellers, and spent very little time in the water.

Yet after all the dust has settled, it seems that there is good evidence for only a few possibilities. The likeliest is that they served for sexual display,

Figure 16.4 An array of different kinds of duckbilled dinosaurs: (A) *Tethyshadros*, (B) *Eotrachodon*, (C) *Hadrosaurus*, (D) *Corythosaurus*, (E) *Lambeosaurus*, (F) *Parasaurolophus*, (G) *Edmontosaurus*, (H) *Kritosaurus*, (I) *Shantungosaurus*.

and to advertise their status in the herd, and for display not only to let others in their herd know who they were, but also to help in recognition of other unrelated hadrosaurs. This would explain why the crests grew as they did, and why there seems to be a significant difference between male and female crests, and between adults and juveniles.

The second line of evidence came from looking at the detailed internal anatomy of the nasal passages. In *Corythosaurus* (**Figures 16.4[D]** and **16.5[B]**) and *Lambeosaurus* (**Figure 16.4[E]**), these passages are wide and roughly "S"-shaped. The long curving crest of *Parasaurolophus* (**Figures 16.4[F]** and **16.5[A]**) is even more remarkable, curving back from the nasals to the tip of the crest in the back, then returning along the lower part of the crest until it reached the throat region. Such a long curving nasal passage linking the throat to the nasals makes the most sense if it was some kind of amplification chamber for sounds generated down in the vocal cords. In fact, the U-shaped loop of the crest in *Parasaurolophus* resembles the long, curved tubes of a trombone. If it generated low-frequency honking and hooting sounds (below 400 Hertz), this is excellent for communication among similar species, since low-frequency sounds travel much further than high-frequency sounds. Not only do these sounds travel further, but also their direction is harder to locate, so the hoots and honks of hadrosaurs could communicate that a predator has been spotted, without the hadrosaur giving their own position away. Living elephants also use low-frequency ultra-sounds (mostly below the range of human hearing) to communicate within their own herd, and among herds long distances away.

Hadrosaurs came to dominate Late Cretaceous herbivorous niches in most of the northern continents of North America and Eurasia, but rarely reached any of the Gondwana continents. The earliest hadrosaurs include

Figure 16.5 Skeletons of some well-known hadrosaurs. (A) *Parasaurolophus*, with the trombone-like crest on its head for sound production. (B) *Corythosaurus*, showing the helmet-like crest, skin impressions, and the truss of crisscrossing tendons in the spine and tail. (C) A pair of the hadrosaurs *Edmontosaurus*. (Courtesy Wikimedia Commons.)

Eotrachodon from the Mooreville Chalk of Alabama (**Figure 16.4[B]**), about 86 Ma, and *Hadrosaurus* itself (**Figure 16.4[C]**) from beds about 80 Ma in New Jersey. Slightly more primitive is *Tethyshadros*, a small form from the Early Cretaceous in the region near Trieste between Italy and Croatia (**Figure 16.4[A]**). The most primitive hadrosaurs are *Telmatosaurus* from the Late Cretaceous of Rumania, and *Jintasaurus* from the Early Cretaceous of China. It appears that hadrosaurs originated in Eurasia before spreading to North America by the middle part of the Cretaceous.

Almost fifty genera of hadrosaurs are now known, not counting all the outdated or invalid genera based on non-diagnostic scraps that have been mostly forgotten (like the genus *Trachodon*). New genera and species are described almost every year now. Their family relationships were once controversial, but are more or less worked out today. Their geographic distribution is also very interesting. Some, like *Corythosaurus*, is only known from Alberta despite the large number of specimens, while *Parasaurolophus* is found in both Alberta and New Mexico. But the closest relative of *Parasaurolophus*, called *Charonosaurus*, comes from the Amur region of Siberia. The close relative of *Corythosaurus*, the hypacrosaurs, largely come from the American West, but *Olorotitan* and *Amurosaurus* comes from the Amur region of Siberia, and *Blasisaurus* from Spain. Primitive lambeosaurines include *Aralosaurus* from the Aral Sea region of Russia, *Canardia* from France, *Jaxartosaurus* from Kazakstan and China, *Tsintaosaurus* from China, and *Pararhabdodon* from Spain.

Among saurolophines, *Edmontosaurus* (**Figures 16.4[G]** and **16.5[C]**), *Prosaurolophus* and *Saurolophus* come from the American Rockies and Alberta but their relative *Kerberosaurus* is from the Amur region of Siberia, and giant *Shantungosaurus* is from China (**Figure 16.4[I]**). *Kritosaurus* (**Figure 16.4[H]**) is found in in New Mexico and Texas, while the kritosaurs *Secernosaurus* and *Willinakaqe* are from the latest Cretaceous of Argentina but not further north. Yet the kritosaur *Wulagasaurus* is found only in China.

Thus, hadrosaurs seemed to be very wide ranging and mobile, switching back and forth between Eurasia and North America freely during the Late Cretaceous. Most remained in the northern Laurasian continents with the exception of the two kritosaurs *Secernosaurus* and *Willinakaqe* that show up in the latest Cretaceous of Argentina, one of the few Cretaceous groups that reached Argentina from North America.

PACHYCEPHALOSAURS: THE "BONEHEADS"

The two remaining branches of the ornithischians are the pachycephalosaurs and the horned dinosaurs, or ceratopsians. They are combined into a group known as the Marginocephalia, or "frill heads", because both groups have some sort of frill or just small ledge sticking out of the back of the skull just above the neck.

Of these two groups, the less familiar to the public are the pachycephalosaurs (although one made an appearance in *Jurassic World: Fallen Kingdom*, battering down the walls and door of the heroes' jail cell). Their name means "thick headed lizard", and that is their outstanding characteristic—a thick dome of bone above their tiny brain (**Figure 16.6**). They were first discovered in 1902, when the only fossils found were the thick bony dome from the skull, and the earliest paleontologists thought that it was a base for a large horn, so it was known as the "unicorn dinosaur". This early find, known as *Stegoceras*, was still a mystery until 1924 when a complete unbroken skull and jaws were found, and gave a much better sense of what these dinosaurs looked like. The skull not only proved that the bony dome sat on top of the head from the eyes to the back of the skull, but also showed many other interesting features. It had large eyes,

Figure 16.6 Reconstructions of a variety of pachycephalosaurs: (A) *Homalocephale*, (B) *Stygimoloch*, (C) *Dracorex*, (D) *Pachycephalosaurus*.

roofed by a shelf of bone protruding above them and below the dome of the skull. This ridge of bone continued to the back of the skull, producing a short "frill" over the neck that is found in all pachycephalosaurs and ceratopsians. This is one of many features that demonstrate these two groups are closest relatives, now known as the Marginocephalia. The entire frill around the back and side of the skull was covered with bumps and ridges of bones. The skull went from broad in the back to a short narrow pointed snout, with large forward-facing nostrils. CAT scans of the large olfactory bulbs of the brain showed that these dinosaurs had a good sense of smell. The jaw contains small leaf-shaped teeth, with a gap between the front teeth and the cheek tooth row. The front teeth were conical nipping teeth with a small set of ridges and cusps on them, while the cheek teeth were triangular in cross section.

The skeleton is like that of many other small bipedal ornithischians, with no obvious specialized armor or other features typical of the advanced groups like stegosaurs, ceratopsians, or duckbills. Charles Gilmore reconstructed some of the fossils he found as belly ribs, or gastralia, but they are now known to be the ossified trusswork of intermuscular bones in the tail. Like most other dinosaurs, *Stegoceras* held its tail straight out in the back. Because *Stegoceras* is known from a partial skeleton, it is one of the most completely known pachycephalosaurs. Most of the rest are known only from skulls. *Stegoceras* was about 2.0–2.5 meters (6.6–8.2 feet) long counting the long tail, and weighed about 10–40 kg (22–88 lb), about size of a goat. The front limbs were quite small, so the dinosaur was completely bipedal, unlike many larger ornithischians. The animal must have run in a bird-like fashion, with its tail stuck straight out behind it, and the body balanced horizontally over the long hind limbs. Then in 1931, a much larger and more complete fossil was found, which received the name *Pachycephalosaurus*

(**Figure 16.5[B]**), and showed that the group was distinct from any other group of dinosaurs. Although no skeleton was known from the original specimens of *Pachycephalosaurus*, the skull is very impressive. The bulging dome on the skull was 25 cm (10 inches) thick, and cushioned a tiny brain inside (the brain cavity is also preserved). All around the rear and sides of the dome are bony knobs, and there are bony spikes on the nose and snout. Like *Stegoceras*, it had large eyes covered by a rim of bone above the eye sockets. The snout was short, with a pointed beak. *Pachycephalosaurus* had tiny leaf-shaped teeth, similar to those of others in the group.

Even though the rest of the skeleton of *Pachycephalosaurus* was unknown from the original skull fossils, a partial skeleton has since been found in the Hell Creek Formation in South Dakota. It was about 1.5 meters (5 feet) tall and 3 meters (10 feet) long. Its partial skull includes most of the back, side and snout region, but not the dome. In addition, the fossil includes the hind limbs and hips, plus some neck and back vertebrae and ribs. On the basis of this specimen and other related pachycephalosaurs known from skeletons, *Pachycephalosaurus* was a medium-sized bipedal animal that weighed about 450 kg (990 lb), with a fairly short, thick, S-shaped neck, very short fore limbs, a bulky body with a large gut for fermenting and digesting plants, long thick hind legs, and a heavy tail held out straight behind it by ossified tendons.

In the 1950s and 1960s, additional pachycephalosaurs like *Prenocephale*, *Homocephale*, *Goyocephale*, and *Tylocephale* (**Figure 16.6[A]**) were discovered by the Polish-Mongolian expedition to the Gobi Desert, which found many Late Cretaceous dinosaurs. In 1974, the primitive form *Wannanosaurus* was found from the early Late Cretaceous of China. The flat skull roof has almost no dome, and still has the large openings in the sides and back of the skull found in most dinosaurs. These features are lost in advanced pachycephalosaurs as they developed their thick bony domes and covered up the holes in the skull. Not only is *Wannanosaurus* older and more primitive than the rest of the pachycephalosaurs, but also it is one of the smallest dinosaurs known, with an estimated body length of only 60 cm (2 feet). This transitional fossil shows how the weird pachycephalosaurs evolved from much more primitive ancestors.

The first describers of *Stegoceras* and *Pachycephalosaurus* did not speculate much about how they behaved or what the thick dome of bone was used for. In 1955, Edwin Colbert was the first to suggest that the pachycephalosaurs were like dinosaurian rams, head butting with their heavily armored skulls. In addition to the solid helmet of bone, the shape of the neck suggested that they had strong neck muscles, with an "S"-shaped curve to absorb the shock of each blow. Others have suggested that they used their bony helmets to head-butt the flanks of other members of their herd, giving a less lethal glancing blow. They had wide trunks and bellies, which would protect the internal organs from a head blow. One genus, *Stygimoloch*, had horns on the side of its face, which would have been even more effective in flank butting (**Figure 16.6[C]**).

In 2004, Mark Goodwin and Jack Horner argued that pachycephalosaurs could not have endured direct head butting because the bone structure allegedly could not have absorbed such stresses. In their opinion, the dome for was for species recognition only. This has been debunked by numerous analyses since then, which established that the spongy bone of the skull supporting the solid bone of the dome is indeed capable of absorbing head-to-head collisions. Their bone structure is much like that of rams and muskoxen, which also engage in head-to-head impacts. In addition, the domes do not appear to differ much among adults. Such differences between males and females would be expected if they were used for species recognition or mate recognition.

Figure 16.7 (A) Skeleton of *Yinlong*. (J. Clark). (B–C) Specimens of *Psittacosaurus* have been found with soft-tissue preservation, and even their colors fossilized. (D) *Protoceratops* was a slightly more advanced ceratopsian with a large frill but no horns. (Courtesy Wikimedia Commons.)

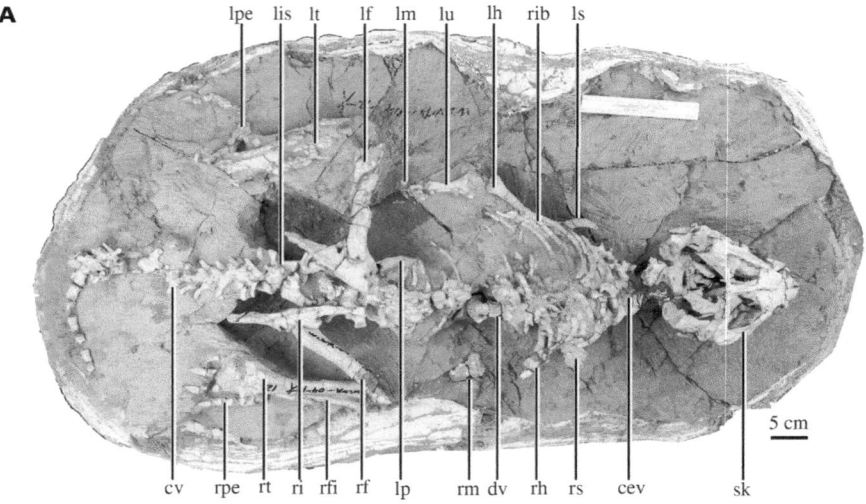

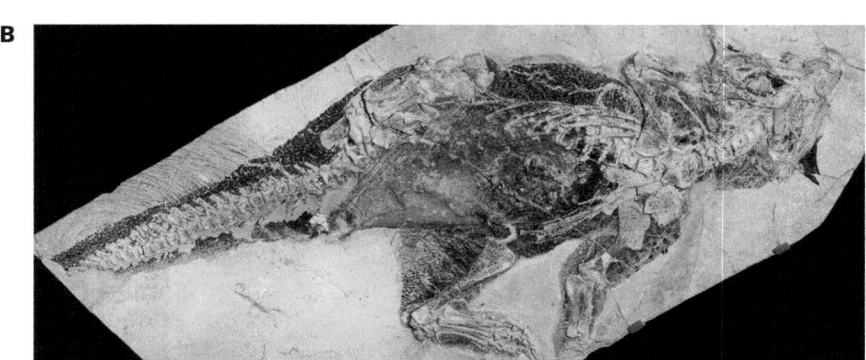

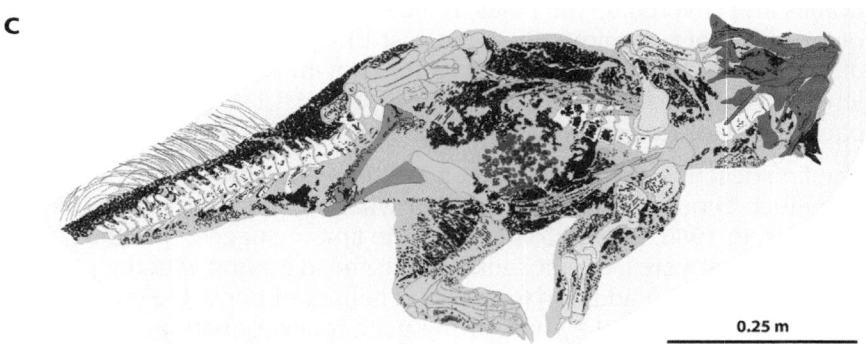

The battering ram model was further supported by a 2013 study of the pathologies of the specimens that had been injured. About 22% of the domes had damage or lesions consistent with osteomyelitis, a bone infection caused by penetrative trauma. By contrast, flat-skulled pachycephalosaurs like *Homalocephale* (**Figure 16.6[A]**) show no such rate of injury, suggesting they did not engage in head-to-head combat. This would make sense if they were females or juveniles, who did not have to compete to become masters of their herds.

New pachycephalosaurs continue to be discovered and named, although they are much less diverse and abundant that groups like hadrosaurs or ceratopsians. The weird horned genus *Stygimoloch spinifer* was a great puzzle when it was first described (**Figure 16.6[C]**). It got its name from the River Styx, the river at the entrance to Hades, while Moloch was a Canaanite god who demanded child sacrifice; "spinifer" means "spiny". (This is the dinosaur in *Jurassic World: Fallen Kingdom* that breaks down the prison cell door and walls.) Then there was the even smaller horned genus *Dracorex hogwartsi* ("dragon king of Hogwarts"), which was described in 2006 as another small horned adult pachycephalosaur (**Figure 16.6[D]**). Some have argued that both of these are juvenile specimens of *Pachycephalosaurus*, but this is still unclear.

Where did marginocephalians come from? In 2004, the most amazing fossil in this lineage was discovered with the description of *Yinlong* (**Figures 16.7[A]** and **16.8[A]**) from much earlier in the Late Jurassic of China. Its name means "hidden dragon" in Mandarin, in reference to the popular movie *Crouching Tiger, Hidden Dragon*, part of which was filmed close to the locality where the fossil was found. *Yinlong* consists of a beautifully preserved skeleton of a bipedal dinosaur not too different in proportions from the primitive ceratopsian *Psittacosaurus* (**Figure 16.7[B,C]**). *Yinlong* has the rostral bone, a feature that is unique to ceratopsians, in its upper beak. However, its skull roof has a unique configuration of bones found in the pachycephalosaurs, which are famous for having a thick dome of bone in their skulls protecting their tiny brains. Paleontologists have long thought that ceratopsians and pachycephalosaurs were closest relatives, based on the fact that they both have a frill of bone around the back of the skull (hence their name, "Marginocephalia"). Like all marginocephalians (pachycephalosaurs plus ceratopsians), there is a frill in the back of the skull of *Yinlong*. But *Yinlong* shows features of both ceratopsians and pachycephalosaurs before their lineage split into the two families. Thus, it forms a transition between more primitive ornithischians in the Jurassic, and the most primitive pachycephalosaurs like *Wannanosaurus* and the earliest ceratopsians like *Psittacosaurus*.

CERATOPSIA: THE HORNED DINOSAURS

Next to the sauropods and big theropods like *T. rex*, one of the most famous and popular dinosaurs is *Triceratops*. But that dinosaur is just the most famous member of a large group, the Ceratopsia, that includes dozens of different genera. Some had one horn, some had three, many had none at all. Most of them had large frills on their skulls, although the primitive ones had only a short frill in the back. But beyond these specializations, all ceratopsians have certain features in common. The most distinctive is a unique not found in any other animal, the rostral bone in the tip of the upper snout. It supported a horny beak over the toothless snout, complementing the beak that sat over the tip of the lower jaw (supporting by the predentary bone, a feature unique to ornithischians). In addition, the frills of many ceratopsians have unique bones attached to their edges, called epoccipitals, another bone found in no other group of animals. Finally, the bones of the cheek region form a "horn" which is pointed down and sideways from the eye socket, another distinctive feature.

The earliest ceratopsians from the Early Cretaceous of Asia are just slightly more advanced than *Yinlong*, and did not have all these advanced features, only the rostral bone and a very short frill on the back of the skull. The most famous is *Psittacosaurus*, a small bipedal dinosaur from China and Mongolia, which a distinctive parrot-like beak (**Figures 16.7[B]** and **16.8[B]**). The name *Psittacosaurus* means "parrot lizard" in Greek, in reference to this beak. The original Mongolian specimens were about 2 meters (6.5 feet) in length, and weighed about 20 kg (44 pounds). Like all ornithischians, the beak also had a horny covering to it, as many reptilian and bird beaks do over their jaw bones. The front of the beak was toothless, but there are small cheek teeth that were self-sharpening,

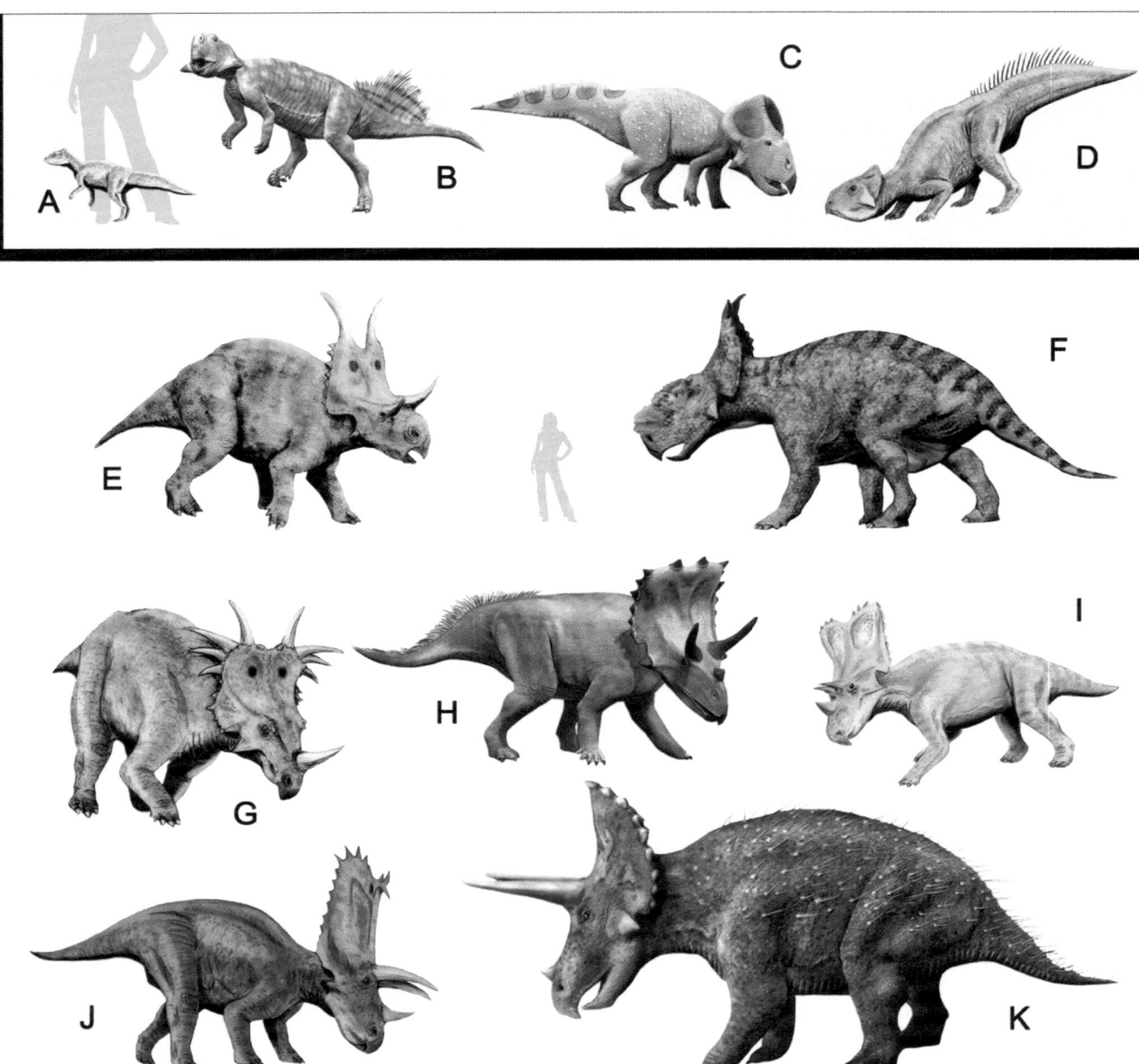

Figure 16.8 Reconstruction of the diversity of the ceratopsians, including: (A) *Yinlong*, (B) *Psittacosaurus*, (C) *Protoceratops*, (D) *Leptoceratops*, (E) *Diabloceratops*, (F) *Pachyrhinosaurus*, (G) *Styracosaurus*, (H) *Agujaceratops*, (I) *Chasmosaurus*, (J) *Pentaceratops*, (K) *Triceratops*.

and had a prominent crest across the top, presumably for chopping up plants, seeds, and nuts for its herbivorous diet. The type skeleton even preserved a pile of small stones in the gut cavity that were probably gastroliths, grinding stones for use its stomach or gizzard.

Another striking feature is the flaring cheekbones, which stick out almost like small horns on the side of the face. The eyes of *Psittacosaurus* were rather large, with a ring of bone (sclerotic ring) protecting the large eyeball. The size of the eyeball suggest that it might have been nocturnal. There was no large ceratopsian frill, but the back of the skull still has a significant shelf of bone that is true of all the marginocephalians, including pachycephalosaurs. It was long thought to have a small brain, but a 2007 study showed the brain was larger and more advanced than most other herbivorous dinosaurs, and with a brain/body size ratio close to that of *Tyrannosaurus rex*.

Psittacosaurus was a bipedal dinosaur with small forelimbs, less than 60% the length of the long hindlimbs. Not only were the arms too short for quadrupedal locomotion, but also it could not rotate its hand to bring the palms flat on the ground, so it was almost certainly bipedal as adults. However, a study of juvenile specimens showed that the limb sizes and bone structure indicated a quadrupedal posture as young animals, gradually becoming fully bipedal as the legs grew faster than the arms. The arms were too short to reach the mouth so they were only good for two-handed grasping of near object, and for scratching and fighting. *Psittacosaurus* had only four fingers on the hand, compared to five in most other ornithischians, and four toes on the feet like most other dinosaurs.

There is one complete specimen (**Figure 16.8[B]**) from the Yixian Formation, in the Liaoning Province of China, which preserves all the soft tissues and even the melanosomes, or fossilized pigment cells, indicating its color. The body was mostly covered by scales rather than feathers, but along the back were an array of bristles which glow like feathers under UV light. The melanosomes preserved on this specimen indicate it was countershaded, with dark on the top and light on its belly, so it was easier for it to hide in the shade of the forests in which it lived. There were patches of color on the face for display, as well as around the cloaca and on the membranes of its hind legs.

Slightly younger early Late Cretaceous specimens from Asia show the next stage in ceratopsian evolution: *Protoceratops*. It was not a big dinosaur, reaching about 1.8 meters (6 feet) long and 0.6 meters (2 feet) high at the shoulder (**Figures 16.7[C]** and **16.8[C]**). It may have weighed about up to 180 kg (400 pounds), about the size of a pig. Like other ceratopsians, *Protoceratops* had a sharp beak on its snout made of the rostral bone, a feature unique to the ceratopsians and some of their distant relatives like *Yinlong*. The large frill over the back of the head and neck were perforated with large holes or "windows" (fenestrae in anatomical terms), which made them lighter and may have added attachment points for jaw muscles. The development of this frill from a tiny ledge of bone in *Psittacosaurus* to the broad flaring structure of adults is one of the most striking features of their growth and development. Initially, it was thought that the frill was mainly to protect the neck, but more recent analysis has argued it wasn't very effective as neck protection, and more likely served as a display structure to communicate with its own herd and advertise its age and status. Certainly, the fact that the frill grew dramatically from juveniles to adults is consistent with its ability to indicate the age and strength of adults, comparable to the way larger horns or antlers in adult antelopes and deer show who was boss.

There is good evidence that *Protoceratops* had a powerful bite and was able to chew tough vegetation, especially with its dental battery packed with small wedge-shaped teeth (a feature distinctive to all ceratopsians). It had large eyes, consistent with the idea that it could see well in the dark and may have been nocturnal. We know that it coped with another night-dweller, the turkey-sized dromaeosaur *Velociraptor*, because of the famous fossil of a *Protoceratops* with a *Velociraptor* attacking it, then both dinosaurs dying as they fought, buried in sand.

Triceratops and *Protoceratops* and *Psittacosaurus* were part of a huge Late Jurassic through Late Cretaceous radiation of horned dinosaurs. Currently, there are over 80 genera recognized in the group (**Figures 16.8**, **16.9**, and **16.10**). Not only were they very diverse, but also many of them were extremely abundant, so a high percentage of the specimens from the Lower Cretaceous of China and Mongolia are *Psittacosaurus*, while *Protoceratops* dominated the Upper Cretaceous Djadokhta Formation of Mongolia, and *Triceratops* makes up 5/6 of the dinosaurs recovered from the Upper Cretaceous Lance and Hell Creek formations. Some, like *Centrosaurus*, were almost certainly herding animals, while others, like *Triceratops*, were apparently loners. As the most common large herbivores through much of the Cretaceous, they must have been important prey for the large predators like the tyrannosaurs.

After starting in the Late Jurassic with *Yinlong*, *Chaoyangsaurus*, and *Xuanhuaceratops* from China, ceratopsians became more common and diverse in the Early Cretaceous with *Psittacosaurus* and numerous other

Figure 16.9 The "wall of skulls" family tree of ceratopsians at the Utah Museum of Natural History in Salt Lake City. The right branch (numbers 08–12) is the chasmosaurines. The large skull on the upper right (number 11) is *Coahuilaceratops magnacuerna*. The two large skulls above it and to the left are (08) *Anchiceratops ornatus* and (09) *Triceratops horridus*. Number 10 is *Kosmoceratops richardsoni* and 12 is *Chasmosaurus belli*. The branch on the left are centrosaurines. In the top row on the left are (left to right) *Einiosaurus procurvicornis*, *Pachyrhinosaurus lakustai*, and *Achelousaurus horneri*. In the bottom row on the left (04–07) are (left to right) *Styracosaurus albertensis*, *Centrosaurus apertus*, *Nasutoceratops titusi*, and *Diabloceratops eatoni*. In the lower left corner are basal ceratopsians *Protoceratops* and *Zuniceratops*. (Photo by the author.)

A

Figure 16.10 Two mounted skeletons of *Triceratops*. (A) The old traditional way of mounting it with the front legs sprawling out to the side like a lizard's legs, and the tail dragging. (B) The modern conception of the posture of *Triceratops*, with the legs fully beneath the body. [(A) Courtesy Wikimedia Commons. (B) Photo by the author.]

B

genera (**Figures 16.7[B]** and **16.8**), almost all restricted to Asia until the very end of the Early Cretaceous. In the early Late Cretaceous, there were Protoceratopsidae in Asia, while *Zuniceratops* and the Leptoceratopsidae were typical of North America (**Figure 16.8[D]**). By the middle Late Cretaceous, ceratopsians apparently vanished from their original homeland in Asia, while they became increasingly more diverse and common in North America. During this last great evolutionary flowering of the group, there were increasingly bizarre and elaborate combinations of horns, frills, and spikes on various parts of the skull (**Figure 16.8**). One group, the Centrosaurinae, include taxa with a single nose horn, brow horns which are small or absent, and a broad frill with various ornamentations on its edges, like the spiked frill of *Styracosaurus* (**Figure 16.8[G]**), or the pair of "devil horns" on *Diabloceratops* (**Figure 16.8[E]**), or the blunt thick boss of bone instead of a nose horn on *Pachyrhinosaurus* (star of the CG movie *Walking with Dinosaurs*) and *Einiosaurus* (**Figure 16.8[F]**). The other group, the Chasmosaurinae, include genera with prominent brow horns, and long frills with few or no big spikes or other ornamentation on the edge of the frill. These include not only the familiar genera like *Triceratops* and *Torosaurus* (**Figure 16.8[K]**) but also the triangular-frilled *Pentaceratops* (**Figure 16.8[J]**) and the bizarre-looking *Medusaceratops* and *Mojoceratops*.

Although the ceratopsians were primarily Asian and North American, there are a handful from Europe, including *Ajkaceratops* from Hungary, *Craspedodon* teeth from Belgium, and a possible leptoceratopsid from Sweden. Then there are claims for possible ceratopsians in other continents. The Australian genus *Serendipaceratops* is known only from an isolated lower arm bone, so it is not certain that it is from a true ceratopsian, and if so, to what ceratopsian group it belonged. *Notoceratops* from Argentina was based on a single toothless jaw that has since been lost.

Triceratops (**Figure 16.8[K]**) is now known from dozens of skulls and many partial skeletons, although no complete skeleton of a single individual has ever been found. Barnum Brown claimed he had seen as many as 500 skulls in the field (in various states of completeness), and Bruce Ericson of the Science Museum of Minnesota reported over 200. As John Scannella commented, "It is hard to walk out into the Hell Creek Formation and not stumble upon a *Triceratops* weathering out of a hillside". The dinosaur is now so familiar to us that it's hard to appreciate how startling its appearance is (**Figures 16.8[K]** and **16.10**). Most of the largest specimens were about 9 meters (30 feet) in length, 3 meters (10 feet) high at the shoulder, and are thought to have weighed between 6 and 12 tonnes (6.5–13 tons). The skulls of *Triceratops*, like that of most ceratopsians, was enormous, the largest skulls known for any group of land animals.

In addition to the famous combination of three horns on their face, *Triceratops* has many other anatomical peculiarities. The nose horn is sometimes compressed and narrow, but in other skulls it is more rounded and conical in cross section, located above a remarkably large opening for the nostrils. Like all ceratopsians, *Triceratops* had a prominent upper beak made of the rostral bone, which occluded with a lower beak made of the predentary bone found in all ornithischians. Both had a sheath of keratin on them to produce a sharp beak that was self-sharpening with wear. *Triceratops* also had dozens of small teeth arranged in stacked rows called a dental battery. However, their teeth looked very different from the dental battery of hadrosaurs, which were built of closely packed tall polygonal prisms (**Figure 16.3**). In *Triceratops*, each battery was made of 35–40 tooth columns, and each column was built of 3–5 stacked teeth, which shed teeth off the top as they wore out. Each battery occluded against the inclined surface of the battery in the opposite jaw. Altogether, they had between 432 and 800 teeth in their mouth at once (hadrosaurs had more than a thousand tiny prisms in their mouth). Their teeth sheared in a vertical plane and were able to chop up even the toughest vegetation. The likeliest food would be palms and cycads and ferns, since grasses were not yet common, and *Triceratops* could not reach its head high enough to eat tall shrubs or trees.

The function of the frill has long been a source of speculation among paleontologists. The conventional story was that the frill served to protect its neck against the bite of *Tyrannosaurus rex*, an idea first proposed by Charles H. Sternberg in 1917 and revived many times. There is some evidence of tyrannosaur bite marks on their brow horns and cheekbones, although not on the frill. In some cases, there are horns that were bitten off, then regrew while the *Triceratops* was still alive. But there is no direct evidence that tyrannosaurs bit the frill in an effort to reach down and bite the neck of their prey. There are, however, many *Triceratops* carcasses that show bite marks and even shed teeth of tyrannosaurs, so they were definitely scavenged. And there is a *Styracosaurus* specimen with damage to one side of the frill, which may have been inflicted by a predator.

Instead, many paleontologists think that the frill was more important for display and dominance within the species. This would explain the wide variation in shapes and sizes of frills, as well as the dramatic change in size and shape of the frill as *Triceratops* grew up. Juvenile *Triceratops* even had a small frill before they reached the age of sexual reproduction, so the frill helped aid in communication visually, and in species recognition.

As far as social behavior goes, *Triceratops* is never found in large herd assemblages like those discovered for *Centrosaurus*. Although they were among the most common dinosaurs of the latest Cretaceous, they were apparently lived in small family groups or were solitary. However, there is evidence of combat between adult *Triceratops*. A study of skulls showed that about 14% showed some kind of damage from intraspecific combat, although this is low compared to other ceratopsians. There is little evidence, however, *Triceratops* engaged in direct head-to-head jousting, because we find at most one or two wounds caused by the horns puncturing the faces of their opponents. It is more likely that they engaged in head-to-head wrestling, based on injuries the skull and face.

Triceratops was clearly quadrupedal, with four sturdy legs with toes tipped by short hooves. However, most of the early reconstructions were based on the "sprawling lizard" way of visualizing dinosaurs, so their front limbs are shown bent in a crouch, like a crocodile or lizard, and their tails shown dragging on the ground (**Figure 16.10[A]**). More recently, their limbs have been reconstructed as being more upright beneath the body with their limbs only slightly flexed, and their tails were held straight out (**Figure 16.10[B]**). This is confirmed by trackways, which show their footprints close together indicating an upright columnar posture, and much too narrow for a stance if their front legs splayed to the side; in addition, there are no tail drag marks. The hands of *Triceratops* did not have the ability to rotate to allow their fingers to face forward, unlike other quadrupeds like stegosaurs and sauropods. Instead, *Triceratops* walked with most of their fingers pointing outward to the side and front. Their hands had three large functional fingers, with only vestigial remnants of the ring finger and pinky.

Finally, at the peak of their success, *Triceratops* and a few other genera were among the last non-bird dinosaurs still around when the Cretaceous came to an end 66 Ma. *Triceratops* bones can be tracked right to within a meter or so of sediment below the boundary itself. The media gives the oversimplified and incorrect impression that the extinction of the non-bird dinosaurs was simply due to the impact of an asteroid in Yucatan. However, for the past 40 years, there has also been strong evidence that huge volcanic eruptions, the Deccan lavas of Pakistan and western India, were changing the climate well before the rock from space hit the earth. Among paleontologists (especially vertebrate paleontologists), the idea that the asteroid impact caused the extinction all by itself is not very popular, since the Deccan lavas clearly were changing the earth's climate before the impacts occurred. Most view the complex biological signal (such as the near total survival of crocodilians, frogs, salamanders, and many turtles through the extinction) as evidence that the event was much more complex than the media and the general public think.

FURTHER READING

Alexander, R.M. 1989. *Dynamics of Dinosaurs and Other Extinct Giants*. Columbia University Press, New York.

Barrett, P.M. 2001. Did dinosaurs invent flowers? Dinosaur-angiosperm coevolution revisited. *Biological Reviews*. 76 (3): 411–447.

Brett-Surman, M.K. 1979. Phylogeny and paleobiogeography of hadrosaurian dinosaurs. *Nature*. 277 (5697): 560–562.

Butler, R.J.; Barrett, P.M.; Kenrick, P.; Penn, M.G. 2009. Diversity patterns amongst herbivorous dinosaurs and plants during the Cretaceous: Implications for hypotheses of dinosaur/angiosperm co-evolution. *Journal of Evolutionary Biology*. 22: 446–459.

Dodson, P. 1976. Quantitative aspects of relative growth and sexual dimorphism in *Protoceratops*. *Journal of Paleontology*. 50: 929–940.

Dodson, P. 1996. *The Horned Dinosaurs*. Princeton University Press, Princeton.

Dodson, P.; Forster, C.A.; Sampson, S.D. 2004. Ceratopsidae, pp. 494–513. In Dodson, P.; Weishampel, D.B.; Osmolska, H., eds. *The Dinosauria* (2nd ed.). University of California Press, Berkeley.

Eberth, D.A.; Evans, D.C., eds. 2015. *Hadrosaurs*. Indiana University Press, Bloomington, IN.

Fastovsky, D.E.; Weishampel, D.B. 2021. *Dinosaurs: A Concise Natural History* (4th ed.). Cambridge University Press, Cambridge.

Han, F.-L.; Forster, C.A.; Clark, J.M.; Xu, X. 2016. Cranial anatomy of *Yinlong downsi* (Ornithischia: Ceratopsia) from the Upper Jurassic Shishugou Formation of Xinjiang, China. *Journal of Vertebrate Paleontology*. 36 (1): e1029579.

Holtz, T.R., Jr.; Rey, L.V. 2007. *Dinosaurs: The Most Complete, Up-to-Date Encyclopedia for Dinosaur Lovers of All Ages*. Random House, New York.

Horner, J.R.; Goodwin, M.B. 2009. Extreme cranial ontogeny in the Upper Cretaceous dinosaur *Pachycephalosaurus*. *PLoS One*. 4 (10): e7626.

Horner, J.R.; Makela, R. 1979. Nest of juveniles provides evidence of family structure among dinosaurs. *Nature*. 282 (5736): 296–298.

Lucas, S.G. 2005. *Dinosaurs, the Textbook* (5th ed.). W. C. Brown, Dubuque, Iowa.

Maryańska, T.; Chapman, R.E.; Weishampel, D.B. 2004. Pachycephalosauria, pp. 464–477. In *The Dinosauria* (2nd ed.). University of California Press, Berkeley.

Maryańska, T.; Osmólska, H. 1974. Pachycephalosauria, a new suborder of ornithischian dinosaurs. *Palaeontologica Polonica*. 30: 45–102.

Mcdonald, A. 2012. Phylogeny of basal iguanodonts (Dinosauria: Ornithischia): An update. *PLoS ONE*. 7 (5): e36745.

McGowan, C. 1983. *The Successful Dragons: A Natural History of Extinct Reptiles*. Samuel Stevens, Toronto.

McGowan, C. 1991. *Dinosaurs, Spitfires, and Sea Dragons*. Harvard University Press, Cambridge.

Norman, D.B. 1985. *The Illustrated Encyclopedia of the Dinosaurs*. Crescent Books, New York.

Norman, D.B.; Weishampel, D.B. 1990. Iguanodontidae and related ornithopods. In Weishampel, D.B.; Dodson, P.; Osmólska, H., eds. *The Dinosauria*. University of California Press, Berkeley.

Ostrom, J.H. 1964. A reconsideration of the paleoecology of the hadrosaurian dinosaurs. *American Journal of Science*. 262 (8): 975–997.

Peterson, J.E.; Dischler, C.; Longrich, N.R. 2013. Distributions of cranial pathologies provide evidence for head-butting in dome-headed dinosaurs (Pachycephalosauridae). *PLoS ONE*. 8 (7): e68620.

Ryan, M.J.; Chinnery-Allgeier, B.J.; Eberth, D.A., eds. 2010. *New Perspectives on Horned Dinosaurs: The Royal Tyrrell Museum Ceratopsian Symposium*, Indiana University Press, Bloomington, IN.

Sereno, P.C. 1986. Phylogeny of the bird-hipped dinosaurs (order Ornithischia). *National Geographic Research*. 2 (2): 234–256.

Sereno, P.C. 1998. A rationale for phylogenetic definitions, with application to the higher-level taxonomy of Dinosauria. *Neues Jahrbuch für Geologie und Paläontologie, Abhandlungen*. 210: 41–83.

Sereno, P.C. 1999. The evolution of dinosaurs. *Science*. 284 (5423): 2137–2147.

Snively, E.; Cox, A. 2008. Structural mechanics of pachycephalosaur crania permitted head-butting behavior. *Palaeontologia Electronica*. 11 (1).

Sullivan, R.M. 2006. A taxonomic review of the Pachycephalosauridae (Dinosauria: Ornithischia). *New Mexico Museum of Natural History and Science Bulletin*. 35: 347–365.

Weishampel, D.B.; Dodson, P.; Osmolska, H., eds. 2007. *The Dinosauria* (2nd ed.). University of California Press, Berkeley.

You, H.; Dodson, P. 2004. Basal Ceratopsia, pp. 478–493. In Weishampel, D.B.; Dodson, P.; Osmolska, H., eds. *The Dinosauria* (2nd ed.). University of California Press, Berkeley.

SAUROPODS

LONG-NECKED GIANTS

17

When it walked the earth trembled under the weight of 120,000 pounds, when it ate it filled a stomach enough to hold three elephants, when it was angry its terrible roar could be heard ten miles, and when it stood up its height was equal to eleven stories of a skyscraper.

—*New York Herald*, 1898

THE LARGEST LAND ANIMALS

When most people hear the word "dinosaur", the first images that come to mind are the enormous sauropods with their long necks with small heads, long tails, and huge bodies and legs like tree trunks. For years, the Sinclair Oil Company had a green sauropod silhouette as their logo, and the names of dinosaurs like *Brontosaurus*, *Diplodocus*, and *Brachiosaurus* are familiar even to young children. Many large natural history museums have a mounted skeleton (or a replica) of a sauropod on display in their exhibit halls, guaranteed to attract visitors who marvel at the enormous size of these creatures.

Yet these animals have a long history of misunderstanding and misinterpretation as well. The first ever found was *Cetiosaurus* from the Middle Jurassic of England. It was known from only a handful of bones of enormous size. *Cetiosaurus* eventually was named by legendary paleontologist Richard Owen in 1841, but he considered to be a gigantic crocodile or marine reptile. Not until the 1870s and 1880s were nearly complete skeletons discovered that showed the true nature of sauropods, giving us our first look at their incredible necks and small heads and long tails, as well as their enormous size. By the early twentieth century, several complete skeletons were mounted in major eastern museums and drew huge crowds.

One of the oldest animations ever made was the cartoon of Gertie the dinosaur, hand drawn by Winsor McCay in 1914, who became a show biz sensation. Gertie was drawn as a sauropod, but acted more like a puppy. Originally it was projected on a screen on the vaudeville stage and appeared to respond to McCay, who was pretending to be its master and giving in commands from the stage. Since that time, sauropods have been featured in films and television, in giant sculptures and robotic animated figures, and still is the first thing people think of when they hear the word "dinosaur".

But much of what people think about sauropods is wrong, or at least badly outdated. At the beginning of the twentieth century, sauropods were portrayed as huge sluggish sprawling reptiles. Some reconstructions had their legs sprawling out from the side of their bodies like the legs of a crocodile, which was problematic, because this posture would cause their bellies to drag on the ground or even require them to drag their bellies in ruts. Many paleontologists could not imagine them supporting their enormous weight on land, so sauropods were usually portrayed as swamp-dwellers, living a sluggish reptilian life eating water

DOI: 10.1201/9781003128205-17

plants. But in the 1960s and 1970s, dinosaur paleontology underwent a renaissance, and most of these images were debunked. The idea that dinosaurs were warm-blooded and faster moving and active became popular, and eventually even sauropods were portrayed this way. Track-ways proved that sauropods moved easily on dry land and did not drag their tails, but held their tails straight out behind them. These tracks also suggested that sauropods had social structure and herds, with the juveniles moving in the middle of a convoy of adults for protection. Most of their fossils were known from floodplains and even drier habitats, showing that they didn't live in water most of the time and certainly didn't need it for holding up their huge bodies. Finally, dozens of genera of sauropods are now known, showing a much wider variety of body shapes and sizes than anyone could imagine a century ago.

If the giant body size and long necks are not the only criterion for defining a sauropod, what do paleontologists use to recognize the group? In addition to their small heads and long tails, their mouth had a very wide gape and no cheek region, with the upper teeth having tooth-to-tooth occlusion with the tips of the lower teeth. Most sauropods had simple peg-like or spoon-shaped teeth, relatively small for such huge animals. This all suggests that they opened their mouths very wide and stripped off and swallowed large amounts of vegetation without any chewing. Another characteristic is that they were completely quadrupedal (unlike their bipedal ancestors), and their massive limbs were completely upright and beneath their bodies. Like all dinosaurs, they were digitigrade, meaning that they walked on the tips of their fingers and toes, not on the palms of their hands or soles of their feet, as we do. But their compressed finger and toe bones are extremely short and thick and stubby to support their enormous weight, and their thumbs were relatively short, so their hands could not grasp anything (something primitive sauropod relatives could still do). Instead, their entire hand and foot was encased in a thick fleshy pad, giving them distinctive round footprints, much like those of an elephant.

THE ORIGIN OF SAUROPODS

Where did these incredible huge and odd-shaped animals come from? Before the twentieth century, only highly specialized giants from the Jurassic were known (plus *Plateosaurus*, which was not yet well studied), so it was hard to imagine deriving a sauropod from a more primitive form, such as the tiny bipedal dinosaurs known from the Late Triassic. But now there are a number of fossils which show just how this transition occurred. The earliest fossil relatives of the sauropods are placed in the larger group, the Sauropodomorpha, which encompasses the Sauropoda and all their more primitive kin. (The old wastebasket term "Prosauropoda" was often used for these primitive sauropod relatives that were not advanced sauropods.) They include fossils from the Late Triassic of South America, such as *Panphagia*, *Saturnalia*, *Bagualosaurus*, and *Guaibasaurus* (**Figure 17.1**), which are only slightly more advanced than *Herrerasaurus* found in slightly older beds in Argentina (see Chapter 14). All of these creatures were small (total length about a meter long, up to 1.5 meters in *Guibasaurus*) lightly built fast-running bipeds with tiny front limbs. Still, they have some sauropod-like features in the skeleton, and their teeth are not the strictly carnivorous pointed conical teeth of other primitive dinosaurs, but more robust and peg-like for an omnivorous diet. Slightly more advanced are fossils like *Thecodontosaurus* and *Pantydraco*, found in Upper Triassic deposits in England and Wales (**Figure 17.1**). These creatures were about 2.5 to 3 meters long, but otherwise they were very similar, with a short neck but long tail, and short forelimbs but long hind legs for a fully bipedal lifestyle. Their teeth are often leaf-shaped, indicating even further commitment to

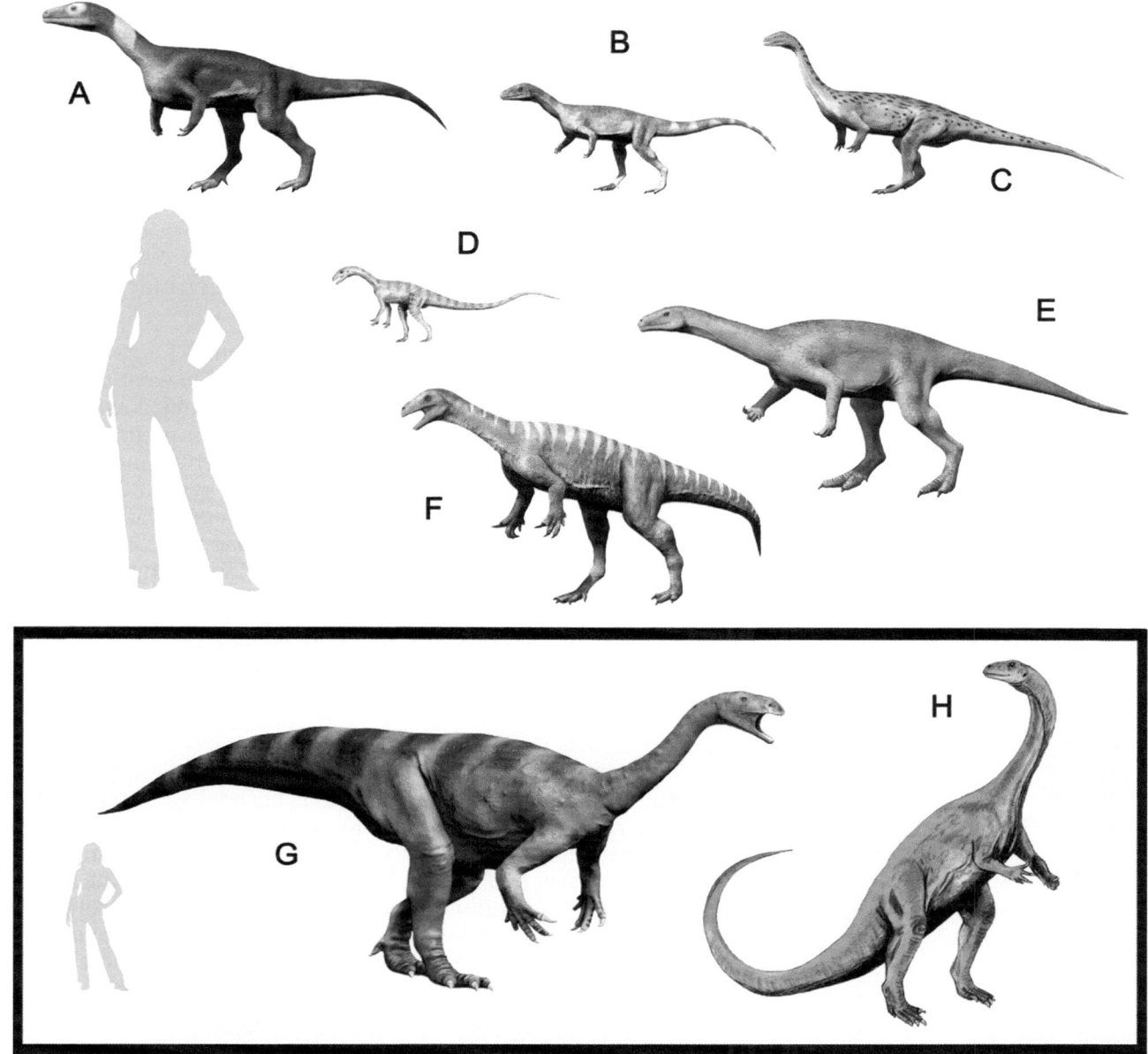

Figure 17.1 Reconstructions of some sauropodomorphs, including: (A) *Guaibasaurus*, (B) *Panphagia*, (C) *Saturnalia*, (D) *Pantydraco*, (E) *Bagualosaurus*, (F) *Thecodontosaurus*, (G) *Plateosaurus*, (H) *Massospondylus*.

a herbivorous diet, rather than the carnivorous or omnivorous diet of their ancestors. Additional taxa, such as *Nambalia* from India, *Efraasia* and *Ruehlia* from Germany, and *Plateosauravus* and *Arcusaurus* from South Africa, show that these very primitive sauropodomorphs were widespread across Pangea in the Late Triassic.

The next stage in sauropodomorph evolution is demonstrated by the well-known Late Triassic dinosaur *Plateosaurus* (**Figures 17.1[G]** and **17.2**). Although it was first named and described in 1837, it was not even recognized as a dinosaur at first. Eventually, it came to be known from dozens of complete skeletons in about 50 localities in the Triassic of Germany, Switzerland, and France, and showed how the transition for tiny slender bipeds to the giant sauropods occurred. *Plateosaurus* was a large-bodied herbivorous dinosaur—in fact, it was the largest dinosaur of its time (the Late Triassic), reaching up to 10 meters (33 feet) long, weighing up to 4000 kg (8800 lb), although *Plateosaurus* had a wide range in adult body sizes. It had a very long neck and long legs but a small head, all sauropod features. It had a deep, narrow box-like skull with eyes facing sideways, good for spotting predators around it. *Plateosaurus* had conical fang-like teeth in the front, and thick, bluntly serrated leaf-shaped teeth in the back of the jaw. These were suitable for shredding vegetation (mainly ferns, cycads, and conifers) that grew in the Late Triassic forests. The sharp teeth in front have led some paleontologists to suggest that *Plateosaurus* was not a strict herbivore, but may have been omnivorous. Its low jaw joint gave it the leverage for a powerful bite, so it could crush even the toughest vegetation. With its large size and long neck, it could reach high vegetation that no other animal at the time could access.

Plateosaurus had rather small but robust front limbs for its size, relative to the long hind limbs, so it was clearly bipedal. The wrist and hand are configured so they cannot rotate and place the palms down or the tips of the fingers down. This completely rules out their putting their hands flat on the ground for a quadrupedal posture. In certain museums, some of the skeletal mounts incorrectly switched the two lower arm bones (radius and ulna) in order to bend the wrist and make their palms lie flat on the ground. Instead, *Plateosaurus* hands had large recurved claws on them, which could have been used for tearing down plant branches, digging up roots, or for defense against predators (such as the primitive theropod *Liliensternus* from the same beds). It also had a large thumb with a claw that was partially opposable, so it could grip branches. This flexible thumb was lost in later sauropods when their hands became specialized for bearing enormous weight.

Plateosaurus was like most bipedal dinosaurs that held their bodies parallel to the ground and their tails out straight behind it (**Figures 17.1[G]** and **17.2**). The large rib cage and the broad hip bones suggest that *Plateosaurus* was quite barrel-chested, consistent with having a large complex gut for fermenting and digesting tough fibrous leaves. But most importantly, *Plateosaurus* had many features that foreshadowed the huge sauropods that were later discovered in North America and elsewhere: the simple small narrow skull with peg-like teeth for stripping and crushing leaves, the relatively long neck, the long tail, and the large trunk for digesting plants. Yet these primitive sauropodomorphs still had some distinctive features as well, such as a "wrap-around overbite", where the upper teeth around the dental arcade overlapped and covered the lower teeth on the inside.

Figure 17.2 Mounted skeleton of the sauropodomorph *Plateosaurus*, known from the Late Triassic of Germany, France, and Switzerland. (Courtesy Wikimedia Commons.)

Plateosaurus was the first of the primitive sauropodomorphs to be discovered, and it is still the best known and best studied. However, sauropodomorphs were widespread in Upper Triassic and Lower Jurassic beds all over the supercontinent of Pangea, although most of the rest of the fossils are very incomplete. Other relatives of *Plateosaurus* include *Euskelosaurus* from Africa, *Jaklapallisaurus* from India, *Unaysaurus* from Brazil, and *Yimenosaurus* from China. The other branch of sauropodomorphs is related to *Massospondylus* (**Figure 17.1[H]**), originally described in 1854 from fossils found in Lower Jurassic beds of southern Africa. Close relatives of *Massospondylus* include *Prahdania* from India, *Coloradisaurus* from Argentina, *Lufengosaurus* from China, and *Glacialisaurus* from Antarctica. Another branch is the Riojasauridae, including *Riojasaurus* from Argentina and *Eucnemesaurus* from South Africa. The third group is related to *Anchisaurus*, known from the Lower Jurassic beds of Connecticut. Other primitive sauropods include the Chinese *Yunnanosaurus* and *Jingshanosaurus*, and *Melanorosaurus* from South Africa.

JURASSIC PARK OF THE SAUROPODS

During the Jurassic, these mostly smaller and bipedal sauropodomorphs were replaced by a radiation of more advanced members of the group Sauropoda, which were fully committed to quadrupedal locomotion, and got too heavy to ever rear back on their hind limbs. (Contrary to images of rearing sauropods in *Jurassic Park* and some museum exhibits, most paleontologists agree that their hind limbs and tail were not strong enough to support their entire mass without breaking.) With their larger size came all the other features that diagnose sauropods discussed earlier, including the massive pillar-like limbs with the stumpy flattened finger and toe bones, and the loss of the ability to grasp with their thumbs. By the end of the Jurassic, sauropods had reached their heyday, and were extremely diverse and numerous in many parts of the world.

The primitive members of the early radiation of true Sauropoda including *Shunosaurus* from the Middle Jurassic of China, which reached about 11 meters (36 feet) in length, and had a club on its tail (**Figure 17.3[A]**). *Cetiosaurus*, the first sauropod ever discovered, is another primitive sauropod from the Middle Jurassic of England, which reached over 16 meters (52 feet) in length (**Figure 17.3[C]**). *Jobaria* from the Middle Jurassic of Niger was 18.2 meters (60 feet) long, so the sauropods were getting bigger and bigger (**Figure 17.3[B]**). But the most unusual of these primitive Sauropoda was *Mamenchisaurus* from the Middle Jurassic of Sichuan, China (**Figures 17.3[D]** and **17.4[A]**). It is very unusual in that its neck was extremely long, over half its total body length. With 19 very long vertebrae in its neck, it has one of the longest necks of any sauropod, and reached up to 35 meters (115 feet) in length and weighed up to 80 tonnes (88 short tons).

From these Middle Jurassic primitive Sauropoda arose a more advanced group of Late Jurassic sauropods, called the Neosauropoda (**Figure 14.1**). Unlike the primitive sauropods, the neosauropods lost the back teeth in their jaws, and focused on the teeth near the front of their mouth for nipping and stripping and tearing off vegetation when feeding. Another feature is the shift of their nasal opening from the tip of the snout to the top of the skull. The Neosauropoda then split into the two main groups of advanced sauropods (**Figure 14.1**): the Diplodocoidea (**Figures 17.4[B–D]** and **17.5**), which tended to have long necks and even longer, whip-like tails, and distinctive pencil-shaped teeth with squared-off snouts, and the nostrils on top of their head merged into a single opening, and the Macronaria (**Figure 17.6**), which include many of the familiar groups, including the brachiosaurs, camarasaurs, and the titanosaurs.

Figure 17.3 Reconstructions of primitive Sauropoda including: (A) the club-tailed *Shunosaurus*, (B) the African sauropod *Jobaria*, (C) the first sauropod ever discovered, *Cetiosaurus*, (D) the long-necked Chinese sauropod *Mamenchisaurus*.

Figure 17.4 (A) Mounted skeleton of the extremely long-necked Chinese dinosaur *Mamenchisaurus*. As the long vertebrae show, it could not flex its neck much, and certainly not curl it in a tight arc. At best, it could only flex it like a fishing rod. (B) Mounted skeletons of the diplodocoids *Diplodocus* (left) and *Apatosaurus* (right) in the Carnegie Museum of Natural History in Pittsburgh. (C) The weird Argentinian Cretaceous diplodocoid *Amargasaurus*, with the spikes along its neck. (Courtesy Wikimedia Commons.)

Figure 17.5 Comparison of some of the diplodocoid sauropods, including: (A) *Diplodocus*, (B) *Apatosaurus*, (C) *Nigersaurus*, (D) *Amargasaurus*, (E) *Dicraeosaurus*.

DIPLODOCOIDS

Diplodocus (**Figures 17.4[B]** and **17.5[A]**) and its relatives are among the more familiar of the dinosaurs, since it has been known since the 1890s, and numerous skeletons have been found in the Upper Jurassic Morrison Formation, especially in Wyoming, Colorado, and at Dinosaur National Monument in Utah. It also became even more famous when the first complete skeleton was named *Diplodocus carnegii* in honor of the benefactor, millionaire Andrew Carnegie. His wealth paid for the founding of the

Figure 17.6 Reconstructions of some of the most famous macronarian sauropods, including: (A) *Camarasaurus*, (B) *Giraffatitan*, (C) *Brachiosaurus*.

Carnegie Museum of Natural History in Pittsburgh and funded the expeditions that found all its dinosaurs. Carnegie was so fond of the dinosaur named after him that he hired a team of Italian plasterers to make dozens of copies of the original and donated them to the natural history museums around the world. In this way, museums in England, France, Germany, Austria, Italy, Russia, Spain, Argentina, and Mexico all had copies of Carnegie's dinosaur on display. Until recently, for example, the central hall in the Natural History Museum in London was decorated, not by any British dinosaur (which were too incomplete to feature in the main halls), but by Carnegie's gift. One of its collectors, Arthur Coggeshall, wrote that "to *Diplodocus carnegii* goes the credit of making 'dinosaur' a household world . . . presidents, kings, emperors, and czars besieged Andrew Carnegie for replicas to be installed in their national museums".

The dinosaur that has been called *Brontosaurus* (**Figures 17.4[B]** and **17.5[B]**) for many years is also a diplodocoid—but its story is complicated. First, the name "*Brontosaurus*" was coined by O.C. Marsh in 1879 based on a nearly complete skeleton which was the pride of the Yale Peabody Museum. A similar mount at the American Museum of Natural History in New York was also called "*Brontosaurus*". But it turns out that a less complete juvenile specimen of this dinosaur from the same locality (Como Bluff, Wyoming) had already been named *Apatosaurus* by Marsh in 1877. By the rules of priority in zoology, if the two animals are the same (as Elmer Riggs showed in 1903), then *Apatosaurus* is the senior and only valid name and *Brontosaurus* must be abandoned. Even though most paleontologists accepted that *Apatosaurus* was the correct name since 1903, the skeletons on display were still labeled *Brontosaurus*, and that name became much more familiar to the public. But by the 1970s, even the books for kids and the toy dinosaur kits had caught up with the science and begun calling it *Apatosaurus*. Then in 2015, Emmanuel Tschopp and colleagues did a study where they claimed that *Brontosaurus* and *Apatosaurus* were different dinosaurs, allowing the name *Brontosaurus* to be resurrected. Many paleontologists do not accept this idea, because it's unlikely that so many distinct genera of dinosaurs with the same size and shape competed for the same resources and all lived in the same time and place. This goes against the principle of competitive exclusion in ecology, where no two species can share the same ecological niche, so there is still debate about whether *Brontosaurus* is valid and should be brought back into use.

The other problem with the popular conceptions of *Brontosaurus* is that the original specimen was headless (for some reason, sauropod skeletons typically lose their skulls before they are fossilized). The early mounted skeletons on display had a short-snouted, tall skull resembling *Camarasaurus* on the neck, and that version is what people imagine as its normal head. But in 1977, Dave Berman and Jack McIntosh argued that a skull found near the neck of a specimen at Carnegie Quarry was the head of *Apatosaurus*, and it looked very much like *Diplodocus*. This made sense, since the rest of the skeleton is very diplodocine. Since then, most of the major museums have replaced the wrong head with the correct skull, often with great fanfare and publicity.

In addition to familiar North American long-necked diplodocines with whip-like tails, there are some other interesting variations on that theme. A slightly shorter-necked form, *Dicraeosaurus* (**Figure 17.5[E]**) comes from the Upper Jurassic Tendaguru beds of Tanzania, the same ones that produced the famous brachiosaur *Giraffatitan*. *Amargasaurus* from Argentina had a row of spines on its neck and its back (**Figures 17.4[C]** and **17.5[D]**). Weirdest of all was *Nigersaurus* (**Figure 17.4[C]**) from the middle Cretaceous of Niger, which had a wide flattened snout that looks like a dinosaurian lawnmower; its head was set on its neck so it mostly faced downward.

There is now a vigorous debate in paleontology about how these diplodocoids fed. Traditionally, the long neck is supposed to have given them the ability to reach to high branches like giraffes. But studies of living giraffes, plus other considerations about how they managed the blood pressure of their heads being so high off the ground, leads to another possibility. As we now know from studying giraffes, a long neck allows them to have a larger "feeding envelope" to reach a lot of vegetation around them without expending the energy to move their body and their legs. Thus, despite their long necks, most diplodocoids are now thought to have fed mostly on low- to middle-level vegetation, not on the tops of trees.

Another interesting thing about their long necks is that most sauropods (especially sauropods like *Mamenchisaurus*—**Figure 17.4[A]**) had very long neck vertebrae. This means that their necks had limited flexibility, and were not able to curl into a tight coil, like the neck of a snake or long-necked birds. Instead, the neck was more like a flexible fishing rod, able to make gentle curves, but not capable of curling back on itself. For generations, sauropods have been drawn with necks as flexible as those of snakes or birds, but this all comes from not knowing the anatomy of their neck vertebrae. This is consistent with the idea that the neck was not built for reaching up very high, but more for swinging from side to side and up and down as it exploits its feeding envelope without moving its legs and body.

MACRONARIANS

The other main branch of Neosauropoda (**Figure 14.1**) is the Macronaria ("big nostrils" in Greek), so named because they have a bony arch on top of their head which surrounds their nasal opening, and these openings are larger than their eye sockets. Instead of the pencil-shaped teeth of diplodocoids, macronarians mostly had short stout spoon-shaped teeth. They also tended to have the wrist and finger bones in their hands that were more elongate than in any other group of sauropods.

Macronarians include the small short-faced sauropod *Camarasaurus* (**Figure 17.6[A]**), but the most spectacular Jurassic group was the brachiosaurs, which had much longer front legs than hind legs, so their shoulders are high and their back slopes down to their hips. Their long necks also made them among the tallest known dinosaurs. The genus *Brachiosaurus* (**Figure 17.6[C]**) was first named and described by Elmer Riggs in 1903 based on just a few huge bones from the Morrison Formation, near Grand Junction, Colorado. Then a larger, more complete brachiosaur was found in the Tendaguru beds of Tanzania by German expeditions of 1910–1913. Originally considered to be *Brachiosaurus*, in 1988 it was renamed *Giraffatitan* (**Figures 17.6[B]** and **17.7**). When the Tendaguru fossils were finally mounted and displayed in Berlin, *Giraffatitan* became the largest dinosaur known from a nearly complete skeletons on display, and remained so until fairly recently when the Argentinian titanosaurs surpassed it as the largest land animal that ever lived.

Some of the early ideas about brachiosaurs were remarkably naïve and even ridiculous. In the early twentieth century, most dinosaur paleontologists viewed sauropods and many other dinosaurs as slow sluggish lizards living in swamps, needing the buoyancy of water to support their enormous bulk. The original mount of *Giraffatitan* actually had the legs mounted in a sprawling lizard-like posture with its limbs flexed and bowed out sideways, rather than with its limbs in the upright vertical posture that we now know they had. The current skeletal mount in Berlin has fixed this mistake (**Figure 17.7**).

Other paleontologists imagined that the long neck and head of sauropods was used like a snorkel, allowing them to submerge their bodies and only have their heads above water. Some even thought that the nostrils on top of their heads allowed them to be almost completely immersed, although we now know that the nostrils faced forward, not upward. A famous old painting by the Czech paleoartist Zdenek Burian shows a brachiosaur walking in a deep fjord completely under water except for its head. This notion is absurd. Any animal this deep below the surface would be subjected to so much water pressure on its body that it could not expand its lungs. Brachiosaurs had no special mechanism for pulling air from the surface down their windpipes against so much hydraulic pressure on the lungs and body. Creatures such as whales can live deep underwater only because they have unique anatomical specializations that allow them to control the air in their lungs when under huge pressures.

Figure 17.7 Mounted skeleton of _Giraffatitan_ from the Tendaguru beds in Tanzania, now in the Museum für Naturkunde in Berlin. (Courtesy Wikimedia Commons.)

This is not the only absurdity of snorkeling brachiosaurs. We now know that nearly all dinosaurs (including birds) have numerous air sacs in their bodies (especially around the backbone), which lighten their weight considerably. This helps decrease the amount of weight they must carry on their limbs (and eliminates the requirement for them to float in water to support their weight), but it would also prohibit them from diving into deep water, since they would immediately float to the surface with all that air trapped inside them. Most of the outdated notions of sauropods were ruled out by the 1980s, and fortunately were incorporated into the first _Jurassic Park_ novel and movie, so modern audiences are familiar with the current version of brachiosaurs.

Paleontologists have done numerous studies to determine the feeding habits of brachiosaurs. They appear to have been specialists on the highest limbs of trees, which were nearly all conifers back in the Late Jurassic. (Flowering plants had not yet evolved.) Their simple small spoon-shaped teeth were sufficient only to strip branches and pine needles off intact, and they had no ability to chew their food. Instead, their fodder was gulped down whole, and must have been ground up in a large gizzard in their chest, then slowly fermented and digested in a huge intestinal tract that took up most of their body cavity. Paleontologists estimate that they needed about 180–240 kg (400–550 pounds) of fodder a day to feed their enormous appetite. That kind of intense browsing would have consumed a lot of conifer trees and ferns in relatively short order, so there could not have been large populations of sauropods living for very long in any one area, nor was it likely that they coexisted with many other species of sauropods in the same place. Instead, they must have constantly migrated long distances to get enough food to survive.

Instead of the idea that brachiosaur nostrils served as a snorkel, we now think that the high arches may have supported a resonating chamber for making sounds. The brain had a volume of only about 300 cc, large enough to control their bodies, but very small relative to their enormous body mass, so these dinosaurs were not smart. Brachiosaurs probably didn't need to be intelligent, since once partially grown they were too large to fear any predators. They must have had enormous hearts to create the blood pressure to pump all that blood uphill to their heads, and probably had a series of valves in their veins of the neck (as giraffes do) to keep their blood from rushing to their head when they bent down to drink, or flowing away from their brain and making them dizzy and pass out (as can happen to humans) when they raised their heads suddenly.

TITANOSAURS

The largest subgroup of the Macronaria are the titanosaurs (**Figures 17.8** and **17.9**). Unlike the diplodocoids, the camarasaurs, and the brachiosaurs that flourished in the Jurassic and then vanished, the titanosaurs were far more diverse and largely diversified through the Cretaceous all over Gondwana and Laurasia. Brachiosaurs are distinctive because of their long forelimbs and high shoulders and long necks adapted for feeding higher than any other animal. Diplodocoids tended to have slender bodies with long necks and extremely long tails. By contrast, some titanosaurs had relatively shorter necks and tails, and relatively small heads with large nostrils and crests formed by their nasal bones. Most of them had small teeth shaped like little spatulas (broad at the tip, narrow at the root), although a number had teeth shaped like pencils (similar to the teeth of diplodocoids). Titanosaurs also had very broad shoulders and hips with a wider stance than other sauropods, and this can be recognized from their trackways. They tended to have stocky forelimbs, sometimes longer than their hind limbs (although not as disproportionately long as in brachiosaurs). Most sauropods have a few stumpy remnants of finger and toe bones, and maybe a claw on their thumb, but many titanosaurs lost their bones in their fingers and toes completely, replacing it with cartilage. Apparently, they walked on the blunt "stumps" of their hand and foot bones, covered by pads of cartilage and keratin, as cover the bones of fingers and toes in most animals. Some of the titanosaurs (*Saltasaurus*) had bony plates on its back shaped like large dishes (**Figure 17.8[A]**), while many of the titanosaurs that are well enough preserved show small dish-like pieces of armor (osteoderms) in their skins that made it harder for a predator to bite into them. The most diagnostic feature of many titanosaurs, however, was how the centra of the tail vertebrae are convex on the rear surfaces. More advanced titanosaurs have a distinctive peg-and-socket joint in their vertebrae. Thus, even a single vertebra can be identified as titanosaur.

Titanosaurs were by far the most diverse and widespread group of sauropods, spreading to all the continents when they began their evolutionary

Figure 17.8 Reconstructions of some of the best-known titanosaurs, including: (A) *Saltasaurus*, (B) *Patagotitan*, (C) *Argentinosaurus*.

Figure 17.9 Skeletal mount of *Patagotitan* in the American Museum of Natural History in New York. (Courtesy Wikimedia Commons.)

radiation in the Early Cretaceous. They seem to have replaced diplodo-cines and brachiosaurs and may have competed with and displaced those groups ecologically as well. Once they began evolving, titanosaurs reached a diversity of over 100 genera known from the Cretaceous (although many of these names are dubious because they are based on very fragmen-tary non-diagnostic fossils). Titanosaurs were found on every continent in the Cretaceous, although they are particularly well known from South America. But they also occur in Africa, as well as India, China and other parts of Asia, Australia, New Zealand, Europe, and even Antarctica. After the Early Cretaceous appearance of *Sauroposeidon*, sauropods of any kind were absent in the Cretaceous of North America until the very latest Cre-taceous, when the titanosaur *Alamosaurus* is found in the southern Rocky Mountains (but not the famous Upper Cretaceous dinosaur beds of the northern Rockies in Wyoming, Montana, or Alberta).

SIZE MATTERS

How big were these giants? Estimates vary tremendously because so many of the specimens are incomplete. The nearly complete skeleton of *Giraffatitan* (**Figure 17.7**) in Berlin was about 22.5 meters (75 feet) long and 12 meters (39 feet) tall. Its weight has been estimated as low as 15 tonnes to as high as 78 tonnes, but most estimates now place the likely mass in the 20–40 tonne range, because brachiosaurs were made much lighter by numerous air sacs throughout their bodies. This is not the limit for *Giraffatitan*, since the mounted skeleton is not fully grown. There is another limb bone of an adult that is 13% larger, suggesting their maxi-mum adult dimensions were considerably bigger.

So how big was *Patagotitan*, supposedly the largest of all dinosaurs (**Fig-ures 17.8[B]** and **17.9**)? The original press releases gave a length esti-mate of 40 meters (131 feet) and a weight estimate 77 tonnes (85 tons), but by the time the specimen was finally published those dimensions had shrunk to 37 meters (122 feet) in length and weight 69 tonnes (76 tons). Other authors estimated that it measured 33.5 meters (110 feet) and weighed about 45.4 tonnes (50 tons).

Meanwhile, another huge titanosaur named *Argentinosaurus* (**Figure 17.8[C]**) was long considered the largest known land animal, yet it is known from much less complete material: a few enormous vertebrae of the back and hip, and some of the hind limb bones. The reconstructed skeleton is 40 meters (130 feet) long. However, it is so incomplete that estimates of its size range from 26 meters (85 feet) to 30 meters (98 feet) to 30–35 meters (98–115 feet) in length, and weight estimates from different scientists range from 60–88 tonnes (66–97 tons), 80–100 tonnes (88–110 tons), and 83.2 tonnes (91.7 tons). These size estimates might put *Argentinosaurus* back at Number One if you use the smaller numbers for *Patagotitan*.

Why are these numbers so widely divergent? This is a problem dealing with partial skeletons, especially with material that consists only of a few vertebrae. Partial skeletons often have no bones in common with other skeletons and other species that would allow direct comparison. The length estimate is highly dependent on how large certain key bones were, and how many vertebrae were in the neck and tail, and none of the spec-imens have a complete neck or tail yet (most don't have a skull, either).

There is an even bigger problem with estimating weight. All we have is skeletons, so we really don't know how fat or skinny the soft fleshy parts of the living animal were. Even with a complete skeleton, we can only get a rough mass estimate, usually by taking a key limb bone dimension and comparing it to the same bones of animals of known weights to get a mass.

The second technique involves constructing a three-dimensional model of the living animal and calculating its volume and thus its mass. In the old days, this meant sculpting an actual physical model and plunging into a graduated cylinder of water to estimate its volume, but more recently several people have developed software for virtual models in the computer than can be quickly estimated in terms of volume and mass. In addition to these problems with weight estimate, there are also the problems with density. Like birds and apparently many dinosaurs, sauropods had many weight-reducing air sacs in the body (especially along their spine, as the bones suggest). In this case they might not have been as heavy as suggested by taking a simple model of uniform density and calculating its weight.

SAUROPOD PHYSIOLOGY

Since the 1970s, there has a been a major controversy about the physiology of dinosaurs, and whether they were "warm blooded" or "cold blooded". We are used to talking about dinosaur metabolism in this way, but vertebrate physiology is more complex than just oversimplifications like "cold-blooded animals" and "warm-blooded animals". There are actually two main components to thermal physiology: the *source* of the heat, and whether it is *regulated* or not. As far as source goes, animals that get their body head from the environment are called *ectotherms*, while those that burn food to create body heat through metabolism are *endotherms*. In the case of regulation, animals that let their body temperatures change with the surrounding temperature are called *poilkilotherms*, while animals that try to keep their body temperature constant are called *homeotherms*.

In the modern world, the boundaries seem pretty clear: all living birds and mammals are homeotherms and endotherms, while the rest of animals are all poikilotherms and ectotherms. Homeothermic endotherms can live in almost any environment, no matter how hot or cold, but pay a heavy price in that they burn most of the food they consume for metabolism. Poikilothermic ectotherms regulate their body temperature by moving in and out of hot and cold areas, but if it gets too cold or too hot, they die. For example, a desert lizard typically has a higher body temperature than a "warm-blooded" mammal like you when they are running or active—but it regulates its temperature by shuttling between sun and shade, or burrowing down into the cool sand, not by burning food.

But even with these broad generalizations, the exceptions are informative. For example, ectotherms like pythons can generate body heat by shivering when they are incubating their eggs, and sea turtles, tuna, sharks, and even some insects are capable of some endothermy. Many homeotherms (such as the platypus, sloths, and certain rodents, shrews, and small birds) let their body temperature fluctuate tremendously, as do animals that go into torpor when they hibernate. These animals allow their body temperature to drop as they go into their suspended animation state. At the other extreme of body size, camels are famous for letting their body temperature cool down during the cold desert night, then slowly heat up as the desert sun warms them. It's all a matter of the ratio of surface area of skin (to gain and lose heat) versus the body mass. Small animals have a relatively large surface area compared to their tiny volume, so they gain and lose heat quickly. But as animals get larger, their surface only increases as a square, but volume increases as a cube. Animals with a large body mass relative to their small surface area take a long time for the heat of the desert to warm them up. Camels can even let their bodies reach unusually high temperatures at the end of the day, because they are about to cool down in the cold desert night.

During the peak of the controversy, there were lots of arguments back and forth about the possible evidence for dinosaur endothermy. When dinosaurs were found in polar regions like Alaska and eventually Antarctica, it was thought to be proof of endothermy—until we came to realize that polar regions in the greenhouse world of the Jurassic and Cretaceous were temperate in climate. Since the 1950s, anatomists pointed out that dinosaur bones have large canals for blood vessels inside them called Haversian canals, a feature found in mammals but not in reptiles. But it turns out that the presence of these canals is also affected by body size and rate of growth, because some large ectotherms (like sea turtles, tortoises, and some crocodilians) also have them while some small birds, bats, shrews, and rodents don't always have them. The presence of Haversian canals seems to be more an indicator of rapid growth to large body size, not of endothermy. Dinosaurs are now known to have extremely rapid growth after they hatched, which better explains the Haversian canals.

Another line of evidence was the ratios of biomass of predators in a food pyramid to the biomass of prey. For an endothermic predator like a lion, there needs about ten times as much biomass of prey species as total biomass of lions, because most of its food is burned to produce body heat. In other words, the predator/prey biomass ratio is about 1:10. An ectothermic predator, like a crocodile, eats very rarely and doesn't use its food for body heat, but for activity and growth, so there can be an almost equal biomass of predatory crocodiles to the biomass of prey species (predator/prey biomass ratio is 1:1). It was claimed that when you look at the predator/prey ratios in Early Permian fossil assemblages from northern Texas (preyed upon by the fin-backed protomammal *Dimetrodon*), the ratio is about 1:1, but for most dinosaur faunas, the prey biomass is about ten times that of predators.

This sounds convincing at first, but on closer examination, it breaks down. There are too many factors which bias which animals get fossilized and which do not, so you cannot interpret fossil collections as perfect reflections of what was originally living. Museum collections tend to be highly biased because the fossil hunters are after only the spectacular skulls and other diagnostic parts, and don't take an unbiased sample of what was actually present in the field area. Lots of things just don't fossilize well, or are overabundant or rare for factors having nothing to do with biology. For example, there are numerous examples of dinosaur quarries that are almost nothing but predators (such as the *Falcarius* quarry in Utah, or the Cleveland-Lloyd Quarry in Utah which is full of allosaurs and other theropods). The famous La Brea tar pits in Los Angeles have far more predators and scavengers (primarily dire wolves and saber-toothed cats, plus vultures and predatory birds) than they do prey species. If you took this overabundance of predators at face value, then their world was entirely carnivores with almost no prey species to eat, and they were mostly cannibals.

Many other arguments were debated back and forth, and the "hot-blooded" dinosaur controversy raged for several decades, but now seems to be resolved. So what's the answer? Were dinosaurs endotherms or ectotherms? The answer is: "It's complicated". Certainly, the smaller predatory dinosaurs (like the "raptors" of *Jurassic Park* fame) were endotherms, because at their small body size and high levels of activity, they would need a high metabolism to be successful. Indeed, there is good evidence that "raptors" and most predatory dinosaurs (including even *T. rex*) were covered by a downy coat of feathers for insulation, so these animals were not slow and stupid, but active and smart and warm-blooded.

But for huge dinosaurs like the sauropods and the larger ornithischians, size presents a different problem. Sauropods had no obvious ways of gaining or losing heat from their bodies rapidly. The largest living land endotherms, the elephants, are a good example of this physiological dilemma. At its huge size, an elephant must spend much of its daytime in mud or water or resting in the shade to dump excess body heat. Its huge ears are primarily used as radiators to shed heat from its body. Most sauropods would have had even greater difficulties if they were endotherms and generating body heat from metabolism of food. Instead, such large beasts didn't need to use metabolic body heat at all, but could keep warm thanks to the stable warm climates around them during the Jurassic and Cretaceous. It was a greenhouse world, with no polar ice caps, and indeed there was almost no ice anywhere on the planet, since even the polar regions were lush and temperate. With their large size, sauropods would have gained or lost body heat only very slowly, so they could obtain a stable warm body temperature by sheer size alone. This strategy is known as "inertial homeothermy" or "gigantothermy". Many scientists think it probably characterized all of the larger non-predatory dinosaurs, including sauropods, and possibly stegosaurs, horned dinosaurs, duckbills, and many others.

The sauropods were truly amazing as the largest land animals the world had ever known. But they too vanished at the end of the Cretaceous like all the other non-bird dinosaurs. The world is left only with their enormous bones, and a sense of awe and wonder about these incredible animals.

FURTHER READING

Alexander, R.M. 1989. *Dynamics of Dinosaurs and Other Extinct Giants*. Columbia University Press, New York.

Buffetaut, E.; Suteethorn, V.; Cuny, G.; Tong, H.; Le Loeuff, J.; Khansubha, S.; Jongautchariyakul, S. 2000. The earliest known sauropod dinosaur. *Nature*. 407 (6800): 72–74.

Cobley, M.J.; Rayfield, E.J.; Barrett, P.M. 2013. Inter-vertebral flexibility of the ostrich neck: Implications for estimating sauropod neck flexibility. *PLoS ONE*. 8 (8): e72187.

Fastovsky, D.E.; Weishampel, D.B. 2021. *Dinosaurs: A Concise Natural History* (4th ed.). Cambridge University Press, Cambridge.

Holtz, T.R., Jr.; Rey, L.V. 2007. *Dinosaurs: The Most Complete, Up-to-Date Encyclopedia for Dinosaur Lovers of All Ages*. Random House, New York.

Klein, N. 2011. *Biology of the Sauropod Dinosaurs: Understanding the Life of Giants*. Indiana University Press, Bloomington, IN.

Lucas, S.G. 2005. *Dinosaurs, the Textbook* (5th ed.). W. C. Brown, Dubuque, Iowa.

McGowan, C. 1983. *The Successful Dragons: A Natural History of Extinct Reptiles*. Samuel Stevens, Toronto.

McGowan, C. 1991. *Dinosaurs, Spitfires, and Sea Dragons*. Harvard University Press, Cambridge.

Norman, D.B. 1985. *The Illustrated Encyclopedia of the Dinosaurs*. Crescent Books, New York.

Rogers, K.C.; Wilson, J.A. 2005. *The Sauropods: Evolution and Paleobiology*. University of California Press, Berkeley.

Sander, P.M.; Christian, A.; Clauss, M.; Fechner, R.; Gee, C.T.; Griebeler, E.E.; Gunga, H.-C.; Hummel, J.; Mallison, H.; Perry, S.F.; et al. 2011. Biology of the sauropod dinosaurs: The evolution of gigantism. *Biological Reviews*. 86 (1): 117–155.

Sellers, W.I.; Margetts, L.; Coria, R.A.B.; Manning, P.L. 2013. March of the titans: The locomotor capabilities of sauropod dinosaurs. *PLoS ONE*. 8 (10): e78733.

Sereno, P.C. 1998. A rationale for phylogenetic definitions, with application to the higher-level taxonomy of Dinosauria. *Neues Jahrbuch für Geologie und Paläontologie, Abhandlungen*. 210: 41–83.

Sereno, P.C. 1999. The evolution of dinosaurs. *Science*. 284 (5423): 2137–2147.

Seymour, R.S. 2009. Raising the sauropod neck: It costs more to get less. *Biological Letters*. 5 (3): 317–319.

Seymour, R.S.; Lillywhite, H.B. 2000. Hearts, neck posture and metabolic intensity of sauropod dinosaurs. *Proceedings of the Biological Society of Washington*. 267 (1455): 1883–1887.

Stevens, K.A.; Parrish, J.M. 1999. Neck posture and feeding habits of two Jurassic sauropod dinosaurs. *Science*. 284 (5415): 798–800.

Taylor, M.P.; Wedel, M.J. 2013. Why sauropods had long necks: And why giraffes have short necks. *Peer Journal*. 1: e36.

Tschopp, E.; Mateus, O.; Benson, R.B.J. 2015. A specimen-level phylogenetic analysis and taxonomic revision of Diplodocidae (Dinosauria, Sauropoda). *Peer Journal*. 3: e857.

Upchurch, P.; Barrett, P.M.; Dodson, P. 2004. Sauropoda, pp. 259–322. In Weishampel, D.; Dodson, P.; Osmólska, H., eds. *The Dinosauria* (2nd ed.). University of California Press, Berkeley.

Wedel, M.J. 2009. Evidence for bird-like air sacs in Saurischian dinosaurs. *Journal of Experimental Zoology*. 311A: 18.

Yates, A.M.; Kitching, J.W. 2003. The earliest known sauropod dinosaur and the first steps towards sauropod locomotion. *Proceedings of the Royal Society B: Biological Sciences*. 270 (1525): 1753–1758.

THEROPODS
CARNIVOROUS DINOSAURS

18

I propose to make this animal the type of the new genus, Tyrannosaurus, *in reference to its size, which far exceeds that of any carnivorous land animal hitherto described . . . This animal is in fact the* ne plus ultra *of the evolution of the large carnivorous dinosaurs: in brief it is entitled to the royal and high sounding group name which I have applied to it.*
—Henry Fairfield Osborn, 1905

THEROPODA

Tyrannosaurus rex. Velociraptor. Spinosaurus. Dilophosaurus. Carnotaurus, the "flesh bull". The media, movies like Disney's *Dinosaur* and the *Jurassic Park/Jurassic World* series, and the dinosaur merchandising industry have made these predators famous. They are portrayed as terrifying killing machines, the stuff of nightmares. They are members of a major branch of dinosaurs known as Theropoda, Greek for the "beast footed" dinosaurs. Once they arose in the Late Triassic, nearly all the carnivorous and fish-eating dinosaurs of the Mesozoic were members of the Theropoda (**Figure 18.1**). Until recently, it was also assumed that not only were theropods the only carnivorous group of dinosaurs, but also that theropods were all carnivorous. But in the 1990s and 2000s, a number of startling finds revealed groups of oddly shaped giant theropods which were omnivorous or even herbivorous.

But being carnivorous is not a good definition of Theropoda, because nearly all the primitive dinosaur relatives were also carnivorous, or at least fed on insects or fish (Chapter 14). So what anatomical features tell us that Theropoda is a natural group? Certainly, the blade-like teeth for cutting flesh are important in their definition, but in addition Theropods usually have fine serrations on the cutting edges of their teeth, like the edge of the blade of a steak knife. But there are other features as well. All theropods were bipedal and never became fully quadrupedal. Originally, the most primitive theropods had well-developed hands with some grasping ability, although later in their evolution, some groups (like *Tyrannosaurus* and *Carnotaurus*) greatly reduced their hands and arms until they were virtually useless. Theropod hands bore at most only four fingers (the pinky is completely lost), and in some groups like tyrannosaurs they reduced it down to only two fingers (thumb and index finger). One group, the alvarezsaurids, are down to a single finger with a hooked claw on their hands. Most theropod jaws had a joint midway back at the rear end of the tooth-bearing dentary bones, that hinged the jaw and allowed it more flexibility in opening and biting. The collarbones, or clavicles, are seldom preserved in theropod skeletons, but where they do occur, they are fused into a boomerang-shaped "wishbone" or furcula, a feature that birds inherited from their theropod ancestors. Theropod feet bore their weight on the middle three toes, with the big toe and pinky

DOI: 10.1201/9781003128205-18

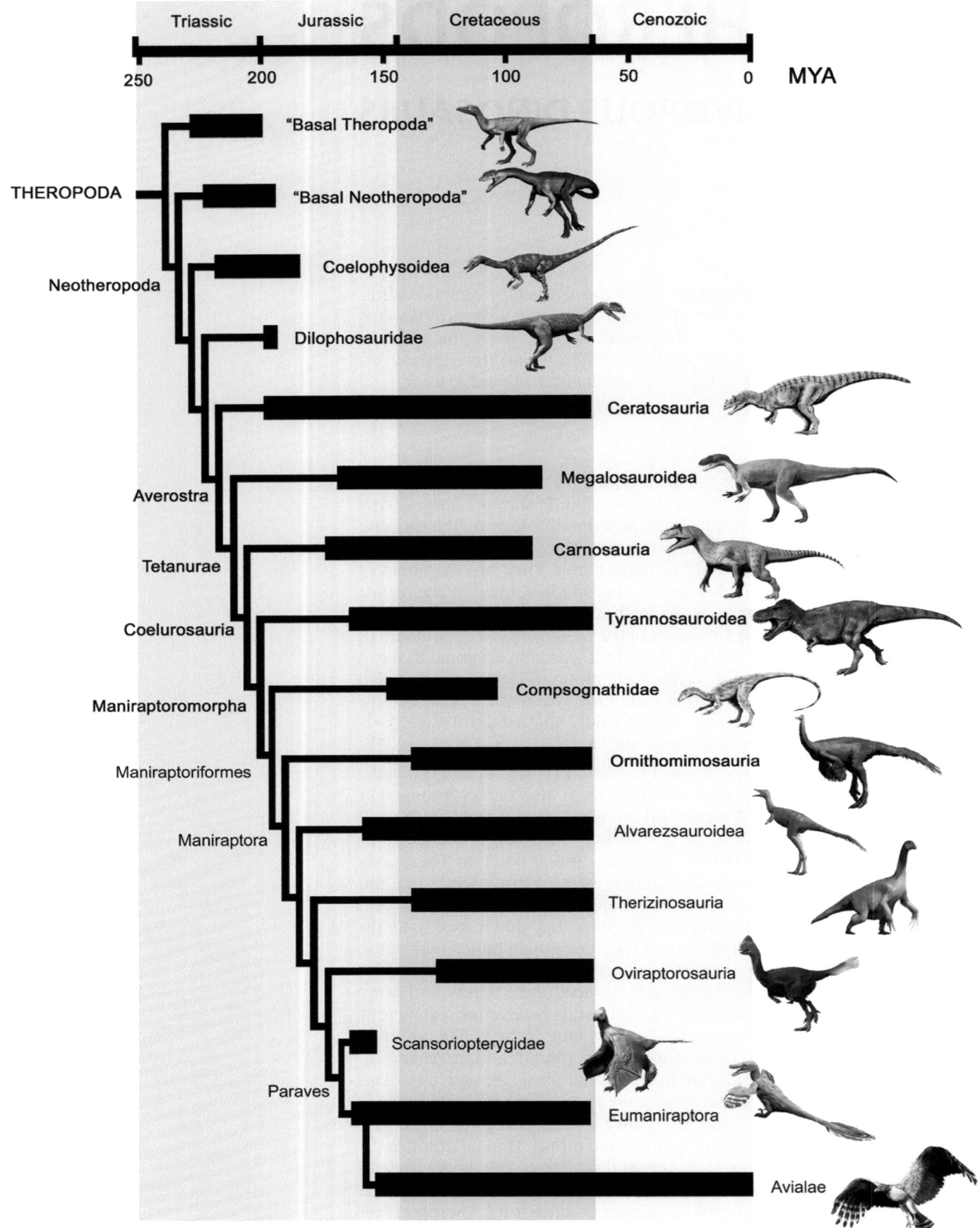

Figure 18.1 Family tree of theropod dinosaurs (including birds).

very reduced or lost in some theropods (although some retained the big toe and birds have developed a perching foot with it). Finally, the snout region of more advanced theropods (but not the coelophysids described next) bears an additional hole in the side of the skull known as the pro-maxillary fenestra, which lies just in front of the antorbital fenestra, a feature found in the skulls of all archosaurs.

Finally, by definition a natural group must include all its descendants. As we shall discuss in Chapter 19, birds are descended from the thero-pods, especially the group that includes *Velociraptor*. In a very real sense, theropods are still alive and among us—we just call them birds.

EARLY THEROPODS

As discussed in Chapter 14, the primitive dinosaurs of the Late Triassic include a number of small delicate long-necked long-tailed forms like *Herrerasaurus* and *Eoraptor* (**Figure 14.5**). According to some special-ists, these dinosaurs are actually within the Theropoda, although other paleontologists regard them as primitive relatives of the theropods, or slightly closer to sauropods.

The first group of dinosaurs that are universally regarded as theropods are the Late Triassic coelophysids (**Figures 18.1**, **18.2[A]**, and **18.3[A]**). Although several genera of coelophysids are known from South Africa, Germany, Argentina, and the southwestern United States, the best known of these is *Coelophysis* itself. It is represented by dozens of complete skel-etons recovered from the legendary Ghost Ranch bone beds north of Abiquiu, New Mexico. *Coelophysis* was about 3 meters (10 feet) in length, including its long tail, and weighed about 15–20 kg (33–44 pounds). It was lightly built with long running legs and a long tail (**Figures 18.2[A]** and **18.3[A]**). Even though it looks superficially similar to the small bipedal fossils like *Eoraptor* and *Herrerasaurus* from the Late Triassic of Argentina (Chapter 14), *Coelophysis* is a much more advanced dinosaur. The shoulder girdle is fully theropod in its anatomy, and it is the earliest dinosaur known to preserve its collarbones fused into a furcula, or boomerang-shaped "wishbone". Unlike more advanced theropods, *Coelophysis* still had four fingers on its hand (although it used only three, and the fourth was tiny and embedded in tissue). By contrast, most theropods have only three fingers, and some just two. The feet of *Coelophysis* have only three toes with a tiny vestigial fourth toe, and are configured like the classic theropod foot, although very slender. *Coelophysis* had a long narrow head that was lightly built with thin struts of bone. The eyes faced forward, so it clearly had good stereovision for running and catching prey. Combined with the bony ring in the eye to protect it (sclerotic ring), the eyes suggest that *Coe-lophysis* was mostly a daytime predator. Further research showed that its vision was much better than that of most lizards, but more like that of an eagle or hawk. *Coelophysis* had dozens of small sharp recurved teeth with serrated edges on the leading and trailing edge of the tooth, showing that it was a vicious predator that could rip open smaller prey. Together with its well-developed front limbs with a wide range of motion, it clearly could reach out and grab fast prey of many sizes.

Theropods continued to evolve in the Early Jurassic, becoming larger and more specialized. The best-known Early Jurassic theropods are the dilophosaurids. Those who have seen the first *Jurassic Park* movie might remember odd-looking dinosaur known as *Dilophosaurus* (**Fig-ures 18.2[B]** and **18.3[B]**). Unfortunately, the movies got it all wrong when it came to *Dilophosaurus*. For one thing, the actual fossil is about twice the size of the movie monster, yet they make it a cute creature smaller than a kangaroo. According to *Jurassic Park*, *Dilophosaurus* spat venom, but this is not supported by any evidence. The spitting cobras of

Africa are among the few animals that spit venom. They squirt the fluid from their mouths using a tiny hole in the front of their tubular fangs, so under pressure the venom is sprayed forward. But the fossils of actual *Dilophosaurus* have normal thick blade-shaped serrated theropod teeth, with no internal canals for injecting venom, let alone spraying it forward.

Figure 18.2 Reconstructions of examples from some of the primitive branches of the Theropoda: (A) *Coelophysis*, (B) *Dilophosaurus*, (C) *Ceratosaurus*, (D) *Elaphrosaurus*, (E) *Carnotaurus*.

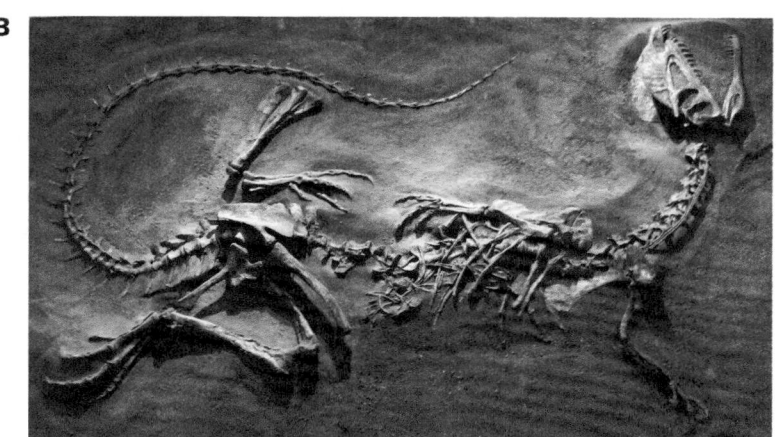

Figure 18.3 Fossils of some of the more primitive theropods: (A) *Coelophysis.* (B) *Dilophosaurus.* (C) *Cryolophosaurus.* (D) *Carnotaurus.* (Courtesy Wikimedia Commons.)

Finally, the moviemakers added the frill around its head on a whim, just to make it scarier, but there is no evidence for this feature either. It was inspired by *Chlamydosaurus*, the frilled dragon lizard of Australia, which has the flap of leathery skin around its neck that can create a frightening display when threatened. These frills are held out by giant modified hyoid bones of the throat region, so if *Dilophosaurus* had a frill, we would see those bones. But it doesn't have them, so the frill is fictional.

The actual specimens of *Dilophosaurus* not only have the paired bony crests on their heads, but also have a distinctive hinge in the upper jaw between the premaxilla and front teeth and the maxillary bone that bears the rest of the back teeth of the jaw. This hinge shows up in other dilophosaurs as well, including the peculiar dinosaur *Cryolophosaurus* (**Figure 18.3[C]**) from the Early Jurassic of Antarctica. *Cryolophosaurus* ("frozen crested lizard") refers to the icy conditions where it was found, and to the odd-shaped crest on its head. It resembled the Spanish comb that Spanish and Mexican señoritas used to wear on the top of their heads, or maybe a lock of hair sticking up like a cowlick. On the basis of the dimensions of the preserved parts, *Cryolophosaurus* was a relatively large theropod for the Early Jurassic, reaching 6.5 meters in length (21.3 feet) and weighing about 465 kg (1025 pounds). The function of the weird crests in dilophosaurs is not known for sure, but since they are thin and fragile, they probably served to advertise the age and maturity of the individual to its rivals and potential mates, as most horns and antlers do for many living mammals.

In the Middle and Late Jurassic, the next branch (**Figure 18.1**) of the Theropoda, known as the Ceratosauria (**Figure 18.2[C]**), took over the large predator niche from the dilophosaurids. This large group includes more than two dozen Middle Jurassic to Late Cretaceous predators, found on every part of Pangea during that time. It derives its name from *Ceratosaurus*, a primitive theropod from the Upper Jurassic Morrison Formation with distinctive horns over its snout (**Figure 18.2[C]**). Another group of ceratosaurs are the elaphrosaurs, which include *Elaphrosaurus* (**Figure 18.2[D]**) from the Upper Jurassic Tendaguru beds of Tanzania, and *Limusaurus* from Mongolia, as well as the Noasauridae, which include genera from the Jurassic and Cretaceous of India (*Laevisuchus*), North Africa (*Deltadromeus*), Argentina (*Noasaurus*, *Velocisaurus*), and Madagascar (*Masiakasaurus*). But the most distinctive of all the ceratosaurines were the abeliasaurs, a short-snouted group of theropods found mostly in South America (*Abeliasaurus*, *Ilokelesia*, *Pycnomemosaurus*, *Skorpiovenator*, *Aucasaurus*, *Quilmesaurus*, *Ekrixanatosaurus*) but also in Europe (*Arcovenator*), Africa (*Rugops*), Madagascar (*Dahalokely*, *Majungasaurus*), and India (*Indosaurus*, *Rahiolisaurus*, *Rajasaurus*).

The most remarkable of the abeliasaurs was *Carnotaurus*, whose name means "flesh bull" in Latin (**Figures 18.2[E]** and **18.3[D]**). Its name refers to the large robust horns over its eye sockets, which somewhat resemble the horns of a bull. But this creature is very distinct from other predatory dinosaurs in lots of ways besides the horns. It is the most specialized and distinctive member of a group of South American dinosaurs called abeliasaurs, which occupied one of the mid-sized predator roles during the latest Cretaceous in the Gondwana continents (especially South America, Madagascar, and India) that tyrannosaurs occupied in North America and Mongolia at the same time.

Like most abeliasaurs, *Carnotaurus* had a shorter snout than did tyrannosaurs, making it look very snub-nosed when compared to *T. rex*. Like tyrannosaurs, it had a powerfully built bulldog-like neck, so it could snap its head side-to-side to rip flesh from its prey, or wrestle with powerful prey items as they struggled. Its skull is lightly built and full of bony struts, so it was highly flexible. Its lower jaw was quite thin and shallow, not nearly as thick and robust as the jaw of tyrannosaurs. Its teeth were smaller and slenderer than those of tyrannosaurs, so it was not as likely to bite as hard, but instead it used them more for nipping and slashing. Although the bite force of *Carnotaurus* was stronger than any living alligator, it was not as strong as that of tyrannosaurs. Thus, many paleontologists think *Carnotaurus* may have been adapted to quick, rapid biting of smaller prey, or possibly making quick slashing wounds to larger prey, rather than bone-crushing power bites like tyrannosaurs used. On the other hand, *Carnotaurus* was one of the largest predators in the latest Cretaceous of Argentina, and lived alongside some remarkable large dinosaurs including some huge titanosaurs, so it seems likely that it preyed on dinosaurs close to its own size (or at least on their young).

Carnotaurus was smaller and more lightly built than *T. rex* or the dinosaurs we shall describe next. The complete skeleton of *Carnotaurus* is about 9 meters (30 feet) long, and so its body weight is estimated at about 1.35 metric tons (1.5 tons). Other abeliasaurs, such as *Ekrixinatosaurus* and *Abeliosaurus*, also were found in the Upper Cretaceous beds in Argentina. They were probably larger, but their fossils are too incomplete to be sure.

Another feature of both tyrannosaurs and *Carnotaurus* were the ridiculously small arms. Even though the arms of *T. rex* are tiny, those of *Carnotaurus* are even smaller proportionally. Most paleontologists think they were vestigial arms that were in the process of being lost, and were completely useless. The lower arm bone of *Carnotaurus* was much shorter than the upper arm bone, and its wrist bones never developed. Although it still had four fingers (*T. rex* had only two), *Carnotaurus* had only two functional digits (index and middle finger), which were short and stubby. Its pinky was gone entirely, and only the middle finger still had a claw. The tip of the ring finger was also missing, and in its place the hand bone was a long bony spur that stuck out of the hand. These tiny limbs clearly had no function whatsoever, and remind paleontologists of other dinosaurs that were the process of losing their forelimbs. Like tyrannosaurs, *Carnotaurus* relied on its mouth and feet to catch prey, but it outdid even *T. rex* in reducing its arms.

The long slender hind limbs of *Carnotaurus* suggest that it was among the fastest runners of all the large predatory dinosaurs. This is supported by the structure of the thigh bone which could withstand lots of bending and twisting, and the vertebrae of the hips and tail, which are the attachment points for its powerful leg muscles. Estimating its speed is difficult to obtain, but it could certainly run faster than a human (something *T. rex* could not do, despite what you have seen in the movies), but it was probably not as fast as an ostrich, which can top 43 mph.

Unlike most dinosaurs, *Carnotaurus* is known from a nice nearly complete articulated skeleton (**Figure 18.3[D]**). Most dinosaurs are known from a few broken bones, with most of the parts missing. Even fewer are known with their bones articulated together as they were in life. Amazingly, *Carnotaurus* was preserved with skin impressions on several parts of its body. The skin impressions show a mosaic of polygon-shaped scales that did not overlap, scored with pairs of parallel grooves in several places. The scales of the head are less regular than those of the body. Unlike tyrannosaurs, there is no evidence of feathers in *Carnotaurus*, so if it had any, they were in areas of the body for which no skin impressions are known.

TETANURAE: CARNOSAURIA

The next more advanced grouping of theropods was named the Tetanurae, or "stiff tails" in Greek (**Figure 18.1**). The Tetanurae got that name because some or all of the vertebrae in the tail are interlocking to form a partially or completely rigid tail that stuck out horizontally behind the dinosaur as it ran. Tetanurae have features of the ribcage that indicates they had the complicated air sac and lung ventilation system that is found in all living birds. They have also completely lost the fourth finger ("ring finger") of the hand, and have an additional hole in the snout region of the skull called the maxillary fenestra. Finally, their teeth are found only in the front part of their jaws, and there are no teeth in the back of the jaws.

Tetanurae is divided into two main branches (**Figure 18.1**): the Carnosauria and the Coelurosauria. The Carnosauria includes the Spinosauridae (*Spinosaurus*, *Irritator*, *Baryonyx*) (**Figures 18.4** and **18.5**); the Megalosauridae, including the very first dinosaur ever formally named, *Megalosaurus* (**Figure 18.4[A]**); the Allosauridae; the Metriacanthosauridae; and the Carcharodontosauridae.

Figure 18.4 Reconstructions of the some of the megalosaurs and spinosaurs: (A) *Megalosaurus*, (B) *Baryonyx*, (C) *Suchomimus*, (D) *Spinosaurus*.

A

B

C

D

Figure 18.5 Fossils of ceratosaurs, spinosaurs, and carcharodontosaurs. (A) Skeleton of *Spinosaurus*. (B) Skeleton of *Allosaurus*. (C) Skull of *Carcharodontosaurus*, compared to a human skull. (D) Skeleton of *Giganotosaurus*. [(A and D) Courtesy Wikimedia Commons. (C) Courtesy P. Sereno. (B) By the author.]

Carnosauria: Spinosauridae

Spinosaurus is now very familiar to the public, thanks to the highly outdated and inaccurate version of it that appeared in *Jurassic Park III*. Another spinosaurid, the British Cretaceous fossil *Baryonyx*, also appears in *Jurassic World: Fallen Kingdom*, chasing the characters out of the computer control room as lava pours around them. There are a number of additional spinosaurids known, including *Irritator* and *Oxalaia* from the Cretaceous of Brazil, *Suchomimus*, *Cristatusaurus*, and *Sigilmassasaurus* from the Cretaceous of northern Africa, *Ostafrikasaurus* from Tanzania, *Siamosaurus* from Thailand, and *Ichthyovenator* from Laos. Although these dinosaurs vary widely in size and overall shape, they all have elongate narrow snouts that resemble that of a crocodile with robust conical teeth, and powerful arms with large recurved claws, suggesting that they fed on mostly aquatic prey like fish or turtles with a semi-aquatic lifestyle, and probably didn't hunt larger land prey like other dinosaurs very often (although *Baryonyx* has been found with iguanodont bones inside it).

The original fossils of *Spinosaurus* itself were found in the western part of Egypt by German scientists. It was only known from a few isolated bones (including the spines that stuck out of the back) in the natural history museum in Munich, which were destroyed by bombing during World War II. *Spinosaurus* has been completely rethought as new specimens have been found in the past two decades. We now know that it wasn't much like the outdated version seen in the third *Jurassic Park* movie, which was a huge biped that terrified even a tyrannosaur. Modern reconstructions suggest that it was a quadrupedal swimmer that acted more like a crocodile, and spent very little time walking on its hind legs on land (**Figures 18.4[D]** and **18.5[A]**). Its long narrow snout was filled with conical teeth for catching fish, not blade-like teeth with serrated edges suitable for ripping flesh, as found in most theropods. And it was nowhere near as big as *Jurassic Park III* had rendered it. At best it was much more slender and only slightly longer than a *Tyrannosaurus*, but without the huge body mass or the flesh-ripping teeth or powerful neck and jaws for biting large prey. If it encountered a large theropod, it would almost certainly have backed off or fled for the nearest water.

Almost all its features of *Spinosaurus* are adaptations for the aquatic lifestyle as well. Its nostrils were located midway back on the snout, suitable for breathing while partially submerged. The snout also had channels for nerves that would have helped it sense changes in water pressure caused by motion of prey in the water. Its stumpy subequal quadrupedal limbs were not optimal for walking on land, but good for paddling in the water, and the long thin finger and toe bones suggest it had webbed feet. Even the geochemistry of the bones and teeth suggest it was an aquatic animal. Finally, the limb bones are very dense and solid, typical of animals like hippos that need dense limb bones to help as ballast.

The biggest mystery was the enormous sail on the back of *Spinosaurus* that gave it the name. It was clearly not big enough to be a true "sail" for propelling it through the water under wind power, because it was much too small for the huge bulk of the dinosaur. In fact, it's so large and conspicuous that it would prevent the dinosaur from completely submerging underwater and sneaking up on prey, as crocodilians do. Others have argued that it was a big heat-gathering surface for regulating body temperature, but then why does it not occur in any other theropod (but for some reason, does occur in the African iguanodontid *Ouranosaurus*)? Most paleontologists point to the large conspicuous nature of the sail and consider it some sort of device to advertise its huge size and dominance in competing with other spinosaurs in its territory, as the horns and antlers of many deer and antelopes do today.

Some people claim *Spinosaurus* is the largest predatory dinosaur known, but that is debatable. The size of *Spinosaurus* is hard to estimate given how incomplete the fossils are, but the most recent estimates places its length at 15.2 meters (51 feet). If *Tyrannosaurus rex* was about 13 meters (43 feet) and about 10 tonnes (11 tons), it was only slightly shorter than *Spinosaurus*, but probably weighed more. Clearly, the scene in *Jurassic Park III* where *Spinosaurus* picks up a *T. rex* and tosses it around like a toy is impossible, based on what we now know of *Spinosaurus*. It was much more likely the other way around: a bulldog predator with the crushing bite of a *T. rex* would easily kill a lightly build fish-eater like *Spinosaurus* in a fight, even if the latter was slightly longer.

Carnosauria: Megalosauridae

In 1837, *Megalosaurus* was described by the famous naturalist William Buckland, making it the very first dinosaur ever to be formally named (**Figure 18.4[A]**). Originally known only from skull and jaw fragments and some broken limbs and vertebrae from the Middle Jurassic of England, it was originally reconstructed as a giant elephantine quadrupedal predatory lizard. Eventually, better specimens of megalosaurs were found with the discovery of *Eustreptospondylus* in England, as well as theropods on other continents, and we now recognize that *Megalosaurus* was a typical bipedal theropod. Despite its name which means "giant lizard" in Greek, as more specimens were found, *Megalosaurus* turned out to be a medium-sized theropod, reaching only about 6 meters (20 feet) long and weighing only about a metric tonne. Now there are almost a dozen genera of Megalosauridae known, nearly all larger than *Megalosaurus*. They include not only the much more complete *Eustreptospondylus*, *Magnosaurus*, and *Duriavenator* from England (all long misassigned to *Megalosaurus*), *Dubreuillosaurus* and *Piveteausaurus* from France, *Wiehenvenator* from Germany, *Torvosaurus* from the Late Jurassic of the Rocky Mountains and Portugal, *Afrovenator* from the Middle Jurassic of North Africa, and *Leshansaurus* from China. Thus, the group was spread across nearly all of Pangea in the Middle and Late Jurassic, but particularly in Europe; it was apparently absent from South America, Australia, and Antarctica, which may be partly a consequence of the limited Jurassic exposures on those continents.

Carnosauria: Metriacanthosauridae

Metriacanthosaurids were very similar to allosaurs in many aspects, but they are distinguished by a series of anatomical details that define them as a natural group (**Figure 18.6[A]**). The most obvious is that neural spines on the middle of the backbone tend to be very tall, although they were never developed into a tall sail like in *Spinosaurus*. Five or six genera are known, including *Metriacanthosaurus* from the Middle Jurassic of England, several genera (*Yangchuanosaurus*, *Sinraptor*, *Shidaisaurus*) from the Middle Jurassic through Early Cretaceous of China, and *Siamotyrannus* from the Early Cretaceous of Thailand.

Carnosauria: Allosauridae and Carcharodontosauridae

The last two groups within the Carnosauria are the Allosauridae and Carcharodontosauridae (**Figure 18.1**). *Allosaurus* (**Figures 18.5[B]** and **18.6[B]**) is one of the best known of all theropods, with dozens of complete skeletons from the Upper Jurassic Morrison Formation of the Rocky Mountains (especially from the famous Cleveland-Lloyd Quarry in central Utah, and Carnegie Quarry at Dinosaur National Monument). These quarries also yield competitors like ceratosaur *Ceratosaurus* and the megalosaur *Torvosaurus*, which are from very different branches of the Theropoda (**Figure 18.1**). *Allosaurus* was typically 8 meters (26 feet) long, with mostly generalized theropod anatomy and only a handful of specializations.

Figure 18.6 Reconstructions of some metriacanthosaurs, allosaurs, and carcharodontosaurs: (A) *Metriacanthosaurus*, (B) *Allosaurus*, (C) *Carcharodontosaurus*, (D) *Giganotosaurus*.

The carcharodontosaurs, on the other hand, tended to be huge and highly specialized (**Figures 18.5[C,D]** and **18.6[C,D]**). Some of them were even slightly bigger than not only the tyrannosaurs but also the abeliasaurs and spinosaurs as well. From the same beds in North Africa that yielded *Spinosaurus* came another giant dinosaur named *Carcharodontosaurus* (**Figures 18.5[C]** and **18.6[C]**). It was so named because its teeth were about the size and shape of those of the great white shark, *Carcharodon*. *Carcharodontosaurus* had a relatively long, tall skull that is much more lightly built than the skulls of tyrannosaurs, abeliasaurs, or spinosaurs. The skull roof is composed of high bony arches, with big openings on both sides of the skull to make it lighter and to increase the area for attaching powerful jaw muscles. Unfortunately, only the skull of *Carcharodontosaurus* is known, plus a few other bones. On the basis of this limited evidence, the body of *Carcharodontosaurus* was about 12–13 meters (39–43 feet) long, and weighed about 6–15 tonnes, making it about the same size as *Spinosaurus* and large *T. rex*.

If the skeletons of *Carcharodontosaurus* and *Spinosaurus* are too incomplete to reliably estimate their size, what was the biggest land predator of all time? Currently, the title goes to a South American carcharodontosaur named *Giganotosaurus carolinii* (**Figures 18.5[D]** and **18.6[D]**). In Greek, its name translates as *Giga* ("big"), *-noto* ("southern"), and *-saurus* ("lizard"). Many people fail to read the name properly and mispronounce it "GIGAN-TO-saurus". It's actually GIG-a-NO-to-saur-us. (There was already a different dinosaur named *Gigantosaurus*, but it is an invalid name now.) About 70% of the skeleton of *Giganotosaurus* has been found, making it much more complete than the other contenders for "biggest land predator ever". Like its close relative *Carcharodontosaurus*, the skull of *Giganotosaurus* is built of high arches of bony struts, with lots of openings on the sides. Such a light skull is a big contrast from the bulldog-like massive skull of *T. rex*, so the bite force of *Giganotosaurus* was probably only a third as strong. Its broad shark-like teeth were better suited for producing slashing wounds, rather than biting down and crushing as in tyrannosaurs. It probably gashed and disabled its prey from ambush, then disemboweled them before gorging itself on the prey, while the unfortunate victim was slowly bleeding to death.

What did it eat? *Giganotosaurus* comes from lower Upper Cretaceous beds of South America, which were dominated by sauropods like the titanosaur *Andesaurus*, diplodocids *Nopcsaspondylus* and *Limaysaurus*, as well as an array of iguanodonts, and small predatory dinosaurs related to *Velociraptor*.

The nearly complete skeletons of *Giganotosaurus* were up to 14.2 meters (53 feet) long, and this suggests that they weighed as much as 13.8 tonnes (30,420 pounds). This is a bit longer than the largest *T. rex* which was about 13.0 meters long and weighed only 8 tonnes. Thus, until some other dinosaur is found which is both larger and more complete, *Giganotosaurus* is the current champion and holder of the crown of the largest land predator that ever evolved.

TETANURAE: COELUROSAURIA

The other great branch of the tetanurine theropods is now known as the Coelurosauria. It includes most of the advanced and uniquely specialized theropods, including the tyrannosaurs, the ostrich dinosaurs or ornithomimids, the weird therizinosaurs and oviraptors, the *Velociraptor* group, and of course, the birds. Even though these creatures look very different from one another, they have some important anatomical features in their skeletons that show they are a natural group. These include the fact that their shin bone is longer than their thighbone (an adaptation for running), a tail that is stiffened at the tip, and a very long set of sacral vertebrae fused to the ilium of the pelvis. In addition, feathers seem to be found in nearly all members of this group. This is the main group of theropods that evolved herbivory (independently in the therizinosaurs and oviraptorids), and omnivory in a number of groups.

Coelurosauria: Tyrannosaurs

Tyrannosaurus rex is probably the most famous and popular of all dinosaurs, and the subject of numerous books and TV shows and movies, especially as the main terror of the *Jurassic Park/World* series. But it is not the only tyrannosaur. Over 35 genera (**Figure 18.7**) are currently placed

Figure 18.7 Reconstructions of some coelurosaurs, including tyrannosaurs: (A) *Dilong*, (B) *Compsognathus*, (C) *Sinosauropteryx*, (D) *Yutyrannus*, (E) *Tyrannosaurus*.

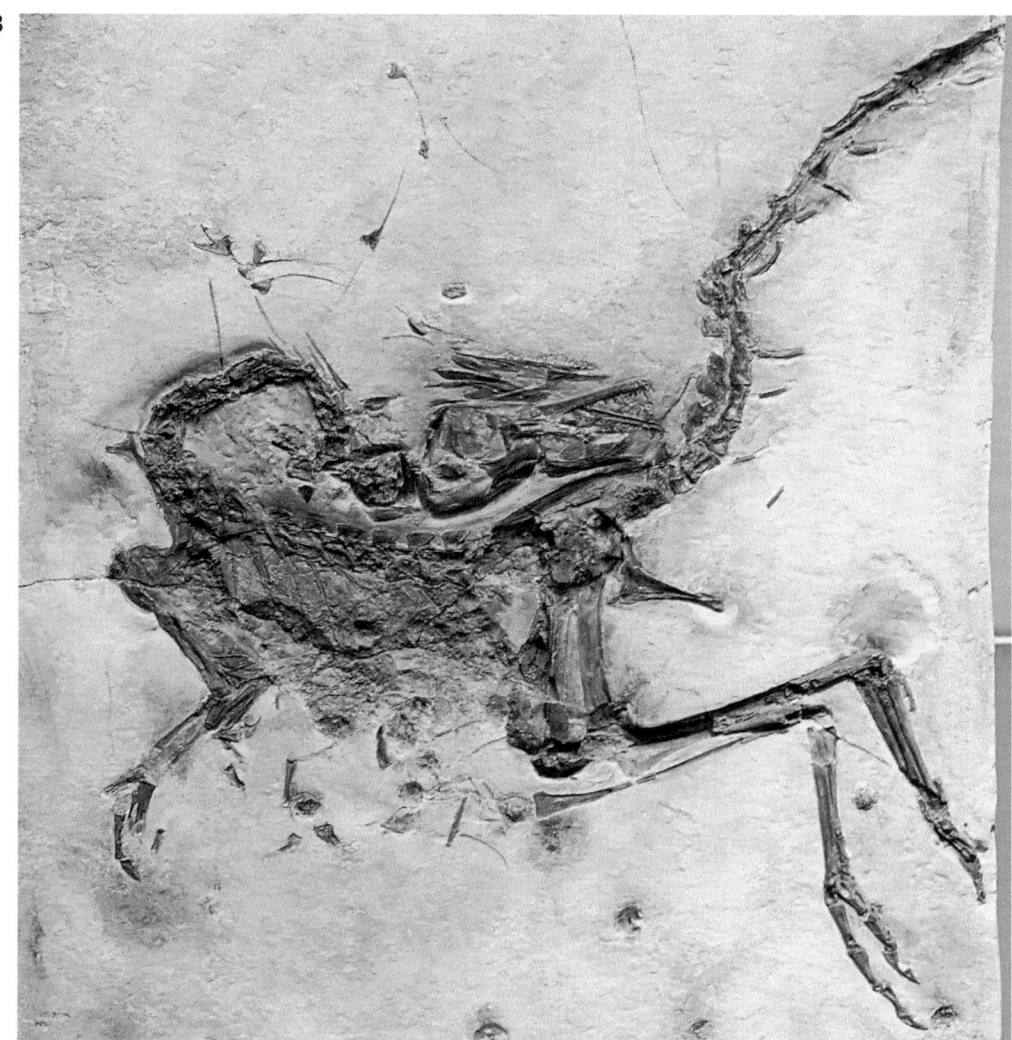

Figure 18.8 Fossils of some more advanced theropods. These include (A) The famous skeleton named "Sue", a *Tyrannosaurus* at the Field Museum of Natural History in Chicago. (B) The first complete specimen of *Compsognathus* from the Solnhofen Limestone of Germany. [(A,B) Courtesy Wikimedia Commons. (C–E) By the author.]

Figure 18.8 (Continued) (C) A modern ostrich skeleton shown next to the ostrich-dinosaur *Struthiomimus*. (D) The long arms of *Deinocheirus*. (E) The weird crested beaked skull of an oviraptorid is shown by the bizarre *Hagryphus*, from the Cretaceous of Utah.

in the Tyrannosauridae, from the very primitive form *Proceratosaurus* of the Middle Jurassic of England, then a huge radiation of tyrannosaurs in Asia (mostly China and Mongolia) through most of the Cretaceous, spreading to North America in the Late Cretaceous, where they underwent another evolutionary radiation. There is not enough room in this book to discuss all of these genera, so let us focus on the most famous, most popular, and best studied of them: *Tyrannosaurus rex*.

When Barnum Brown found the first five specimens of *Tyrannosaurus* in the Hell Creek beds of eastern Montana, the fossils were originally mounted in the old "tail-dragging lizard" model of dinosaurs, standing in a kangaroo pose with its tail on the ground. But during the Dinosaur Renaissance of the 1970s and 1980s, paleontologists realized that tyrannosaurs were completely bipedal with their tails sticking straight out in the back to balance their body in horizontal position over the hind legs. That has been the accepted pose for most theropods ever since.

Tyrannosaurus rex (**Figures 18.7[E]** and **18.8[A]**) had an extremely robust skull and huge teeth, and could generate an enormous bite force. The skull of the biggest *Tyrannosaurus rex* was about 1.5 meters (5 feet long), but it was made lighter with numerous air pockets and holes in the solid bone, so the weight of the head was reduced. In cross-section, the snout was shaped like an upside-down "U", making it more rigid and stronger than a typical theropod skull, which has a cross-section shaped like an upside-down "V". The snout was also narrow enough that the eyes could face fully forward, giving *Tyrannosaurus rex* excellent binocular stereovision and depth perception for hunting. The enormous teeth, (about 30 cm, or 12 inches, from tip to root) curve backward, giving it strength for the teeth to pull back as they rip out hunks of flesh. They were shaped a bit like steak knives but they were as big as a banana and had serrated ridges on the cutting edges. The front teeth were thicker and deeply rooted, with a "D"-shaped cross-section, giving them strength so they didn't break when the *Tyrannosaurus rex* bit down and pulled flesh backward. Instead of being a slashing predator, *Tyrannosaurus* was more like a bone-crushing "bulldog dinosaur". Modern techniques of modeling of bite forces suggested that *Tyrannosaurus rex* could produce 35,000–57,000 newtons of force (7900 to 13,000 pound-force). This is three times more powerful that the bite of the great white shark, 3.5 times as strong as the Australian saltwater crocodile, seven times more powerful than *Allosaurus*, and 15 times as powerful as the bite of a lion. Recently, that estimate has been revised upward to 183,000–235,000 newtons (41,000–53,000 pound-force), stronger than the bite of even the giant extinct shark *Otodus megalodon*.

So what did *Tyrannosaurus rex* do with those powerful jaws? We know that they fought with each other, because several skulls show healed wounds on their faces and other bones that could have only been caused by the bite of another *Tyrannosaurus rex*. Some even have broken teeth embedded in their bones. Clearly, they were capable of killing and eating nearly any dinosaur that lived during the Late Cretaceous with them. Unfortunately, the media have given a lot of publicity to the silly argument that *Tyrannosaurus rex* was a scavenger and not a carnivore. In reality, modern large predators are not very picky. Lions, which mostly hunt their food, will gladly eat carrion when they are hungry, while hyenas, which are famous for breaking down carcasses and crushing bones, are actually very efficient pack hunters who prefer to kill their own meals when they get the chance. There is no reason to think that *Tyrannosaurus rex* was a picky eater, but instead was opportunistic and ate anything it could, especially when it was hungry!

One of the most famous features of *Tyrannosaurus rex* was its tiny arms. Dozens of cartoons, gags, and even greeting cards have made fun of the fact that its arms seem useless. If you look closely, *Tyrannosaurus rex* also had only two functioning fingers, while most other theropods still had three fingers. Paleontologists have debated why the forelimbs were so small ever since *Tyrannosaurus rex* was first found. Pioneering early twentieth-century paleontologist Henry Fairfield Osborn, who named and described *T. rex*, suggested that they might have been useful to hold a mate during copulation. Others argued that it would help them rise from a prone position, and recent digital models have shown that it was plausible. More recent research has shown that the actual bones of the arms are quite strong and robust, and would have had powerful muscles, capable of lifting 200 kg (440 pounds), so they were not weak arms. This suggests they could hold a smaller prey animal more easily than was once believed, and certainly were capable of slashing a prey animal or another *Tyrannosaurus rex* in close combat. Actually, compared to dinosaurs like *Carnotaurus*, which has even tinier stunted arms with vestigial fingers, the arms of *Tyrannosaurus rex* are not that small. More importantly, *Tyrannosaurus rex* focused on the powerful head and neck and jaws as its primary weapon. Along with its strong hind feet with long sharp claws, it had a different way of feeding than animals that rely on strong arms to catch prey. The arms of tyrannosaurs are mostly likely vestigial and less important in a predator that primarily used it head and legs.

One of the biggest changes in how we think about *Tyrannosaurus* concerns its body covering. For almost a century, it was rendered as a big scaly reptile, a sort of lizard on steroids. The only known skin impressions of *Tyrannosaurus* have a mosaic of small scales preserved. But in the 1990s, discoveries in China produced many different fossils of dinosaurs, birds, and mammals, especially in lake shales, which are low in oxygen and formed in stagnant water, so soft tissues were preserved. These produced a small tyrannosaur (**Figure 18.7[A]**) called *Dilong paradoxus*, which clearly showed filamentous feathers or fluff on its body. When a larger tyrannosaur, *Yutyrannus halli*, was found in China (**Figure 18.7[D]**), it too was covered with a coat of feathers. Given that these animals are very closely related to *Tyrannosaurus rex* and all other tyrannosaurs, it is extremely likely that the iconic dinosaur of the *Jurassic Park/World* movies was not the scaly lizard that the moviemakers created, but a bird-like creature with at least some coating of down or at least feather tufts over parts of its body.

Coelurosauria: Compsognathidae

In 1861, the very first complete dinosaur skeleton ever described was the small chicken-sized theropod *Compsognathus* (**Figures 18.7[B]** and **18.8[B]**) from the Upper Jurassic Solnhofen Limestone of Bavaria. (They are familiar in the *Jurassic Park* movies as the tiny "compies" which attacked humans in packs.) *Compsognathus* had a long neck, small head with a pointed snout, delicate small arms and long slender legs, and a long tail. Although it looks much like the most primitive theropod dinosaurs from the Triassic, it is actually quite advanced in many features, especially in the ankle and foot bones. In 1868, Darwin's advocate Thomas Henry Huxley redescribed the original specimen, and noticed its similarities to *Archaeopteryx*, and then proposed that birds evolved from dinosaurs. Since then, at least 11 other genera of compsognathids have been described, found from Europe to Asia to South America in the Late Jurassic and Early Cretaceous. One of these new discoveries, *Sinosauropteryx* from the Early Cretaceous of China, is known from complete articulated specimens with feathers preserved, and even some pigment cells (melanosomes), which showed that they not only were completely feathered, but also had black-and-white striped feathers on their tails (**Figure 18.7[C]**).

Coelurosauria: Ornithomimids

The "ostrich dinosaurs" or ornithomimids are familiar from *Gallimimus* in *Jurassic Park* as the long-necked, long-legged fast runners that stampede in one of the crucial scenes. The first specimen discovered was called *Ornithomimus* ("bird mimic") in 1890, because of its bird-like feet and legs, and it was about 3.8 meters (12 feet) in length. The first nearly complete skeleton that showed the ostrich-like neck and legs and general build was *Struthiomimus* ("ostrich mimic") from the Upper Cretaceous beds of Alberta (**Figures 18.8[C]** and **18.9[A]**). Described in 1917, it was about 4.3 meters (14 feet) long counting its long slender tail. These are the oldest named genera and some of the best known, but there are at least 10 other genera in the family Ornithomimidae, and another 9 genera of primitive ornithomimosaurs that are not members of the family Ornithomimidae. Thus, the ostrich mimics were a diverse group of dinosaurs found all over Eurasia and North America during most of the Cretaceous.

The weirdest of the ornithomimids were first discovered in the Gobi Desert in the 1960s and only consisted a set of huge arms with long claws (**Figures 18.8[D]** and **18.9[B]**) known as *Deinocheirus*. It was a complete mystery until 2013, when most of the skeleton was found in Mongolia (part of it poached and recovered from the black market). Deinocheirids were huge animals all right (**Figure 18.9[B]**), but they were not predators at all, but herbivores or omnivores—despite having the claws of a theropod! The largest known specimen was 11 meters (36 feet) long, and may have weighed 6.4–12 tonnes (7–13 tons), but the smaller specimens were only about 75% the size of the largest. Even though it had a huge bulky body, the bones were hollow, which made it lighter and caused less of a strain on its relatively short legs and toes with blunt claws (which bore all the weight, because the hands were for grasping, not walking). The oddest feature is the long spines on the top of its backbone from the lower back to the base of the tail, which may have given it a tall "sail" or possibly supported a large fleshy "hump". The tail ended with a fusion of most of the tail vertebrae into a pygostyle (like the "parson's nose" in birds), which apparently supported a fan-like array of tail feathers as in birds. The huge skull was long and narrow, and over a meter long (3.36 feet). Yet the skull was nothing like a typical predatory theropod, but more like the toothless ostrich-like heads of ornithomimids. It had a wide bill and deep lower jaw, resembling a duckbill dinosaur snout rather than a predator (but it was toothless, unlike duckbills or most theropods). The eyes were relatively small with a ring of bone around the pupil (sclerotic ring). Since they were herbivores, it is thought that they were mostly daytime feeders. *Deinocheirus* had a relatively small brain, with a ratio of brain size to body mass more like the huge sauropods than its own group, the more intelligent theropods.

So what did this weird creature eat? The beak suggests a herbivorous or omnivorous diet of plants, but fish scales were found in one specimen, so they ate at least some fish, if not meat. One specimen had hundreds of gastroliths (gizzard stones) in its gizzard to grind down its plant diet. They had enormous bellies, which would be expected for a plant eater that needed a big gut and a long digestive tract to process and ferment large volumes of vegetation. The huge clawed hands were apparently for grasping and pulling down branches, not attacking prey, and possibly for digging for roots and tubers, or maybe defending against predators.

As bizarre as *Deinocheirus* seems, it was not the only big-handed heavy-bodied herbivorous ornithomimosaur from the Cretaceous of Asia. There is also *Garudimimus*, named after the Garuda bird in Hindu and Buddhist mythology, which was about 2.5 meters (8.2 feet) long.

Figure 18.9 Reconstructions of some ornithmimids and therizinosaurs, including: (A) *Struthiomimus*, (B) *Deinocheirus*, (C) *Therizinosaurus*, (D) *Gigantoraptor*.

Larger still is *Beishanlong* from the Early Cretaceous Ghost Castle site in the White Mountains of Gansu Province, China. Its name means "White Mountain dragon" in Mandarin. It reached about 7 meters long (23 feet) and about 550 kg (1200 pounds) in weight. Thus, deinocheirids were apparently widespread in Asia through most of the Cretaceous.

Coelurosauria: Maniraptora: Therizinosaurs

The higher coelurosaurs are known as the Maniraptora, which are distinct from most other advanced theropods in that their arms and hands are relatively large, with a highly specialized wrist with the "half-moon"-shaped (semilunate) carpal bone (see Chapter 19). They also have their pubic bone pointed backward, a condition that occurs not only in birds but also in all other maniraptorans, as well as a bunch of other skeletal specializations that show they are a natural group.

One of the most recently solved puzzles of theropods, and some of the weirdest of all dinosaurs, were the therizinosaurs, also known as segnosaurs (**Figure 18.9[C]**). They were a second example of carnivorous theropods becoming herbivorous or omnivorous. Although the first fragmentary fossils were named in 1948, it wasn't until the 2000s that good specimens gave us a complete picture of these odd creatures. *Segnosaurus* (described in 1979) was the first to show what therizinosaurs were like. It had lower jaw with a downturned snout and leaf-shaped teeth, suggestive of a herbivore. *Segnosaurus* had powerful forelimbs with the long claws that were strongly curved and flattened, like the blade of a sickle. The pelvis was very broad, the hind limbs robust and short, and still retained four toes pointing forward (most theropods had only three, plus a tiny "big" toe or hallux that points backward), suggesting it was a very heavy slow-moving biped with a huge gut. The feet are unusual in that the first toe is large and contacts both the ankle joint and the ground, whereas the first toe of most theropods is extremely reduced and doesn't touch the ankle joint. Even more oddly, the pubic bone of the hip points backwards, but in a different way than it does in the ornithischian hips, or in the hips of true birds—yet all its other features are indicative of a non-avian theropod dinosaur, which had a saurischian pelvis.

Coelurosauria: Maniraptora: Oviraptorosauria

If weird creatures like deinocheirids and therizinosaurs were not odd-looking enough, even stranger are the oviraptorosaurs. Most of them had a flat snout with a parrot-like beak (with no teeth or only a few teeth). They often had a weird crest on the top of their skulls over their large eyes (**Figures 18.8[E]** and **18.9[D]**). Their long arms with only three fingers apparently bore feathers, and they have long legs for running and a short tail. They were first described in 1924 based on *Oviraptor philoceratops*, whose name means "egg thief who loves horned faces", because the first specimen was found in the Gobi Desert near a nest of eggs thought to belong to the ceratopsian *Protoceratops*. But this is a slander, because later specimens showed *Oviraptor* brooding right on the nest, and the eggs produce *Oviraptor* embryos, so it was not an egg thief, but the mother of the nest. Many more have been discovered, so now there are at least 35 genera, mostly from the Late Cretaceous of Mongolia and China, although a few are found in North America.

Coelurosauria: Eumaniraptora: Dromaeosaurs

Velociraptor is perhaps one of the most popular dinosaurs today, and certainly the most famous member of the group called dromaeosaurs (**Figures 18.10** and **18.11[A]**). It was made famous by the *Jurassic Park* books and movies, and yet almost everything the public "knows" about it is wrongly portrayed in the same books and movies. First, the

Figure 18.10 Reconstructions of some dromaeosaurs, including: (A) *Deinonychus*, (B) *Velociraptor*, (C) *Microraptor*, (D) *Utahraptor*.

name is wrong. The terrifying human-sized dinosaur in the movies is based on *Deinonychus* (**Figures 18.10[A]** and **18.11[A]**), the first complete dromaeosaur, found by John Ostrom in 1963. *Velociraptor* was a much smaller dromaeosaur, the size of a turkey, and would never have terrorized anyone in a movie theater. Author Michael Crichton followed a mistake in a book by an amateur which falsely claimed *Velociraptor* and *Deinonychus* were the same thing, so that the senior name for both dinosaurs would be *Velociraptor*. That mistake has propagated ever since, so "raptors" are everywhere in the culture, and now there is even an NBA team called the Toronto Raptors. Second, all dromaeosaurs had feathers, and there are even quill knobs for the attachment for feathers on the arms of some specimens of dromaeosaurs to confirm this—yet the movies persist in giving us naked dinosaurs. These major mistakes, along with many minor ones (*Velociraptor* only comes from Mongolia, and is not found in Montana—but *Deinonychus* was found in Montana) can drive paleontologists crazy.

Fortunately, the movies have made the general public aware that dinosaurs such as dromaeosaurs were active and intelligent, and most of the anatomy of *"Velociraptor"* in the movies is accurate except for the size. When Ostrom described *Deinonychus*, he recognized that it had a large brain and wicked-looking mouth full of sharp teeth, strong forelimbs with sharp claws, a rigid tail that stuck straight out to balance their bodies, and most importantly, an enlarged claw on their foot that helped slash other creatures when they would leap up and attack. The rest of the dromaeosaurs are highly diverse, with over 50 genera known now, mostly from the Late Cretaceous of Asia but also from the Cretaceous of North

Figure 18.11 (A) Comparison of the turkey-sized skeleton of *Velociraptor* (foreground) with the human-sized dinosaur *Deinonychus* (background). (B) The fossil of the crow-sized dromaeosaur *Microraptor gui* from the Cretaceous of China, with flight feathers on both the arms and legs. (Courtesy Wikimedia Commons.)

America, Europe, Africa, and South America. Many were crow-sized feathered flying creatures (like *Microraptor gui*, which had wing feathers on its hands and legs—18.11B) that looked and acted much like birds, but were not anatomically birds yet. Others were non-flying ground animals like turkey-sized *Velociraptor*, the human-sized *Deinonychus*, and the even bigger *Utahraptor*, which was up to 4.8 meters (16 feet) long, not counting the longer feathers in its tail (**Figure 18.10[D]**).

Dromaeosaurs are so much like birds in so many ways that it is just a small step to the next branch of the theropods: the Avialae, which includes the birds and many of their extinct relatives. These are discussed in the next chapter.

FURTHER READING

Alexander, R.M. 1989. *Dynamics of Dinosaurs and Other Extinct Giants*. Columbia University Press, New York.

Bonaparte, J.; Novas, F.; Coria, R. 1990. *Carnotaurus sastrei* Bonaparte, the horned, lightly built carnosaur from the Middle Cretaceous of Patagonia. *Contributions in Science (Natural History Museum of Los Angeles County)*. 416: 41–50.

Farlow, J.O.; Gatesy, S.M.; Holtz, T.R., Jr.; Hutchinson, J.R.; Robinson, J.M. 2000. Theropod locomotion. *American Zoologist*. 40 (4): 640–663.

Fastovsky, D.E.; Weishampel, D.B. 2021. *Dinosaurs: A Concise Natural History* (4th ed.). Cambridge University Press, Cambridge.

Hendrickx, C.; Hartman, S. A.; Mateus, O. 2015. An overview of non-avian theropod discoveries and classification. *PalArch Journal of Vertebrate Palaeontology*. 12 (1): 1–73.

Holtz, T.R., Jr.; Rey, L.V. 2007. *Dinosaurs: The Most Complete, Up-to-Date Encyclopedia for Dinosaur Lovers of All Ages*. Random House, New York.

Lucas, S.G. 2005. *Dinosaurs, the Textbook* (5th ed.). W. C. Brown, Dubuque, Iowa.

McGowan, C. 1983. *The Successful Dragons: A Natural History of Extinct Reptiles*. Samuel Stevens, Toronto.

McGowan, C. 1991. *Dinosaurs, Spitfires, and Sea Dragons*. Harvard University Press, Cambridge.

Norman, D.B. 1985. *The Illustrated Encyclopedia of the Dinosaurs*. Crescent Books, New York.

Ostrom, J.H. 1969. Osteology of *Deinonychus antirrhopus*, an unusual theropod from the Lower Cretaceous of Montana. *Peabody Museum Natural History Bulletin*. 30: 1–165.

Rauhut, O.W. 2003. *The Interrelationships and Evolution of Basal Theropod Dinosaurs*. Blackwell Publishing, New York.

Sereno, P.C. 1998. A rationale for phylogenetic definitions, with application to the higher-level taxonomy of dinosauria. *Neues Jahrbuch für Geologie und Paläontologie, Abhandlungen*. 210: 41–83.

Sereno, P.C. 1999. The evolution of dinosaurs. *Science*. 284 (5423): 2137–2147.

Therrien, F.; Henderson, D.M. 2007. My theropod is bigger than yours . . . or not: Estimating body size from skull length in theropods. *Journal of Vertebrate Paleontology*. 27 (1): 108–115.

Weishampel, D.B.; Dodson, P.; Osmólska, H., eds. 1990. *The Dinosauria*. University of California Press, Berkeley, Los Angeles, Oxford.

BIRDS

THE FLYING DINOSAURS

19

And if the whole hindquarters, from the ilium to the toes, of a half-hatched chick could be suddenly enlarged, ossified, and fossilised as they are, they would furnish us with the last step of the transition between Birds and Reptiles; for there would be nothing in their characters to prevent us from referring them to the Dinosauria.

—Thomas Henry Huxley, 1870, *Further Evidence of the Affinity between Dinosaurian Reptiles and Birds*

BIRDS ARE DINOSAURS

The story of the origin of the birds is one of the first great success stories in the history of evolutionary biology, and one of its most famous. In his first edition of *On the Origin of Species* in 1859, Darwin was quite apologetic about the apparent lack of transitional fossils between major taxa. Then in 1861 the first fossils of *Archaeopteryx* (**Figure 19.1**) were discovered in the Upper Jurassic Solnhofen Limestone of Bavaria, famous for its fine-grained limestones that were quarried to make lithographic stone. The scientific world was galvanized, and Darwin's supporters were energized that there was the first good fossil demonstrating the transition from reptiles to birds.

Almost every skeletal feature of *Archaeopteryx* was dinosaurian, except for two important bird-like characters: the fused collarbones forming a wishbone, and the presence of feathers (and we now know that these features also evolved in theropod dinosaurs). In the following century and a half, six more specimens were found that added further details to the story. The "Berlin specimen" (**Figure 19.1**), found in 1877, was the most complete and well preserved of the 13 known specimens. It was fossilized in a "death pose" with all its bones in place and clear feather impressions. In 1996, ornithologist Alan Feduccia (p. 29) called it "the most important natural history specimen in existence, comparable perhaps in scientific and even monetary value to the Rosetta Stone. Beyond doubt, it is the most widely known and illustrated fossil animal—a perfectly preserved Darwinian intermediate, a bird that has anatomical features of a reptile, feathers, and a long, lizard-like tail". Other specimens (at least 12 in total are now known) revealed the presence of a broad breastbone for the attachment of flight muscles. Another specimen of *Archaeopteryx* was misidentified as a pterosaur, and one specimen was long misidentified as the small dinosaur *Compsognathus* before its faint feather impressions were discovered. These specimens, and a careful analysis of Mesozoic reptiles, have given us a startling perspective: birds are most closely related to certain types of theropod dinosaurs.

The fact that one specimen of *Archaeopteryx* could be so easily misidentified for a different dinosaur clearly proved this. But it is not a new

DOI: 10.1201/9781003128205-19

Figure 19.1 The most of famous of the 13 known specimens of *Archaeopteryx*. Discovered in 1877, it is now on display at the Museum für Naturkunde in Berlin (so it is called the "Berlin specimen", even though all known specimens of *Archaeopteryx* come from the same Solnhofen Limestone quarries in southern Germany). It is the most complete specimen of this bird, and was fossilized in a natural pose with the neck pulled backward due to the contraction after death of the nuchal ligament which holds the neck straight. (Courtesy Wikimedia Commons.)

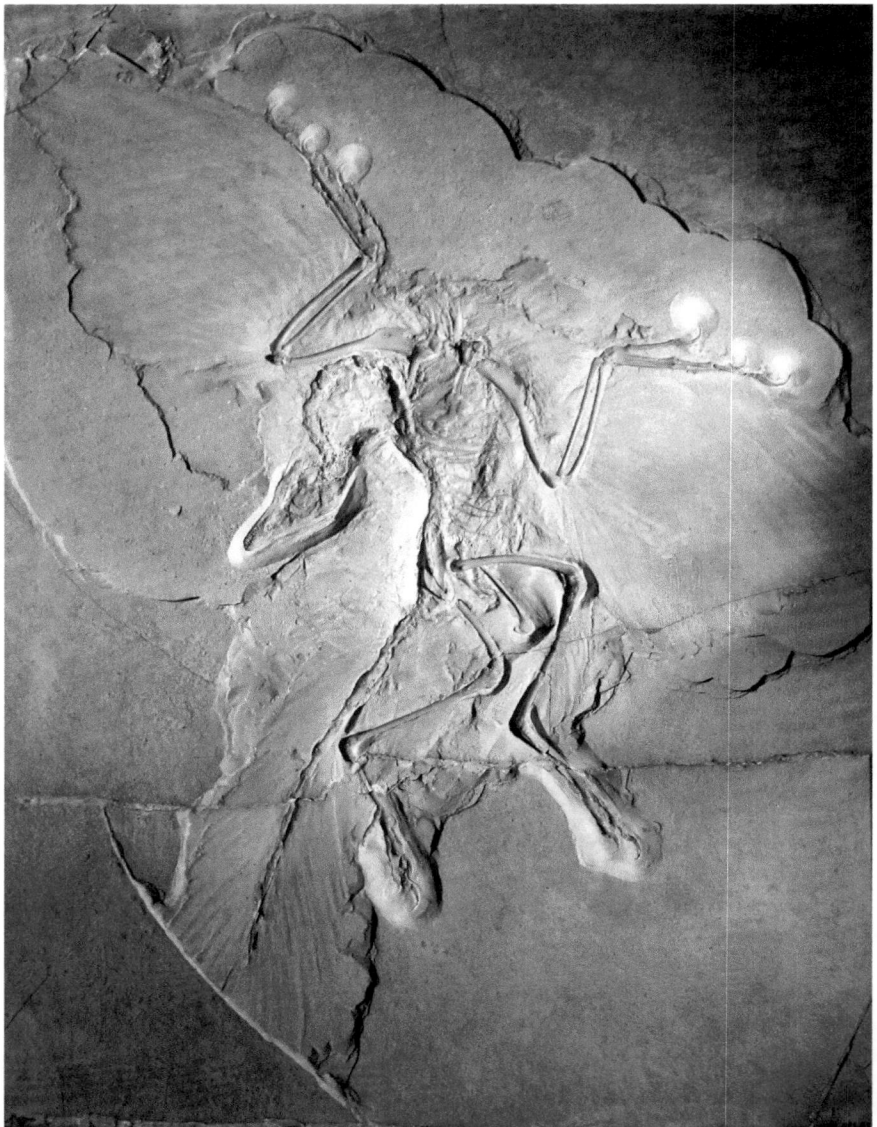

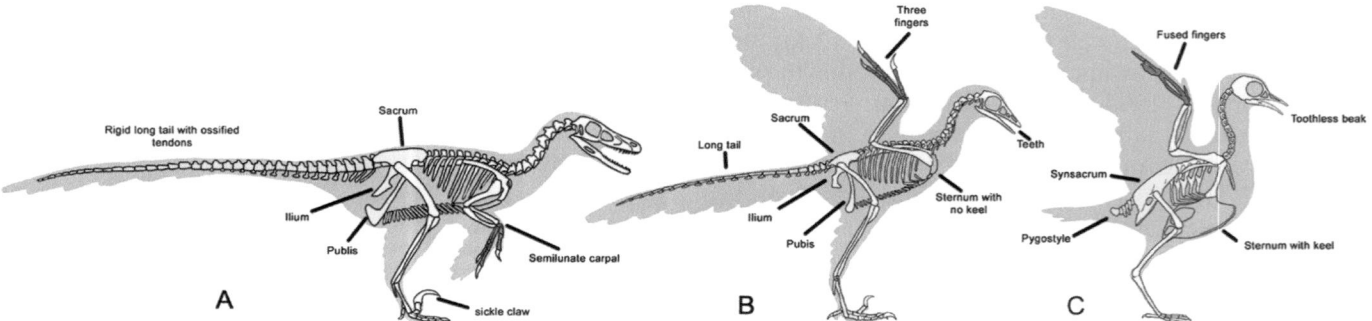

Figure 19.2 Diagram comparing the anatomic similarities and differences of a (C) bird, (B) *Archaeopteryx*, and (A) small theropod.

idea. Thomas Henry Huxley first suggested it soon after *Archaeopteryx* was described. Dozens of evolutionary specializations support the close relationship of birds and theropods known as dromaeosaurs, such as *Deinonychus* and *Velociraptor*. *Archaeopteryx* has a long bony tail, a dinosaurian feature found in no living bird, which have all fused their tail bones up into a tiny reduced nub of bones called the pygostyle or "parson's nose" (**Figure 19.2**). The skull of *Archaeopteryx* has the same arrangement of holes in the side that dinosaurs have, especially similar to that of the predatory dinosaurs like *Velociraptor*, and very different from the highly modified skulls of modern birds. The vertebrae are also like those of dinosaurs, and not arranged in as flexible a configuration that is seen in modern birds. The hip bones are intermediate in the condition between that of dinosaurs and birds, as is the strap-like shoulder blade. *Archaeopteryx* has gastralia, or belly ribs, found in many predatory dinosaurs but no modern birds.

Its most striking feature is the configuration of the hand and wrist. *Archaeopteryx* had long claws like those of predatory dinosaurs. It still had a fully functional hand with fingers 1–3 (thumb, index finger, middle finger) like most of the theropod dinosaurs. By contrast, a modern bird's hand is fused into just a few bones (called the carpometacarpus) plus tiny reduced finger bones. These bones form the small triangular pointed bit of bone at the end of the chicken wings you order for a meal, which you never eat because there is no meat (muscle) on them. Instead of fingers supporting their wings (as in bats), birds have greatly reduced their fingers and instead form their wings with feather shafts rather than skin membranes. In the wrist, birds and *Velociraptor* have a unique configuration of wrist bones fused into a half-moon shape, called the semilunate carpal (**Figure 19.2**). With this kind of wrist configuration, *Velociraptor* and its relatives can strike downward and forward quickly with their hands—but they cannot easily rotate their palms downward, so commonly seen on incorrect reconstructions of dinosaurs. In other words, they could catch a basketball between their hands, but they could not rotate them palms down to dribble. That rapid downward and forward snap of the wrist is the same motion that you see in the downstroke of the wing of a bird during flight—and it's all due to the semilunate carpal in the wrist.

The clincher is found in the hindlegs of *Archaeopteryx*. They have a unique ankle configuration called the mesotarsal joint (**Figure 11.9**), found only in dinosaurs, birds, and their close relatives, the pterosaurs. Most vertebrate ankles (including yours) are made of a series of rows of ankle bones, and they have a hinge between the tibia and fibula (shin bones) and the first row of ankle bones (calcaneum and astragalus). However, in birds, dinosaurs, and pterosaurs, the hinge is between the first and second row of ankle bones, so the astragalus and calcaneum actually can fuse to the end of the tibia. Next time you eat a chicken or turkey drumstick, notice the little cap of cartilage and bone on the "handle" end of the drumstick. This is the first row of ankle bones, a dinosaurian feature found in every bird. In addition, all birds and dinosaurs have a bony spur called the ascending process of the astragalus, that sticks up in the front of the tibia, another unique dinosaurian feature. The toes and feet of *Archaeopteryx* are like those of dinosaurs rather than most birds. *Archaeopteryx* even had one toe claw that was enlarged like the slashing toe claws seen in *Velociraptor* and its kin. In most respects, *Archaeopteryx* is just another dromaeosaur relative with feathers. Only the reversal of the big toe to point backwards, the lack of steak-knife serrations on the edges of the teeth, the asymmetrical flight feathers, and the relatively large arms distinguish it from dinosaurs like *Velociraptor*.

Or have birds completely lost their teeth? One would think so, since they are never seen on living birds, which have a horny beak instead. The idea is reflected in the phrase "as scarce as hen's teeth" (as in, they are so scarce that they are never found). But in a famous pioneering experiment in embryology and genetics in 1980, E.J. Kollar and C. Fisher grafted the mouth epithelium of a lab mouse into the mouth of an embryonic chick's beak. They let the chick develop, and were stunned to find that somehow it had grown the tooth buds that could become teeth again (although they never mature enough to develop enamel or dentin)! But they were not mouse teeth at all, but the simple conical teeth of predatory dinosaurs and the Cretaceous birds that still had teeth. Apparently, birds still have in their genes the information to make dinosaurian teeth, but this has been disabled or broken or suppressed by their regulatory genes so it is never expressed—except when tampered with by scientific experiments.

Since the famous Kollar and Fisher experiment, scientists have found lots of other dinosaurian genes that were repressed in birds, but can be expressed if the shut-off command is eliminated. One study managed to manipulate the chick genome so it developed a long bony dinosaurian tail like that in *Archaeopteryx*, not the short stubby pygostyle of modern birds. Yet another experiment tampered with the chick genes so their feet look dinosaurian, not bird like. Even more amazing is genetic manipulation of the genes for the mouth of a bird, so they ended up with a dinosaurian mouth with tooth buds rather than a beak of modern birds.

MESOZOIC BIRD EVOLUTION

Besides *Archaeopteryx*, there are now a number of other Jurassic close relatives of birds, known as the Scansoriopterygidae. These include *Xiaotingia*, *Yi qi*, *Anchiornis*, and *Epidexipteryx* from the Late Jurassic of China (**Figure 19.3**). They mostly look much like *Archaeopteryx* except for small details in the skulls shape, the elongate third finger and a few other features. The Cretaceous bird *Confuciusornis* from China (**Figures 19.3** and **19.4[A]**) already had a toothless beak, but still had long fingers with claws. By the Early Cretaceous, however, bird fossils show that a tremendous diversification was occurring. In the past few decades, hundreds of amazing discoveries (especially from the Lower Cretaceous lake beds of Liaoning Province, China) have produced almost 100 species of primitive toothed birds, beautifully preserved with their feathers intact, and some with the original coloration visible, as well as rare specimens with stomach contents or internal organs preserved. In addition, these same beds produce a spectrum of non-bird dinosaurs, showing that feathers are found in most groups of dinosaurs. Most of these feathered non-bird dinosaurs do not have flight feathers, but the feathers are there for their original purpose, insulation, and also for display, since they were brightly colored. This is just as it is for modern birds, who use only a small percentage of their feathers (mainly wing feathers and tail feathers) for flight, but most of their feathers are body feathers and down, which hold in their body heat.

Another very primitive bird was *Rahonavis*, from the Cretaceous of Madagascar (**Figure 19.3**). It also had the dromaeosaur-like sickle claw on the hind feet, the long bony tail, teeth, and several other dinosaurian features—but like more advanced birds, its hips fused to the lower back vertebrae to form a synsacrum. In addition, it had holes in its vertebrae for the air sacs found in living birds, and it even had quill knobs (bumps where the feathers attached to the bone) on its arms and fingers, showing it bore robust flight feathers and was probably a better flier than *Archaeopteryx*. One of its most bird-like features was that the fibula, the tiny bone that runs parallel to the shin bone, no longer reached all the way down to the ankle like in *Archaeopteryx*, but tapered down into nothing. If you've ever eaten a chicken or turkey drumstick, you will find this tiny toothpick of a bone, and it does not reach down to or connect to the ankle but is embedded in the muscles of the leg.

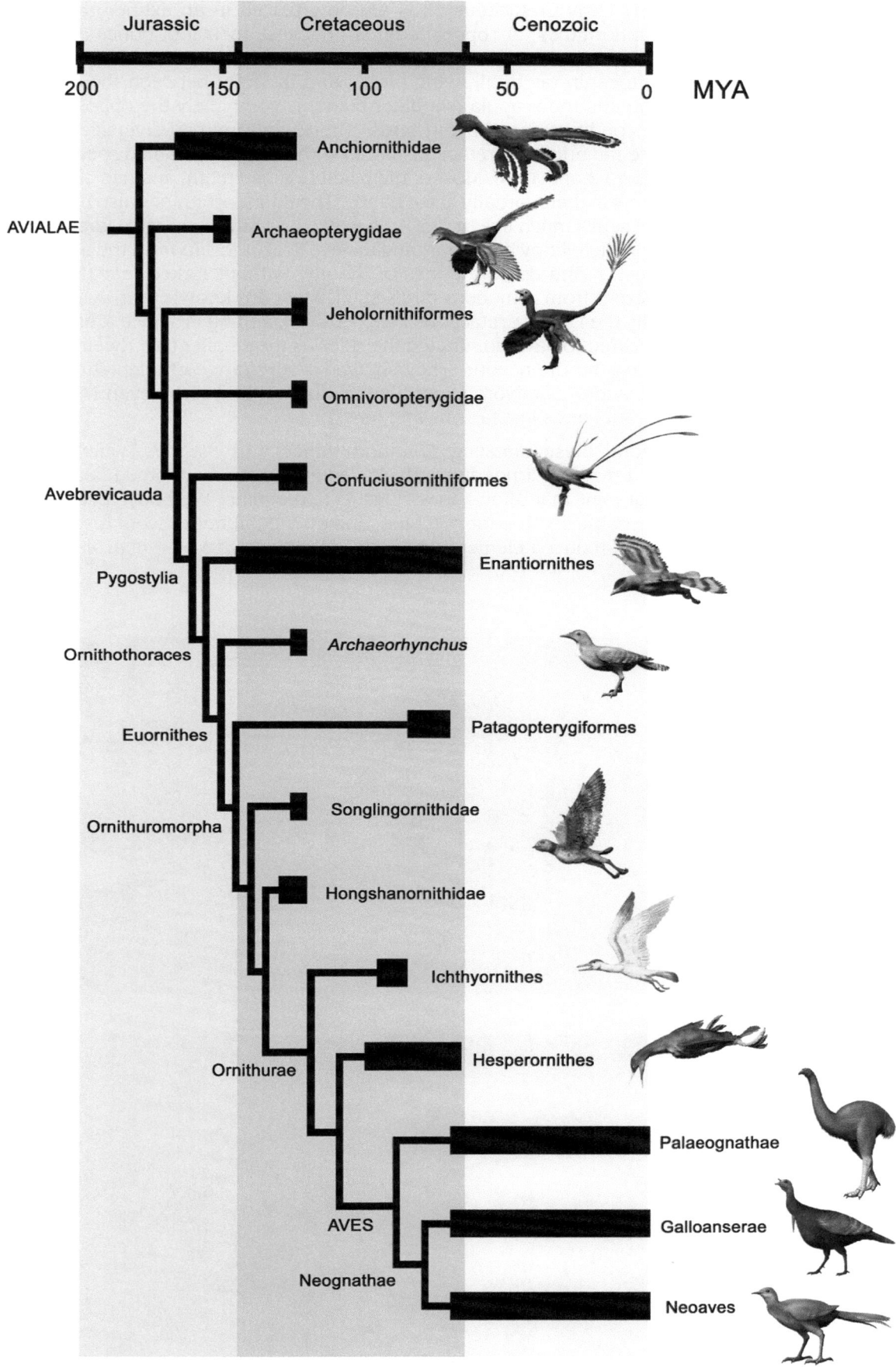

Figure 19.3 Family tree of Mesozoic birds, with reconstructions of their appearance at the branch tips.

Many Early Cretaceous birds are now placed in an extinct group, the Enantiornithes, or "opposite birds", because their foot bones (metatarsals and tarsals) fuse from the ankle down to the toes, rather than from the toes up (as in all living birds), and the joint between the coracoid bone and the scapula (shoulder blade) is completely the opposite from the condition in all living birds. The Enantiornithes diversified into a huge radiation of Cretaceous birds, with at least 80 named species (**Figures 19.3** and **19.4**). Most of them still retained teeth, and had claws and fingers in their partially fused hands. The Enantiornithines also had primitive skulls much like that of *Archaeopteryx*, with a simple quadrate bone, a complete bony bar separating the eye socket (orbit) from the antorbital fenestra, and dentary bones of the jaw without forked rear tips—very different from a modern bird's skull. Most are known from Asia (especially the Lower Cretaceous lake beds of Liaoning Province, China), and appeared to have dominated the skies as the smaller tree-dwelling birds during the Cretaceous. They showed a very range of adaptations such as waders, granivores, insectivores, fishers, and some even resembled raptorial birds like falcons and hawks.

Among these amazing Enantiornithinae was *Sinornis*, which had a hand made of long, unfused, clawed fingers, and a toothed beak, but a wrist joint that allows the wings to fold against the body, feet with an opposable first toe for perching, and all the tail bones were fused into a single, reduced element, the pygostyle (**Figure 19.3**). *Cathayornis*, from

Figure 19.4 Fossils of some Cretaceous birds. (A) *Confuciusornis*, (B) *Gansus*. (Courtesy Wikimedia Commons.)

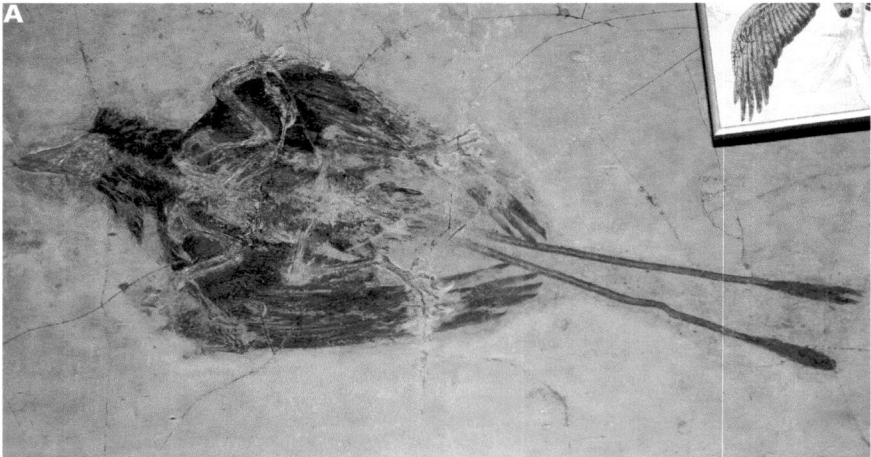

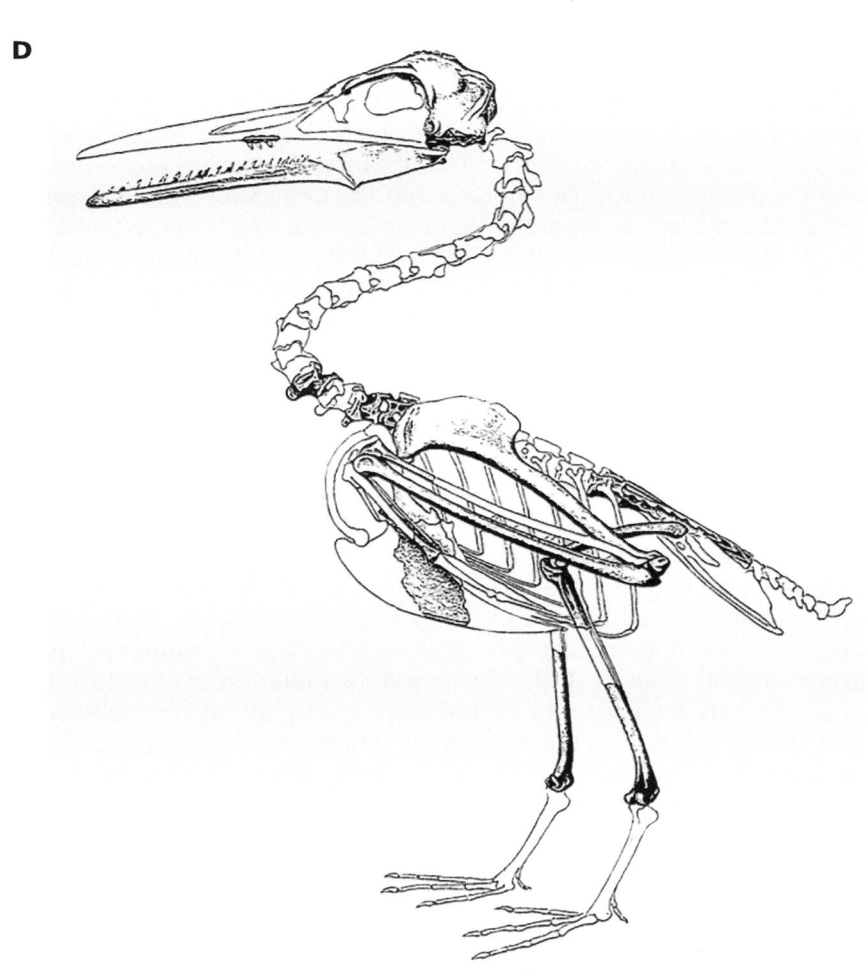

Figure 19.4 (Continued) (C) *Hesperornis*.
(d) *Ichthyornis*.

the Lower Cretaceous of Jiufotang, China, was like *Sinornis* in retaining teeth in its beak, and a primitive pelvis and hindlimb, but was a powerful flier. *Iberomesornis* from the Lower Cretaceous of Las Hoyas, Spain, was among the early birds with a pygostyle (reduced and fused tail bones), larger wings, and a well-developed breastbone, but still retaining an unfused hand and unfused ankle.

A second major clade of Mesozoic birds was the ornithuromorphs, which are Cretaceous relatives of the living bird radiation and their living descendants (**Figures 19.3** and **19.4[C,D]**). They are represented by *Ambiortus* from the Lower Cretaceous of Mongolia, with a fully modern wishbone, fused hand bones and reduced fingers, a keel on its breastbone for the flight muscles, and even feather impressions. *Gansus* from the Lower Cretaceous of Gansu Province, China, has a hind limb that suggests that it was an early swimming and diving bird (**Figure 19.4[B]**). *Chaoyangia* from the Lower Cretaceous of China is known primarily from a pelvis, but it is modern in many aspects, as are the vertebrae and ribs.

By the Late Cretaceous, a more advanced subgroup of the ornithuromorphs, known as the ornithurines, are represented from many good specimens of marine birds known from the chalk beds of Kansas, Texas, and Alabama. These include (**Figure 19.4[E,F]**) the loon-like *Hesperornis*, and the tern-like *Ichthyornis*, which were described by O.C. Marsh in 1880 in his famous monograph on toothed birds, *Odontornithes*. This monograph was originally published by the U.S. Geological Survey, and caused a scandal on the floor of Congress, when fundamentalist Congressman Hilary Herbert of Alabama was outraged that taxpayer dollars were spent on studying birds with teeth. To his mind, these were a biblical impossibility.

Because the Mesozoic bird fossil record is so scrappy (except for the Early Cretaceous of China), it is difficult to say how abruptly these archaic birds died out at the end of the Cretaceous. Nevertheless, the fact remains that few lineages survived into the Cenozoic. All of the enantiornithines vanished, as did most of the ornithurines. Some think that the ornithurines, which tended to be adapted to a semi-aquatic lifestyle, were able to survive the end-Cretaceous catastrophe by sheltering in water, while the songbird-like enantiornithines were more vulnerable in their treetop habitats.

THE CENOZOIC RADIATION OF AVES

Finally, we come to the earliest members of the living class Aves, or modern birds (**Figure 19.5**). There are many anatomical features in Aves that are not found in their ancestors, including the complete loss of the teeth, and the complete fusion of the foot and ankle bones to form the tarsometatarsus. In the Paleocene, there was a renewed radiation of modern bird families from a few survivors related to the ornithurines, so by the Eocene there was a huge diversity of birds, many from living groups. In this respect, the Paleocene bird radiation event resembles the enormous evolutionary radiation of mammals in the Paleocene, after the extinction of the non-avian dinosaurs cleared the way for large terrestrial vertebrates. In other words, the avian dinosaurs also underwent a huge evolutionary radiation after the non-avian dinosaurs died out.

The study of most fossil birds is a real challenge, because their relatively fragile thin-walled hollow bones break up very easily, so only

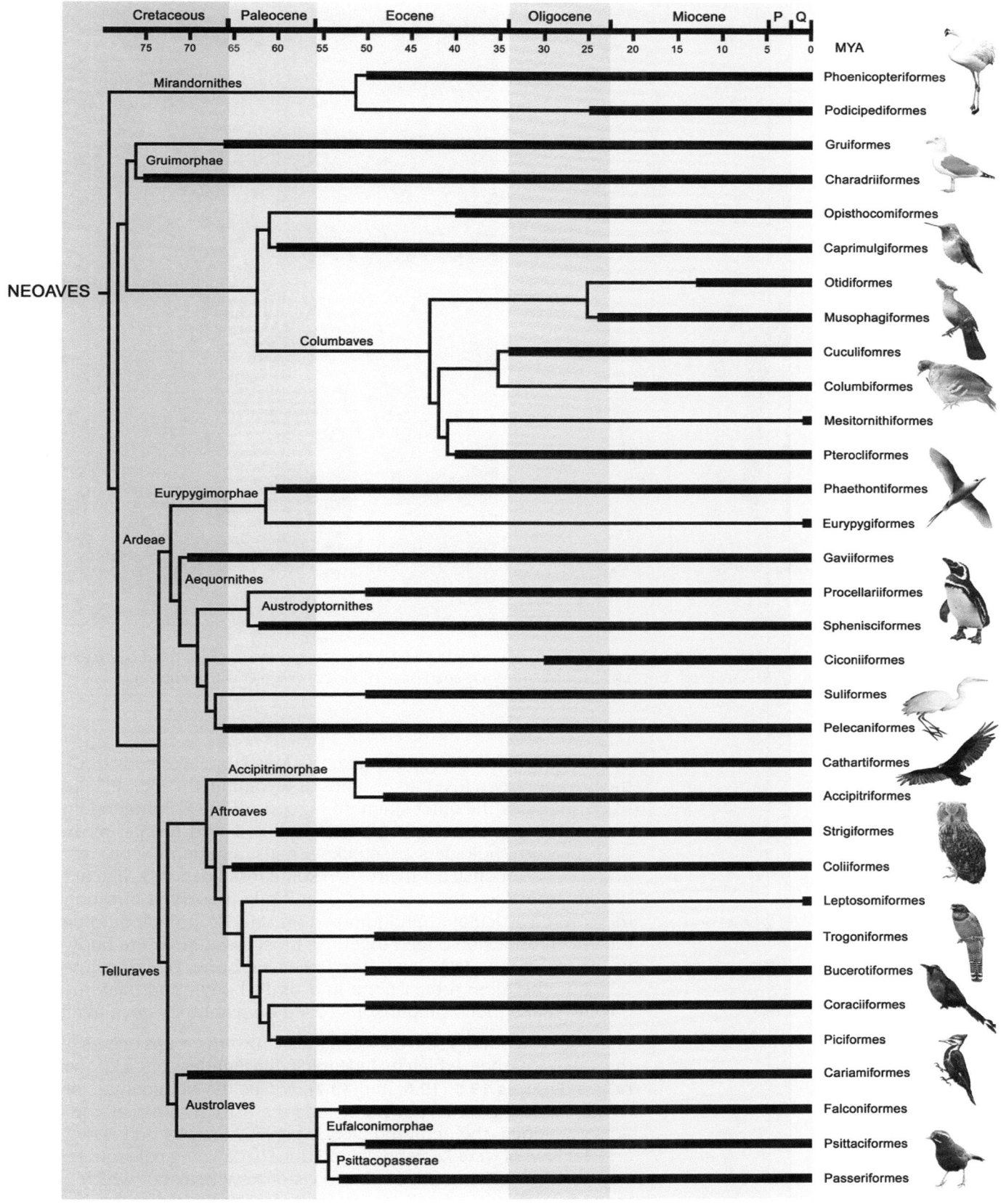

Figure 19.5 Phylogeny of Neoaves, showing the interrelationships of the modern bird families. [Modified from Kuhl et al. (2021).]

A

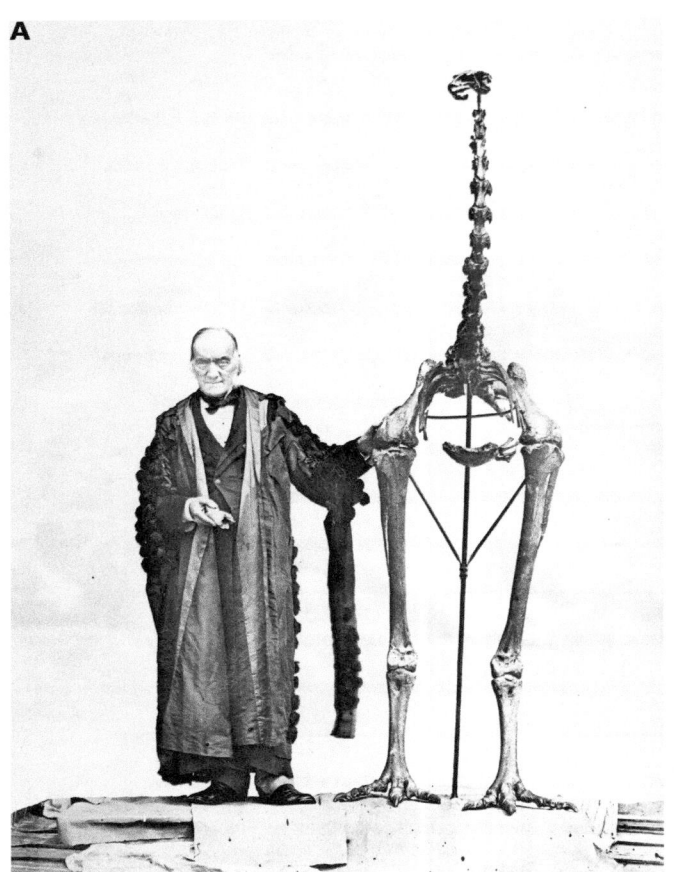

B

Figure 19.6 Some of the paleognath birds grew enormous on islands. Shown here are fossils of (A) the recently extinct moa from New Zealand, *Dinornis*; (B) *Aepyornis*, the "elephant bird" from Madagascar. (Courtesy Wikimedia Commons.)

extraordinary fossil deposits (like the Mesozoic lake beds of China, or the Eocene Green River lake shales of Wyoming, the Messel lake shales of Germany, or the quiet stagnant lagoons of the Solnhofen Limestone in Germany) preserve complete articulated bird fossils. Most of the time, paleornithologists are stuck with just a handful of broken bones, and have to reconstruct the rest. In some fossil deposits, like at La Brea tar pits, there are thousands of beautifully preserved bird bones, but they are disarticulated and jumbled around, so no paleontologist can tell which bones belonged together. Most fossil birds are known from just a few bones (especially the lower leg bone, or tarsometatarsus, which is the most robust bone in a bird's body), and paleontologists have to make their comparisons based on what they have available.

The first evolutionary split between the living bird groups occurs between the Palaeognathae, which include the living flightless birds known as the ratites (**Figures 19.5**, **19.6**, and **19.7**), versus the Neognathae, or all the remaining living birds. Ratites include the ostrich of Africa, the rhea of South America, the emu and cassowary of Australia and New Guinea, and the kiwi of New Zealand. In addition to the living paleognaths, there were extinct giants, such as the moas of New Zealand, which were up to 3.7 meters (12 feet) tall, and were hunted to extinction by the Maoris only 400 years ago (**Figure 19.6[A]**). The biggest birds of all, however, were the famous "elephant bird", *Aepyornis*, and its relatives *Vorombe* (the biggest of all), and the smaller *Mullerornis*, from the Pleistocene and Holocene of Madagascar. *Vorombe* weighed close to 450 kg (1000 pounds), stood almost 3 meters tall, and laid an egg almost 1 foot long

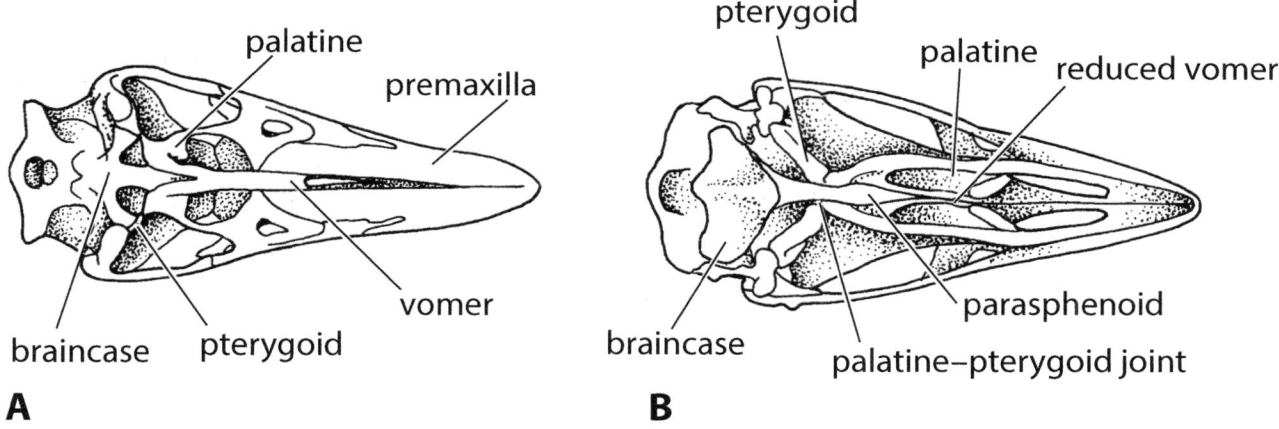

Figure 19.7 (A) Palaeognath vs. (B) neognath palates. Palaeognaths, like the ratites, have a solid bony palate made of broad plates of the premaxillary and palatine bones, with only small openings between the bones. Neognath palates, found in all living birds except ratites, are much more open with the flat plates of the premaxillaries, palatines, and pterygoids reduced to narrow struts and splints of bone. (Redrawn from several sources.)

with a 2-gallon capacity (**Figure 19.6[B]**). They, too, were hunted to extinction when humans first arrived on Madagascar.

Ratites are members of the most primitive group of living birds, the Palaeognathae, because they have a number of unique features besides their flightlessness (except for tinamous, which can still fly a bit). The palaeognathous palate (**Figure 19.7**) is distinctively different and more primitive compared to the palate seen in most modern birds (neognathous), but it does have a number of specializations in the palate showing that palaeognathous birds are a natural group, and not simply a wastebasket of large primitive flightless birds. Ratites have additional unique features that define them, especially in the configuration of the pelvis: the pubis and ischium are longer than the ilium, so these bones stick out beneath the tailbones, and there is no large fenestra between the ilium and ischium, which is a feature of the neognaths.

The striking occurrence of the living ratites on Gondwana continents today has led some authors to suggest that this pattern is an old relict distribution, implying that their divergence began before these continents rifted apart in the Late Cretaceous. However, more recent evidence shows that their close relatives, the lithornithids, are known from the Paleocene of Europe and Asia, and there are fossil ostriches from the Eocene of Germany and the Miocene of central Europe, and fossil ratites from the Paleocene of France, South America, Antarctica, and Australia. Apparently, ratites had a worldwide distribution during the early Cenozoic, so that only the surviving relicts happen to be confined to Gondwana continents.

NEOGNATH BIRDS

All the remaining living birds have the derived neognathous palate (**Figure 19.7**). They comprise an enormous radiation of some 9000 species, nearly all of which originated and diversified during the Cenozoic. The evolution and relationships of these living orders of birds is still highly controversial although there now seems to be a consensus regarding the molecular family tree of birds (**Figure 19.5**). A number of family trees of the living birds based on their anatomy were published over the years, but without any consensus. The first molecular approach to bird phylogeny in 1990 produced a completely different geometry of bird relationships using DNA hybridization, but these results have also been criticized, and now they are outdated as direct sequencing of the genome has replaced DNA hybridization. Even today there is some conflict between different DNA phylogenies, because many of the orders of birds apparently diverged so rapidly in the early Cenozoic that the molecular differences among them are slight.

Although there are some conflicting results, what the DNA consistently shows is that there is a clear split between an early group, the Galloanserae, consisting of the Galliformes (the various land fowl, including chickens, turkeys, grouse, quail, and their kin) and Anseriformes (ducks, geese, swans, screamers, and their kin) versus all the rest of the birds (**Figure 19.5**). The most primitive fossils of Galloanserae are a series of fragmentary specimens which cannot be assigned to any specific group of galliforms or anseriforms. The best studied and most complete fossil is the latest Cretaceous bird known as *Vegavis*, which came from Vega Island in Antarctica. *Vegavis* was apparently a long-legged bird with webbed feet like anseriform birds, but is really a primitive relative of both branches of Galloanserae. CT analysis of the voice box suggested that it could honk like a goose, but could not make more complex sounds. Other similar Late Cretaceous birds are *Australornis* from New Zealand, *Neogaeornis* from Chile, and *Polarornis* from Seymour Island in Antarctica, which have been grouped in the family Vegavidae. These birds were clearly diverse in the southern hemisphere before the end-Cretaceous extinctions, and then recovered and diversified into the Galloanserae in the Cenozoic.

By the middle Eocene, there are definite fossils that can be assigned to the Galliformes and the Anseriformes. The duck clade includes many fossils that clearly belong to living groups of ducks, geese, swans, screamers, and other waterfowl, but also some spectacular birds that don't look duck-like at all. One of the best known is the Eocene bird *Presbyornis*, known from dozens of complete skeletons from the lower to middle Eocene Green River lake beds of Wyoming and Utah (**Figures 19.8[A]** and **19.9[D]**). Its long neck and very long legs originally made paleontologists think it was a flamingo, but when the head with the duck-like bill was found it was clear it was not that bird at all; scientists are not sure to what group within the Anseriformes it belongs. Its bill is shaped much like that of dabbling ducks, so it apparently filtered small food from the lake waters. The large number of specimens suggested that it waded in huge colonies on the shores of the ancient Green River lake system. The bonebeds of these birds are deposited in such a way that it suggests large numbers of them died from botulism poisoning when lake waters became stagnant and toxic, as often happens to colony-nesting birds and shorebirds today.

One of the earliest branches of the Anseriformes in the early-middle Eocene of North America and Europe was a gigantic bird known as *Diatryma* (North America) or *Gastornis* (Eurasia) (**Figures 19.8[B]** and **19.9[A]**). Most paleornithologists consider the two so similar that they regard *Diatryma* as a junior name for *Gastornis*. These birds stood over 2 meters (6.5 feet) tall, and had a head with a huge, deep, sharp beak over 0.5 meters (1.5 feet) long. They may have weighed as much as 175 kg (385 pounds). Their wings were vestigial (as in most flightless birds), while they had robust hind limbs with robust claws. However, their thick limbs did not allow for much rapid running, so some paleontologists thought they were ambush predators, who struck out of hiding. Their huge thick beaks had an enormous crushing

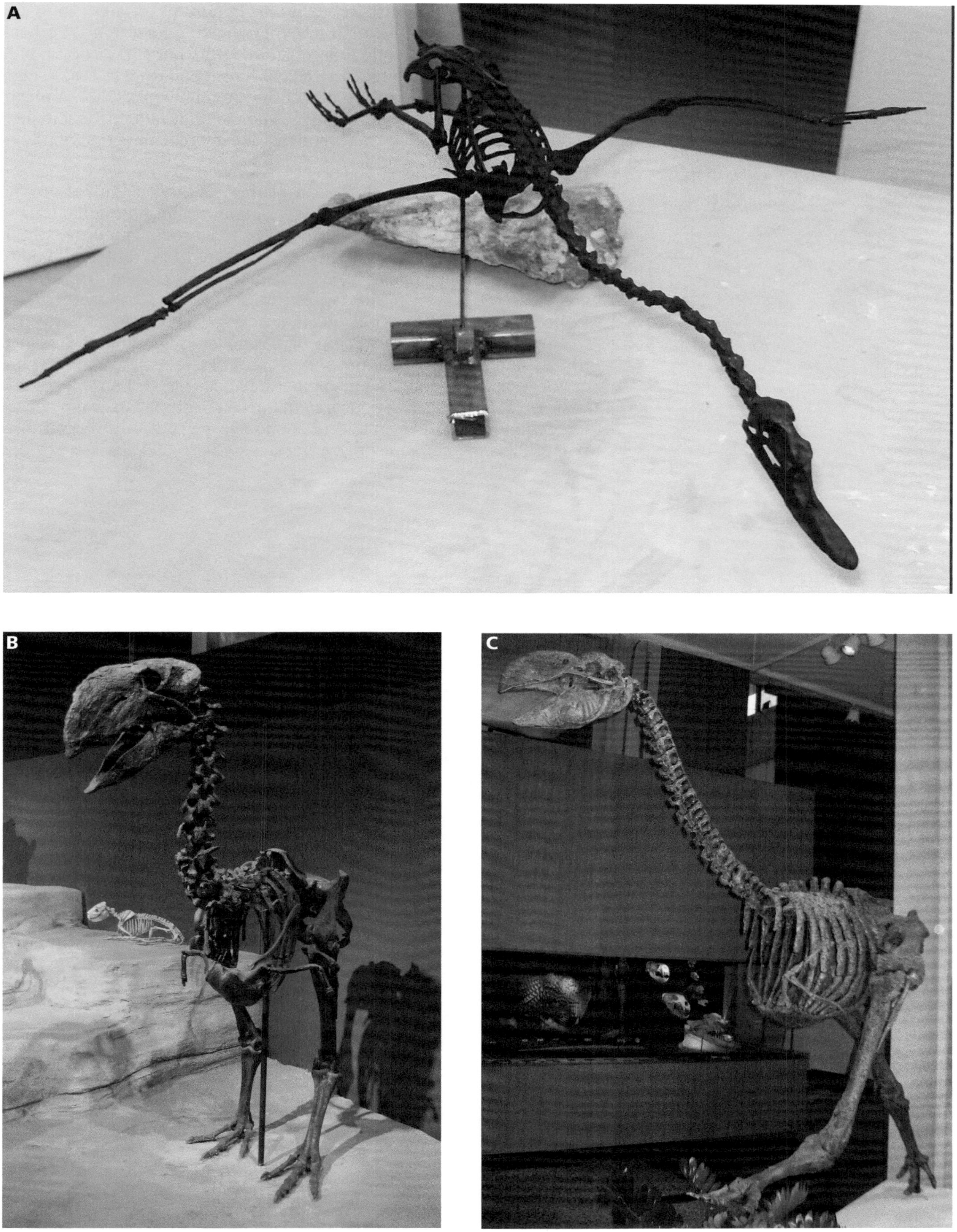

Figure 19.8 Fossil specimens of (A) *Presbyornis*. (B) *Gastornis*. (C) *Dromornis*. (Courtesy Wikimedia Commons.)

force, which some scientists thought was important for a crushing bite on their prey when they caught it. Others have argued that their beak did not possess a hooked tip that most predatory birds have (to rip prey apart), and thought the huge beak might have been used to crush seeds and nuts, suggesting a herbivorous or maybe omnivorous diet. A recent chemical analysis of their bones suggested that they had no meat in their diet at all, because their bone chemistry doesn't resemble that of predators like theropod dinosaurs.

Meanwhile, Australia had its own radiation of giant flightless ground birds, known as the Dromornithidae, or *mihirungs* in Aboriginal tongue (**Figures 19.8[C]** and **19.9[B]**). They have been nicknamed the "thunder birds" or "demon ducks". Formerly thought to be ratites related to ostriches, now they are classified as members of the Galloanserae, perhaps close to *Gastornis*. The largest ones, such as *Dromornis stirtoni*, reached 3 meters (10 feet) tall, and may have weighted up to 240 kg (530 lb). They had massing crushing beaks, suggesting an omnivorous diet. But they had hoof-like feet, not sharp claws of a predator, a huge stomach for fermenting food, so these features suggest they ate plants and seeds. Dromornithids first appeared in the Oligocene of Australia (where there is almost no pre-Oligocene Cenozoic record), and lasted until the late Pleistocene, vanishing around 50,000–20,000 years ago along

Figure 19.9 Reconstructions of a variety of extinct birds, including: (A) *Gastornis***, (B)** *Dromornis***, (C)** *Pelagornis***, (D)** *Presbyornis***, (E)** *Kumimanu***.**

with all the megafauna of Australia (possibly due to human hunting, although this is controversial).

Another interesting group is the Pelagornithidae, or "pseudo-toothed birds", which had jagged tooth-like edges in the horny keratin at the edge of their beaks, acting as true teeth would to grip slippery fish and squid with their bills. They were the dominant group of soaring seabirds in the entire Cenozoic, appearing the Paleocene and lasting until the early Pleistocene before vanishing, only to be replaced by marine soarers like albatrosses. The largest of these comes from Oligocene beds in South Carolina, and is known as *Pelagornis sandersi*. (Another species from Chile is almost as big.) It was shaped like a gigantic albatross, only twice as large as the living species, with a wing span of 7.4 meters (24 feet) (**Figure 19.9**). That is the longest wingspan ever known in the birds.

NEOAVES

All the rest of the remaining living non-galloanseran neognathous birds are placed in a group called the Neoaves (**Figure 19.5**). According to the consensus phylogeny of Hackett and others in 2008, and Prum and others in 2015, the next branch point within the Neoaves was a cluster of several well-supported supraordinal groups: a group including the hummingbirds, swifts, and nightjars; a group clustering rails, cranes, cuckoos, and bustards; a clade of shoebills, pelicans, herons, ibises, cormorants and frigatebirds, storks, penguins, albatrosses, and loons; and a number of unresolved groups within this large polytomy, including pigeons and doves, tropicbirds, hoatzins, and a clade of flamingoes plus grebes (this last grouping was long contentious, but seems supported by the molecules). Finally, there were two large groups clustered at the top of the family tree. One group includes the gulls, plovers, sandpipers, snipes, and some other shorebirds and wading birds. The other clade includes the bulk of the familiar birds: the passerine birds (most of the smaller familiar birds, from sparrows and warblers to robins and flycatchers) plus (surprisingly) the parrots and then the falcons. The other subgroup of this larger group included the woodpeckers, kingfishers, hornbills, hoopoes, trogons, owls, mousebirds, and a group consisting of hawks, eagles, and New World vultures.

These comprehensive efforts to use DNA to decipher bird relationships by Hackett and others, and by Prum and others, has not yet been fully assimilated and critiqued by the avian research community, although there are criticisms of specific parts of it, and different topologies of key groups emerging from other molecular studies. It is not possible in a book like this to review this in detail, especially since the arrangement of the bird groups is still being debated. Nor is there space to discuss the fossil record of 31 orders and about 9000 species of neoavian birds. Instead, let us look at a few unusual and spectacular examples of Neoaves in the fossil record.

One remarkable group of birds are the penguins, the only birds that not only are flightless but also have turned their wings into flippers for swimming. Some of them get really large, such as the King Penguin, and the Emperor Penguin, which reaches 1.2 meters (4 feet) in height, and weighs up to 45 kg (100 pounds). But they are dwarfed by the gigantic penguin *Kumimanu* from the Paleocene of New Zealand (**Figure 19.9[E]**). Its name comes from the Maori for "monster bird". *Kumimanu* was up to 1.8 meters (6 feet) tall, and weighed about 91 kg (200 pounds), yet it was already a true penguin committed to diving and swimming with its wings. New Zealand in the Paleocene was much warmer and milder than it is today, since there was no Antarctic ice cap, and warm tropical waters flowed down past the southern continents. The waters were rich in fish and sea turtles, and *Kumimanu* must have been able to capture much larger prey than any living penguin. Over two dozen species of fossil penguin are now known, and they have a long history of diversifying in the Paleocene and Eocene around the southern continents.

TERROR FROM THE SKIES

We discussed the flying birds with the largest wingspan, the albatross-like *Pelagornis* (**Figure 19.9[C]**), but a close second to them was the gigantic condor-like bird *Argentavis* (**Figure 19.10[A]**). Image a bird the size of a small airplane soaring above you, getting ready to dive down and attack from above. That bird was *Argentavis magnificens*, the "magnificent Argentine bird", and it was probably the heaviest bird ever to fly, with one of the largest wingspans. It was originally discovered in the 1970s by Argentine paleontologists Rosendo Pascual and Eduardo Tonni, working in the Miocene badlands in the eastern foothills of the Andes. Its upper arm bone (humerus) was longer than the entire arm of man! It was formally published in 1980, and it was a shock to imagine a flying bird this large. Most estimates suggest that it had a wingspan of 7 meters (23 feet), a body length of 1.26 meters (4.1 feet), and weighed about 72 kg (160 lb). Compare this to the largest living flying bird, the wandering albatross, which has a wingspan of only 3.6 meters (12 feet).

Figure 19.10 Reconstructions of some extinct birds, including: (A) *Argentavis*, (B) *Teratornis*, (C) *Paraphysornis*, (D) *Brontornis*, (E) *Kelenken*.

Since *Argentavis* was a vulture-like bird, it is best to compare to the largest flying land bird, the Andean condor, which soars above the beds yielding *Argentavis* even today. The Andean condor has a wingspan of 3.2 meters (10 feet) and weighs 15 kg (33 lb), only about a fifth the size of *Argentavis*. The only heavier flying birds today are bustards, which weigh as much as 21 kg (46 lb)—still less than a third the size of *Argentavis*.

Many scientists have speculated about the aerodynamics of a flying animal this large. Although it is not beyond the physical limits for a flying creature (after all, some pterosaurs were larger), it is near the limits for birds. Its broad wings compared to seabirds would have made its wing loading large enough for powered flight although it probably flew like most eagles and vultures and condors, using thermal currents and updrafts in the mountains to soar for miles without flapping their wings. Some have suggested it would have needed a headwind to get off the ground, although its powerful legs could also have given it a running or jumping start. But its long wings would not have been able to flap while they were standing, and it would have needed to get a good take-off before it could flap a full downstroke of its wings.

The size of this bird also suggests other things about its paleobiology. Birds that are this big must lay large eggs (probably weighing almost a kg), and they tend to have only one or two per clutch each year. Birds this large develop slowly, so they would have not have been fledged and become independent until about 16 months, and not fully mature until they were 12 years old. Such large birds do not fear predators, so they have a slow reproductive rate but a high survival rate and long lifespan, usually dying from accidents, disease, or old age, rather than from larger predators. On the basis of the age spans of living birds, *Argentavis* would have lived 50–100 years barring accidents, but lived in small numbers mostly in the mountainous regions of the Andes where the thermals provide lift.

They probably soared above huge territories (at least 500 square km) to find enough food for such a large body. *Argentavis* is neither an eagle nor a vulture or condor, but a member of an extinct group of birds called teratorns, which have features of both eagles and condors. Since they are extinct, we don't know exactly what *Argentavis* ate, but with their size they would have eaten a lot of carrion as condors and vultures do, driving off the predators from their kill when necessary. Their eagle-like beaks and their powerful legs show that they certainly were aerial predators as well, being able to grab and kill smaller prey in their talons as eagles and hawks do. Their large skulls have structures that suggest they usually ate their prey whole, gulping it down without tearing it apart as eagles and hawks do.

One of the last of the teratorn species was the biggest bird from the La Brea tar pits, *Teratornis merriami* (**Figure 19.10[B]**). It had a wingspan up to 3.8 meters (12.5 feet), with a wing area of about 17.5 square meters, and weighed about 15 kg (33 lb), so it was a third again larger than the largest living flying bird today, the Andean condor, and twice the size of the modern California condor. As the largest bird at La Brea, its powerful bill could have ripped holes in large carcasses and opened them up for smaller scavengers to reach the edible parts. *Teratornis* was part of an enormous fauna of predatory and scavenging birds at La Brea that fed upon animals stuck in the tar, which made it a deathtrap for birds of prey and carrion feeders. In addition to the teratorns, there are large ancestors of the California condor, and 23 other species of predatory and scavenging birds, including relatives of the black vulture, Egyptian vulture, American vulture, the caracaras, turkey vultures, rough-legged buzzards, four species of eagles, 15 other species of hawks, falcons, and kites, and 10 species of owl, out of about 140 species of birds.

TERROR BIRDS

In some places, the avian descendants of the dinosaurs did not completely yield the role of large terrestrial predators to carnivorous mammals. Whenever there were no large mammalian predators during the Cenozoic, birds often occupied that role. As discussed already, the early-middle Eocene of North America and Europe was ruled by the huge *Gastornis*, which might have been a predator. If it was a predator, it had no mammalian carnivores larger than a dog to compete with in the middle Eocene.

In the Cenozoic of South America, there was an entire family of giant predatory birds, the phorusrhacids, or "terror birds", which also had large sharp beaks and stood 2 to 3 meters tall (**Figures 19.10[C–E]** and **19.11[A–C]**). They had huge skulls with long, deep, but narrow beaks with a hooked tip, ideal for catching mammalian prey and ripping it open. Their heads were proportionally the largest heads ever to evolve in a bird, which makes sense if they were taking large prey and only had their huge beaks and powerful feet to attack with. They had long necks with an S-shaped bend, that could flex and strike out during a lunge after prey.

These birds had tiny vestigial wings that could not have lifted them in flight, but they had no need to fly with their fast-running long hind legs. Like the ratites, phorusrhacids surrendered the advantages of flight in order to get larger and focus on running. Their feet were powerful and robust with sharp claws, so they undoubtedly used them to kick and slash at prey, and pin the prey animal down as they ripped it open with their hooked beaks.

We can get a sense of how phorusrhacids might have lived and fed by watching their closest living relatives, a South American ground bird known as the seriema. Seriemas act much like the reptile-hunting secretary birds of Africa. Both have long powerful legs with a fringe of feathers so they can kick and slash at lizards and venomous snakes without fearing getting bitten. Both birds grab prey with their powerful hind legs, then tear it apart with their beaks, or grab a snake or lizard in their mouth, crush it, or shake it to stun it or break its neck or back. Then they slam it to the ground before jumping on it and slashing it with their beaks. But some phorusrhacids were large enough (over 3 meters or 10 feet tall) that they would not have settled for just small mammals and reptiles, but many could attack sheep-sized or larger mammals as well.

Phorusrhacids were a very diverse group through the entire Cenozoic in South America, with at least 18 species in 5 different subfamilies. They first appeared with the meter-tall *Paleopsilopterus* from the middle Paleocene (60 Ma), and increased in diversity through the Eocene and Oligocene. By the middle Miocene, all five subfamilies had appeared, and they came in a variety of sizes and shapes. Some were huge and heavy boned (the Brontornithinae), yet reached up to 2.8 meters (9.2 feet) in height, and weighed as much as 400 kg (880 pounds) (**Figure 19.10[D]**). On the other extreme, the subfamily Phorusrhacinae were taller and more slender, reaching 3.3 meters (10 feet) in height, but not weighing nearly as much. The Patagornithinae were of medium height (1.7 meters or 5.6 feet tall), with a slender build and longer legs suggesting they were nimble specialized runners. The Mesembriornithinae were relatively small phorusrhacids, reaching only 1.5 meters (5 feet) in height.

A

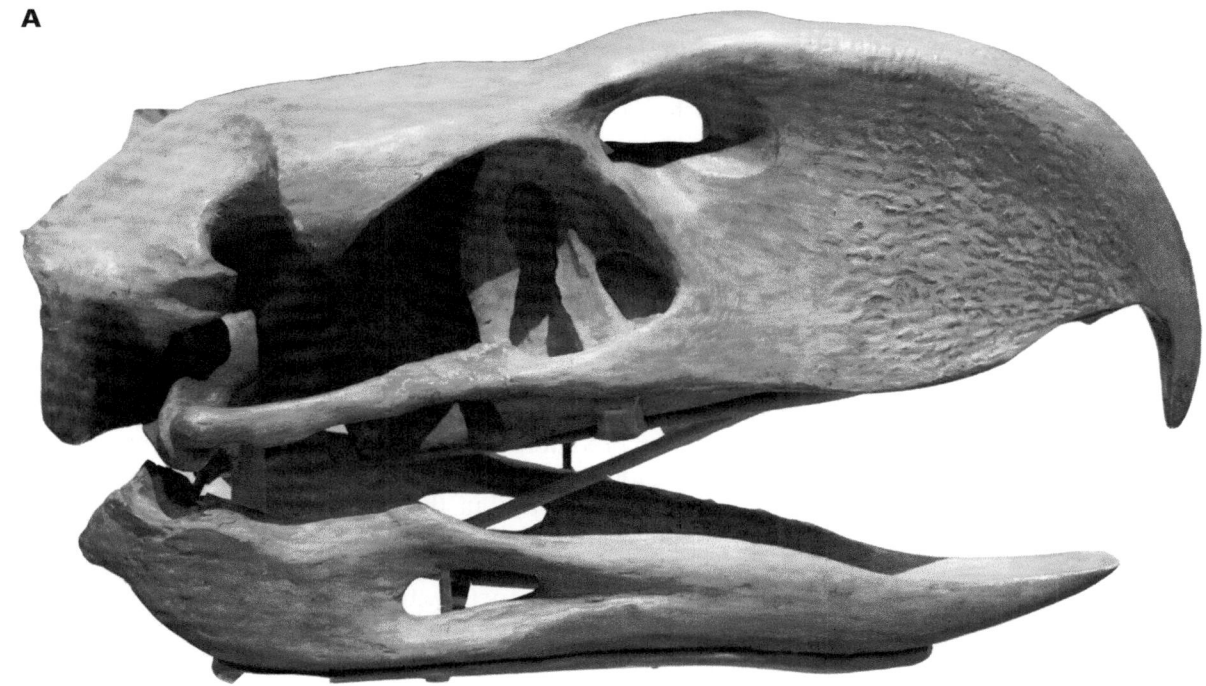

Figure 19.11 Fossils of some different phorhusrhacids. (A) The skull of *Phorhusrhacos*, with its hooked beak and powerful jaws. (B) The skeleton of *Kelenken*, a larger but more lightly built phorhusrhacid. (C) The skeleton of *Titanis*, the phorhusrhacid that reached North America in the Pliocene. (Courtesy Wikimedia Commons.)

Figure 19.11 (Continued)

Finally, the primitive Psilopterinae were the lineage from which the rest evolved, and never exceeded 1 meters (3.3 feet) in height, even though they ranged from the Paleocene to the early Pliocene (60 to 3 Ma), the entire history of the group.

The biggest of all the phorusrhacids was *Kelenken* (**Figures 19.10[E] and 19.11[B]**), from the middle Miocene of Patagonia. Discovered in 2006, it had the largest skull of any bird ever known. The skull was over 71 cm (28 inches) long, of which the beak portion was 48 cm (18 inches) all by itself! However, instead of the massive deep skull of some phorusrhacids, *Kelenken* had a relatively long narrow beak with a hooked end that resembles an eagle's beak. It reached over 3.3 meters (10 feet) in height, and with its long legs, it is estimated that it could run 48 km/h (30 mph).

Thus, the top predators in South America for most of the last 65 million years were not the dog-like and hyaena-like opossums, but huge terror birds that could kill all but the largest mammals. Near the lakes and rivers, small animals had to contend with huge alligators, and snakes as long as a bus, but on the dry land, the terror birds were the kings. No mammal could compete with them, and most had to fear them and hide from them or run from them—which was difficult, since some phorusrhacids could run faster than any animal at the time. The dinosaurs had not relinquished the role of top predators in South America. The phorusrhacid birds, their direct descendants, kept their spot at the top of the food chain.

Phorusrhacids ruled South America unchallenged from 60 Ma until only 3 Ma, then they rapidly began to vanish in the middle Pliocene. The reason seems apparent: this was the time when large mammalian predators from North America, such as saber-toothed cats, jaguars, cougars, wild dogs, bears, and many other advanced predators, came down across the Panama land bridge. The last of the phorusrhacids was found in the late Ice Age deposits of Uruguay, dying out with the last of the Ice Age megamammals.

But one phorusrhacid managed to buck the trend of being overwhelmed by Northern invaders. Known as *Titanis walleri* (**Figure 19.11[C]**), it managed to walk north up through Central America, showing up in Texas and Florida in deposits about 5 Ma, much earlier than most of the other creatures that walked across Panama land bridge during the Pliocene "Great American Interchange". Like the other phorusrhacids, it was a huge predator that could hold its own against most mammalian carnivores. The biggest specimens were 2.5 meters (8.2 feet) tall, and weighed 150 kg (330 pounds), so they were no bird to mess with! It was also very fast; its speed has been estimated at 65 km/h (40 mph), faster than nearly all mammals of that time. However, it was different from many phorusrhacids that stayed behind in South America in having a shorter, thicker neck, bulkier head, and overall a heavier build than the faster-running phorusrhacids that remained in South America. For a long time, it was thought to have died out at the end of the last Ice Age with the extinction of the megamammals, but more recent redating of the specimens show that *Titanis* only survived until the beginning of the Ice Ages, about 2 Ma. Still, it managed to hold its own in the Gulf Coast for 3 million years, fighting back challenges from saber-toothed cats and dogs and bears that eventually wiped out all its southern relatives.

This is just a small sampling of the incredible fossil record of birds, especially among the 9000 species of Neoaves that have evolved in the Cenozoic. They may be the last survivors of the great radiation of dinosaurs, but in many ways, they still dominate the planet. There are more species of birds than there are of mammals, or reptiles or amphibians, and they are almost as diverse as the great radiation of teleost fish in the Cretaceous and Cenozoic. Our mammalian chauvinism declares the Cenozoic to be the "Age of Mammals" but in reality, the birds still rule the land and the air in terms of total diversity, so we are still in an "Age of Dinosaurs".

FURTHER READING

Brown, J.W.; Rest, J.S.; Garcia-Moreno, M.D.; Sorenson, J.; Mindell, D.P. 2008. Strong mitochondrial DNA support for a Cretaceous origin of modern avian lineages. *BMC Biology.* 6: 6.

Chiappe, L.M. 1995. The first 85 million years of bird evolution. *Nature.* 378: 349–355.

Chiappe, L.M. 2007. *Glorified Dinosaurs: The Origin and Early Evolution of the Birds.* Wiley-Liss, New York.

Cracraft, J. 1973. Continental drift, paleoclimatology, and the evolution and biogeography of birds. *Journal of Zoology (London).* 169: 455–545.

Cracraft, J. 1974. Phylogeny and evolution of the ratite birds. *Ibis.* 116: 494–521.

Cracraft, J. 1985. The origin and early diversification of birds. *Paleobiology.* 12: 383–389.

Cracraft, J. 1988. The major clades of birds, pp. 339–361. In Benton, M. J., ed. *The Phylogeny and Classification of the Tetrapods, vol. 1: Amphibians, Reptiles, Birds.* Clarendon Press, Oxford.

Dyke, G.; Kaiser, G., eds. 2011. *Living Dinosaurs: The Evolutionary History of Modern Birds.* Wiley-Blackwell, London.

Feduccia, A. 1994. Tertiary bird history: Notes and comments, pp. 178–188. In Prothero, D. R.; Schoch, R.M., eds. *Major Features of Vertebrate Evolution.* Paleontological Society Short Course 7. Paleontological Society, Lawrence, KS.

Feduccia, A. 1995. Explosive radiation of Tertiary birds and mammals. *Science.* 267: 637–638.

Feduccia, A. 1999. *The Origin and Evolution of Birds* (2nd ed.). Yale University Press, New Haven.

Gauthier, J.A. 1986. Saurischian monophyly and the origin of birds. *Memoirs of the California Academy of Sciences.* 8: 1–55.

Gauthier, J.A.; Padian, K. 1985. Phylogenetic, functional, and aerodynamic analyses of the origin of birds, pp. 185–197. In Hecht, M. K., J. H. Ostrom, G. Viohl, and P. Wellnhofer, eds. *The Beginnings of Birds.* Freunde des Juras-Museums, Eichstätt.

Hecht, M.K.; Ostrom, J.A.; Viohl, G.; Wellnhofer, P., eds. *The Beginnings of Birds.* Freunde des Juras-Museums, Eichstätt.

Hou, L.; Zhou, Z.; Martin, L.D.; Feduccia, A. 1995. A beaked bird from the Jurassic of China. *Nature.* 377: 616–618.

Houde, P. 1988. Palaeognathous birds from the early Tertiary of the Northern Hemisphere. *Publications of the Nuttall Ornithological Club.* 22: 1–148.

Ksepka, D.T., Clarke, J.A.; Grande, L. 2011. Stem parrots (Aves, Halcyornithidae) from the Green River Formation and a combined phylogeny of Pan-Psittaciformes. *Journal of Paleontology.* 85: 835–852.

Kuhl, H., Frankl-Vilches, C.; Bakker, A.; Mayr, G.; Nikolaus, G.; Boerno, S. T.; Klages, B.; Timmermann, B.; Gahr, G. 2021. An unbiased molecular approach using 3′-UTRs resolves the avian family-level tree of life. *Molecular Biology and Evolution.* 38 (10): 108–127.

Long, J.A. 2008. *Feathered Dinosaurs: The Origin of Birds.* Oxford University Press, New York.

Marsh, O.C. 1880. Odontornithes: A monograph on the extinct toothed birds of North America. *Report on the Geological Exploration of the Fortieth Parallel.* 7: 1–201.

Marshall, L.G. 1994. The terror birds of South America. *Scientific American.* 270 (2): 90–95.

Martin, L.D. 1991. Mesozoic birds and the origin of birds, pp. 485–540. In Schultze, H.-P.; Trueb, L., eds. *Origins of the Higher Groups of Tetrapods.* Comstock Publishing Company, Ithaca, NY.

Martin, L.D. 1995. The Enantiornithes: Terrestrial birds of the Cretaceous in avian evolution. *Courier Forschunginstitut Senckenberg.* 181: 23–36.

O'Connor, J., Chiappe, L.M.; Bell, A. 2011. Pre-modern birds: Avian divergences in the Mesozoic, pp. 39–116. In Dyke, G.; Kaiser, G., eds. *Living Dinosaurs: The Evolutionary History of Modern Birds.* Wiley-Blackwell, London.

Olson, S.L. 1985. The fossil record of birds. *Avian Biology.* 8: 79–252.

Ostrom, J.H. 1973. The ancestry of birds. *Nature.* 242: 136.

Ostrom, J.H. 1975. The origin of birds. *Annual Reviews of Earth and Planetary Sciences.* 3: 55–77.

Ostrom, J.H. 1976. *Archaeopteryx* and the origin of birds. *Biological Journal of the Linnean Society (London).* 8: 91–182.

Ostrom, J.H. 1979. Bird flight: How did it begin? *American Scientist* 67: 46–56.

Ostrom, J.H. 1991. The question of the origin of birds, pp. 467–484. In H.-P. Schultze, Trueb, L., eds. *Origins of the Higher Groups of Tetrapods.* Comstock Publishing Company, Ithaca, NY.

Ostrom, J.H. 1994. On the origin of birds and of avian flight, pp. 160–177. In Prothero, D.R.; Schoch, R.M., eds. *Major Features of Vertebrate Evolution.* Paleontological Society Short Course 7. Paleontological Society, Lawrence, KS.

Pickrell, J.; Currie, P. 2014. *Flying Dinosaurs: How Fearsome Reptiles Became Birds.* Columbia University Press, New York.

Reilly, J. 2019. *The Ascent of Birds: How Modern Science Is Revealing Their Story.* Pelagic Publishing, New York.

Sanz, J.L., Bonaparte, J.F.; Lacasa, A. 1988. Unusual Early Cretaceous birds from Spain. *Nature*. 331: 433–435.

Sanz, J.L.; Buscalioni, A.D. 1992. A new bird from the early Cretaceous of Las Hoyas, Spain, and the early radiation of birds. *Palaeontology*. 35: 829–845.

Sanz, J.L., Chiappe, J.M.; Buscalioni, A.D. 1995. The osteology of *Concornis lacustris* (Aves: Enantiornithes) from the Lower Cretaceous of Spain and a re-examination of its phylogenetic significance. *American Museum Novitates*. 3133: 1–23.

Sereno, P.C.; Chenggang, R. 1992. Early evolution of avian flight and perching: New evidence from the Lower Cretaceous of China. *Nature*. 255: 845–848.

Sibley, C.G.; Ahlquist, J.E. 1990. *Phylogeny and Classification of Birds*. Yale University Press, New Haven, CT.

Walker, C. A. 1981. New subclass of birds from the Cretaceous of South America. *Nature*. 292: 51–52.

Witmer, L.M. 1991. Perspectives on avian origins, pp. 427–466. In Schultze, H.-P.; Trueb, L., eds. *Origins of the Higher Groups of Tetrapods*. Comstock Publishing Company, Ithaca, New York.

Witmer, L.M.; Rose, K.D. 1991. Biomechanics of the jaw apparatus of the gigantic Eocene bird *Diatryma*: Implications for diet and mode of life. *Paleobiology*. 17: 95–120.

Xu, X.; You, H.; Du, K.; Han, F. 2011. An *Archaeopteryx*-like theropod from China and the origin of Avialae. *Nature*. 475: 465–470.

Zhou, Z.; Jin, F.; Zhang, J. 1992. Preliminary report on a Mesozoic bird from Liaoning, China. *Chinese Science Bulletin*. 37: 1365–1368.

SYNAPSIDS

THE ORIGIN OF MAMMALS

20

Of all the great transitions between major structural grades within vertebrates, the transition from basal amniotes to basal mammals is represented by the most complete and continuous fossil record, extending from the Middle Pennsylvanian to the Late Triassic and spanning some 75 to 100 million years.

—James Hopson, "Synapsid Evolution and the Radiation of Non-Eutherian Mammals", 1994

THE ORIGIN OF MAMMALS

Mammals dominate most of the terrestrial habitats on earth today, and include the largest animals in the sea (whales and dolphins). Mammals such as bats (plus a number of different gliding mammals) also have taken to the air. Ever since the end of the Cretaceous, mammals have ruled all the large-animal roles once occupied by the non-avian dinosaurs. For a long time, the Cenozoic has been called "the Age of Mammals", even though we know that birds, teleost fishes, and especially insects were much more diverse in the Cenozoic. And now one species of mammal, *Homo sapiens*, has overwhelmed the planet and driven thousands of species to extinction and is destroying the environment for all life.

Where did mammals come from? The evolutionary sequence from the earliest mammal relatives to the appearance of the first true mammals is one of the most remarkable and complete in all the fossil record. At the very beginning of their history, amniotes split into two lineages, the synapsids and the reptiles (**Figures 8.1** and **20.1**). Traditionally, the earliest synapsids have been called the "mammal-like reptiles" but this is an obsolete term. The earliest synapsids had nothing to do with reptiles as the term is normally used (referring to the living reptiles and their extinct relatives). Early synapsids are "reptilian" only in the sense that they initially retained a lot of primitive amniote features. Furthermore, the earliest reptiles (*Hylonomus* and *Westlothiana* from the Early Carboniferous) and the earliest synapsids (*Protoclepsydrops* from the Early Carboniferous and *Archaeothyris* from the Middle Carboniferous—see **Figure 20.2**) are equally ancient, showing that their lineages diverged at the beginning of the Carboniferous, rather than synapsids evolving from the animals we call reptiles. For all these reasons, it is no longer appropriate to use the term "mammal-like reptiles". If one must use a nontaxonomic term, "protomammals" or "stem mammals" is an alternative with no misleading implications.

DOI: 10.1201/9781003128205-20

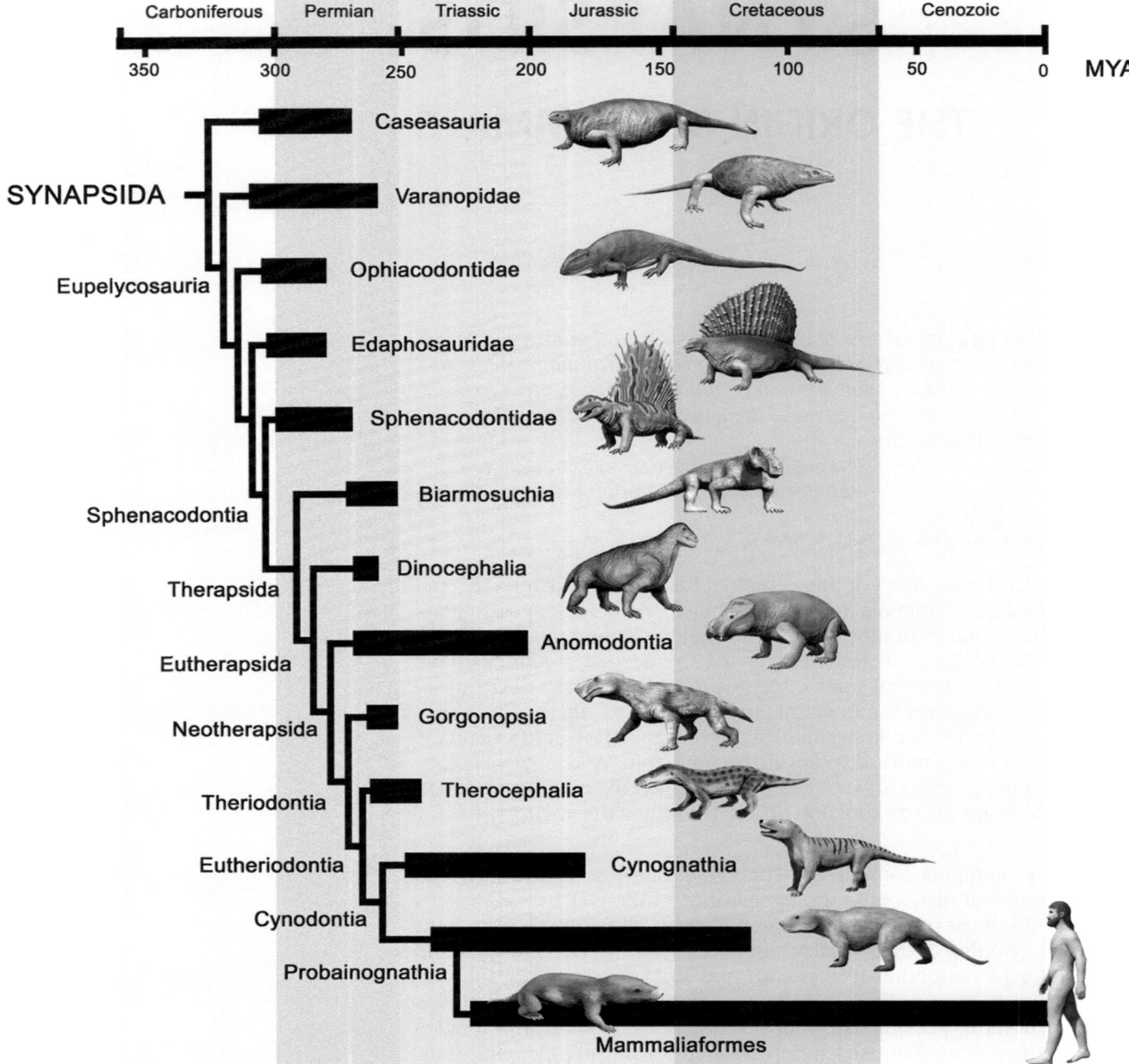

Figure 20.1 Family tree of synapsids and mammals.

TO BE A MAMMAL

From their origin in the early Late Carboniferous, an amazing array of synapsid fossils shows the transition from early amniote to mammal in remarkable detail. Yet though the fossil record is excellent, many features that distinguish mammals from reptiles do not fossilize (**Figure 20.3**). Some distinguishing features include:

1. Physiological characters—Mammals are usually diagnosed as homeothermic (having constant body temperature) endothermic amniotes with hair. They also have other features related to their high metabolism and active lifestyles, such as a four-chambered

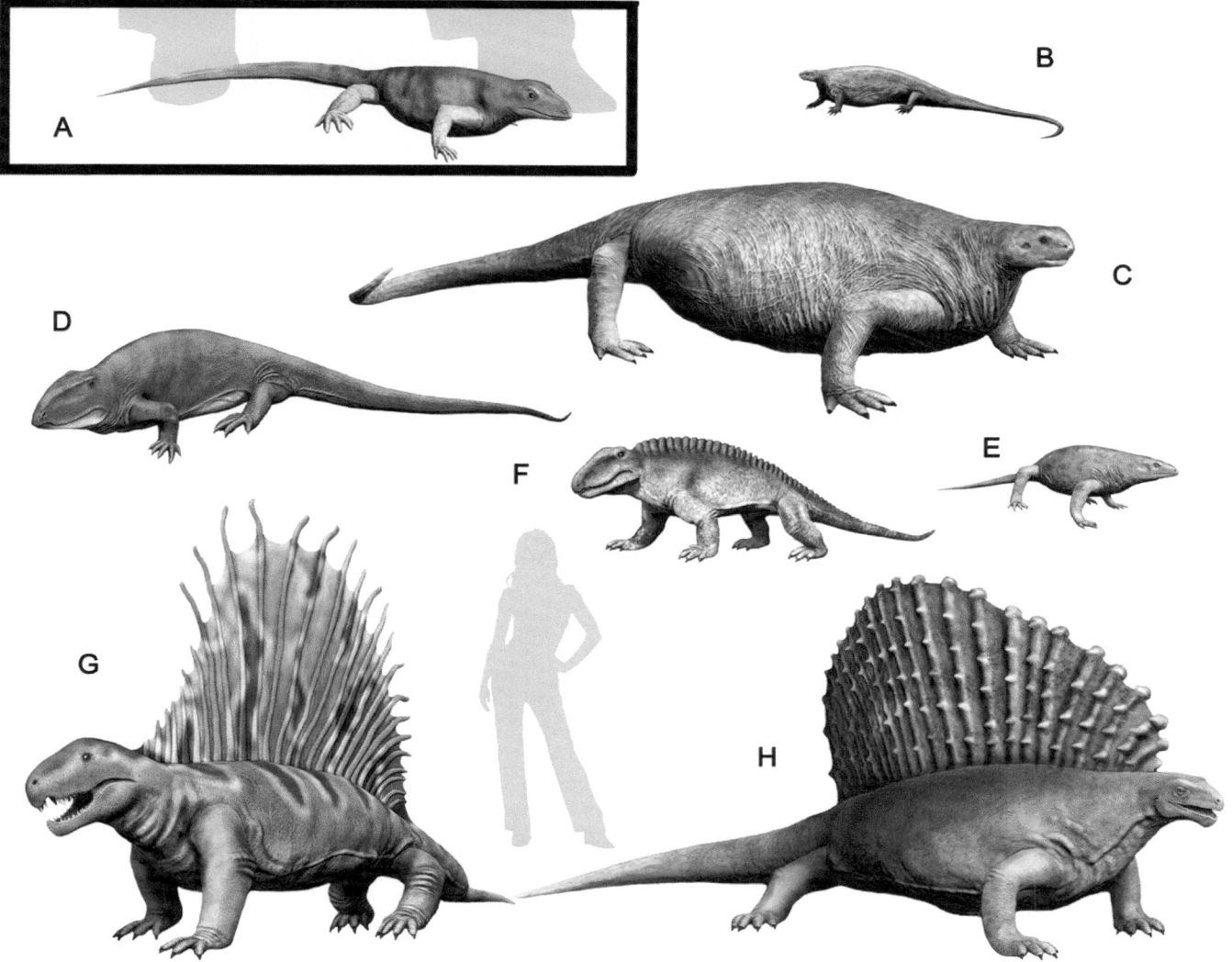

Figure 20.2 Reconstructions of some of the better-known primitive synapsids ("pelycosaurs"), including: (A) *Archaeothyris*, (B) *Casea*, (C) *Cotylorhynchus*, (D) *Ophiacodon*, (E) *Varanops*, (F) *Sphenacodon*, (G) *Dimetrodon*, (H) *Edaphosaurus*.

heart, a diaphragm for actively pumping air in and out of the lungs, and a sophisticated brain with an enlarged neocortex. Most of these characters do not preserve in fossils, especially in the skeleton. The internal molds of the brain cavity are known from many synapsids, so it is possible to determine when the enlargement of the neocortex occurs.

2. Reproductive characters—Another distinctive characteristic of mammals is their mode of reproduction. Most mammals (except the egg-laying platypus and echidna) give birth to live young, which the females then nurse with milk from their mammary glands. Instead of laying eggs and then abandoning them, most mammals invest a lot of parental care into each offspring, so that fewer are born, and they are born more helpless than hatchling reptiles or amphibians. Young mammals grow rapidly after birth, but their growth slows down to a terminal, adult growth stage (in contrast to most other animals, which grow continuously throughout their lives). The best way to detect a pattern of terminal growth in fossil bones is by the presence of bony caps (epiphyses) on the ends of the long bones of juveniles, indicating

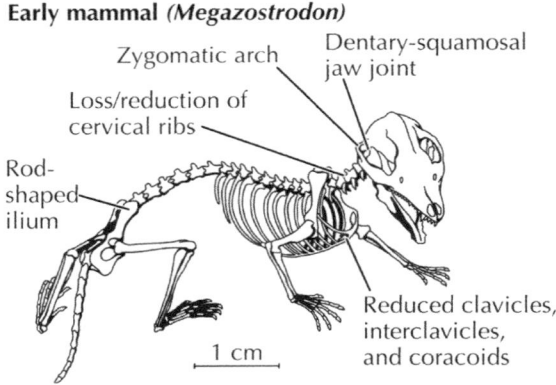

Early mammal (Megazostrodon)

Zygomatic arch

Dentary-squamosal jaw joint

Loss/reduction of cervical ribs

Rod-shaped ilium

Reduced clavicles, interclavicles, and coracoids

1 cm

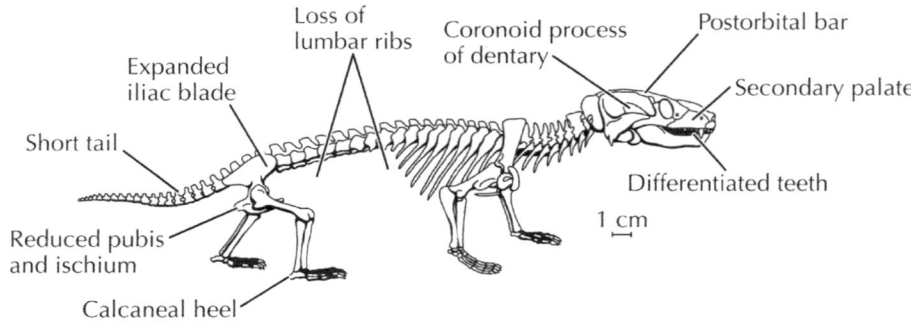

Cynodont therapsid (Thrinaxodon)

Loss of lumbar ribs

Coronoid process of dentary

Postorbital bar

Expanded iliac blade

Secondary palate

Short tail

Differentiated teeth

Reduced pubis and ischium

Calcaneal heel

1 cm

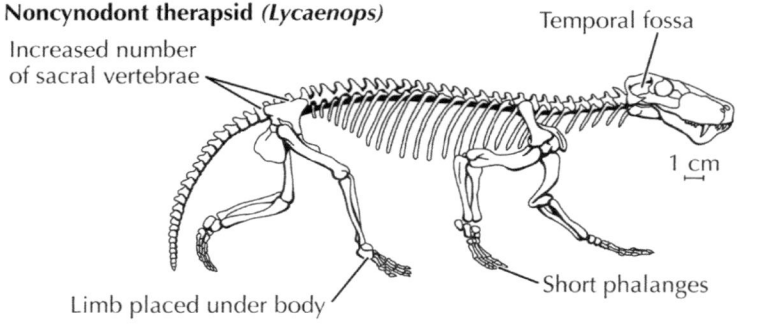

Noncynodont therapsid (Lycaenops)

Increased number of sacral vertebrae

Temporal fossa

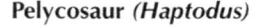

Limb placed under body

Short phalanges

1 cm

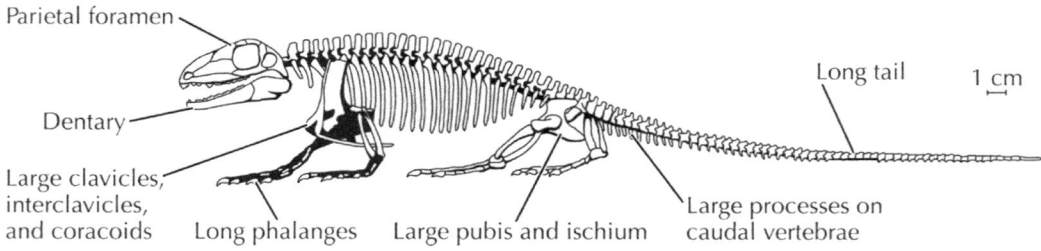

Pelycosaur (Haptodus)

Parietal foramen

Long tail

1 cm

Dentary

Large clavicles, interclavicles, and coracoids

Long phalanges

Large pubis and ischium

Large processes on caudal vertebrae

Figure 20.3 Changes in the synapsid lineage from a primitive "pelycosaur" like *Haptodus* through a more advanced therapsid (*Lycaenops*) to a small cynodont (*Thrinaxodon*) to the most primitive mammaliform (*Megazostrodon*). Through this evolutionary sequence, the limbs go from sprawling to completely underneath the body, and the shoulder girdle and hip girdle are modified to hold that posture; the reptilian bones of the shoulder girdle, such as the interclavicles and coracoids, reduce and eventually disappear; the rib cage gets shorter until the lumbar region of the back is completely without ribs, and the neck ribs (cervical ribs) also vanish; the skull becomes more open with a larger and larger temporal fenestra on the side, until it becomes a broad opening with no bone behind the eye socket, but a zygomatic arch in the cheeks instead; the dentary bones that carried the teeth in primitive synapsids grow backward and crowd out all the non-dentary jaw bones until the dentary touches the squamosal bone of the skull and develops a new hinge, replacing the old quadrate-incus part of the reptilian skull. In addition to these features, not shown are the development of the secondary palate in the mouth and the splitting of the hinge between the skull and first neck vertebra (occipital condyle) from a single ball joint below the neural canal to a double ball joint below and on each side of the nerve cord.

that the animals underwent rapid growth as a juvenile, and then stopped growing when these caps fused to the shaft of the bone. Unfortunately, the other reproductive features have a very low fossilization potential, although there are indirect means of detecting some of them.

For paleontologists, the transformation to mammals must be detected in skeletal features that have at least some fossilization potential. Most of these give indirect evidence for mammalian physiology and reproduction. For example, there are many modifications of the skull and jaws for chewing and eating food more rapidly and efficiently, which is required for an animal with high metabolism. The teeth in early synapsids are simple cones or pegs for catching and puncturing prey, but later in synapsid evolution, the teeth become differentiated into nipping incisors in front, a large stabbing canine on each side of the jaw to catch and hold the prey, and multi-cusped cheek teeth (premolars and molars) for chewing up the food. Reptiles replace their teeth continuously throughout their lives, but mammals replace their deciduous teeth ("baby teeth") only once, and the molars are never replaced. Many primitive amniotes have teeth on the palate and in the throat region for holding a struggling prey item, but mammals have teeth only on the margin of their jaws.

In reptiles, the nasal passage opens into the front of the mouth cavity, so that when a lizard or snake slowly swallows a large prey item, it must hold its breath while there is food in its mouth. Clearly, the high metabolism of mammals would not allow them to hold their breath for long while eating or chewing. For this reason, the bones of the upper jaw grow toward the midline and form a secondary palate that roofs over the original amniote palate, so that the internal nasal passage is enclosed, separated from the mouth cavity, and opens in the back near the throat. (If you feel the roof of your mouth with your tongue, you can detect the suture along the midline of your palate. Some babies have a birth defect called cleft palate, where the two halves of the secondary palate fail to grow completely together, making it difficult for them to eat and breathe.)

The primitive synapsid jaw was a simple snap-trap mechanism, with a strong temporal muscle pulling up on the jaw and inserting on the top of the skull behind the eyes. Numerous bones made up this primitive jaw: the dentary in front, which bore the teeth; the articular, which formed the jaw hinge with the quadrate bone of the skull; the coronoid, forming a ridge on the top of the back of the jaw; the angular and surangular, on the back lower corner of the jaw, and several others (**Figure 20.3**). Such a jaw was suitable for grabbing and crushing prey, but not for extensive chewing (although a few herbivorous synapsids, like *Edaphosaurus* and also the labidosaurine reptile *Moradisaurus*, developed a tooth plate on the palate for chewing). A single-element jaw is mechanically much stronger against the pressures and torques of the chewing motion than one with numerous elements that are sutured together; the sutures are lines of weakness under stress. Through synapsid evolution, the post-dentary elements of the jaw become smaller and smaller as the dentary becomes the primary, and eventually the only, bone of the jaw. As the post-dentary elements reduced in size and most of them disappeared, the dentary extended back and took their place as the main area of muscle attachment. Eventually, the dentary developed a tall coronoid process to which the temporal muscles attached, replacing the amniote coronoid bone, and allowing them to have even greater bite strength. In addition, a pair of new muscles, the masseters, arose between the outer edge of the cheekbones and the outer side of the jaw, allowing front-back and side-to-side motion in chewing (**Figure 20.3**).

Finally, the non-dentary bones of the jaw were lost completely (although some persisted even in the earliest mammals) as the dentary expanded backward and took their place. In advanced synapsids, the dentary reaches far enough back to touch the squamosal bone of the skull and develop a dentary/squamosal jaw joint, replacing the old reptilian quadrate/articular jaw joint. In some specimens of synapsids, such as *Diarthrognathus* ("double jaw joint" in Greek), both jaw joints operated side-by-side on each side of the head. Eventually, however, the dentary/squamosal joint took over completely, and then the quadrate and articular no longer functioned as a jaw joint. Instead of vanishing, however, they took over a new function. In reptiles, they not only served as a jaw hinge, but are also able to transmit sound to the ear, since most reptiles hear with their lower jaws. (The snake charmer's flute is for the spectators, not for the cobra, since snakes cannot hear well when their jaw is up off the ground in a threat posture.)

Once the quadrate and articular became detached from the jaw hinge, they took up a different role as bones of the middle ear. The quadrate became the incus, or "anvil" bone, and the articular became the malleus, or "hammer" bone. (The "stirrup" bone, or stapes, has been part of the middle ear since the early tetrapods.) When sound vibrates your eardrum, the chain of bones—"hammer", "anvil", and "stirrup", or malleus, incus, and stapes—that transmits this vibration to the inner ear is actually a remnant of your reptilian jaw apparatus. This amazing story is apparent not only in synapsid fossils, but also in mammalian embryology. When you began your development, your ear bones started out as part of your jaw, but were transferred entirely to your ear later in development of the embryo.

Other skeletal modifications are apparent as synapsids became mammals. The early amniotes had a sprawling posture, resting on their bellies with the legs held out from the side of the body, but early in synapsid evolution, the body adopted an erect posture, with the limbs held under the body and moving rapidly fore and aft. These skeletal changes are particularly evident in the shoulder blade, which flares out into a broad triangle with a ridge down the middle for more complex muscle insertions. The hips became long and narrow for greater flexibility, with forward expansions of the ilium bone for stronger leg muscles, and eventually the three bones of the pelvis fused into a single bone (not the multiple bones of the primitive amniote hip). In advanced synapsids and mammals, the free ribs of the chest are linked together with a breastbone, forming a solid rib cage. This means that advanced synapsids did not breathe primarily by flexing their ribs (as occurs in most reptiles), but must have had a muscular wall in their chest cavity called a diaphragm to pump their lungs within the rigid rib cage. The ribs of the lower back, on the other hand, were lost, allowing the trunk to become more flexible, and are thought to be related to the development of the diaphragm as the end of the thoracic rib cage. The small lower temporal opening of primitive synapsids became larger and larger as the jaw muscles expanded, until only a thin cheekbone, the zygomatic arch, remained. In many advanced synapsids and in most mammals, the temporal opening is so large that the bony bar between it and the eye is lost. The single ball joint that connects the skull to the vertebral column (the occipital condyle) in reptiles split into two small ball joints on either side of the spinal column, allowing much greater strength, flexibility, and stability in moving the head to catch and hold prey.

EARLY SYNAPSIDS: "PELYCOSAURS"

Most of these skeletal features can be traced through the course of synapsid evolution (**Figures 20.1**, **20.2**, and **20.3**). For example, the earliest synapsids (mostly from the Pennsylvanian-Early Permian) are known as the "pelycosaurs" (which is a wastebasket artificial group unless it includes the rest of the synapsids, even the mammals), and include such fossils as *Casea*, *Cotylorhynchus*, *Ophiacodon*, and *Varanops* (**Figures 20.1**, **20.2**, and **20.4**). *Casea* (**Figures 20.2[B]** and **20.4[A]**) was among the most primitive of the synapsids from the early Middle Permian, but at up to 1.2 meters (4 feet) in length and weighing 200 kg (400 lb), it was much larger than the tiny Pennsylvanian fossils like *Archaeothyris*. Its head was unusually small and short for its huge body, and it had a blunt snout with a rounded front, large eyes, and simple peg-like teeth not suited for catching prey. Instead, it appears to be one of the very first herbivorous amniotes to evolve.

The weirdest of all the caseids was *Cotylorhychus* (**Figures 20.2[C]** and **20.4[B]**). It had an enormous barrel-shaped body up to 6 meters (20 feet) long, the largest animal of its time. Despite this huge body, the head was ridiculously small and looked completely out of place on its body. Yet the small skull was filled with peg-like teeth with a large overbite on the snout, suggesting that it was also an herbivore which had a long digestive tract in its barrel-shaped body, serving as a huge fermenting vat for breaking down vegetation. It also had large nasal openings, but relatively small eyes. The massive sprawling limbs and the clawed fingers and toes suggested that it might have had webbed feet, and supported its weight in water with an aquatic lifestyle—basically, a primitive synapsid version of a hippopotamus.

Another primitive synapsid was *Ophiacodon* (**Figures 20.2[D]** and **20.4[C]**), which reached 3 meters (10 feet) in length and weighing 230 kg (500 lb), larger than *Casea*. The skull of *Ophiacodon* alone was the longest of any synapsid, reaching 50 cm (almost 2 feet). This skull was also very deep and narrow, and bore dozens of teeth like those found in snakes for catching smaller prey (the name *Ophiacodon* means "snake toothed" in Greek). At one time its broad claws, thin lower jaws, and simple teeth were interpreted as evidence that it was an aquatic predator, but since then it has been shown that these features are just typical of primitive synapsids, and there is now evidence that it was primarily terrestrial predator. In addition, *Ophiacodon* has fibro-lamellar bone, which has been interpreted as evidence that they were at least partially warm-blooded. *Varanops* (**Figures 20.2[E]** and **20.4[D]**) was another typical Early Permian synapsid, reaching up to 1.2 meters (4 feet) in length, or about the size and proportions of large monitor lizard (indeed, its name means "like a monitor"). It had wicked recurved teeth in a robust deep skull with a triangular-shaped snout.

Besides the caseids and ophiacodontids, another major group of Early Permian synapsids were the sphenacodontids. One of the most primitive of these was *Sphenacodon* itself (**Figures 20.2[F]** and **20.3[E]**), which gave its name to the group. It had a very large, long, deep, narrow skull with sharp conical teeth, including some that were enlarged and shaped like stabbing canines. The upper jaw had a notch in the area where the lower caniniform teeth would insert when the jaw closed. *Sphenacodon* had a short neck, robust trunk, and short front and hind limbs, and a tapering tail. However, the depictions of its sprawling posture may be mistaken, because trackways have been found which suggest that its legs were more upright and directly beneath the body. Large individuals

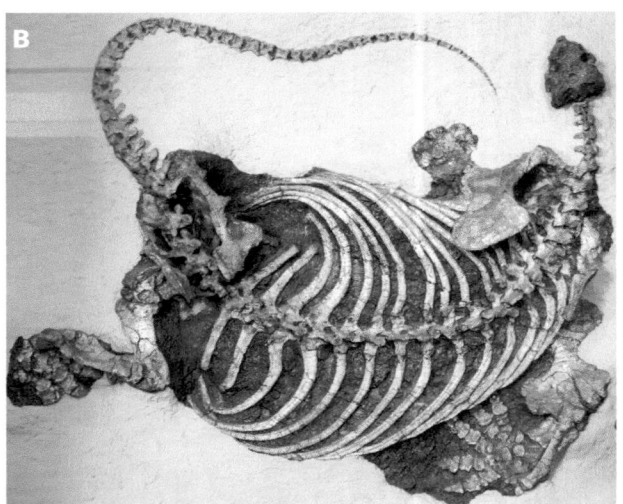

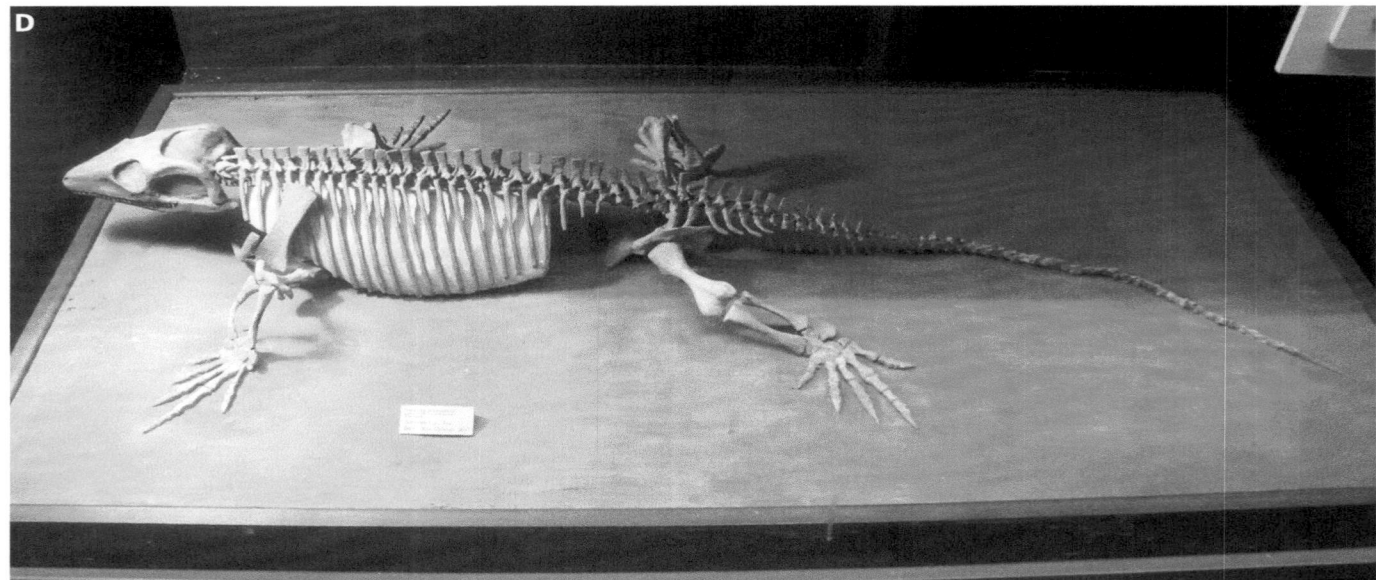

Figure 20.4 Photos of important synapsid ("pelycosaur") fossils, including: (A) *Casea*, (B) *Cotylorhynchus*, (C) *Ophiacodon*, with *Edaphosaurus* in the background, (D) *Varanops*.

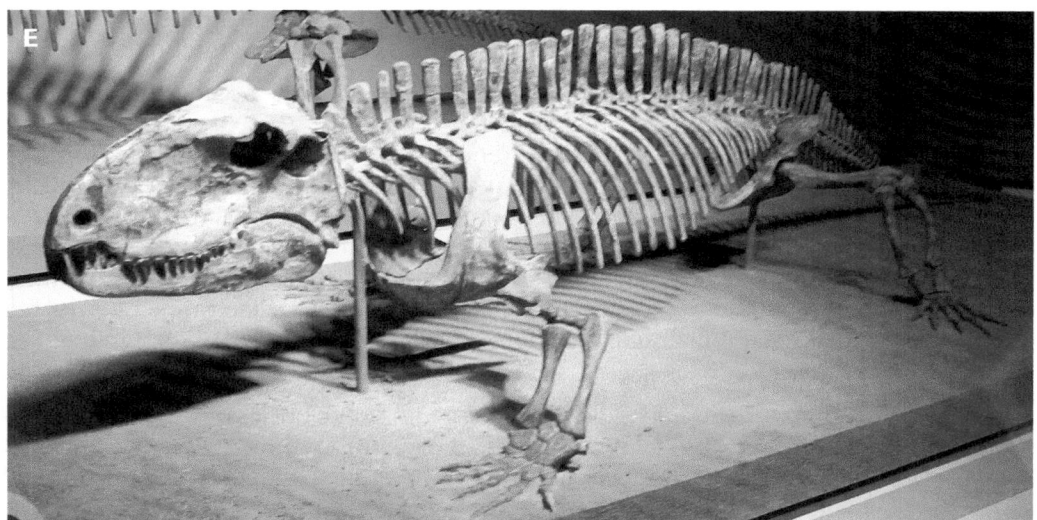

Figure 20.4 (Continued)
(E) *Sphenacodon*, (F) *Dimetrodon*,
(G) *Edaphosaurus*.

were up to 3 meters (10 feet) long. More importantly, it has a short sail on its back with moderately long spines on its back vertebrae.

The most famous and spectacular of these sphenacodontids was the predatory finback *Dimetrodon* (**Figures 20.1**, **20.2[G]**, and **20.4[F]**). It is often mistakenly called a "dinosaur", or put in the toy kits with real dinosaurs, even though it vanished 40 million years before dinosaurs evolved; and it's a synapsid, more closely related to us than it is to dinosaurs. *Dimetrodon* was the top predator of the Early Permian ecosystem, bearing wicked-looking recurved pointed teeth, and even larger caniniform teeth in the front of the mouth. They had a wide range of sizes but the largest individuals may have reached 4.6 meters (15 feet) in length.

Closely related to the sphenacodonts was *Edaphosaurus* (**Figures 20.2[H]** and **20.4[G]**), which also had a large fin on its back with short crossbars of bone sticking out up and down the lengths of the long bony spines that supported a sail. Its sail was roughly oval in shape, with steep front and back edge, very different from the bell-curve profile of the sail of *Dimetrodon*. *Edaphosaurus* was clearly herbivorous, because it had large grinding tooth plates in its mouth, as well as blunt peg-like teeth on the edge of its jaws for cropping plants. Large individuals of *Edaphosaurus* were up to 3.5 meters (11.5 feet) in length and weighed about 330 kg (660 pounds).

Dimetrodon and *Edaphosaurus* both bore large "sails" along their backs supported by long spines extending from the top of their vertebrae. Many ingenious ideas have been proposed for the function of these fins. One plausible suggestion is that they served as heat gathering and dumping devices for thermoregulation. They have the appropriate surface area for an animal of their body volume to allow them to dump heat when the sail is out of the sun, and pick up heat when it is exposed broadside to the sun. This suggests that the earliest synapsids were not yet endotherms, but used sunning behavior to regulate their body temperature (as do most living reptiles). However, since most other synapsids at that time did not have a sail for thermoregulation and apparently didn't need it, other paleontologists argue that it was a display device for recognizing their own species, and for signaling their size and strength to other animals, just as large horns and antlers serve today in antelopes and deer.

The early synapsids that have long been put in the wastebasket group known as "pelycosaurs" were primitive in many other aspects. They had a sprawling posture with a simple shoulder blade, small iliac blade on the pelvis, and simple thigh bone. Their teeth were typically simple conical pegs (although those in the canine position were a bit larger) replaced multiple times, and they had no secondary palate. Instead, they had many teeth on their original reptilian palate and in the throat region. They had a single ball joint (occipital condyle) in the back of the skull, a small brain, as well as a jaw composed of a small dentary and many accessory jaw bones. Most of these conditions are primitive for amniotes, so the primary advanced feature that earmarks the "pelycosaurs" as synapsids is the presence of the lower temporal opening below the postorbital and squamosal bones, although it is small, indicating relatively small jaw muscles.

LATER SYNAPSIDS: "THERAPSIDS"

Upper Permian red beds, especially in South Africa and Russia, produced an incredible diversity of synapsids, and demonstrated their evolution over about 30 million years. Gone were the archaic fin-backed synapsids like *Dimetrodon* and all of its "pelycosaur" kin, replaced by a new radiation of more advanced synapsids, called by the wastebasket group name "Therapsida" (**Figures 20.1**, **20.5**, and **20.6**). They include at least two large groups of herbivores that dominated the Late Permian, and a wide variety of huge wicked-looking predators that could eat any animal alive at the time.

Biarmosuchids: The most primitive of the "therapsids" were lightly built predators called the biarmosuchids (**Figure 20.5[A]**). They were intermediate in shape and size between sphenacodonts and more advanced synapsids. *Biarmosuchus*, for example was about 2 meters (6.6 feet) long. Its skull was much like that of sphenacodonts, except that the temporal fenestra was much larger, and it had a much wider flare of the back of the skull. This suggests that biarmosuchids had much stronger and larger muscles closing its jaws than did *Dimetrodon*, and thus a more powerful bite. The caniniform teeth of biarmosuchids were reduced to a single large tooth on each side, but were nowhere near as larger as the caniniform teeth in later "therapsids". Some biarmosuchids, especially burnetiomorphs like *Proburnetia* (**Figure 20.5[B]**) had bumps and bosses

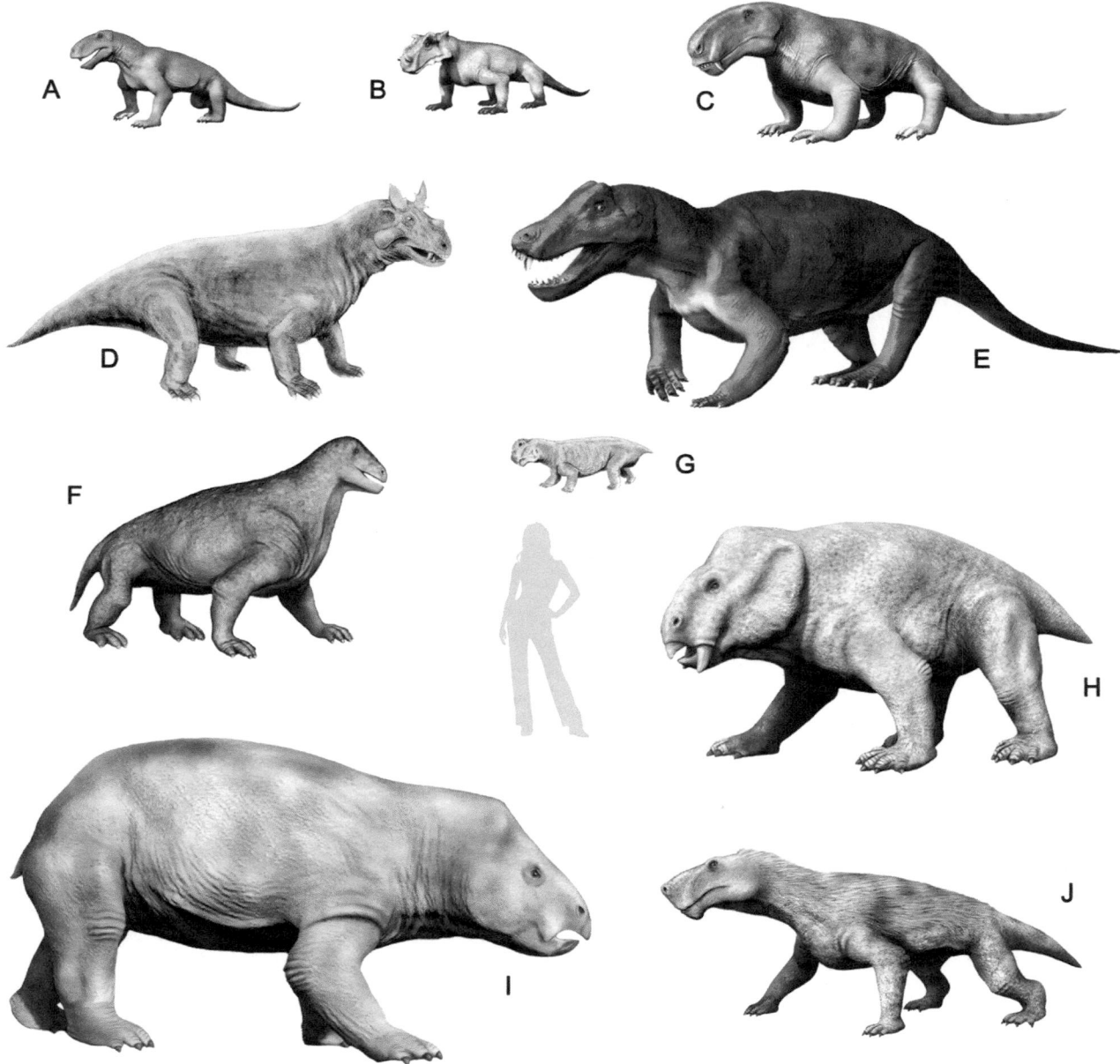

Figure 20.5 Reconstructions of some of the more advanced synapsids ("therapsids") including: (A) *Biarmosuchus*, (B) *Proburnetia*, (C) *Eotitanosuchus*, (D) *Estemmenosuchus*, (E) *Anteosaurus*, (F) *Moschops*, (G) *Lystrosaurus*, (H) *Placerias*, (I) *Lisowicia*, (J) *Inostrancevia*.

A

B

C

D

E

F

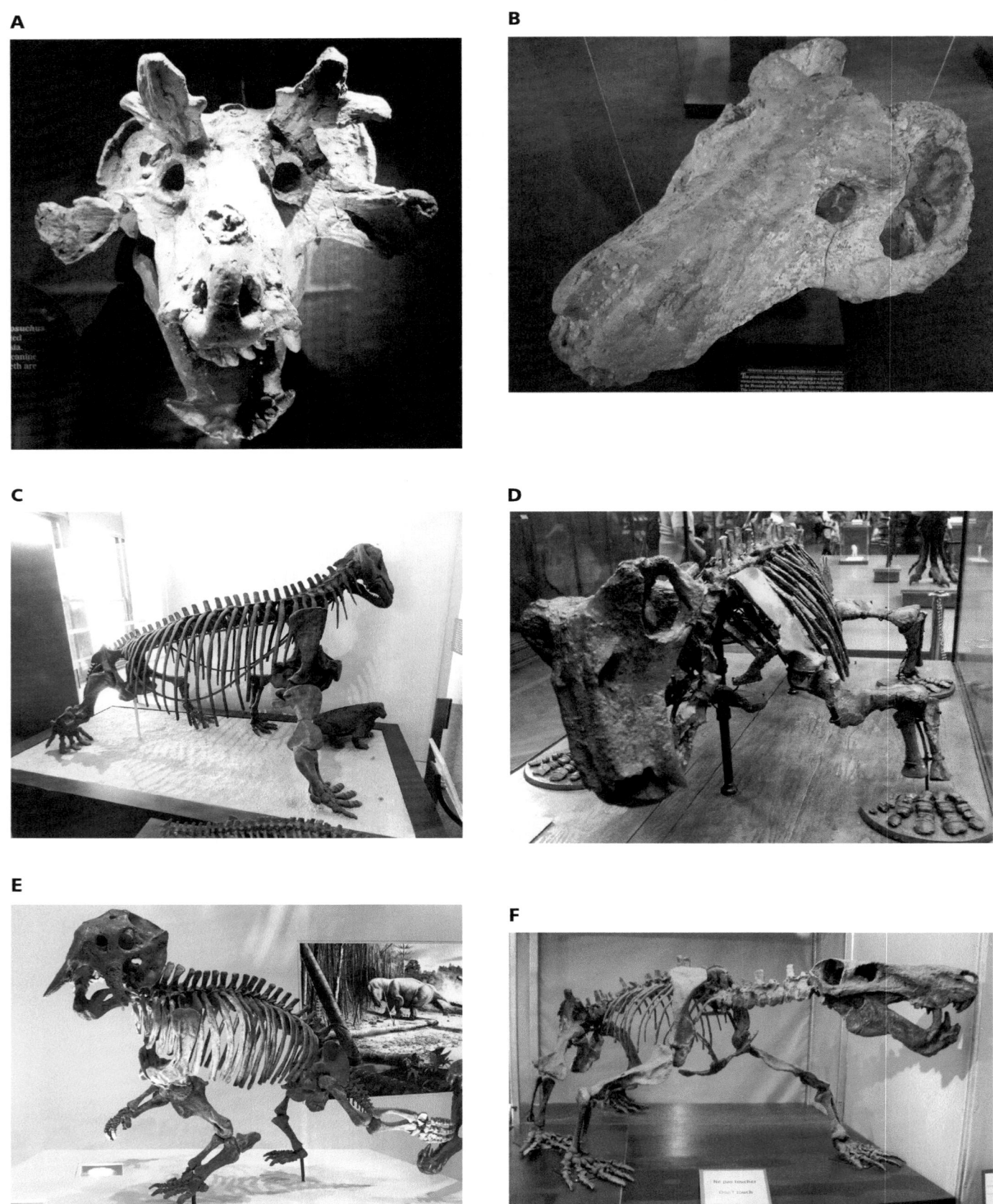

Figure 20.6 Photos of some "therapsids" including: (A) The bizarre bumpy skull of the dinocephalian *Estemmenosuchus*, (B) the huge predatory dinocephalian *Anteosaurus*, (C) the herbivorous dinocephalian *Moschops*, (D) the small dicynodont *Lystrosaurus*, (E) the giant dicynodont *Placerias*, (F) the huge predatory gorgonopsian *Inostrancevia*. (Courtesy Wikimedia Commons.)

on their skulls, and others that thick domes of bone on their skulls, similar to those found in some of the more advanced therapsids. The skeleton (especially the shoulder girdles and hip) was much more advanced in posture as well, with limbs virtually vertical beneath the body. Biarmosuchids had smaller more symmetrical hands and feet with reduced toes, suggesting that they moved in the front-to-back plane and did not flare outwards. Even larger and more advanced were biarmosuchids such as *Eotitanosuchus*, which was probably over 2.5 meters (6.6 feet) long, and had huge upper canine teeth in its skull (**Figure 20.5[C]**).

Dinocephalians: Some of the Late Permian "therapsids" were among the early herbivorous land animals, and others were among the largest herbivores known on land up until then. One of the most remarkable group of therapsids were the dinocephalians ("terrible heads"), which included not only carnivores and omnivores, but also herbivores. They sported an array of warts and bumps and thick bony battering rams on their heavily armored skulls. One branch of dinocephalians was represented by *Anteosaurus* (**Figures 20.5[E]** and **20.6[B]**). They were huge predators up to 6 meters (20 feet) in length and 600 kg (1300 lb) in weight, with a long, heavily built body with a long tail and relatively short weak legs. These suggest that *Anteosaurus* was crocodile-like in its habits, waiting in the water to ambush its prey with a quick lunge. *Anteosaurus* had not only huge canine teeth in front, but also long sharp incisors in the front of the mouth, and additional teeth on the roof of the mouth. The skulls had large flaring cheekbones, and a huge bony crest on the top, and some anteosaurs had additional bumps and knobs on their heads.

Even more bizarre are the estemmosuchids, which had skulls covered with weird horns, bumps, and knobs (**Figures 20.5[D]** and **20.6[A]**). These creatures could reach 3 meters (10 feet) in length, but still had the primitive sprawling posture of primitive synapsids. The condition of the horns and knobs vary among the different species and genera, suggesting they were used for display and species recognition, and possibly for sparring between males or for courtship. Skin impressions of estemmosuchids suggest that they had a smooth, hairless, warty hide.

Another odd-looking dinocephalian was *Moschops* (**Figures 20.5[F]** and **20.6[C]**). These creatures were heavily built herbivores with thick bodies, short thick necks, and heads that were covered with a thick dome of bone, presumably for some sort of head-butting or head-wrestling with other members of their population. Large specimens of *Moschops* were up to 2.7 meters (9 feet) in length. They had short hindlimbs and much longer forelimbs, so their backs sloped down abruptly to the rear, but their limbs were also upright beneath their bodies, while their forelimbs bent only slightly at the elbow, so they could move more quickly and did not sprawl like other synapsids. The teeth of *Moschops* were very stout and high-crowned, suggesting that they chewed up tough vegetation like cycads and ferns. Their chubby bodies and herbivorous diets have suggested a semi-aquatic hippo-like lifestyle.

Dicynodonts: The dinocephalians were creatures of the Middle-Late Permian. They vanished during the great Permian extinction, but another group of herbivorous therapsids flourished in the Late Permian and persisted into the Triassic. Over 70 genera are known of these squat creatures with a toothless beak and big canine tusks were known as dicynodonts ("double dog teeth" in Greek, in reference to their two canine tusks), which ranged from rat sized to about 3.5 meters (11 feet) long and weighed up to 1000 kg (2200 pounds). Like dinocephalians, they were stout-bodied with short hind limbs and longer forelimbs bent slightly at the elbow. With only a toothless beak and almost no other teeth in their mouth other than their canine tusks, they were clearly herbivores. They employed a unique "cheek-pivot" system of chewing, which allowed the jaw to slide front-to-back because of their unique system of jaw joints, giving them some ability to shred their diets of tough vegetation before they swallowed.

Dicynodonts first appeared in the Middle Permian with small primitive forms such *Eodicynodon*, which was only about half a meter long (about 2 feet). They rapidly evolved during the Permian to become the most diverse and numerous synapsids of the Late Permian in South Africa and Russian, with dozens of genera occupying the roles of small, medium, and large-sized herbivores. Then the great Permian extinction wiped out all but four lineages of dicynodonts. One of these genera, the pig-sized *Lystrosaurus* (**Figures 20.5[G]** and **20.6[D]**), was extremely common in the Early Triassic, and was found on almost all the Pangean continents: South Africa, South America, India, Russia, China, Mongolia, and even in Antarctica. This was one of the key pieces of evidence supporting continental drift almost a century ago. Another group of dicynodonts, typified by *Placerias* (**Figures 20.5[F]** and **20.6[E]**), underwent an evolutionary radiation of at least 24 genera in the Triassic, and are common in Upper Triassic beds such as Petrified Forest in Arizona, as well as China, Mongolia, South Africa, and South America. One of the largest and best known is *Placerias*, which was the size of a hippo, reaching 3.5 meters (11.5 feet) in length, and weighing about 1000 kg. The largest of these Triassic dicynodonts was *Lisowickia* (**Figure 20.5**) from the Late Triassic of Poland, which reached the size of a small elephant, weighing about 6 tons. It is thought to have reached such size to compete with the rise of large dinosaurs in the Late Triassic.

Gorgonopsians and other predatory therapsids: Preying upon these herbivores was a wide array of ferocious carnivorous synapsids, including groups known as the therocephalians, and the bauriamorphs. The most impressive were the terrifying gorgonopsians ("Gorgon appearance" in Greek). They had huge skulls with sharp stabbing canine teeth, strong jaw muscles for chewing, and powerfully built bodies. The largest, such as *Inostrancevia*, were bigger than bears, with a skull 45 cm long, saber teeth over 12 cm long, and a long sprawling crocodile-like body up to 3.5 meters (11 feet) in length, weighing about 300 kg (660 pounds) (**Figure 20.6[F]**).

Throughout the evolution of these more advanced synapsids in the Late Permian, we see more and more mammal-like features appearing. The small opening on the side of the skull in *Dimetrodon* became a large expanded arch behind the eye for powerful jaw muscles to bulge, and allow powerful bite forces and even some chewing. The original reptilian palate began to be covered by a secondary palate, which grew over it and enclosed the nasal passages.

Instead of the single ball joint on the back of the skull just below the spinal cord connecting to the neck, therapsids had a double ball joint, allowing for greater strength and flexibility in their neck muscles. Therapsids also have many modifications of the skeleton (**Figure 20.3**) that make them more mammalian in appearance, including a gait which no longer sprawls on the belly like a crocodile, but held the body in a semi-sprawling to nearly upright position.

THE THIRD WAVE: "CYNODONTS"

The greatest extinction in earth history occurred at the end of the Permian (about 252 Ma), wiping out about 70% of the species on land (mostly therapsids), and 95% of animal species in the ocean. The causes of the great Permian extinction ("the mother of all mass extinctions") were complex, but the event was apparently triggered by enormous volcanic lava flows pouring across most of northern Siberia, the biggest volcanic eruptions in earth history. These injected huge amounts of greenhouse gases (especially carbon dioxide, as well as sulfur dioxide) into the atmosphere and oceans. The climate became a "super-greenhouse", and the oceans then became supersaturated in carbon dioxide, making them too hot and acid and killing nearly everything that lived there. The atmosphere was also too low on oxygen, and too loaded with carbon dioxide, so land animals above a certain size nearly all vanished, and only a few smaller lineages of synapsids, reptiles, amphibians, and other land creatures made it through the hellish planet of the latest Permian, and survived to the aftermath world of the earliest Triassic.

Once most of the Permian therapsids vanished at the great end-Paleozoic extinction, the synapsids started all over again with a third great evolutionary radiation (**Figure 20.1**) of at least 60 genera of much more mammal-like synapsids called cynodonts ("dog toothed" in Greek). These included forms as big as a bear called *Cynognathus* ("dog jaw" in Greek), which was 1–2 meters long, with a head over 60 cm (2 feet) in length, and many smaller forms in the size range of raccoons and weasels. Most cynodonts had upright postures (**Figure 20.3**), with their limbs completely under their bodies for rapid running. Their non-dentary jawbones were tiny, and reduced to mere splints in the inside back part of the jaw near the hinge. They had secondary palates going all the way back to the throat, as in mammals, and many other indicators of active living and rapid metabolism. Most had multi-cusped cheek teeth instead of the simple conical pegs of the primitive synapsids, suggesting that they were capable of complex chewing motions, rather than gulping the foot down whole, as do reptiles and primitive synapsids.

The transition from primitive amniotes to mammals is demonstrated *Thrinaxodon* (**Figures 20.7[A]** and **20.8[A]**), which represents the start of the cynodont radiation of synapsids after the Early Permian finbacks, and the Middle-Late Permian therapsids of the Karoo beds of South Africa. *Thrinaxodon* was one of the earliest cynodonts, the first fossil to show many of the advanced features of the final phase of the evolution of synapsids into mammals. It was quite common in the Early Triassic of the Beaufort Group in South Africa, so there are lots of nearly complete specimens and we know its anatomy and behavior better than most other synapsids.

Thrinaxodon was about the size and shape of a weasel, with a long narrow snout on the skull, and a long slender low-slung body with short legs. *Thrinaxodon* were typically 30–50 cm (1 foot to almost 2 feet) in length. The dentary bone of *Thrinaxodon* dominates the entire jaw, so the non-dentary bones were tiny splints—although it still had the reptilian quadrate/articular jaw joint (**Figure 20.3**). *Thrinaxodon* had a complete secondary palate, so it could breathe and eat at the same time. It had large eyes (for seeing in the dark, or in its burrows), and a relatively large head. Like its descendants, the cheek teeth were not simple conical pegs, but had complex cusps and could be rightfully called molars and premolars. In fact, its name means "trident tooth" in Greek, referring to the three-cusped molar teeth in its mouth. The temporal opening for the muscles on the side and top of its head was unusually large, allowing for complex chewing motions of the jaw. Yet unlike most mammals, *Thrinaxodon* still had a bony bar that separated the temporal jaw opening from the eye socket. On each side of its snout were tiny pits in the bone, suggesting that it had whiskers. If *Thrinaxodon* had hair on its snout, it's a good bet that it had hair all over. Hair normally doesn't fossilize, so this may be the first evidence of its appearance in the mammalian lineage.

Figure 20.7 Reconstructions of some of the better-known advanced synapsids or "cynodonts". These included: (A) *Thrinaxodon*, (B) *Cynognathus*, (C) *Chiniqudon*, (D) *Tritylodon*, (E) *Oligokyphus*.

Even though *Thrinaxodon* had short legs, its posture placed the legs beneath the body in a semi-upright stance (**Figure 20.8[A]**). It had specialized shoulder bones and broad hip bones (especially the iliac blade, which attaches the hips to the spinal column, and anchors the leg muscles) much like the more advanced cynodonts and mammals. There are ribs only in the chest region around the lungs; all the ribs from the lower back (lumbar region) are lost, as in mammals. This allows them to bend their backs more sharply, and turn around in a small space, or curl up tightly (specimens curled up in their burrows are common). Even more revealing, *Thrinaxodon* had broad flanges on its thoracic ribs that would have made the rib cage fairly solid and rigid. This would have prevented the kind of rib-assisted breathing found in most reptiles (and apparently in primitive synapsids as well). Instead, *Thrinaxodon* must have had a muscular diaphragm between the lung cavity and the abdominal cavity, which pumps air in and out of the lungs. This feature is found in all mammals. Putting all these clues together—complex cheek teeth, whiskers, diaphragm—suggests that *Thrinaxodon* was extremely mammal-like, and probably was covered in fur with a high metabolic rate and warm-blooded physiology.

In contrast to the small cynodonts like *Thrinaxodon*, the Middle Triassic saw the rise of larger forms like *Cynognathus* (**Figures 20.7[B]** and **20.8[B]**) mentioned previously. Its body was up to 2 meters (6.6 feet) long counting the tail, so it approached the size of a small bear or large wolf, but its skull was much bigger than that of any bear and equipped with huge stabbing canines and numerous cheek teeth for chewing up meat. It had a full secondary palate, and its skull was almost fully mammalian

A

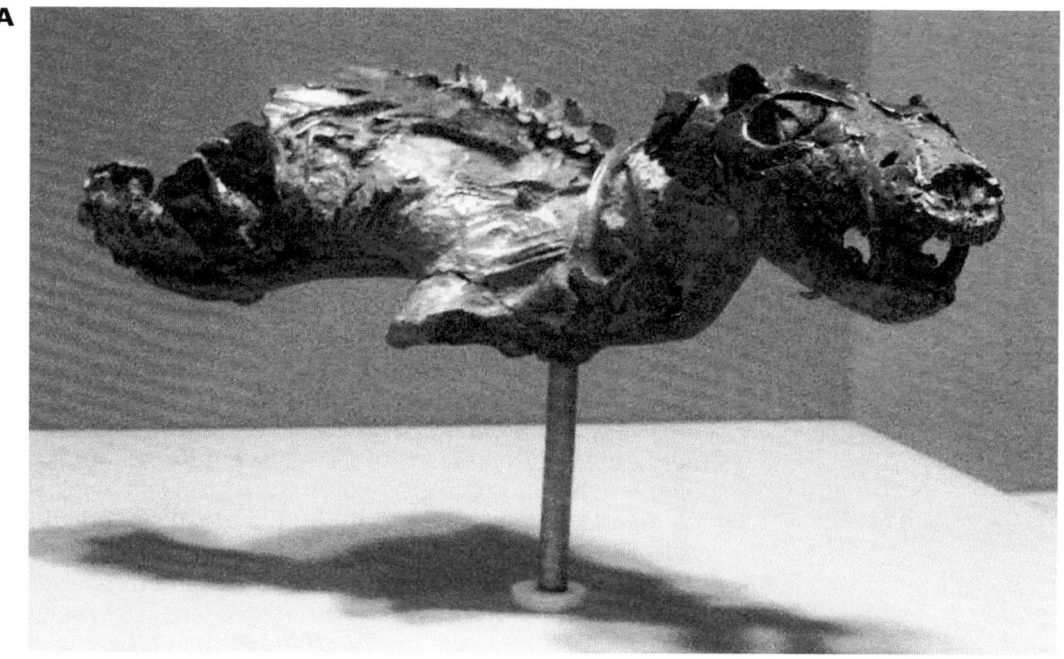

B

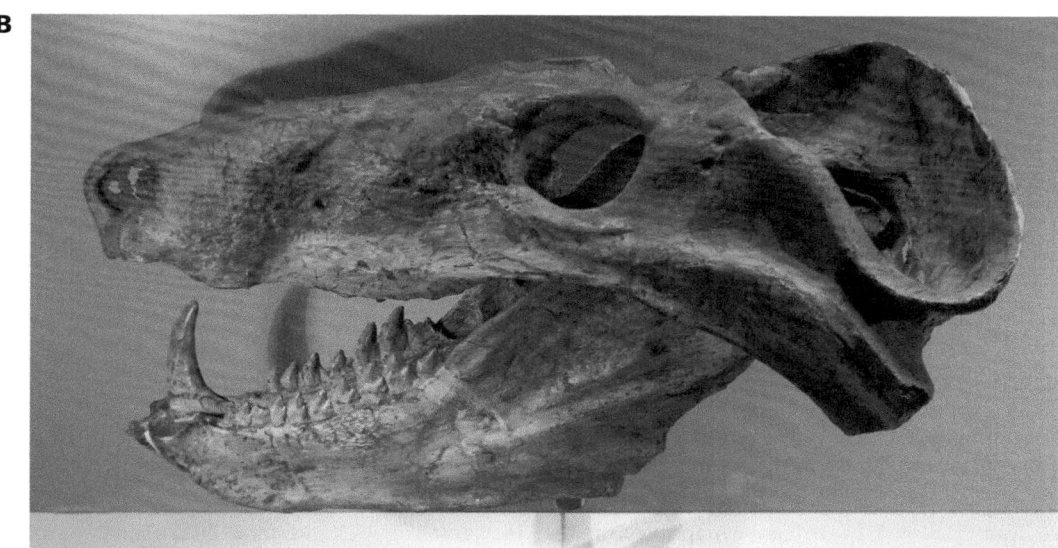

C

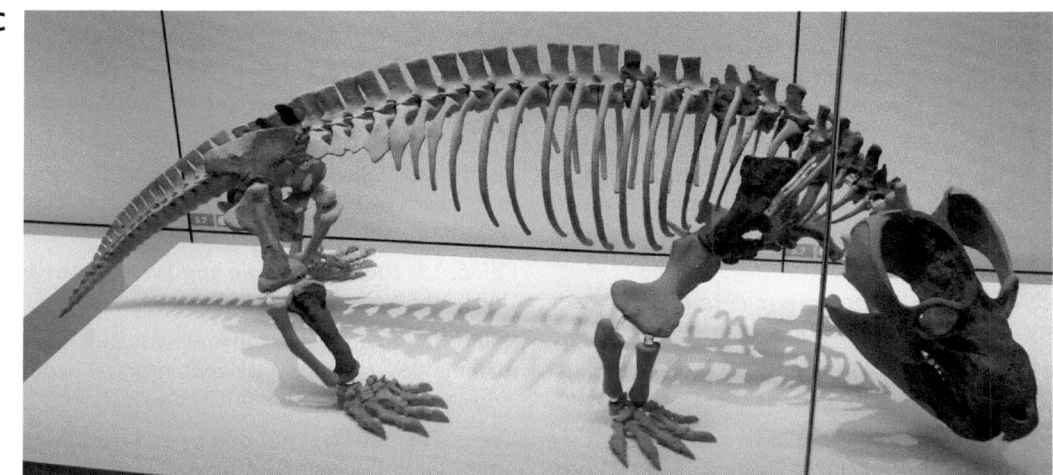

Figure 20.8 Fossils of some of the "cynodont" synapsids, including: (A) the weasel-sized Triassic form *Thrinaxodon*, (B) the bear-sized predator *Cynognathus*, (C) the more advanced cynodont *Chiniqudon*. (Courtesy Wikimedia Commons.)

in most aspects. It had flanges on the ribs, suggesting it had a diaphragm to aid in breathing. Likewise, there were pits on the snout suggesting whiskers, so it was probably covered in some sort of body hair.

Cynognathus vanished by the Late Triassic, but its more advanced cynodont descendants continued to dominate the Triassic, even as other groups of animals (especially the primitive relatives of crocodiles, and the earliest dinosaurs) began to appear. One important group of cynodonts were the relatives of *Probainognathus* from the Middle and Late Triassic. Genera like *Probainognathus*, *Pachygenelus*, *Diarthrognathus*, and *Chiniquidon* (**Figures 20.7[C]** and **20.8[C]**) were dog-sized predators from South Africa and South America with completely mammalian skeletons, and skulls that were extremely mammal-like as well. As mentioned with *Diarthrognathus* previously, they had the first examples of double jaw joints where the old quadrate-articular jaw joint was still functioning, side-by-side with the new dentary-squamosal jaw joint. The teeth were highly complex, with multiple conical cusps aligned on their molars, and large stabbing canines in a broad skull with relatively short snout. Like all cynodonts, the secondary palate extends all the way to the back of the mouth. The wide arches behind the eyes allowed for not only large temporalis muscles for the jaw, but also branches of the masseter muscles to pull the jaw back and forth in chewing.

Some of the last and most advanced of the cynodonts were small rat-sized creatures like *Tritylodon* (**Figure 20.7[D]**) from the Lower Jurassic beds of South Africa, which had very advanced teeth and jaws capable of grinding seeds and nuts. Others, like *Oligokyphus* (**Figure 20.7[E]**) from the Late Triassic and Early Jurassic, were about 50 cm (20 inches) long, and had a skull much like that of a modern rodent, with chisel-like front incisors, a gap between the incisors and cheek teeth, and grinding molars suitable for seeds, nuts, and vegetation. Bit by bit, the last of the cynodonts were becoming extremely similar to the earliest mammals, so that deciding where to draw the line between them is problematic.

FURTHER READING

Angielczyk, K.; Kammerer, C.F.; Frobisch, J. 2013. *Early Evolutionary History of Synapsida*. Springer Science & Business Media, New York.

Benson, R.J. 2012. Interrelationships of basal synapsids: Cranial and postcranial morphological partitions suggest different topologies. *Journal of Systematic Palaeontology*. 10 (4): 601–624.

Botha-Brink, J.; Modesto, S.P. 2007. A mixed-age classed 'pelycosaur' aggregation from South Africa: Earliest evidence of parental care in amniotes? *Proceedings of the Royal Society B*. 274 (1627): 2829–2834.

Brocklehurst, N.; Reisz, R.; Fernandez, V.; Fröbisch, J. 2016. A re-description of '*Mycterosaurus*' *smithae*, an Early Permian eothyridid, and its impact on the phylogeny of pelycosaurian-grade synapsids. *PLoS ONE*. 11 (6): e0156810.

Chinsamy-Turan, A., ed. 2011. *The Forerunners of Mammals*. Indiana University Press, Bloomington, IN.

Hopson, J.A. 1987. The mammal-like reptiles: A study of transitional fossils. *The American Biology Teacher*. 49 (1): 16–26.

Hopson, J.A. 1994. Synapsid evolution and the radiation of non-eutherian mammals, pp. 190–219. In Prothero, D.R.; Schoch, R.M., eds. *Major Features of Vertebrate Evolution*. Paleontological Society Short Course 7. Paleontological Society, Lawrence, KS.

Hotton, N.; MacLean, P.D.; Roth, J.J.; Roth, E.C., eds. 1986. *The Ecology and Biology of Mammal-Like Reptiles*. Smithsonian Institution Press, Washington, DC.

Huttenlocker, A.K.; Rega, E. 2012. The paleobiology and bone microstructure of pelycosaurian-grade synapsids, pp. 90–119. In Chinsamy-Turan, A., ed. *Forerunners of Mammals: Radiation, Histology, Biology*. Indiana University Press, Bloomington, IN.

Kemp, T.S. 1982. *Mammal-Like Reptiles and the Origin of Mammals*. Academic Press, London.

Kemp, T.S. 1988. Interrelationships of the synapsida, pp. 1–22. In Benton, M.J., ed. *The Phylogeny and Classification of the Tetrapods, vol. 2: Mammals*. Clarendon Press, Oxford.

Kemp, T.S. 2005. *The Origin and Evolution of Mammals*. Oxford University Press, Oxford, UK.

Kemp, T.S. 2006. The origin and early radiation of the therapsid mammal-like reptiles: A palaeobiological hypothesis. *Journal of Evolutionary Biology*. 19 (4): 1231–1247.

Kemp, T.S. 2011. The origin and radiation of therapsids, pp. 3–30. In Chinsamy-Turan, A., ed. *Forerunners of Mammals*. Indiana University Press, Bloomington, IN.

Kielan-Jaworowska, Z.; Luo Z.; Cifelli, R.L. 2004. *Mammals from the Age of Dinosaurs: Origins, Evolution, and Structure*. Columbia University Press, New York.

King, G. 1990. *The Dicynodonts: A Study in Paleobiology*. Chapman & Hall, London.

Laurin, M.; Reisz, R.R. 1996. The osteology and relationships of *Tetraceratops insignis*, the oldest known therapsid. *Journal of Vertebrate Paleontology*. 16 (1): 95–102.

Liu, J.; Rubidge, B.; Li, J. 2009. New basal synapsid supports Laurasian origin for therapsids. *Acta Palaeontologica Polonica*. 54 (3): 393–400.

McLoughlin, J.C. 1980. *Synapsida: A New Look into the Origin of Mammals*. Viking, New York.

Modesto, S.P.; Anderson, J.S. 2004. The phylogenetic definition of Reptilia. *Systematic Biology*. 53 (5): 815–821.

Modesto, S.P.; Smith, R.M.H.; Campione, N.E.; Reisz, R.R. 2011. The last "pelycosaur": A varanopid synapsid from the *Pristerognathus* Assemblage Zone, Middle Permian of South Africa. *Naturwissenschaften*. 98 (12): 1027–1034.

Romer, A.S.; Price, L.I. 1940. Review of the Pelycosauria. *Geological Society of America Special Papers*. 8: 1–538.

PRIMITIVE MAMMALS

MESOZOIC MAMMALS, MONOTREMES, AND MARSUPIALS

21

With malleus
Aforethought
Mammals
Got an earful
of their ancestors'
Jaw.

—John Burns, *Biograffiti*, 1975

MESOZOIC MAMMALS

By the latest Triassic and earliest Jurassic, cynodonts were vanishing, and the first undoubted mammal fossils (with a dentary-squamosal joint and complex molar teeth) had appeared. They were only shrew-sized creatures, but they were now living in a world dominated by the rise of the huge dinosaurs. For the next 120 million years (two-thirds of the history of the mammals), there were a number of very small-bodied, mammal-like "cynodont" lineages that approach mammals in most features, yet most paleontologists are unwilling to call them mammals. Some, like the tritylodonts (**Figure 20.7[E]**), had a very rodent-like skull with long incisors, no canines, rows of molars with multiple cusps for grinding; they had a long body shaped like a weasel. Others, such as the ictidosaurs or trithelodonts (**Figure 20.7[F]**), had very advanced jaws—*Diarthrognathus* (mentioned in Chapter 20) had both a dentary/squamosal and quadrate/articular jaw joint operating side-by-side. However, most paleontologists do not regard a fossil as mammalian unless it had a robust dentary-squamosal jaw joint; others use the presence of an incus and malleus in the middle ear as their criterion for which fossil is a mammal. This condition first appears in the latest Triassic and Early Jurassic with tiny, shrew-sized animals such as *Morganucodon* and *Sinoconodon* (**Figures 20.3**, **21.1**, and **21.2**). Although they have a robust dentary/squamosal jaw joint, they retain tiny vestiges of some of the other non-dentary jaw bones on the inside and back of the jaw. They had an upright mammalian posture, with a long blade on the iliac portion of the fused pelvis, an advanced thigh bone with several bony ridges for attaching muscles, and a broad shoulder blade with a spine down the middle (although the primitive amniote interclavicle bone was still present in the shoulder). Their teeth are specialized into incisors, canines, premolars, and molars, and the premolars had only a single replacement. However, they did not yet have the precise occlusion of the teeth seen in more advanced mammals.

DOI: 10.1201/9781003128205-21

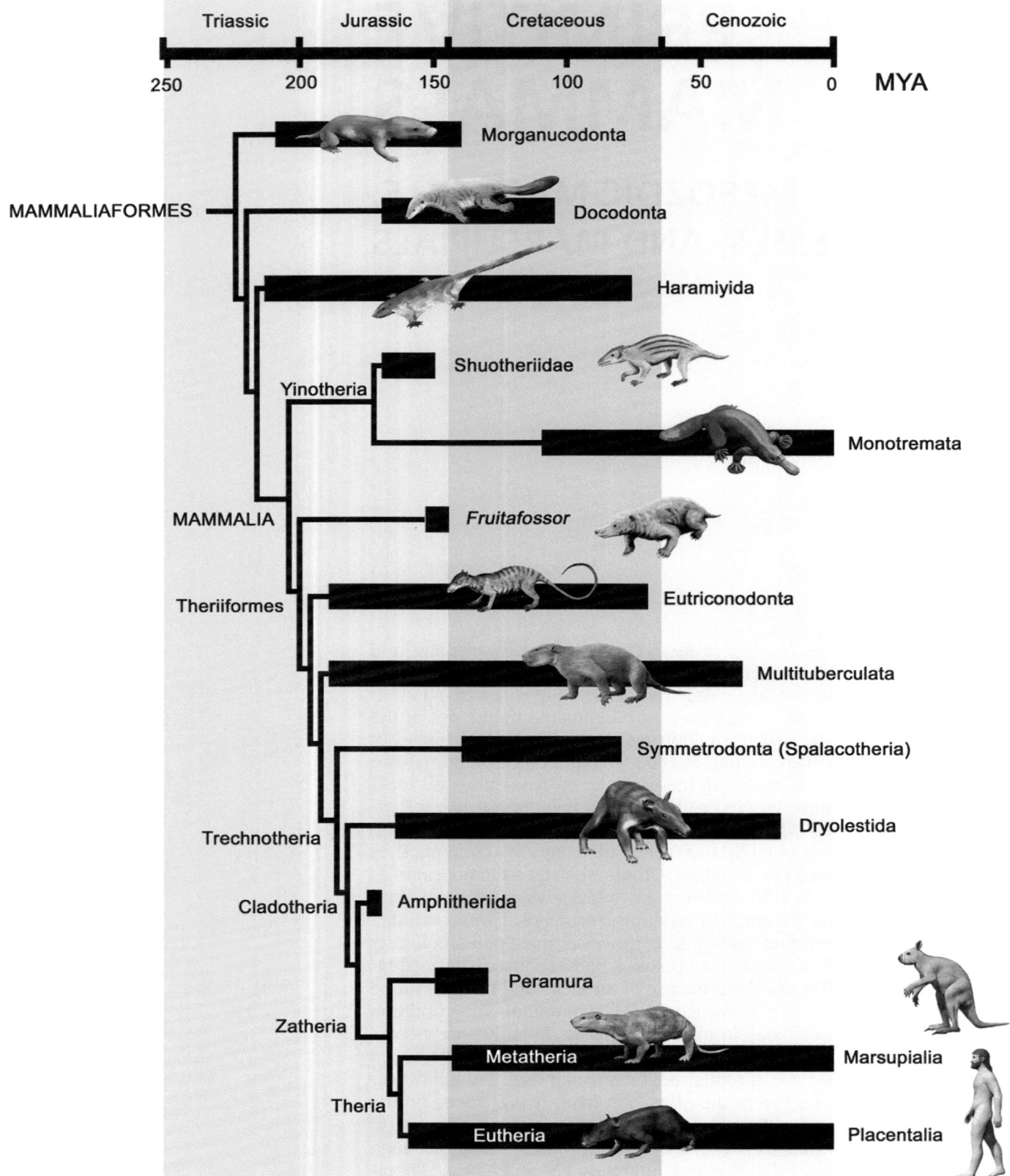

Figure 21.1 Phylogeny of Mesozoic mammals.

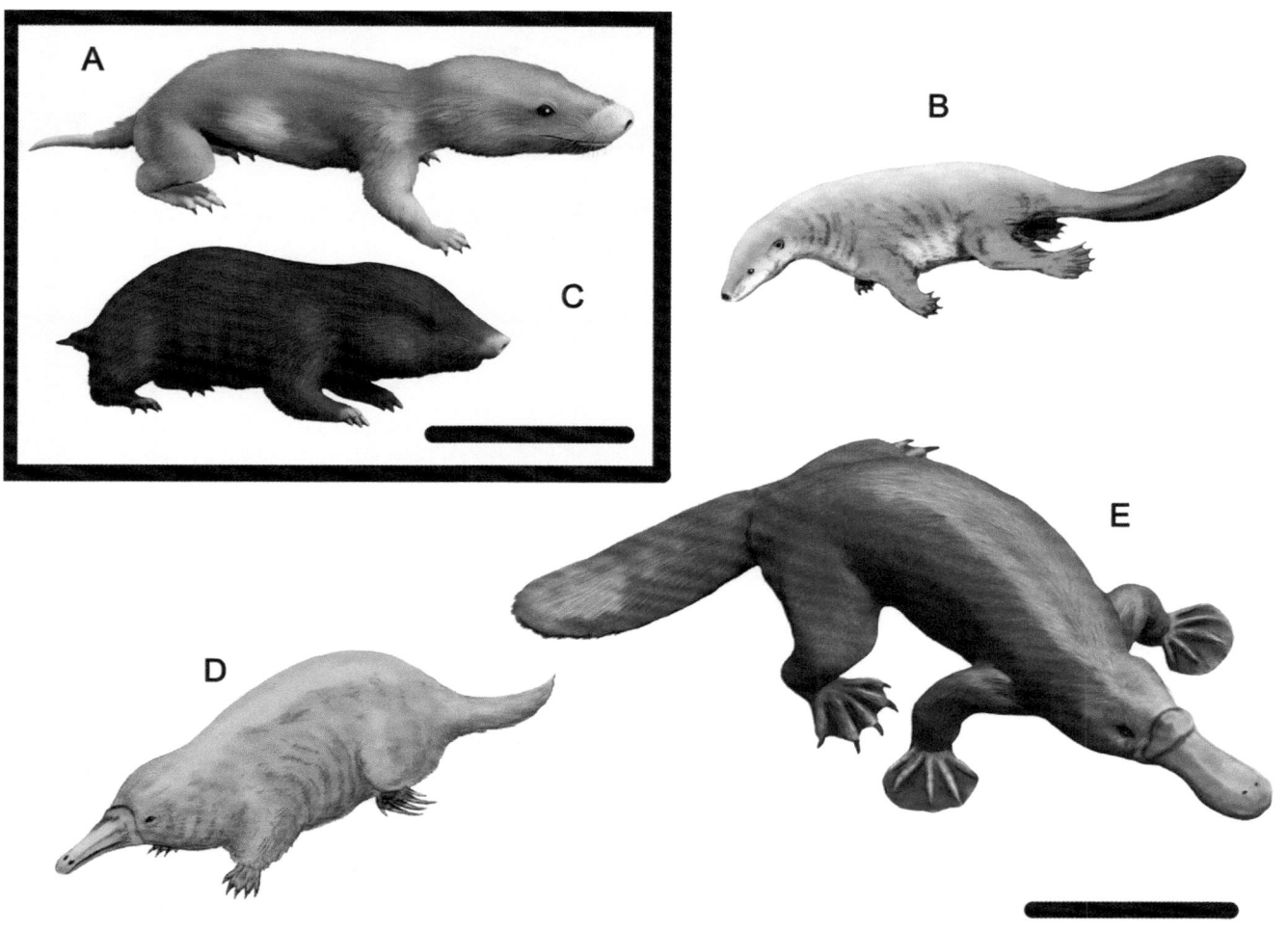

Throughout the Jurassic and Cretaceous, most mammals remained small (rat- to shrew-sized) animals that may have hidden from the dinosaurs in the undergrowth, or may have been mostly active at night (like a lot of smaller mammals are today). They remained as tiny creatures under the feet of the dinosaurs, or in the trees above them. Consequently, Mesozoic mammal fossils are also tiny, and tend to be fragmentary and hard to find. Most of what is known about Mesozoic mammals comes from tooth and jaw fragments, although in recent years, skeletal remains have been discovered for many major groups.

Figure 21.2 Reconstructions of some of the earliest mammals, including: (A) the morganucodont *Morganucodon*, the docodonts, (B) *Castorocauda* and (C) *Docofossor*, the monotremes, (D) *Steropodon*, and (E) *Obdurodon*. Scale for (A) and (C): 5 cm; scale for (B), (D), and (E): 20 cm.

MORGANUCODONTS

The most primitive and earliest known mammals were the morganucodonts, including *Morganucodon*, *Eozostrodon*, *Megazostrodon*, and about a dozen other genera from the latest Triassic and earliest Jurassic (**Figures 20.3**, **21.1**, **21.2[A]**, and **21.3[A]**). Most are known only from teeth and jaws, but a few are represented by nearly complete skeletons. They would have looked very much like modern shrews in both their size and shape, and their tiny pinhead-sized cheek teeth have triangular three-cusped crowns suitable for shearing up insects, as do the teeth of many modern insectivorous mammals. Even though they had a mammalian jaw joint between the dentary and squamosal bones, they still had some of the ancestral jaw bones as tiny vestiges in the inside back part of the jaw. They had large brains compared to most of the later mammal-like cynodonts, another mammalian feature. Unlike most primitive

Figure 21.3 Photos of (A) a model of the shrew-sized morganucodontid *Megazostrodon*. (B) The cat-sized triconodont *Gobiconodon*. (C) The skull of the Late Cretaceous multituberculate *Meniscoessus*, showing the cheek teeth with multiple tubercles on the grinding surface. (D) The skull of the squirrel-like Paleocene *Ptilodus*, showing the chisel-like incisors, large diastema or gap between the front teeth and the cheek teeth, and the large slicing blade-like lower fourth premolar in the lower jaw. [(A–C, E) Courtesy Wikimedia Commons. (D) Redrawn by N. Tamura, based on Romer (1966).]

A

B
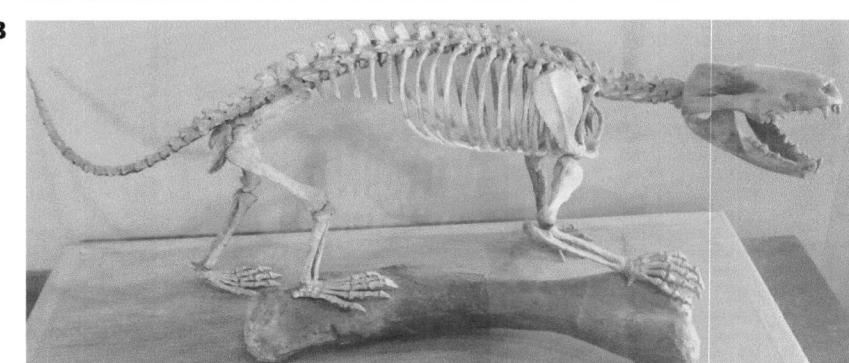

C
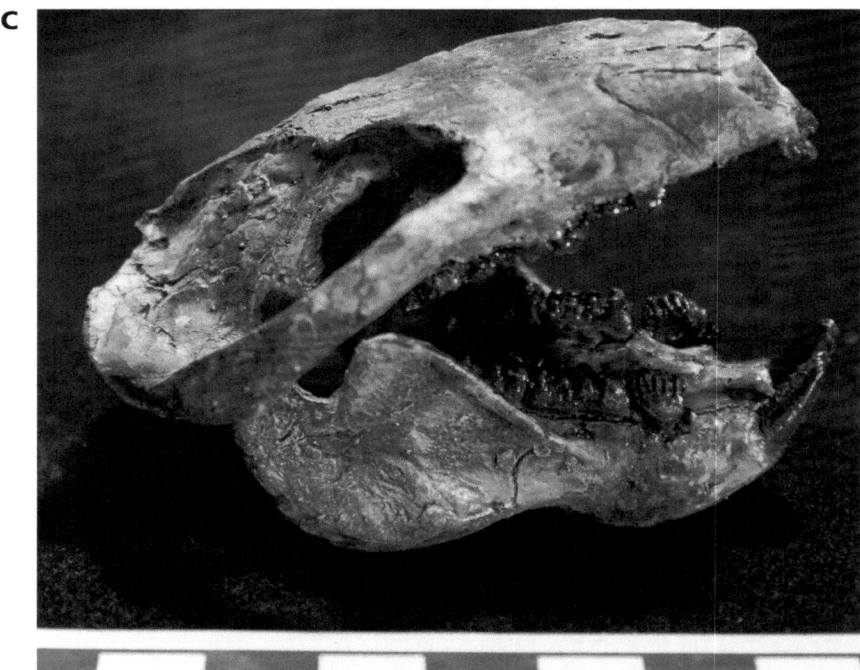

D
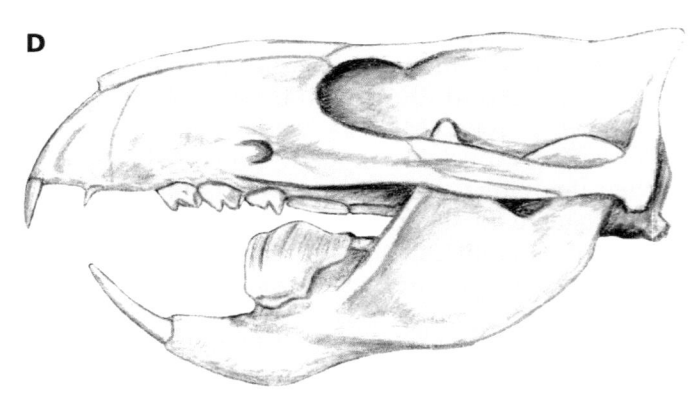

E

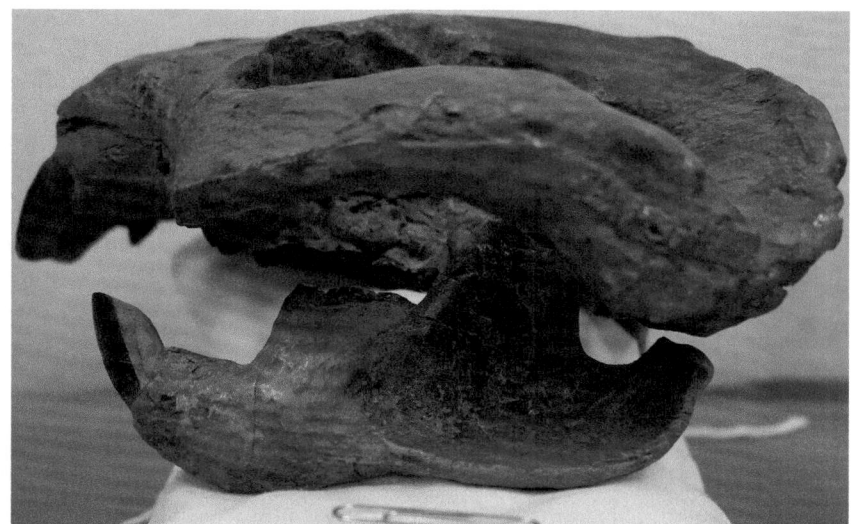

synapsids, which keep replacing all their teeth through their lives, the morganucodonts (and all other mammals) only have one cycle of tooth replacement, so only one round of baby teeth preceded their adult teeth.

DOCODONTS

For a century, jaws and teeth with strangely squared cheek teeth with saddle-shaped crowns (rather than the typical triangular three-cusped pattern of morganucodonts and triconodonts) were common from the same Upper Jurassic beds that produced other Mesozoic mammals as well as huge sauropod dinosaurs. Known as docodonts (**Figure 21.1**), these creatures had teeth suggesting they were slightly more omnivorous, not fully insectivorous like most Mesozoic mammals. Nothing more was known about the skeleton of these animals until the past decade or so, when more complete skeletons were found of fossils such as *Haldanodon* and *Castorocauda* (**Figure 21.2[B]**).

Haldanodon, from the Upper Jurassic Guimarota coal beds of Portugal, was about 38 cm (15 inches) long, and had a mole-like body with powerful forelimbs with short curved fingers and claws for digging. The snout bones had a roughened area for some sort of shield or nail-like hardening on the nose, consistent with the digging lifestyle. Its jaws were very robust, with features showing that they had relatively powerful bite forces.

Even more surprising was *Castorocauda*, known from a remarkable complete skeleton the Middle Jurassic (164 Ma) Daohugou lake beds of Inner Mongolia (**Figure 21.2[B]**). The specimen is so beautifully preserved that even the hair and soft tissues are visible. It is the oldest mammalian fossil to have hair preserved, although fossils such as *Thrinaxodon* suggest that hair occurred much earlier in synapsid evolution (see Chapter 20). The body of *Castorocauda* was about 43 cm (17 inches) long and it weighed about 800 g (almost two pounds), making it one of the largest mammals of its time. It was built a bit like otter, beaver, or platypus, with a broad swimming tail covered with scales like the tail of a beaver. This is how it got its name *Castorocauda lutrasimilis* ("beaver tail like an otter"). Its vertebrae were flattened like those of an otter or beaver and it even had webbed feet. Its forelimbs are very robust for digging or strong swimming, and look like those of the living platypus, which both digs and swims. Its docodont teeth have been highly modified for fish catching, so they resemble the teeth of seals and other fish eaters. Several recently described Chinese Jurassic docodonts were based on beautiful complete articulated skeletons with hair impressions from the Ganggou site in Hebei Province, China. One of them, *Agilodocodon*, was built much like a squirrel, and spent its life in trees. Another one from the same beds is known as *Docofossor* (**Figure 21.2[C]**), had skeletal features suggesting that it burrowed like a mole.

Thus, docodonts were very different from most of the shrew-like insectivorous Mesozoic mammals in that they had distinctively different teeth for an omnivorous diet, and some skeletons show they lived and acted more like moles, or squirrels, or otters.

MONOTREMES AND THEIR RELATIVES

When the platypus was first brought from Australia and shown to European scientists, they thought it was a hoax. A furry creature with a duck's bill and webbed feet? Surely, some prankster had sewn together parts of these animals! But the platypus is indeed a real animal. Even stranger, it is one of only two groups of mammals that lay eggs. When the eggs hatch, the females (like other mammals) nurse their young with milk, although their milk glands have no nipples. The mother simply "sweats" milk from the glands in the fur on their bellies so the babies can lap it up (Mammary glands are just highly modified sweat glands). Platypuses also have retained many other primitive reptilian features. There is no separation between their reproductive tract and their anus and urinary openings, but have a common cloaca like that of reptiles. In their shoulder girdle, they still have reptilian bones such as the coracoids and interclavicle.

The only other mammals with this odd arrangement are the "spiny anteaters" or echidnas of Australia and New Guinea. Even though they look slightly like other ant-eating mammals with a long sticky tongue and toothless snout, echidnas are encased in sharp spines that remind you of those of a hedgehog—but they are primitive egg-laying mammals and their spines and ant-eating mouth are due to convergent evolution. Together, the platypus and echidnas form a group known as the monotremes.

For the longest time, the relationships of monotremes to other mammals and to fossil Mesozoic mammals were a mystery. This was mainly because adult monotremes have no teeth, so it was impossible to compare them to Mesozoic mammals consisting only of tooth fossils. Only juvenile platypuses develop teeth that they shed later in development, and those juvenile teeth were really weird looking. Until recently, no fossil monotremes were known, but more and more fossils of ancient monotremes, whose teeth do resemble those of the platypus pups, have come to light. The oldest known monotreme fossil is *Teinolophos*, from the Early Cretaceous of Australia. The next youngest fossil was found in an Australian opal mine, and the entire bone of the fossil has been replaced with opal (**Figure 21.2[D]**). Named *Steropodon* ("lightning tooth", after the Lightning Ridge locality where it was found), it is also Early Cretaceous, but only about 105 Ma in age. Another similar monotreme from Lightning Ridge is *Kollikodon*. Each has teeth that somewhat resemble a juvenile platypus, and suggest an omnivorous diet. Both of these creatures were very large compared to other Cretaceous mammals, reaching length of 1 meter (3.3 feet) or more. Monotreme fossils remained rare through most of the Cenozoic, although the early-middle Miocene Riversleigh beds produce the platypus *Obdurodon* (**Figure 21.2[E]**), which looked very much like a modern platypus.

MULTITUBERCULATES

The next branch point in mammal evolution is a group known as the allotherians, which include the Triassic-Early Jurassic haramiyidyans, and the late Jurassic to late Eocene multituberculates (**Figure 21.1**). Unlike most of the groups mentioned so far (except for living monotremes), which had a low diversity through only part of the Mesozoic and vanished, the earliest allotherians (the haramiyidians) show up in the Late Triassic (200 Ma), and the multituberculates lasted through the entire early Cenozoic, vanishing near the end of the Eocene (35 Ma). Thus, their lineage lasted for 165 million years, the longest duration of any mammal group, living or extinct. In addition, they survived the mass extinction event at the end of the Cretaceous (which decimated the marsupials), and then were incredibly abundant, diverse, and successful in the jungles of the Paleocene and early Eocene of North America and Europe. In fact, they are so common that they are one of the most abundant fossils recovered from these beds, and so diverse that there are at least 200 species known. Recently, the haramiyidians have been considered to be unrelated to multituberculates.

So what were multituberculates? In shape and size, many of them resembled squirrels, although some were ground dwellers and may have lived more like marmots or beavers (**Figures 21.3[C–E]** and **21.4[A–D]**). Some multituberculates had long prehensile tails, and their wrists and ankles allowed their feet and hands to scale trees and even climb down a trunk head first. Most had a rodent-like skull with long chisel-like front incisors and a toothless gap (diastema) between the incisors and grinding cheek teeth (**Figure 21.3[C–E]**). These molar teeth of multituberculates, however, look like those of no other group of mammals, since they are relatively long and narrow with multiple rows of cusps or tubercles on them (hence their name). Even more remarkable is that the last premolar on the lower jaw on many multituberculates was developed into a distinctive large chisel-like slicing blade (**Figure 21.3[D]**), apparently used for slicing open seeds and nuts. This single blade-like tooth is enough to identify many species of multituberculates in the Paleocene.

Allotherians first originated with a poorly known Late Triassic group known as the haramiyidians. Much better specimens include the first multituberculates, such as the Jurassic Chinese form *Rugosodon* (**Figure 21.4[A]**). Through the Jurassic and Cretaceous, multituberculates flourished on the northern continents, especially North America and Eurasia. However, there was another group of poorly known mammals called Gondwanatheria from the Cretaceous of South America that may be relatives of the multituberculates of Eurasia and North America. And the Lower Cretaceous beds of Australia yield a single tooth, *Corriebataar*, that may establish the group in that continent as well.

After the Cretaceous extinctions wiped out the dinosaurs (other than birds) and decimated the marsupials, the multituberculates dominated the "squirrel" niche of small seed- and nut-eating tree-climbing mammal in the jungles of the Paleocene of North America and Europe (**Figure 21.4[C]**). They are so abundant that they can be found in almost any beds of this age, with their distinctive blade-lower premolars occurring in large numbers. Some of them, such as *Taeniolabis*, were large robust ground dwellers as large as a modern beaver (**Figures 21.3[C]** and **21.4[D]**).

By the end of the early Eocene, however, they declined in numbers and diversity. They were rare in the middle Eocene, and vanished completely by the end of the Eocene. The reasons for this are not known. There was a climatic change where the dense jungles of the middle Eocene were replaced by patchy forests and scrubland in the Oligocene, so their

Figure 21.4 Reconstructions of some Mesozoic mammals, including the multituberculates: (A) the Jurassic Chinese archaic multituberculate *Rugosodon*, (B) *Meniscoessus*, (C) the squirrel-like *Ptilodus*, and (D) the beaver-sized *Taeniolabis*; the euriconodonts, (E) *Yanoconodon*, and (F) *Repenomamus*; the early placental mammals (G) *Eomaia*, and (H) *Juramaia*. (Scale bar = 20 cm for A–D, F; 5 cm for E,G,H.)

habitat may have been disappearing. This is consistent with the decline in lemur-like primates at the same time, which also depended on a dense forest canopy. Others have suggested that they were outcompeted for the niche by the rapidly diversifying rodents, which might have been better adapted as the small seed- and nut-eating herbivores. Thus, after a long diverse run lasting over 165 million years, the longest-lived group of mammals finally vanished.

EUTRICONODONTS

Another group of Late Jurassic and Cretaceous mammals that were originally known from only teeth and a few jaws were the triconodonts. Described as early as 1861, they were distinctive in having cheek teeth with the crowns formed into a triangle with three conical cusps (hence their name, "Euriconodonta", "three-cone toothed"). Since this is the primitive mammalian tooth condition seen even in the earliest mammals, the morganucodonts were once included in the group as well.

Then a number of specimens with complete skeletons were found in China starting in the 2000s. They include the tiny (5 inches long) *Yano-conodon*, from the Yan Mountains of China, about 122 Ma (**Figure 21.4[E]**). It is remarkably primitive for a Cretaceous mammal, with middle ear bones that are not fully mammalian, and ribs in its lower back (a feature lost in nearly all mammals). Even better known in *Jeholodens*, from the Lower Cretaceous Jehol beds (about 125 Ma) of China, based on a complete specimen with hair and soft parts preserved. It was a small furry shrew-like creature with a long skinny tail, much like other Mesozoic mammals, but unlike *Yanoconodon* it was more advanced in having completely mammalian middle ear bones, and no ribs in its lower back. Even larger was *Sinoconodon*, from the Early Cretaceous of China.

A big departure from the normal Mesozoic mammal body plan was the discovery of the eutriconodont *Volaticotherium* from the Jurassic of Mongolia. It is based on a nearly complete skeleton that showed very long delicate limbs and body and the dark carbon film of a patagium, suggesting that it had a gliding membrane between its arms and legs like that of a flying squirrel. Just like docodonts, the triconodonts seemed to come in variety of shapes and ecological niches, all convergent on body forms that placentals would rediscover many millions of years later.

But most surprising of all was the discovery of *Repenomamus giganticus* (**Figure 21.4[F]**) from the Lower Cretaceous beds of Liaoning, China, about 123–125 Ma in age. It is one of several genera of triconodonts known as gobiconodonts, largely known from good specimens from the Cretaceous of Asia. Gobiconodonts tended to be large by Mesozoic mammal standards, with *Gobiconodon* itself (from the Early Cretaceous of the Gobi Desert in Mongolia, and also from Russia, China, and the Cloverly Formation of Wyoming) reaching 12 pounds (5.4 kg) and 20 inches (510 mm), about the size of a large cat (**Figure 21.3[B]**). It was much more robust and short-legged and probably not a good climber, but purely a ground predator.

Repenomamus giganticus is known from complete articulated specimens with soft parts and fur impressions, and as the name implies, it was a gigantic mammal by the standards of the Mesozoic. About the size of a very large dog, big specimens reached 1 meter (3 feet) in length and weight about 12–14 kg (26–31 lb). They had a massive, hulking body with robust sprawling limbs and a long tail and snout, and large hands and feet that allowed them to walk on their palms and soles, not just their toes. Most surprisingly, they were larger than some of the small dinosaurs (like *Graciliraptor*) from the same beds, and a specimen of *Repenomamus* was found with baby *Psittacosaurus* (a distant relative of *Triceratops*) in their stomachs! Although most Mesozoic mammals were tiny and hid from the dinosaurs during the Jurassic and Cretaceous, *Repenomamus giganticus* managed to turn the tables on the dinosaurs.

The relationships of triconodonts are controversial. Most evidence shows that they are more advanced than morganucodonts, docodonts, or monotremes, but not as advanced as multituberculates or therian mammals (**Figure 21.1**). However, other scientists place them just outside the branch point for the monotremes, or outside the monotreme-therian mammalian group altogether.

THERIAN ANCESTORS

Monotremes, multituberculates, docodonts, triconodonts, and morganucodonts were mostly extinct side branches of the mammal tree; only one group (monotremes) still survives. The main surviving lineage of living mammals fall into two groups: the Metatheria (marsupials) and the Eutheria (placentals), both of which are known from excellent complete specimens with fur and soft parts from Chinese beds about 125 Ma in age. One of these fossils, *Sinodelphys*, is the oldest known marsupial, and it is found in the same beds as *Eomaia*, one of the oldest known placentals (**Figures 21.4**, **21.5**, and **22.1**). Together, the marsupials and placentals form a group known as the Theria, which are distinguished by losing the last of the reptilian bones in the shoulder (coracoids and interclavicles) found in many Mesozoic groups, as well as having a

5 cm

Figure 21.5 Photo of the complete articulated skeleton of *Sinodelphys*, the oldest known marsupial fossil, from the Early Cretaceous of China. (Courtesy Z.-X. Luo.)

distinctive type of ankle structure, and cheek teeth with are formed into a "reversed triangle" pattern of three cusps (the tribosphenic tooth). These teeth work something like pinking shears, whose blades have triangular "teeth" sliding in the V-shaped valleys between each other. Such teeth are common in insectivorous mammals and excellent for chopping up insect cuticle. All living therian mammals are also defined by the ability to give birth to live young, mammary glands with distinct nipples, and many other features not found in monotremes or most other Mesozoic groups.

The earliest known relatives of the therians occurred in the Jurassic, with two large groups of therians that were more primitive than either placentals or marsupials appeared. One group, the spalacotherioid symmetrodonts, which had three-cusped cheek teeth that vaguely resembled those of triconodonts, but clearly had the "reversed triangle" pinking-shears style of occlusion. Although over 30 genera are known from the Jurassic and Early Cretaceous of Eurasia and North America, most are based on just fragmentary jaws and teeth. Only *Zhangeotherium* is known from a complete skeleton, complete with fur and soft tissues, from the Lower Cretaceous beds of Liaoning Province, China.

The second group, the Dryolestoidea, was among the most common mammals in the Upper Jurassic beds of North America and Eurasia, and possibly the Cretaceous and Cenozoic of South America. Dryolestoids were very distinctive. Known from several dozen genera (mostly Late Jurassic of Europe and North America), their upper teeth had a triangular crown pattern and the teeth were very narrow and compressed front-to-back, so there were as many as 8–9 molars (most mammals have only 3–4), and as many as 11–12 cheek teeth (most mammals have no more than 7–8). Their lower cheek teeth also had very narrow crowns, compressed in a front-to-back direction, with a hook-like shelf in the back of each. The only dryolestoid known from a skeleton is *Henkelotherium*, from the Upper Jurassic Guimarota beds of Portugal. It was a shrew-sized creature with a long tail, and limbs that suggest it was largely arboreal.

By the Early Cretaceous, mammals with more advanced teeth are known, living alongside the surviving archaic groups, such as the triconodonts, symmetrodonts, and dryolestoids. These more advanced mammals have added a new cusp (the protocone) to the inside corner of the upper cheek teeth, making it essentially a modern mammalian molar (known as the tribosphenic tooth). This basic tribosphenic prototype would be highly modified in later mammals, but the position and homologies of the primary cusps are the same, no matter what the tooth is used for. In the Jurassic, the tribosphenic therian mammalian lineage had split into the two major living groups, the marsupials (Metatheria) and the placentals (Eutheria). Most mammals from the uppermost Cretaceous beds that produce *Tyrannosaurus* or *Triceratops*, or the lowermost Paleocene beds just above them, are clearly opossum-like marsupials, or placentals that seem to be ancestral to hoofed mammals (*Protungulatum*), primates (*Purgatorius*), carnivore relatives (*Cimolestes*), and lots of insectivorous forms. Few are known from more than teeth or jaws, however, so it is difficult to place these fossils within any living orders of mammals. Multituberculates were also extremely abundant, and all three groups made it through the end-Cretaceous extinctions (although marsupials were decimated).

THE MARSUPIALS OR METATHERIA

Most people are familiar with the opossum, kangaroo, koala bear, and (thanks to *Looney Tunes*) the Tasmanian devil. All these animals are marsupials (Metatheria), or pouched mammals, so called because females carry their young in a pouch on their belly. Today, marsupials are the dominant group of mammals only on the island continent of Australia, where there are few native placental mammals.

In many instances in the geological past where marsupials have been forced to compete with placental mammals, they have lost that race for survival, and this has led people to argue that they are more primitive and inferior to placentals. However, marsupials are not inferior; they are just very different in their body plan and evolutionary strategies. The most obvious difference is in their reproduction, which works very differently from the reproduction of a placental. A female marsupial has a pair of uteri (unlike the single uterus of a placental) that open into a vagina with three different branches, a central medial vagina (the birth canal), and two lateral vaginas that lead into the uterus. The penis of many male marsupials is forked, so that it can deliver sperm to both lateral vaginas. Once fertilization occurs, the embryo develops for only a few weeks, after which the young is born prematurely, with only its forelimbs and mouth functional. These limbs are important, because after they are born, the embryo must crawl up the fur of its mother's belly to find the opening of the pouch. In most marsupials, the newborn embryo is the size of a bee, or even smaller. Once it reaches the pouch, it crawls in and latches onto a nipple, where it completes its development.

By contrast, a placental embryo is surrounded by an organ, the placenta (also known as the "afterbirth"). The placenta develops from the chorionic and amniotic membranes which protects the embryo during development. The placenta also serves to pass gases, food, hormones, and waste products between mother and embryo. The placenta has another important function—it serves as a barrier against the mother's immune system, so that when the embryo develops its own immune signature, the mother's system will not reject it as a "foreign object". By contrast, marsupials have no such protection, so the young are born prematurely before they can suffer immune rejection.

Marsupial reproduction allows multiple generations of young to be raised at once. A marsupial mother can carry one baby in the pouch, an embryo in the uterus, and take care of a third generation still living in her vicinity, so the generational turnover can be quite rapid. If a marsupial mother is in great danger from a predator or from starvation, she can drop the babies in her pouch at minimal risk to herself, and live to breed again. By contrast, a placental mother cannot abort her fetus without great risk to herself, so she is obliged to carry it to term, even if it means death for her. In other words, marsupial mothers make less parental investment in each young, but suffer less risk as a result. The main disadvantages of marsupial reproduction are that the young are born with a smaller neocortex in the brain, due to their abbreviated development, and therefore require a longer time to mature and be weaned from the mother. By contrast, some mammals (such as rabbits or rodents) can shorten their generation time until the babies are weaned, and thus can produce offspring faster than most marsupials. In other cases, placental babies (such a zebra or wildebeest or giraffes) are born almost fully functional, because they must be able to run with their mother within hours after they are born, or they will be eaten by predators.

Today, marsupials comprise most of the native fauna of Australia, and are restricted to that continent (except for the opossums and their South American relatives). In the Cretaceous, however, marsupials were widespread and found on most of the continents (they were the most common mammals in North America during the reign of *Tyrannosaurus*). After the end-Cretaceous event, however, the balance shifted to placentals on the northern continents, and only opossums persisted through much of the Cenozoic in Europe or North America. By contrast, the marsupials did very well on the southern continents of South America and Australia, where there was little placental competition (and also Antarctica, before it froze over).

In South America, there were no large carnivorous placental mammals during the early Cenozoic, so marsupials occupied that niche. Some of them (the sparassodonts, or borhyaenids) were shaped much like wolves or hyaenas, while another (*Thylacosmilus*) was a saber-tooth that closely resembles the placental saber-toothed cat (**Figures 21.6[A]** and **21.7[B]**). Most of these marsupial predators disappeared as placental carnivorans came from North America when the Panamanian land bridge opened in the Pliocene. However, South America still supports a large diversity of opossum-like marsupials.

In Australia, the situation was even simpler. Only one possible placental fossil is known from that continent (other than bats, who flew there) before humans arrived with their dingo dogs and other animals in the late Pleistocene, so Australia was apparently isolated from placentals during most of the Cenozoic. In the absence of such placental competition, marsupials evolved into a great variety of body forms to fill the niches occupied by placentals on other continents (**Figures 21.6** and **21.8**). Even today, there are marsupial equivalents of moles, mice, cats, flying squirrels, wolves, groundhogs, anteaters, and many other body forms. In addition, there are many body forms that placentals never invented. Kangaroos are the main herbivorous marsupials, but they get along by hopping, an innovation that large hoofed placental mammals

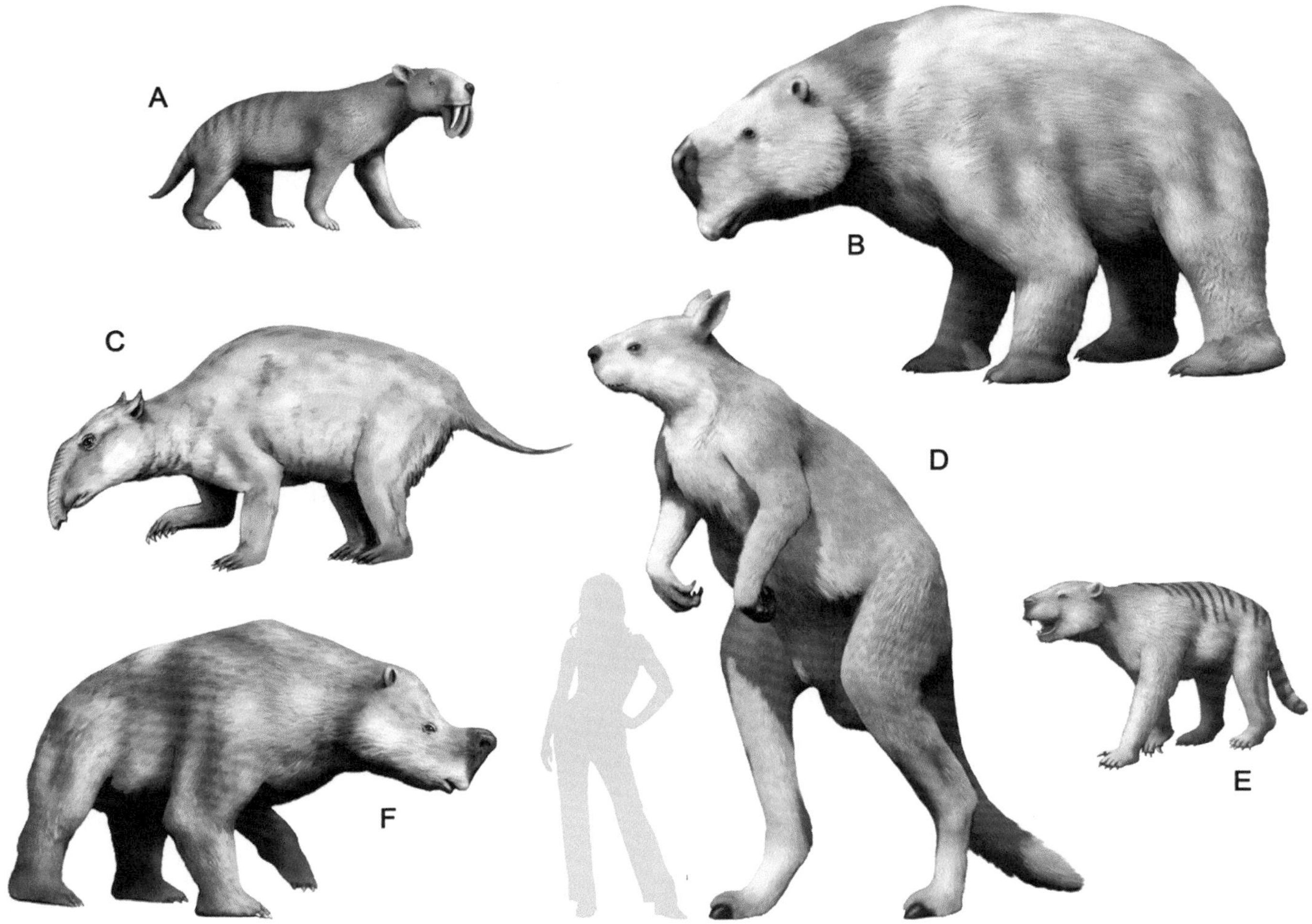

Figure 21.6 Reconstructions of some large extinct marsupials, including: (A) the marsupial "sabertooth" *Thylacosmilus*, (B) the rhino-sized wombat relative *Diprotodon*, (C) the weird *Palorchestes*, with its tapir-like snout and claws like a ground sloth, (D) the giant short-faced kangaroo *Procoptodon*, (E) the "marsupial lion" *Thylacoleo*, (F) the diprotodont *Zygomaturus*.

Figure 21.7 Fossils of some of the more extreme extinct marsupials, including (A) the skeleton of the marsupial "lion", *Thylacoleo*. (Courtesy Wikimedia Commons.)

Figure 21.7 (Continued) (B) the skull of the marsupial "sabertooth" *Thylacosmilus*, showing the convergent evolution on saber-toothed cats. However, there are many differences in detail that show it is a marsupial, and it is the only saber-tooth with a long complete flange on the lower jaw, that protected the saber like a scabbard when the mouth closed. (C) The rhino-sized wombat-like *Diprotodon*.

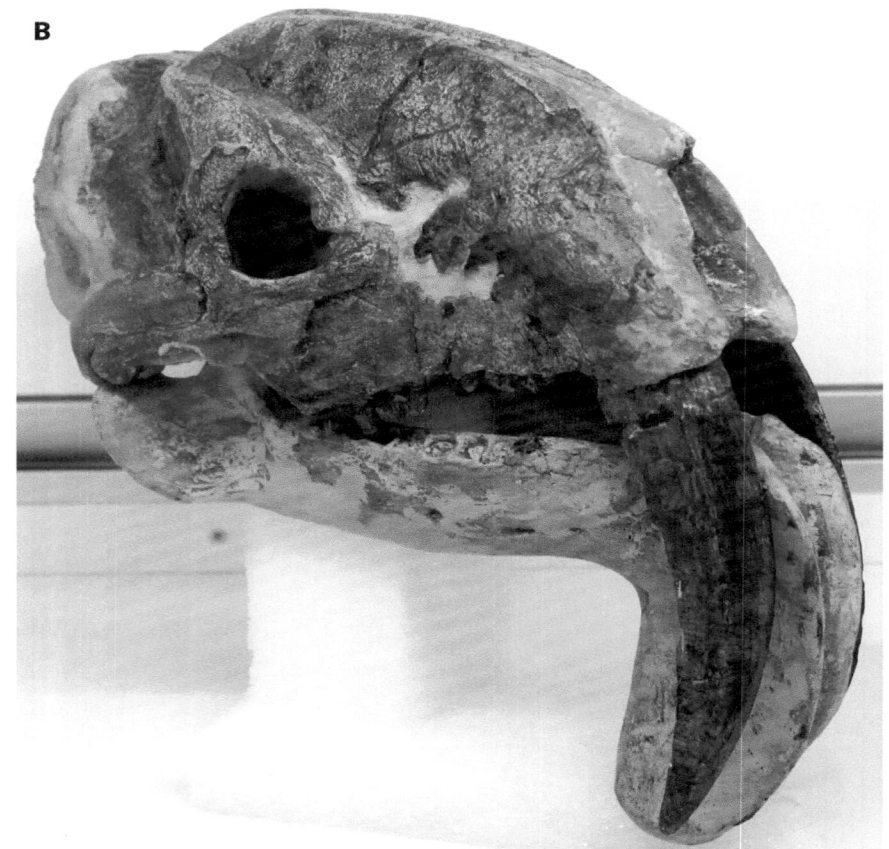

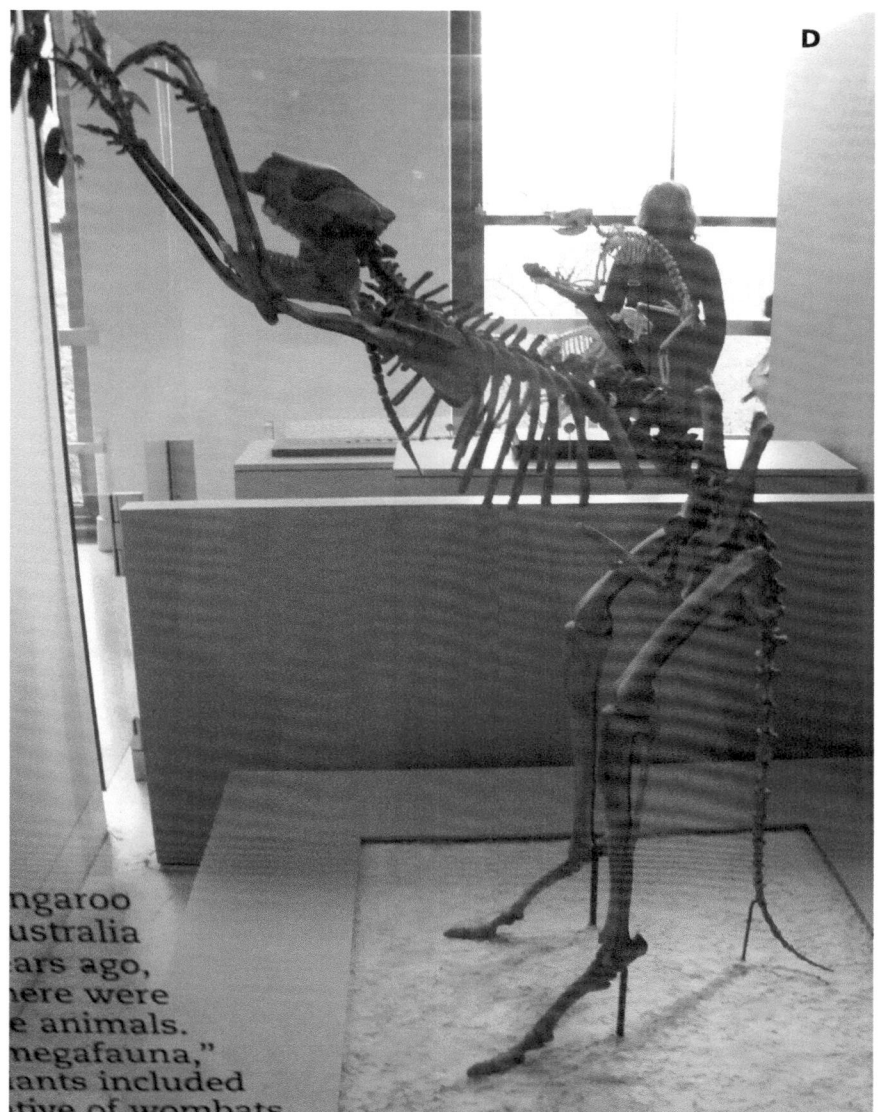

D

Figure 21.7 (Continued) (D) The gigantic short-faced kangaroo *Simonsthenurus*.

never discovered (although many groups of small placentals, such as kangaroo rats and jerboas, did). In the Australian Pleistocene, there were giant wombat relatives, the diprotodonts, which were the size of rhinos (**Figures 21.6[B]** and **21.7[C]**), and kangaroos almost twice the size of any living species (**Figures 21.6[E]** and **21.7[D]**). There was even a marsupial "lion", *Thylacoleo* (**Figures 21.6[E]** and **21.7[A]**) which had a peculiarly short skull with long cutting blades in its jaws instead of multiple shearing teeth, like a placental carnivore.

Many of these giant Pleistocene marsupials vanished as the Ice Ages ended and climate changed. However, the biggest threat for marsupials was the invasion of Aborigines to Australia about 46,000 years ago, and with them their placental dogs (dingoes). When Europeans arrived about two centuries ago, they also brought other destructive placentals, such as goats, rats, and rabbits. Today, many of the native Australian marsupials are endangered as their habitats disappear and placental mammals continue to take over. Australia has long been a "living museum" of unique animals that evolved in isolation through over 70 million years, but that "museum" may vanish within another century.

Placentals

Marsupials

Anteater
(*Myrmecophaga*)

Anteater
(*Myrmecobius*)

Mouse
(*Mus*)

Mouse
(*Dasycercus*)

Groundhog
(*Marmota*)

Wombat
(*Phascolomys*)

Flying squirrel
(*Glaucomys*)

Flying phalanger
(*Petaurus*)

Wolf
(*Canis*)

Tasmanian wolf
(*Thylacinus*)

Mole
(*Talpa*)

Mole
(*Notoryctes*)

Ocelot
(*Felis*)

Native cat
(*Dasyurus*)

Figure 21.8 In the absence of competition from placental mammals, Australian marsupials evolved some remarkable mammals that converged on the body form of placental mammals, even though they are not closely related. Presumably, these Australian forms occupied the niches that placentals occupied elsewhere. These convergent forms include Australian marsupial equivalents of anteaters, mice, groundhogs, flying squirrels, wolves, moles, and cats. To that could be added the wolverine-like Tasmanian devil, the lemur-like cuscus, and the extinct rhino-sized diprotodonts, and the marsupial "lion" *Thylacoleo*. In South America, marsupials such as the "sabertooth" marsupial *Thylacosmilus*, and the hyaena-like borhyaenids or sparassodonts, occupied niches that placentals filled elsewhere. (Modified from several sources.)

FURTHER READING

Agnolin, F.L.; Ezcurra, M.D.; Pais, D.F.; Salisbury, S.W. 2010. A reappraisal of the Cretaceous non-avian dinosaur faunas from Australia and New Zealand: Evidence for their Gondwanan affinities. *Journal of Systematic Palaeontology*. 8 (2): 257–300.

Beck, R.M.D.; Godthelp, H.; Weisbecker, V.; Archer, M.; Hand, S.J. 2008. Australia's oldest marsupial fossils and their biogeographical implications. *PLoS One*. 3 (3): e1858.

Celik, M.A.; Phillips, M.J. 2020. Conflict resolution for Mesozoic mammals: Reconciling phylogenetic incongruence among anatomical regions. *Frontiers in Genetics*. 11: 651.

Chen, M.; Wilson, G.P. 2015. A multivariate approach to infer locomotor modes in Mesozoic mammals. *Paleobiology*. 41 (2): 280–312.

Hand, S.J.; Long, J.; Archer, M.; Flannery, T.F. 2002. *Prehistoric Mammals of Australia and New Guinea: One Hundred Million Years of Evolution*. Johns Hopkins University Press, Baltimore, MD.

Hopson, J.A. 1994. Synapsid evolution and the radiation of non-eutherian mammals, pp. 190–219. In Prothero, D.R.; Schoch, R.M., eds. *Major Features of Vertebrate Evolution*. Paleontological Society Short Course 7. Paleontological Society, Lawrence, KS.

Jackson, S.; Groves, C. 2015. *Taxonomy of Australian Mammals*. CSIRO Publishing, Australia.

Ji, Q.; Luo, Z.-X.; Yuan, C.-X.; Tabrum, A. R.; Yuan, C.-X. 2006. A swimming mammaliaform from the Middle Jurassic and eco-morphological diversification of early mammals. *Science*. 311 (5764): 1123–1127.

Kemp, T.S. 1982. *Mammal-Like Reptiles and the Origin of Mammals*. Academic Press, London.

Kemp, T.S. 2005. *The Origin and Evolution of Mammals*. Oxford University Press, Oxford.

Kielan-Jaworowska, Z.; Hurum, J.H. 2001. Phylogeny and systematics of multituberculate mammals. *Palaeontology*. 44 (3): 389–429.

Kielan-Jaworowska, Z; Luo, Z.; Cifelli, R.L. 2004. *Mammals from the Age of Dinosaurs: Origins, Evolution, and Structure*. Columbia University Press, New York.

Krause, D.W. 1986. Competitive exclusion and taxonomic displacement in the fossil record: The case of rodents and multituberculates in North America. *Contributions to Geology, University of Wyoming, Special Paper*. 3: 95–117.

Krause, D.W.; Hoffmann, S.; Werning, S. 2017. First postcranial remains of Multituberculata (Allotheria, Mammalia) from Gondwana. *Cretaceous Research*. 80: 91–100.

Lillegraven, J.A. 1974. Biological considerations of the marsupial-placental dichotomy. *Evolution*. 29: 707–722.

Lillegraven, J.A.; Kielan-Jaworowska, Z.; Clemens, W.A., eds. 1979. *Mesozoic Mammals: The First Two-Thirds of Mammalian History*. University of California Press, Berkeley.

Luo, Z.; Meng, Q.; Ji, Q.; Liu, D.; Zhang, Y.; Neander, A. I. 2015. Evolutionary development in basal mammaliaformes as revealed by a docodontan. *Science*. 347: 6223, 760–764.

Luo, Z.; Yuan, C.; Meng, Q.; Ji, Q. 2011. A Jurassic eutherian mammal and divergence of marsupials and placentals. *Nature*. 476 (7361): 442–445.

Luo, Z.; Ji, Q.; Wible, J. R.; Yuan, C.-Z. 2003. An early Cretaceous tribosphenic mammal and metatherian evolution. *Science*. 302 (5652): 1934–1940.

Luo, Z.; Martin, T. 2007. Analysis of molar structure and phylogeny of docodont genera. *Bulletin of the Carnegie Museum of Natural History*. 39: 27–47.

McKenna, M.C. 1975. Toward a phylogenetic classification of the Mammalia, pp. 21–46. In Luckett, W. P.; Szalay, F. S., eds. *Phylogeny of the Primates*. Plenum Press, New York.

McKenna, M.C.; Bell, S. K. (1997). *Classification of Mammals above the Species Level*. Columbia University Press, New York.

Meng, Q.; Ji, Q.; Zhang, Y.; Liu, D.; Grossnickle, D. M.; Luo, Z. 2015. An arboreal docodont from the Jurassic and mammaliaform ecological diversification. *Science*. 347: 6223, 760–764.

Moyal, A.M. 2004. *Platypus: The Extraordinary Story of How a Curious Creature Baffled the World*. The Johns Hopkins University Press, Baltimore, MD.

Murphy, W.J.; Pevzner, P.A.; O'Brien, S.J. 2004. Mammalian phylogenomics comes of age. *Trends in Genetics*. 20: 631–639.

Nilsson, M.A.; Churakov, G.; Sommer, M.; Van Tran, N.; Zemann, A.; Brosius, J.; Schmitz, J. 2010. Tracking marsupial evolution using archaic genomic retroposon insertions. *PLoS Biology*. 8 (7): e1000436.

Novacek, M.J. 1992. Mammalian phylogeny: Shaking the tree. *Nature*. 356: 121–125.

Ostrander, G.E. 1984. The early Oligocene (Chadronian) Raben Ranch Local Fauna, northwest Nebraska: Multituberculata; with comments on the extinction of the Allotheria. *Transactions of the Nebraska Academy of Sciences*. 10: 71–80.

Prevosti, F.J.; Forasiepi, A.; Zimicz, N. 2011. The evolution of the Cenozoic terrestrial mammalian predator guild in South America: Competition or replacement? *Journal of Mammalian Evolution*. 20: 3–21.

Prothero, D.R. 1981. New Jurassic mammals from Como Bluff, Wyoming, and the interrelationships of the non-tribosphenic Theria. *Bulletin of the American Museum of Natural History*. 167: 277–326.

Prothero, D.R. 1994. Mammalian evolution, pp. 238–270. In Prothero, D.R.; Schoch, R.M., eds. *Major Features of Vertebrate Evolution*. Paleontological Society Short Course 7. Paleontological Society, Lawrence, KS.

Prothero, D.R. 2016. *The Princeton Field Guide to Prehistoric Mammals*. Princeton University Press, Princeton, NJ, 240 pp.

Rich, T.H.; Vickers-Rich, P.; Flannery, T.F.; Kear, B.P.; Cantrill, D.J.; Komarower, P.; Kool, L.; Pickering, D.; Rusler, P.; Morton, S.; van Klaveren, N.; Fitzgerald, E.M.G. 2009. An Australian multituberculate and its palaeobiogeographic implications. *Acta Palaeontologica Polonica*. 54 (1): 1–6.

Rougier, G.W.; Wible, J.R.; Hopson, J.A. 1996. Basicranial anatomy of *Priacodon fruitaensis* (Triconodontidae, Mammalia) from the Late Jurassic of Colorado, and a reappraisal of mammaliaform interrelationships. *American Museum Novitates*. 3183.

Rowe, T.S. 1988. Definition, diagnosis, and origin of Mammalia. *Journal of Vertebrate Paleontology*. 8 (3): 241–264.

Sánchez-Villagra, M. 2012. Why are there fewer marsupials than placentals? On the relevance of geography and physiology to evolutionary patterns of mammalian diversity and disparity. *Journal of Mammalian Evolution*. 20 (4): 279–290.

Savage, D.E.; Russell, D.E. 1983. *Mammalian Paleofaunas of the World*. Addison Wesley, Reading, MA.

Savage, R.J.G.; Long, M.R. 1986. *Mammal Evolution: An Illustrated Guide*. Facts-on-File Publications, New York.

Shoshani, J.; McKenna, M.C. 1998. Higher taxonomic relationships among extant mammals based on morphology, with selected comparisons of results from molecular data. *Molecular Phylogenetics and Evolution*. 9: 572–584.

Simpson, G.G. 1950. History of the Fauna of Latin America. *American Scientist*. 38 (3): 361–389.

Springer, M.S.; Meredith, R.W.; Janecka, J.E.; Murphy, W.J. 2011. A historical biogeography of Mammalia. *Proceedings of the Royal Society B*. 366: 2478–2502.

Szalay, F.S. 1982. A new appraisal of marsupial phylogeny and classification, pp. 621–640. In Archer, M., ed. *Carnivorous Marsupials*. 2. Royal Zoological Society of New South Wales, Sydney, Australia.

Szalay, F.S.; Novacek, M.J.; McKenna, M.C., eds. 1993. *Mammal Phylogeny*. Springer-Verlag, Berlin.

Turner, A.; Anton, M. 2004. *National Geographic Prehistoric Mammals*. National Geographic Society, Washington, DC.

Van Rheede, T.; Bastiaans, T.; Boone, D.; Hedges, S.; De Jong, W.; Madsen, O. 2006. The platypus is in its place: Nuclear genes and indels confirm the sister group relation of monotremes and therians. *Molecular Biology and Evolution*. 23 (3): 587–597.

Williamson, T.E.; Brusatte, S.L.; Secord, R.; Shelley, S. 2015. A new taeniolabidoid multituberculate (Mammalia) from the middle Puercan of the Nacimiento Formation, New Mexico, and a revision of taeniolabidoid systematics and phylogeny. *Zoological Journal of the Linnean Society*. 177: 183–208.

Wilson, G.P.; Ekdale, E.G.; Hoganson, J.W.; Calede, J.J.; Linden, A.V. 2016. A large carnivorous mammal from the Late Cretaceous and the North American origin of marsupials. *Nature Communications*. 7.

THE PLACENTAL EXPLOSION

THE MAMMALS DIVERSIFY

The placental or eutherian mammals comprise about twenty living orders and several extinct ones. The morphological and adaptive range of this group is extraordinary; diversification has produced lineages as varied as human and their primate relatives, flying bats, swimming whales, ant-eating anteaters, pangolins, and aardvarks, a baroque extravagance of horned, antlered, and trunk-nosed herbivores (ungulates), as well as the supremely diverse rats, mice, beaver and porcupines of the order Rodentia. Such adaptive diversity, and the emergence of thousands of living and fossil species, apparently resulted from a radiation beginning in the late Mesozoic between 65 and 80 Ma. This explosive radiation is one of the more intriguing chapters in vertebrate history.

—Michael J. Novacek, "The Radiation of
Placental Mammals", 1994

PLACENTALS

Placentals make up about 95% of the fossil and living mammals. In Simpson's 1945 classification of mammals, there were over 2600 placental genera, compared to a few hundred marsupials, and a few dozen multituberculates, monotremes, and other Mesozoic forms. The number of described taxa has greatly increased in the last 77 years.

As mentioned in Chapter 21, the fundamental split between most living mammals is the pouched mammals (marsupials) and the placental mammals, which give birth to fully developed young. These differences in reproduction are hard to recognize in fossils, however, so we must resort to features visible in the skeleton. There are quite a few that are useful. Placentals have lost the epipubic bone ("marsupial bone") in the hip joint, and have only three molars but four premolars in their cheek teeth. In their upper molar crowns, the outer portion next to the lips (the stylar area) is greatly reduced compared to those in marsupials, while the inner area of the crown is emphasized. These features, and several others, allow most fossils to be confidently identified as marsupial or placental.

The oldest known fossil that can be confidently related to placentals is *Juramaia* from the Late Jurassic of China, about 160 Ma (**Figure 22.1[A]**). It is a beautifully preserved specimen (although much of the back of it is missing), but it clearly shows the characteristic placental teeth and other anatomical features. The next youngest placental fossil is *Eomaia*, from the Early Cretaceous of China, about 125 Ma (**Figure 22.1[B]**). It is a flattened but complete specimen in an extraordinary state of preservation, with even the fur impressions and other soft tissues. Although the creature was only

DOI: 10.1201/9781003128205-22

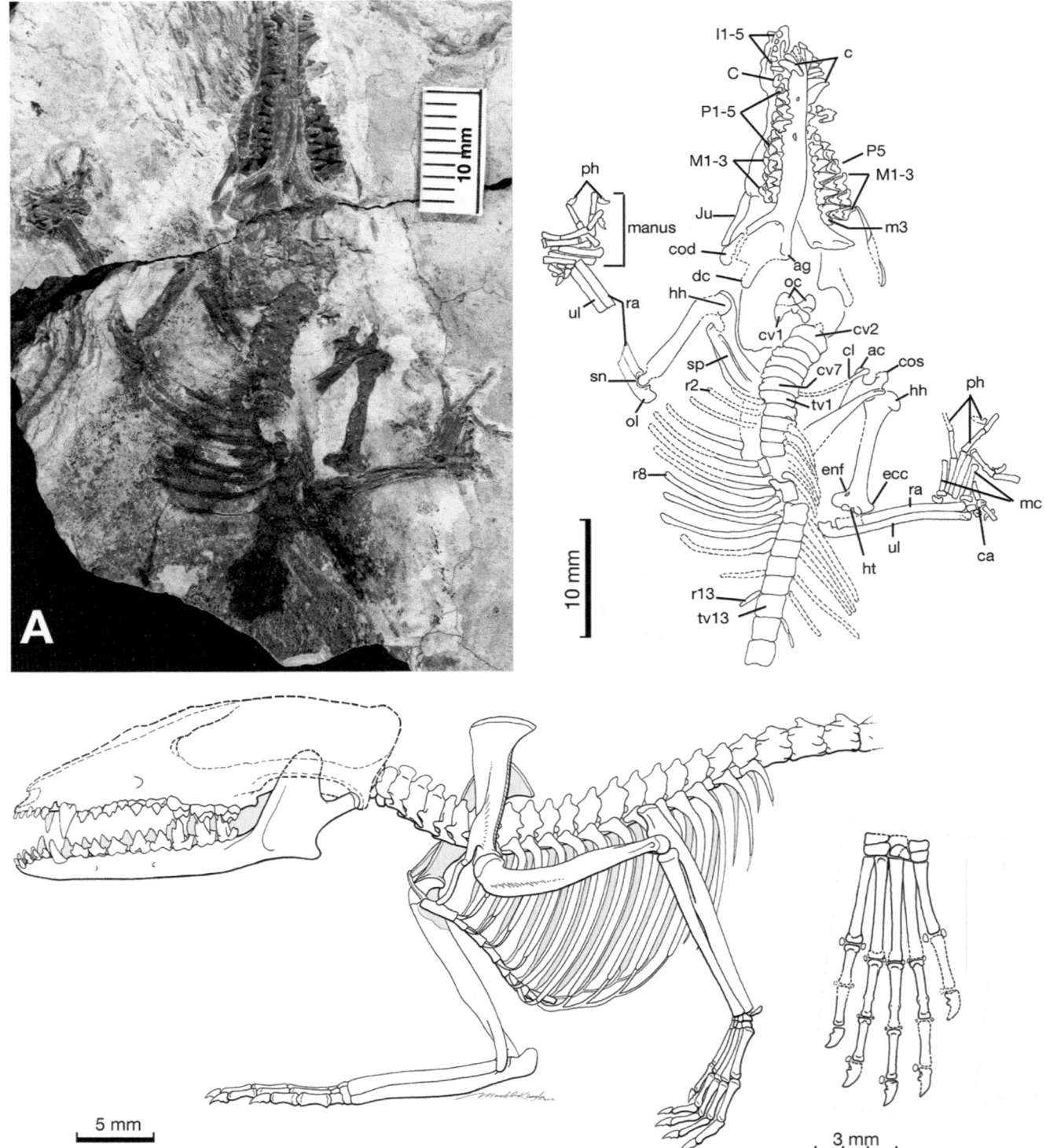

Figure 22.1 (A) The earliest known placental fossil, *Juramaia*, from the Late Jurassic of China. (Courtesy Z.-X. Luo.)

10 cm (4 inches) long and probably weighed only 20–25 grams (0.7–0.8 oz), about the size of a shrew, it was remarkably advanced for its time.

By the Late Cretaceous, placentals began to really evolved rapidly, and there are numerous complete uncrushed skeletons of primitive placentals from Mongolia and China known as kennalestids and asioryctids. In the latest Cretaceous and earliest Paleocene, we begin to see the first

B

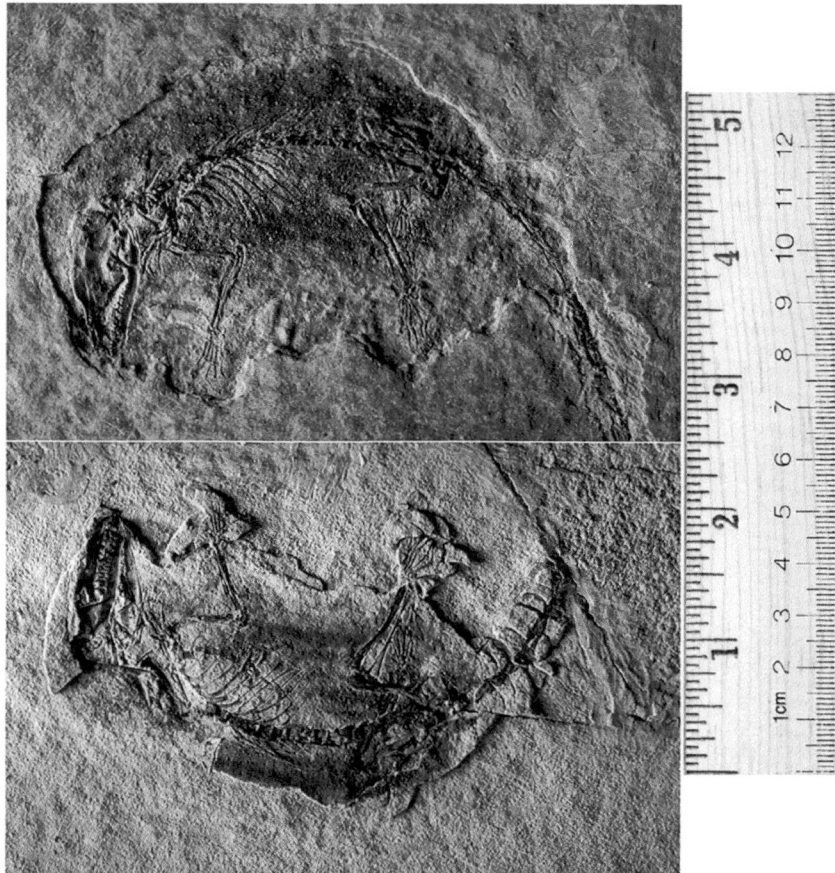

Eomaia scansoria fossil counter parts.   Ji et al. Ms. for Nature (correspondence: Luo)

evidence of some of the major groups, including hoofed mammals (*Pro-tungulatum* from Montana), carnivorous mammals (*Cimolestes*), and primates (*Purgatorius*).

Once the dinosaurs (excluding birds) vanished at the end of the Cretaceous, placental mammals underwent an explosive adaptive radiation, so that by the early Eocene, nearly all of the 20 or so living orders, and numerous extinct ones, had appeared (**Figure 22.2**). These include not only true carnivorans, insectivorans, rodents, primates, and several orders of hoofed mammals, but also animals as different as bats and whales. Evolutionary biologists have long regarded this as one of the most spectacular adaptive radiations ever documented.

For over a century, paleontologists have tried to piece together the origin and early history of each of the orders of placental mammals, primarily by studying the fragmentary teeth and jaws collected from the Cretaceous and Paleocene. Despite all this effort, however, little progress was made from 1910, the date of William King Gregory's massive monograph, *The Orders of Mammals*, until the late 1970s. This was due to several problems. For one thing, a lot of the important evidence is available from anatomy other than the teeth and jaws (especially from the braincase, ear region, and other parts of the skull and skeleton), and yet mammalian paleontologists persisted in trying to trace ancestral-descendant sequences of teeth back through the rocks. They often traced all the orders of mammals back to the "insectivores", long used as a wastebasket group for all mammals that ate insects. This is an ecological feature that evolved many times, not an indication of close phylogenetic relationship. In fact, due to their small size, it is likely that most groups

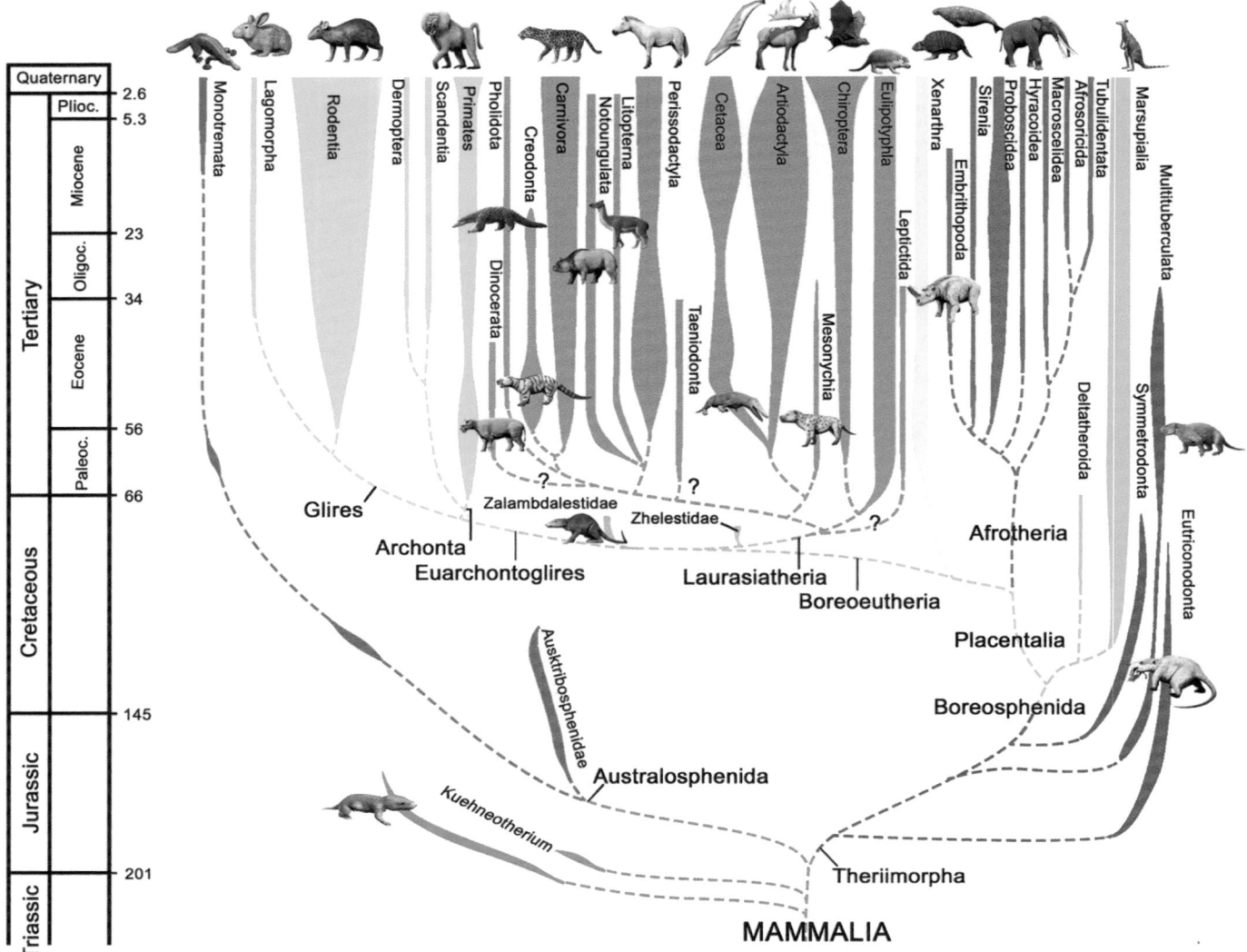

Figure 22.2 The family tree of mammals, including the Mesozoic mammaliaform lineages, the multituberculates, the monotremes, the marsupials, and the enormous evolutionary radiation of placental mammals.

of Mesozoic mammals were insectivorous, but their ecology doesn't make them closely related. There is a natural group known as the order Insectivora (sometimes also called Lipotyphla or Eulipotyphla), composed of shrews, moles, and hedgehogs, that was improperly expanded to include a whole zoo full of unrelated beasts, including tree shrews, elephant shrews, and many extinct Mesozoic and early Cenozoic mammals, sometimes thrown in the wastebasket order "Proteutheria". Bats were supposedly derived from this amorphous cloud of animals because some have an insectivorous diet; it turns out they are members of the Laurasiatheria, a group we will discuss in later chapters. At one time or another, all of the rest of the mammalian orders were also traced to one or more "insectivores" of the Late Cretaceous.

Another problem was that the best-studied collections were primarily from North America and Europe, so paleontologists tended to try to link together fossils found in the same area, neglecting the possibility of immigration from other continents. When the excellent fossil record of the Paleocene of China finally became available for study by international scientists in the late 1970s and 1980s, many of the important "missing links" turned up in Asia, not in North America or Europe.

But the cladistic rethinking of the methods of animal classification, starting in the 1960s and 1970s, spurred a renewed effort to tease out every useful feature of the anatomy of mammals and decipher their interrelationships, using only shared evolutionary novelties, not primitive characteristics, and natural monophyletic groups. By the 1980s and 1990s, a number of mammalian paleontologists had published phylogenies of different groups of mammals and great progress had been made.

Also, in the 1980s and 1990s, molecular biology challenged the traditional ways of sorting out the interrelationships of placental mammals. Many of the groupings proposed by paleontologists in the 1970s through 1990s were confirmed by molecular analysis. However, in some cases the molecular sequence data gave different answers than the anatomy did. In a few cases (such as the suggestion that guinea pigs were not rodents), there were clearly problems with molecular data, most of which were later corrected. But the molecular phylogenies also suggested clusters of mammals that had never been suggested by the anatomy. As more and more studies were published, the molecular evidence became overwhelming, so in most cases the mammalian paleontologists have come to accept these new groups. Keep in mind, however, that nearly all of these groupings are supported only by molecular data, and so far, there is no other evidence (anatomical or otherwise) for them. In a few cases, there are anatomical features with disagree with the molecular data, and this conflict has still not been fully resolved.

Nevertheless, the molecular evidence just seems to get stronger and stronger over the years, so most mammalian paleontologists have begrudgingly come to accept it, and use the groups defined on molecular features only (**Figure 22.2**). They are:

Xenarthra: The sloths, armadillos, and anteaters were long recognized as a natural group, since they have many peculiarities in their skeletons, and very primitive features of the anatomy and metabolism as well. Since 1975, paleontologists have realized that they diverged from the common placental root very early. Some molecular phylogenies place them slightly closer to the majority of the placentals, while others suggest that they are one of the first groups to branch off.

Afrotheria: A group of mammals originally confined to Africa. The core members of the group (elephants and sea cows, plus the extinct arsinoitheres) have long been connected as a group called the Tethytheria, first proposed by Malcolm McKenna in 1975. The hyraxes also tend to cluster with the tethytheres. More recently, a number of insectivorous groups previously lumped into the "Insectivores" wastebasket have proven to be afrotheres. These include the elephant shrews, tenrecs, golden moles, and several other uniquely African mammals, such as the aardvarks.

Boreoeutheria: All the remaining orders of placentals cluster together, excluding the xenarthrans and the afrotheres. These separated into two main branches:

Euarchontoglires: A cluster of the euarchontans (primates, colugos, and tree shrews), and the glires (rodents and lagomorphs).

Laurasiatheria: Almost all the remaining orders of placental mammals are Laurasiatheria, including the hoofed mammals (perissodactyls and artiodactyls, including whales and dolphins), the carnivorous mammals (carnivorans and creodonts) plus pangolins, the true insectivores (shrews, moles, and hedgehogs), and the bats (chiropterans).

Let's look at most of these groups, starting with the most primitive members, the Xenarthra plus the Afrotheria in this chapter. We will discuss the rest in Chapters 23 and 24.

XENARTHRA: SLOTHS, ARMADILLOS, AND ANTEATERS

Of the living placentals, one of the first groups to branch off was the edentates (anteaters, sloths, and armadillos), known as the order Xenarthra. Although the name "edentate" implies that they are toothless, only anteaters fit that description, since sloths and armadillos have simple peg-like teeth with no coating of enamel. In his original 1758 classification of animals, Linnaeus used the "edentates" to include not only the sloths, armadillos, and anteaters, but also other toothless ant-eating groups, like the pangolins, which molecular evidence puts in a very different branch of mammals. Consequently, modern classifications use the next oldest valid name, "Xenarthra" ("strange joints") instead, because the vertebrae of sloths, armadillos, and anteaters all have numerous spines with bizarre extra articulations and joints.

Because they don't have an abundant fossil record of teeth and are known primarily from South America, xenarthrans were long neglected in the analysis. But their anatomical features show that they are very primitive placentals, lacking many of the specializations found in all other eutherians. For example, female xenarthrans have a uterus simplex, which is divided by a septum and has no cervix. The xenarthran metabolism tends to be much slower and less well-regulated than that of other placentals. Xenarthrans still retain a few primitive reptilian bones that all other placentals have lost, and their brain and neural development is also much less advanced. One of the most consistent features found in all placentals except xenarthrans is a stirrup-shaped stapes in the middle ear (the "hammer", "anvil" and "stirrup", or malleus, incus, and stapes, which conduct sound from the eardrum to the inner ear). Xenarthrans have the primitive amniote rod-like stapes with no hole at the base for the stapedial artery. In addition, to the extra articulations between the vertebrae, the Xenarthra have odd fusions of the hip region with the vertebrae of the back and tail, and many other unusual features in the shoulder, ankle, and skull. Molecular phylogenies also place the Xenarthra near the base of the placentals, along with the Afrotheria, or sometimes closer to the Afrotheria than they are to any other placentals. (This group of Xenarthra plus Afrotheria is called the "Atlantogeneta", because the Afrotheria originated and diversified on the Africa side of the Atlantic, and the Xenarthra was restricted to South America for most of their history.)

Part of the reason for the neglect of xenarthrans is that most of their evolution took place in isolation in South America, so Northern Hemisphere paleontologists seldom studied them. Their Cretaceous ancestors were among the earliest mammals to evolve on that continent while it was mostly isolated from the rest of the world, and consequently, through the Cenozoic, South America was home to a wide variety of xenarthrans, including the sloths. Today, there are three species of the three-toed sloth *Bradypus*, and two species of the two-toed sloth *Choloepus*. These small animals are legendary for hanging upside down from trees their entire lives, moving very slowly, and feeding on and digesting leaves slowly between long naps. But they are tiny remnant of a big radiation of huge ground sloths, with four major families, made up of 86 genera and hundreds of species of extinct ground sloths. The biggest were the elephant-sized megatheres, typified by *Megatherium* and *Eremotherium*, which towered over 6 meters tall and weighed 3 tonnes (**Figures 22.3[A]** and **22.4[A]**). In addition, there were the bear-sized mylodonts, such as *Paramylodon*, the biggest sloth at La Brea tar pits, which was about 3.8 meters (10 feet) tall on its hind legs, and weighed about 1000 kg. A third group were the megalonychids, which were about the same size, and represented by 28 genera. Most were South American, but *Pliometanastes* was the first sloth to island-hop through

Central America about 9 Ma, even earlier than did the mylodontid *Thino-badistes*. President Thomas Jefferson himself described *Megalonyx* ("giant claw" in Greek) in 1796 from claw fossils brought to him by Col. John Stuart from a cave in Virginia. He was convinced they belonged to a giant lion that still roamed in the West, and asked Lewis and Clark to look out for it during their expedition in 1803–1805. Only in the 1820s after the description of *Megatherium* and other fossils did people realize the claws were not from a giant lion, but a giant ground sloth. Finally, the smaller sloths were the Nothrotheriidae, which are known from ten genera. Several specimens of *Nothrotheriops* (**Figures 22.3[B]** and **22.4[B]**) are known from mummified specimens with skin and hair found in dry caves in Mexico, New Mexico, and Arizona. Many of these caves yield sloth dung that has been dried and preserved over 10,000 years.

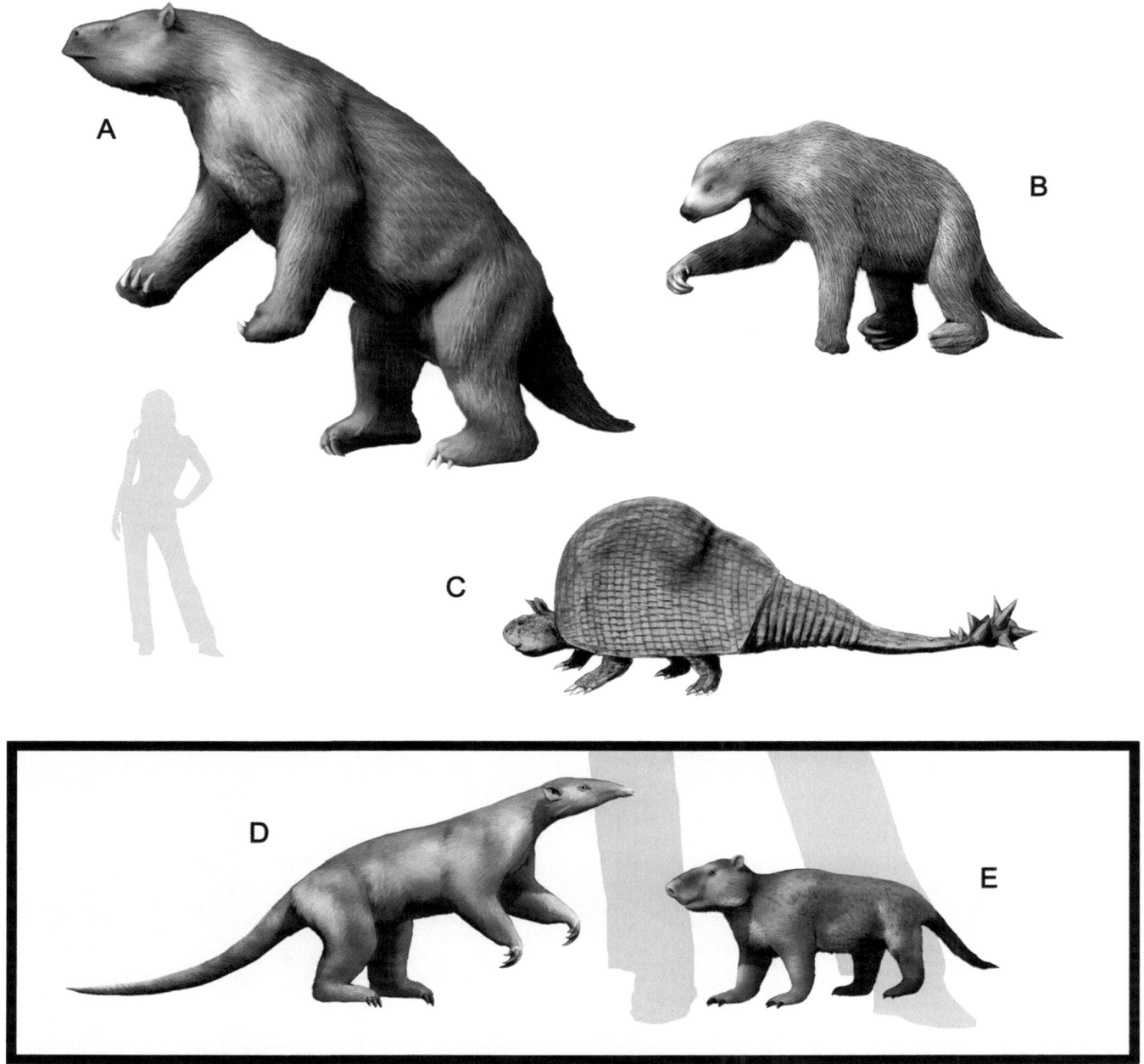

Figure 22.3 Reconstructions of some extinct xenarthrans. (A) The gigantic elephant-sized ground sloth *Megatherium*. (B) The smaller ground sloth *Nothrotheriops*. (C) The armored glyptodont *Doedicurus*, with the spiked club tail. (D) The pangolin from the Eocene Messel beds of Germany, *Eurotamandua*. (E) The odd Chinese fossil *Ernanodon*, which resembled some xenarthrans.

Figure 22.4 Fossils of some extinct xenarthrans. (A) The gigantic ground sloth *Megatherium*. (B) The smaller ground sloth *Nothrotherops*. (Photos courtesy Wikimedia Commons.)

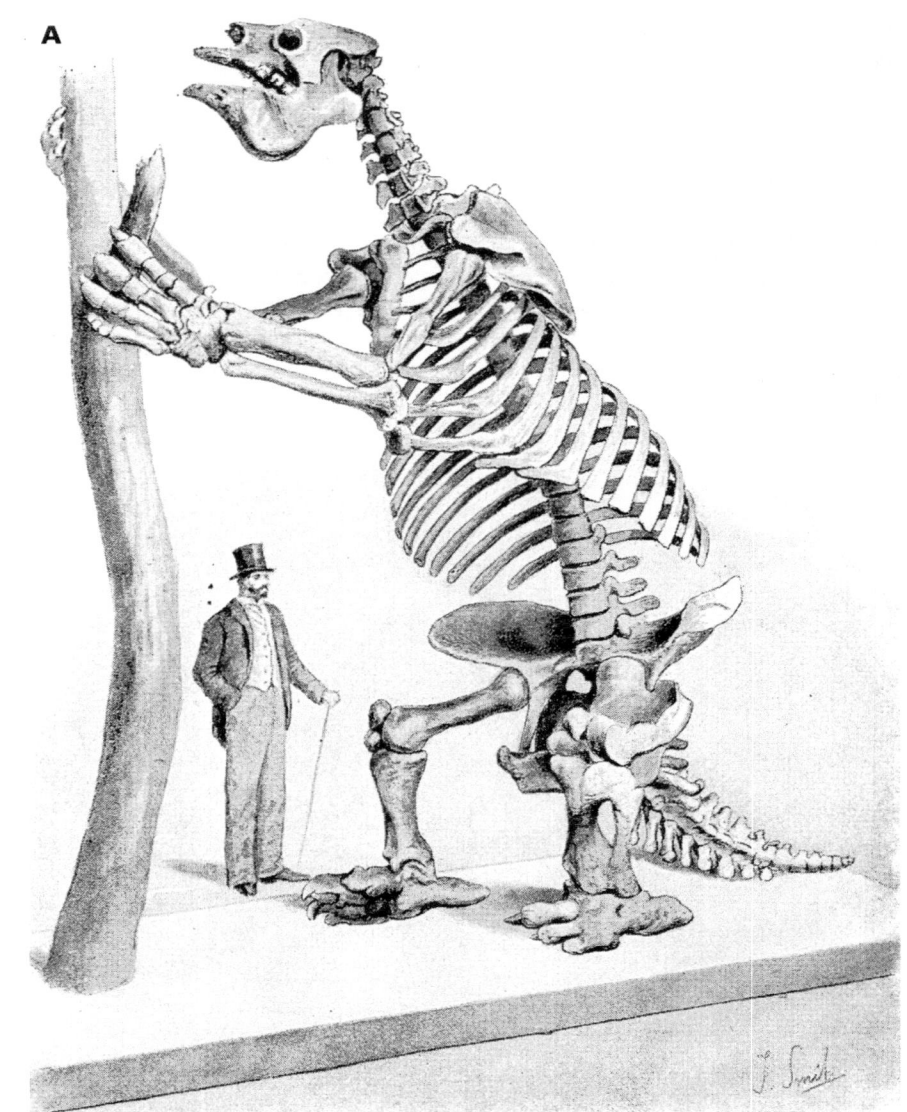

Figure 22.4 (Continued) (C) The pampathere *Holmesina* (left) and the glyptodont *Glyptodon* (right). (D) The gigantic glyptodont *Panochthus*.

The most amazing of the nothrotheriids was *Thalassocnus*, a late Miocene sloth from Peru and Chile that returned to the sea. The oldest species from the late Miocene were apparently semiaquatic, and four successive species show the gradual transition in the skeleton from terrestrial to fully marine modes of life. The bones of each successive species show increasing density necessary for ballast in a marine mammal, and their limbs gradually change shape as they became more committed swimmers who used their long claws to cling to rocks in the surf (as do modern marine iguanas). The teeth show the wear and chemistry of eating vegetation mixed with sand from the nearshore beaches and sea bottom, but later forms show no such wear, and instead are adapted for eating sea grasses and algae.

Even more startling, the two living tree sloths are not closely related to each other. The two-toed sloth is a dwarfed megathere, while the three-toed sloth is most closely related to megalonychids, so they are independently derived from two different groups and secondarily became small as they sought refuge in trees.

In addition to sloths, there were giant relatives of the armadillos, the glyptodonts (**Figures 22.3[C]** and **22.4[C–D]**). Some were the size of a

SmartCar, with the largest species over 2 meters long. They weighed 2 tonnes including 400 kg of bony armor, and some of them, like *Doedicurus* (**Figure 22.3[C]**) had a spiked club at the tip of their tail. The other giant extinct relative of armadillos are the pampatheres (**Figure 22.3[C]**). These creatures had a shell made of bony plates, or osteoderms, with distinctive round or hexagonal shapes, and three broad bands of hinged armor plates across their back that gave them a bit of flexibility. After originating in South America, they got larger through their evolution. They can be traced from the relatively small *Vassalia* (Miocene-Pliocene of South America) to the medium-sized *Kraglievichia* (Pliocene-early Pleistocene, South America and Florida) to the bigger *Pampatherium* (middle-late Pleistocene, South America). Finally, there was the huge *Holmesina* (Pleistocene, North America) (**Figure 22.4[C]**), which was almost the size of a Volkswagen Beetle, reaching 2 meters (7 feet) in length, and weighing about 227 kg (500 lb), almost four times as big as the living giant armadillo.

Even though xenarthrans were confined to South America until late in their evolution, similar creatures like the pangolins have appeared elsewhere (**Figure 22.3[D]**). The strange xenarthram-like animal, *Ernanadon* (**Figure 22.3[E]**), has been described from the Paleocene of China, but whether it is a xenarthran or just an example of convergent evolution is still disputed. Xenarthrans were among the few South American natives to successfully march north across the Panamanian land bridge in the Plio-Pleistocene against to the tide of North American mammals heading south. Ground sloths, armadillos, glyptodonts were all common in the Pleistocene of southern North America, from La Brea tar pits to Texas to Florida

AFROTHERIA

Fossils of mammoths and mastodonts were among the first ever found and described by ancient scholars. Indeed, mummified mammoths complete with stomach contents were known in the 1700s, and the skulls of mammoths (with the single large nasal opening on the front of the skull) might have been responsible for the legend of the Cyclops. Mastodon teeth from Big Bone Lick in Kentucky were attributed to weird creatures, until they were finally correctly related to elephants by Baron Georges Cuvier in the early 1800s. In fact, he famously used mammoths and mastodonts to establish that animals as big as these could not be hiding on earth anywhere, and thus were clearly extinct. This idea that was shocking to religious scholars of the time who could not imagine God letting any of his creatures vanish.

Mammoths and mastodonts were known early, but since then the fossil record has produced almost a complete unbroken sequence of their relatives, collectively known as order Proboscidea (**Figure 22.5**). The earliest proboscideans (*Eritherium*, *Phosphatherium*) are known from the late Paleocene of Africa. This was followed by *Numidotherium*, from the early Eocene of Algeria, which already had the high forehead, the retracted nasal opening (indicating a short proboscis), short upper tusks, mastodont-like teeth, and the lower front jaw is beginning to develop a broad scoop, a diagnostic feature of mastodonts. It was only a meter tall (3 feet) at the shoulder, yet it already had the limb characteristics found in later, larger mastodonts. By the later Eocene, proboscideans were shaped like small hippos, without trunks or tusks (*Moeritherium*, *Barytherium*) (**Figure 22.5[A]**). In the early Oligocene, the famous Fayûm beds of Egypt produce very primitive, small mastodonts with short jaws and even shorter tusks, known as *Palaeomastodon* and *Phiomia* (**Figure 22.5[B]**).

During this time, the various lineages of proboscideans (elephants, mammoths, and mastodonts) are very primitive and hard to tell apart, typical of the early stages of an evolutionary radiation. Soon they diverged into numerous lineages: the deinotheres, with their downward-deflected lower

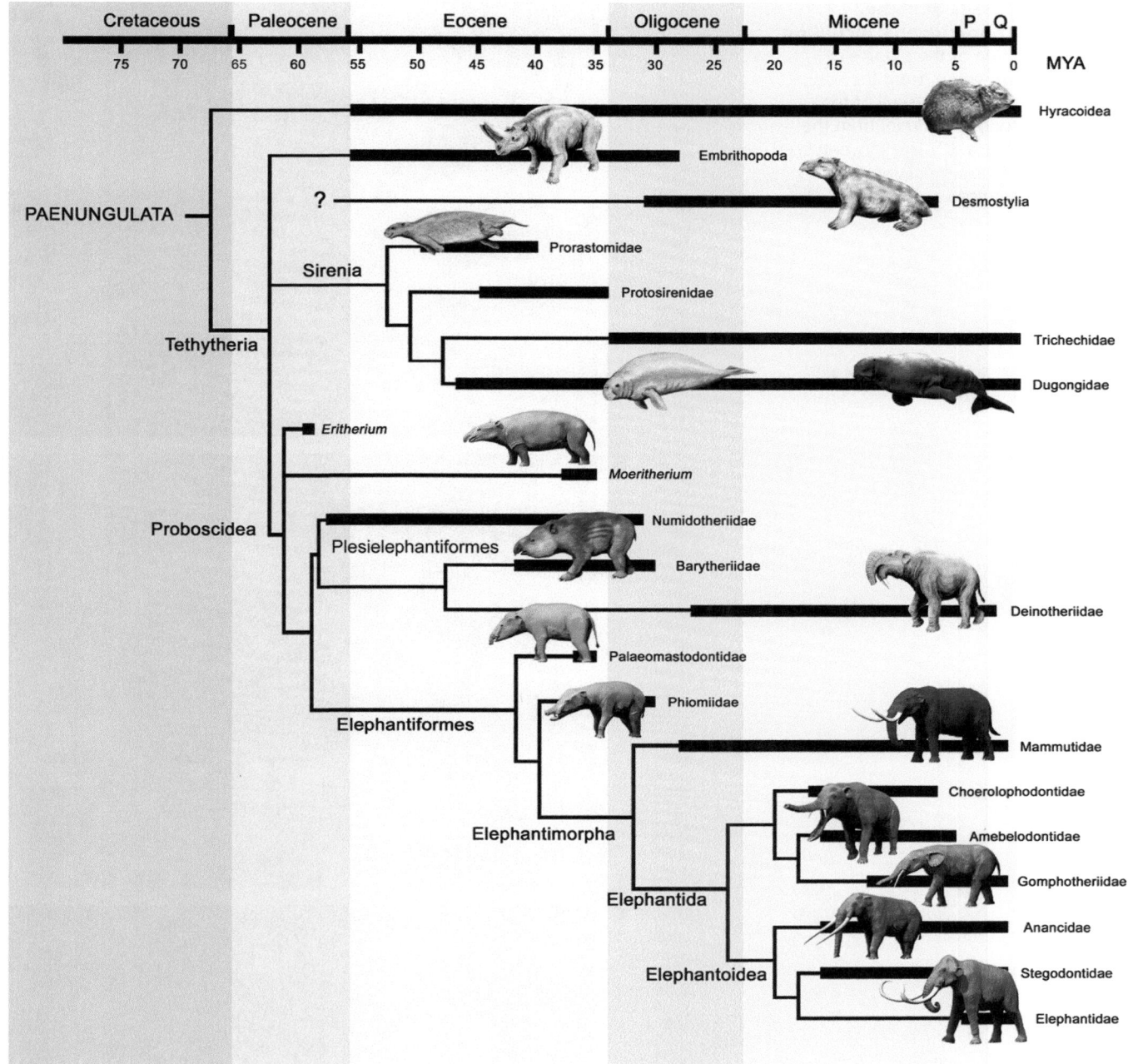

Figure 22.5 Family tree of tethytheres and their relatives, including the proboscideans, sirenians, hyraxes, desmostylians, and arsinoitheres (Embrithopoda).

tusks (**Figure 22.6[D]**); the true mastodons, or family Mammutidae, which focused on living in dense forests and eating conifer needles and leaves, rather than grazing like mammoths did (**Figure 22.6[E]**); and the gomph-otheres, with small upper and lower tusks. Some evolved into beasts with enormous broad tusks shaped like shovels (**Figure 22.6[C]**), while others had various combinations of two and four tusks with different lengths and curvatures. In the late Pleistocene, only the mastodons (**Figure 22.6[E]**) and the diversity of mammoths (**Figure 22.6[F]**) remained, and most of these were driven to extinction at the end of the Pleistocene.

Closely related to the Proboscidea are several other groups that had long been zoological mysteries, placed in their own isolated orders (**Figure 22.5**). In 1975, Malcolm McKenna named them "tethytheres", because

Figure 22.6 Fossils of some important proboscideans, including: (A) Reconstruction of the pig-like or tapir-like *Moeritherium*, from the late Eocene of Africa. (B) The skull of the primitive mastodont *Phiomia*, from the Oligocene of Africa. (C) The shovel-tusked mastodont *Amebelodon*. (Courtesy Wikimedia Commons.)

A

B

C
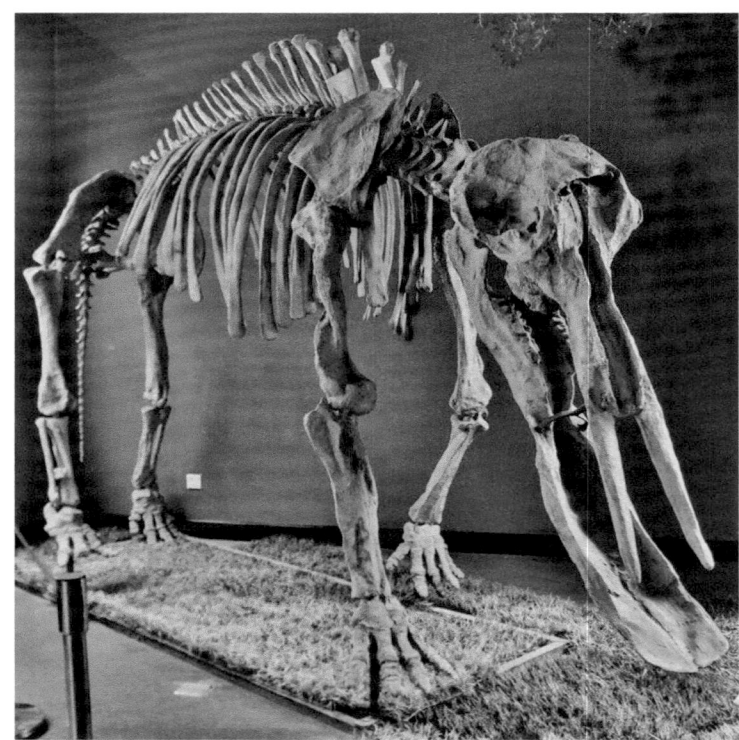

Figure 22.6 (Continued) (D) *Deinotherium*. (E) The American mastodon. (F) The Columbian mammoth.

their earliest fossils are found in Paleocene and Eocene beds around the edge of the Tethys Seaway that ran from Gibraltar to Indonesia (especially Egypt, Morocco, and Pakistan). Living tethytheres have a number of unique specializations found in no other group of mammals. These include a single pair of teats on the breasts (like humans), rather than multiple teats as in most female mammals; eyes that are shifted far forward on the skull; cheekbones that contain a broadly expanded portion of the rear skull bones; and teeth that do not erupt from below, as in most mammals, but from the back of the jaw, pushing the old teeth out the front of the jaw (known as horizontal tooth replacement).

The living tethytheres most closely related to Proboscidea is the order Sirenia, which includes the manatees and dugongs, or "sea cows" (**Figure 22.5**). These animals are completely aquatic, losing their hindlimbs and developing a round fluke on the tail, and paddles for forelimbs. Living manatees have no hind limbs, but a fossil sirenian from the Eocene of Jamaica named *Pezosiren* had not only both front and hind limbs, but also terrestrial hands and feet rather than flippers, a perfect transitional fossil between sirenians and their ancestors. Manatees are restricted to freshwater lakes, rivers, and estuaries, browsing the water plants, but they are so slow and docile that they are now on the endangered species list as a result of hunting and injuries from speedboats hitting them.

There was a third, gigantic species of sirenian alive on this planet just a few hundred years ago. During Vitus Bering's expedition to discover and claim Alaska for the Russian Empire in 1741–1742, the official expedition scientist Georg Steller named and described a huge creature quietly feeding on kelp in the Aleutian Islands (**Figure 22.7**). As large as some whales, it grew to a length of 8–9 meters (26–30 feet), and weighed about 8–10 tons. It was completely docile and unafraid of humans, since no Europeans had ever hunted it. Known as Steller's sea cow (*Hydrodamalis gigas*), the limited population of a few thousand was easily slaughtered for meat, or just for sport by Russian fur trappers hunting sea otter and seal pelts. By 1768, only 27 years after Steller had first seen them and formally described them in the scientific literature, the largest of all sirenians was extinct.

Another longstanding mystery was a peculiar rhino-like group from the African Oligocene known as arsinoitheres, the order Embrithopoda (**Figure 22.8**). These elephant-sized animals had a pair of huge, sharp, recurved bony horns on their noses, and no one had a clue as to what they were related to. However, when more archaic Eocene arsinoitheres were found in Mongolia and Turkey, it was suggested that they were tethytheres, and this has since been confirmed.

Yet another paleontological puzzle were the peculiar Pacific Miocene marine mammals known as desmostylians (**Figure 22.9**). These walrus-sized animals had hoofed feet rather than flippers, with a broad shovel-like tusked jaw containing bizarre molars that look more like

Figure 22.7 Reconstruction of Steller's sea cow, the largest sirenian ever to live, wiped out from its habitat in the northern Pacific by fur trappers and hunters only 27 years after Steller discovered it and described it.

Figure 22.8 (A) The weird skeleton of the tethythere *Arsinoitherium* with its rhino-like body and pair of bony horns on its forehead. (B) Reconstruction of *Arsinoitherium*. [(A) Courtesy Wikimedia Commons.]

a bundle of barrels than anything else. They were long placed in their own order Desmostylia, with no apparent relationships to anything else, until a fossil was found and described in 1986, which was an unusually primitive desmostylian named *Behemotops*. It showed that desmostylians were actually tethytheres, distantly related to sirenians and proboscideans.

Even more distantly related to the Tethytheria are the woodchuck-like hoofed mammals known as hyraxes or conies, the order Hyracoidea (**Figure 22.5**). These little animals are today restricted to rocky outcrops in east African and the Middle East, but during the early Cenozoic, they were among the most common hoofed mammals in Africa, evolving into beasts with hippo-like bodies, and many other shapes as well.

Finally, molecular studies consistently cluster a number of mammalian groups with the Tethytheria to form a larger group known as Afrotheria, since they are all restricted to Africa today or in the past, or came from Africa. Most of these creatures were orphans on the family tree

Figure 22.9 Desmostylians were a weird-looking group of semi-aquatic tethytheres that fed around tidal regions in the North Pacific during the Miocene. (A) The skeleton of *Neoparadoxia*, a large species from the Miocene of southern California. (B) Reconstruction of *Palaeoparadoxia*, from the Miocene of the Bay Area in California. (C) Reconstruction of *Desmostylus*. [(A) Photo by the author.]

of mammals, until the molecular evidence clustered them with living elephants and sirenians. The Afrotheria include a surprising variety of mammals: the long-nosed elephant shrews (Macroscelidea) which had been clustered with insectivores, or even with rabbits, for a long time; the strange African mole-like creatures known as chrysochlorids or "golden moles" (also mistakenly put in the insectivores, like the true moles); the odd insectivorous tenrecs of Madagascar; and last, but not least, one of the weirdest of all creatures, the aardvark (long a mystery as to where it belonged in the Mammalia). So far, there has been no anatomical or fossil evidence to show these animals are related to tethytheres, or to support the idea of Afrotheria, although many people have tried to find it.

This brief discussion is all the room we have for primitive mammals such as the Xenarthra and Afrotheria. Let us now examine the largest group of mammals, the Lauriasiatheria, including the carnivorous mammals and the hoofed mammals in Chapters 23 and 24, and give the Euarchontoglires (rodents and rabbits, plus primates including humans) a detailed treatment in Chapter 25.

FURTHER READING

Agusti, J.; Anton, M. 2002. *Mammoths, Sabertooths, and Hominids: 65 Million Years of Mammalian Evolution in Europe.* Columbia University Press, New York.

Asher, R.J.; Bennett, N.; Lehmann, T. 2009. The new framework for understanding placental mammal evolution. *BioEssays.* 31 (8): 853–864.

Asher, R.J.; Geisler, J.H.; Sánchez-Villagra, M.R. 2008. Morphology, paleontology, and placental mammal phylogeny. *Systematic Biology.* 57: 311–317.

DeJong, W.W.; Zweers, A.; Goodman, M. 1981. Relationship of aardvark to elephants, hyraxes and sea cows from alpha-crystallin sequences. *Nature.* 292 (5823): 538–540.

Delsuc, F.; Catzteflis, F.M.; Stanhope, M.J.; Douzery, E.J.P. 2001. The evolution of armadillos, anteaters and sloths depicted by nuclear and mitochondrial phylogenies: Implications for the status of the enigmatic fossil *Eurotamandua. Proceedings of the Royal Society of London B.* 268 (1476): 1605–1615.

Farina, R.A.; Vizcaino, S.F.; De Iuliis, G. 2013. *Megafauna: Giant Beasts of Pleistocene South America.* Indiana University Press, Bloomington, IN.

Gaudin, T.J.; Croft, D.A. 2015. Paleogene Xenarthra and the evolution of South American mammals. *Journal of Mammalogy.* 96 (4): 622–634.

Gregory, W.K. 1910. The orders of mammals. *Bulletin of the American Museum of Natural History.* 27: 1–524.

Halliday, T.J.D. 2015. Resolving the relationships of Paleocene placental mammals. *Biological Reviews.* 92 (1): 521–550.

Hu, Y.; Meng, J.; Li, C.; Wang, Y. 2010. New basal eutherian mammal from the Early Cretaceous Jehol biota, Liaoning, China. *Proceedings of the Royal Society B.* 277 (1679): 229–362.

Janis, C. 1993. Tertiary mammal evolution in the context of changing climates, vegetation, and tectonic events. *Annual Reviews of Ecology and Systematics.* 24: 467–500.

Janis, C.; Gunnell, G.F.; Uhen, M.D., eds. 2008. *Evolution of Tertiary Mammals of North America. Vol II: Small Mammals, Xenarthrans, and Marine Mammals.* Cambridge University Press, Cambridge.

Janis, C.; Scott, K. M.; Jacobs, L. L., eds. 1998. *Evolution of Tertiary Mammals of North America, vol. I: Terrestrial Carnivores, Ungulates, and Ungulate-Like Mammals.* Cambridge University Press, Cambridge.

Ji, Q.; Luo, Z.-X.; Yuan, C.-X.; Wible, J. R.; Zhang, J.-P.; Georgi, J. A. 2002. The earliest known eutherian mammal. *Nature.* 416 (6883): 816–822.

Kurtén, B. 1968. *Pleistocene Mammals of Europe.* Columbia University Press, New York.

Kurtén, B. 1988. *Before the Indians.* Columbia University Press, New York.

Kurtén, B.; Anderson, E. 1980. *Pleistocene Mammals of North America.* Columbia University Press, New York.

Luo, Z.; Yuan, C.; Meng, Q.; Ji, Q. 2011. A Jurassic eutherian mammal and divergence of marsupials and placentals. *Nature.* 476 (7361): 442–445.

Madsen, O.; Scally, M.; Douady, C.J.; Kao, D.J.; DeBry, W.; Adkins, R.; Amrine, H.; Stanhope, M.J.; de Jong, W.W.; Springer, M.S. 2001. Parallel adaptive radiations in two major clades of placental mammals. *Nature.* 409: 610–614.

McKenna, M.C. 1975. Toward a phylogenetic classification of the Mammalia, pp. 21–46. In Luckett, W.P.; Szalay, F.S., eds. *Phylogeny of the Primates.* Plenum Press, New York.

McKenna, M.C.; Bell, S.K. 1997. *Classification of Mammals above the Species Level.* Columbia University Press, New York.

Murphy, W.J.; Eizirik, E.; Johnson, W.E.; Zhang, Y.P.; Ryder, O.A.; O'Brien, S.J. 2001. Molecular phylogenetics and the origins of placental mammals. *Nature.* 409: 614–618.

Murphy, W.J.; Eizirik, E.; O'Brien, S.J.; Madsen, O.; Scally, M.; Douady, C.J.; Teeling, E.; Ryder, O.A.; Stanhope, M.J.; de Jong, W.W.; Springer, M.S. 2001. Resolution of the early placental mammal radiation using Bayesian phylogenetics. *Science.* 294: 2348–2351.

Murphy, W.J.; Pevzner, P.A.; O'Brien, S.J. 2004. Mammalian phylogenomics comes of age. *Trends in Genetics.* 20: 631–639.

Novacek, M.J. 1992. Mammalian phylogeny: Shaking the tree. *Nature.* 356: 121–125.

Novacek, M. J. 1994. The radiation of placental mammals, pp. 220–237. In Prothero, D.R.; Schoch, R.M., eds. *Major Features of Vertebrate Evolution.* Paleontological Society Short Course 7. Paleontological Society, Lawrence, KS.

Novacek, M.J.; Wyss, A.R. 1986. Higher-level relationships of the recent eutherian orders: Morphological evidence. *Cladistics.* 2: 257–287.

Novacek, M.J.; Wyss, A.R.; McKenna, M.C. 1988. The major groups of eutherian mammals, pp. 31–73. In Benton, M.J., ed. *The Phylogeny and Classification of the Tetrapods, vol. 2: Mammals.* Clarendon Press, Oxford.

O'Leary, M.A.; Bloch, J.I.; Flynn, J.J.; Gaudin, T.J.; Giallombardo, A.; Giannini, N.P.; Goldberg, S.L.; Kraatz, B.P.; Luo, Z.-X.; Meng, J.; Ni, M.; Novacek, M.J., Perini, Z. S.; Randall, G.; Rougier, G.; Sargis, E.J.; Silcox, M.; Simmons, N.B.; Spaulding, M.; Velazco, P.M., Weksler, M.; Wible, J.R; Cirranello, A.L. 2013. The placental mammal ancestor and the post: K-Pg radiation of placentals. *Science.* 339 (6120): 662–667.

Prothero, D.R. 1994. Mammalian evolution, pp. 238–270. In Prothero, D.R.; Schoch, R.M., eds. *Major Features of Vertebrate Evolution.* Paleontological Society Short Course 7. Paleontological Society, Lawrence, KS.

Prothero, D.R. 2006. *After the Dinosaurs: The Age of Mammals.* Indiana University Press, Bloomington, IN.

Prothero, D.R. 2016. *The Princeton Field Guide to Prehistoric Mammals.* Princeton University Press, Princeton, NJ.

Prothero, D.R.; Schoch, R.M. 2002. *Horns, Tusks, and Flippers: The Evolution of Hoofed Mammals and Their Relatives.* Johns Hopkins University Press, Baltimore, MD.

Rose, K.D. 2006. *The Beginning of the Age of Mammals.* Johns Hopkins University Press, Baltimore, MD.

Rose, K.D.; Archibald, J.D., eds. 2005. *The Rise of Placental Mammals: The Origin and Relationships of the Major Extant Clades.* Johns Hopkins University Press, Baltimore, MD.

Sánchez-Villagra, M. 2012. Why are there fewer marsupials than placentals? On the relevance of geography and physiology to evolutionary patterns of mammalian diversity and disparity. *Journal of Mammalian Evolution*. 20 (4): 279–290.

Savage, D.E.; Russell, D.E. 1983. *Mammalian Paleofaunas of the World*. Addison Wesley, Reading, MA.

Savage, R.J.G.; Long, M.R. 1986. *Mammal Evolution: An Illustrated Guide*. Facts-on-File Publications, New York.

Scally, M.; Madsen, O.; Doouady, C.J.; de Jong, W.W.; Stanhope, M.J.; Springer, M.S. 2001. Molecular evidence for the major clades of placental mammals. *Journal of Mammalian Evolution*. 8: 239–277.

Shoshani, J.; McKenna, M.C. 1998. Higher taxonomic relationships among extant mammals based on morphology, with selected comparisons of results from molecular data. *Molecular Phylogenetics and Evolution*. 9: 572–584.

Springer, M.S.; Burk-Herrick, A.; Meredith, R.; Eizirik, E.; Teeling, E.; O'Brien, S.J.; Murphy, W.J. 2007. The adequacy of morphology for reconstructing the early history of placental mammals. *Systematic Biology*. 56: 673–684.

Springer, M.S.; Cleven, G.C.; Madsen, O.; De Jong, W.W.; Waddell, V.G.; Amrine, H.M.; Stanhope, M.J. 1997. Endemic African mammals shake the phylogenetic tree. *Nature*. 388 (6637): 61–64.

Springer, M.S.; Meredith, R.W.; Eizirik, E.; Teeling, E.; Murphy, W.J. 2007. Morphology and placental mammal phylogeny. *Systematic Biology*. 57: 499–503.

Springer, M.S.; Meredith, R.W.; Janecka, J. E.; Murphy, W. J. 2011. A historical biogeography of Mammalia. *Proceedings of the Royal Society B*. 366: 2478–2502.

Springer, M.S.; Murphy, W.J.; Eizirik, E.; O'Brien, S.J. 2003. Placental mammal diversification and the Cretaceous: Tertiary boundary. *Proceedings of the National Academy of Sciences*. 100 (3): 1056–1061.

Springer, M.S.; Stanhope, M.J.; Madsen, O.; de Jong, W.W. 2004. Molecules consolidate the placental mammal tree. *Trends in Ecology and Evolution*. 19: 430–438.

Stanhope, M.J.; Waddell, V.G.; Madsen, O.; de Jong, W.; Hedges, S.B.; Cleven, G.C.; Kao, D.; Springer, M.S. 1998. Molecular evidence for multiple origins of Insectivora and for a new order of endemic African insectivore mammals. *Proceedings of the National Academy of Sciences*. 95 (17): 9967–9972.

Szalay, F.S.; Novacek, M.J.; McKenna, M.C., eds. 1993. *Mammal Phylogeny*. Springer-Verlag, Berlin.

Tabuce, R.; Marivaux, L.; Adaci, M.; Bensalah, M.; Hartenberger, J.-L.; Mahboubi, M.; Mebrouk, F.; Tafforeau, P.; Jaeger, J.-J. 2007. Early Tertiary mammals from North Africa reinforce the molecular Afrotheria clade. *Proceedings of the Royal Society B: Biological Sciences*. 274 (1614): 1159–1166.

Tabuce, R.; Asher, R.J.; Lehmann, T. 2008. Afrotherian mammals: A review of current data. *Mammalia*. 72 (1): 2–14.

Tassy, P.; Shoshani, J. 1988. The Tethytheria: Elephants and their relatives, pp. 283–316. In Benton, M.J., ed. *The Phylogeny and Classification of the Tetrapods, vol. 2: Mammals*. Clarendon Press, Oxford.

Turner, A.; Anton, M. 2004. *National Geographic Prehistoric Mammals*. National Geographic Society, Washington, DC.

Werdelin, L.; Sanders, W.L., eds. 2010. *Cenozoic Mammals of Africa*. University of California Press, Berkeley, CA.

LAURASIATHERIA I

CARNIVORES, BATS, INSECTIVORES, AND THEIR KIN

23

Tho' Nature, red in tooth and claw
With ravine, shriek'd against his creed
                    —Alfred Lord Tennyson, *In Memoriam A.H.H.*, 1850

THE LAURASIATHERES

Molecular phylogenies (**Figure 22.2**) of the living mammals yield a surprising cluster of groups: the Laurasiatheria, so named because they originated in the northern continents (Laurentia, or North America, plus Asia), and most of their history is restricted to the Northern Hemisphere. Today, the Laurasiatheria includes all the flesh-eating mammals (order Carnivora), the hoofed mammals (the odd-toed Perissodactyla, plus the even-toed Artiodactyla and their descendants, the whales), plus the scaly ant-eating pangolins (order Pholidota), the true insectivores (order Lipotyphla), and the bats (order Chiroptera). In addition, there are numerous extinct groups related to these living mammals, including the predatory order Creodonta, numerous groups of archaic hoofed mammals, and a wide range of extinct insectivorous mammals that may or may not be related to the Lipotyphla. This encompasses most of the diversity of living mammals today, from the biggest (whales) to the smallest (shrews), habitats ranging from aquatic (whales) to burrowing (moles) to flying (bats), and a wide range of large predators and well as prey. As discussed in the previous chapter, there is still no strong anatomical evidence for the relationships of these groups, but they are consistently supported by molecular evidence, so this arrangement has gradually won acceptance both by mammalian paleontologists and mammalogists. Consequently, this means that the Lauriasiatheria make up most of the diversity of mammals, other than the rodents. With so much diversity, we cannot cover all the families and genera in detail, but will focus on the more interesting and unusual extinct examples in these groups.

INSECTIVORES

As discussed in Chapter 22, a wide variety of different kinds of mammals with teeth suitable for eating insects have evolved, but most are not closely related in the phylogenetic sense. "Insectivora" was long used as a wastebasket for miscellaneous insectivorous mammals. Today it is no longer considered a natural group. There is a group of insectivorous mammals that are indeed closely related, and they are called the Lipotyphla or Eulipotyphla to avoid confusion with the old wastebasket use of the name "Insectivora". The lipotyphlans include shrews, moles, hedgehogs, and a strange mammal from Cuba known as the solenodon. Most of these mammals are tiny in body size, because their prey is small. The

DOI: 10.1201/9781003128205-23

Figure 23.1 The giant dog-sized hedgehog from Gargano Island, *Deinogalerix*. (A) The complete skeleton on display. (B) Reconstruction. [(A) Courtesy Wikimedia Commons.]

smallest living mammals is the living Etruscan shrew, only 2 g in weight and reaching only 3.5 cm in length (barely over an inch). Even smaller was the extinct *Batodonoides vanhouteni* from the early Eocene (53 Ma) of Wyoming. At 1.3 g (0.05 ounces), it was the smallest mammal that ever lived, and it was barely bigger than the rubber eraser on a pencil.

Once a paleontologist focuses on microvertebrate fossils by washing lots of fossiliferous sediments through a screen, and trap their tiny bones, teeth, and jaws, then shrews, moles, and hedgehogs have an excellent fossil record. Although most remained tiny, some were unusually large. The most impressive is the huge hedgehog *Deinogalerix* from Gargano Island in the Mediterranean (**Figure 23.1**). Lacking competition from other carnivores on that island, it evolved into a dog-sized or badger-sized predator, larger than any lipotyphlan known.

CHIROPTERA (BATS)

Lots of people have negative feelings about bats, but as mammals go, they are quite successful. They are the only mammals that are true fliers, developing wings by spreading a membrane among all five fingers of the hand. The only other true fliers, the birds and pterosaurs, plus the insects, all developed wings in completely different ways. Bats are also extremely diverse, with about 1240 living species in 186 genera and 18 families. Thus, they are the second-most diverse order of mammals after rodents. Bats make up about 20% of all mammal species on earth. But because their skeletons are so small and fragile, bats are rarely fossilized. All but a few fossil bats are known just from their teeth and jaws, and even with these tooth fossils, scientists estimate that 60% to 90% of bat species have never left been preserved. Bats are far more abundant in the tropics, yet these regions are a notoriously poor place for preserving small delicate bones, due to their wet climate and corrosive groundwater which quickly dissolves bone in the sediment. About the only exception are Pleistocene caves, which often preserve lots of cave-dwelling bats, but this is a fraction of the diversity that once lived.

Not all bats are small insect-eaters with echolocation (the suborder Microchiroptera). There also the fruit bats (order Megachiroptera), which can grow quite large. The largest have wingspans of 1.7 meters (5.6 feet) and weigh up to 1.6 kg (3.5 lb). They are often called "flying foxes" because their large heads have a long snout and ears much like a living fox. Unlike microbats, fruit bats are largely daytime animals with excellent sight. They roost in trees, not in caves, and do not use echolocation (except for one species), because they feed on fruits, nectar from flowers, and nuts. Most megabats land on a branch and crawl along with their hooked fingers to reach fruit, but smaller ones can hover in flight to feed on flower nectar with their long tongues. There are about 120 living species of fruit bat, and most live in the tropical jungle regions of Africa, southern Asia, and the eastern Pacific, from Japan to Australia. Many of the tropical islands of the Pacific have their own endemic species of fruit bat, because all it took was one founding population to reach the isolated land, and then they diverged from their mainland ancestors.

Because their bones are very thin and fragile and seldom fossilized, bats have a relatively poor fossil record. For that reason, we get only glimpses about how they evolved from non-flying mammals. The oldest fossil bats (*Wyonycteris*) come from the Paleocene and earliest Eocene, but they are just fossil teeth, since there are no deposits capable of preserving such delicate skeletons. The best fossils of the earliest bats come from lake beds with extraordinary preservation (**Figure 23.2**), such as the late early Eocene of the Green River Formation of Wyoming (*Icaronycteris, Onychonycteris*) and the Messel deposits of Germany (*Archaeonycteris, Palaeochiropteryx, Hassianycteris*). Thus, we are hampered with a poor early fossil record, and when bats finally appear, they already have bat-like wings.

When you look closer, however, you find that these fossils are actually very primitive, and except for the wings, they are not very bat-like. These have primitive tooth crowns, relatively unspecialized skulls and brains, and do not have the inner ear features necessary for echolocation. Their hands are not fully incorporated into the wing, and they have a long tail (not found in modern bats) with no membrane between the tail and legs (uropatagium) seen in most living bats. There is no fusion of the vertebrae and hip bones seen in modern bats, nor is there a large keel on the sternum for strong flight muscles. The recent discovery of another Green River bat, *Onychonycteris*, which is an even more primitive transitional fossil, shows these non-bat-like features and more. It had claws on all five fingers (living bats have them on only two), short broad wings more

Icaronycteris index
bat
cast of holotype YPM18150

Onychonycteris finneyi
bat

Figure 23.2 The oldest well-preserved bat fossils come from the middle Eocene lake shales, where the water is quiet and stagnant and delicate fossils are preserved. These fossils include *Icaronycteris* (left) and Onychonycteris from the Green River Shale of Wyoming. (Photo by the author.)

suited to a mixture of flapping and gliding rather than extended flights, and long hind legs and short forearms, suggesting that it evolved flight from a climbing way of life.

PHOLIDOTA (PANGOLINS)

Pangolins are bizarre-looking mammals that look vaguely like anteaters covered with overlapping plate-like keratinous scales that make them look like a living pine cone; when threatened, they roll up into a ball and the sharp-edged scales form an effective armor. Pangolins also have a long tube-like snout and a sticky tongue almost 40 cm (16 inches) long for snagging ants and termites out of their nests. Their long front claws and powerful limbs are used for digging into those nests and ripping them apart, as well as for climbing trees and for defense when necessary. Some are strictly arboreal, using their prehensile tails to hang from branches as they rip open tree bark and nests for insects, while others live on the ground and dig burrows for protection. They are mostly nocturnal, and they have poor eyesight, relying on sound and smell for getting around instead.

Since they eat ants with their toothless snout, for a long time they were associated with anteaters and other xenathrans as "edentates" (meaning "toothless", a term coined by Linnaeus in 1758). But molecular biology shows they are laurasiatherians, more closely related to carnivorans than any other living mammal.

Rare creatures like pangolins have only a sparse fossil record, and very little was known of their evolutionary history until the 1970s. *Neomanis* from the Oligocene and Miocene of France was known, but it was very incomplete. Then in 1970, a fossil found in the upper Eocene beds of Wyoming (misidentified as a juvenile carnivore skull) was identified by Robert Emry as a pangolin and named *Patriomanis*. This proved that pangolins had once lived in the Americas. In 1978, the beautifully preserved skeleton of a complete pangolin (even including the scales) from the middle Eocene Messel lake beds of Germany was published as *Eomanis*, showing the pangolin body plan was already established 50 Ma (**Figure 23.3**). Another fossil from the same deposits, *Eurotamandua*, was once considered to be related to the tree-dwelling tamandua anteaters (Xenarthra), but is now recognized as a scale-less pangolin since it does not have xenarthrous vertebrae. Thus, good fossils of the Manidae tell us that pangolins were spread across the northern continents by the middle Eocene. Their remaining fossil record is poor, but there are eight species alive today. Sadly, they are being poached intensively because their meat and scales have great value on the Asian black market, so they are likely to become extinct in the near future.

Figure 23.3 Reconstructions of extinct archaic pangolins from the Eocene lake beds of Messel in Germany. (A) *Eomanis*; (B) *Eurotamandua*. Scale bar: 20 cm.

CARNIVOROUS MAMMALS

Predators are essential to the food web, keeping the population numbers of their prey under check, and culling the weak, the young, and the old. But the role of predators has been occupied by many different animals since dinosaurs relinquished that role at the end of the Cretaceous. In the Paleocene and earliest Eocene, most of the mammalian predators were small, and the large predator niche was occupied by avian dinosaurs, the gigantic "terror birds" such as *Diatryma* in North America and *Gastornis* in Europe (**Figures 19.8[B]** and **19.9**) (Most paleontologists now consider both of these birds to be *Gastornis*). In the Miocene and Pliocene in South America, the largest mammalian predators were only wolf-sized, another group of birds, the phorhusrhacids, independently evolved into large land predators (**Figures 19.10[C–E]** and **19.11**).

Among mammalian orders, several different groups have evolved to fill that role. In Australia, where only native mammals were the pouched marsupials, several groups of wolf-like, lion-like, and cat-like predators evolved from possum-like ancestors (Chapter 21). The same was true in South America, where marsupial predators evolved that were remarkable mimics of wolves and hyaenids (the borhyaenids) and saber-toothed cats (the thylacosmilids). In fact, the saber-tooth niche (**Figure 23.4**) was

Figure 23.4 A wide spectrum of mammals has evolved a saber-toothed skulls and teeth, a classic case of convergent evolution. These include the "marsupial saber-tooth" *Thylacosmilus* (see Figure 21.7[A]), as well as: (A) The saber-toothed creodont *Apataelurus*. (B) The saber-toothed "false cat" or nimravid *Hoplophoneus*. (Courtesy Wikimedia Commons.)

A

B

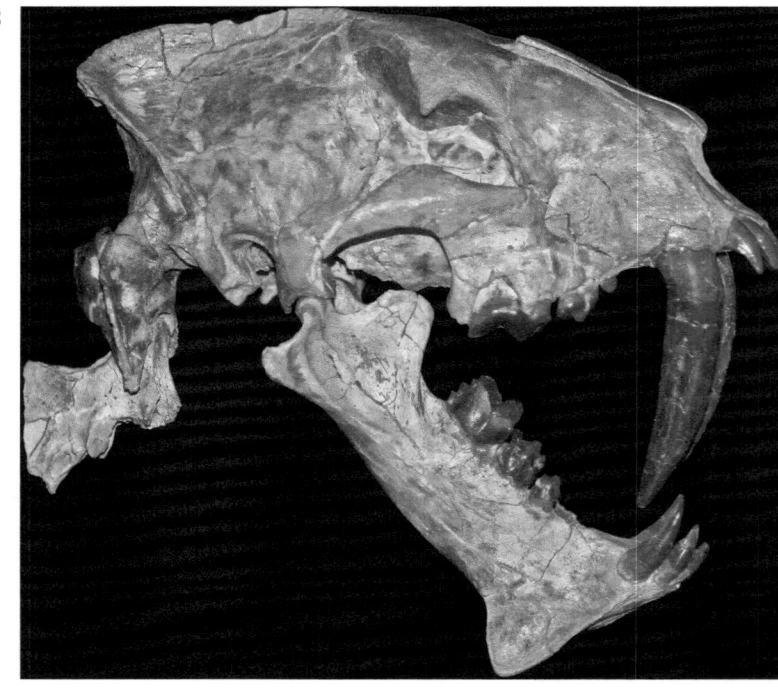

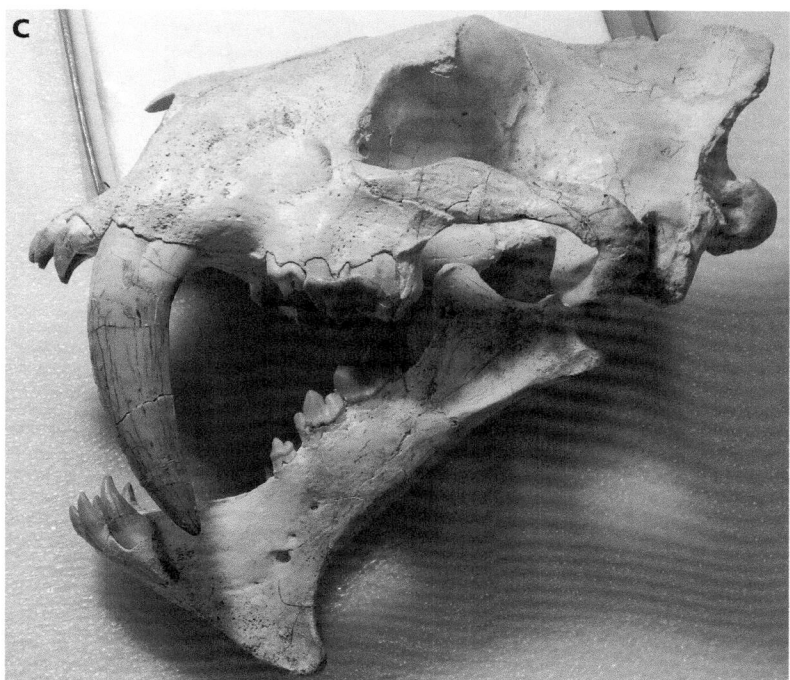

Figure 23.4 (Continued) (C) The sabertooth known as *Barbourofelis*, which may be a nimravid or another independent example of convergent evolution. (D) The true saber-toothed cat, *Smilodon*, a member of the order Carnivora.

occupied not only by marsupials, but also by creodonts and two different groups of carnivorans, including several different kinds of true cats (family Felidae). In the Northern Hemisphere, there were archaic hoofed mammals (the mesonychids) that became the largest predators of the early Cenozoic.

Through most of the Cenozoic, there have been two main groups of mammals that performed the roles of predators (**Figure 23.5**). They include the extinct archaic predators of the order Creodonta, and the living group of predators, the order Carnivora (cats, dogs, bears, hyaenas,

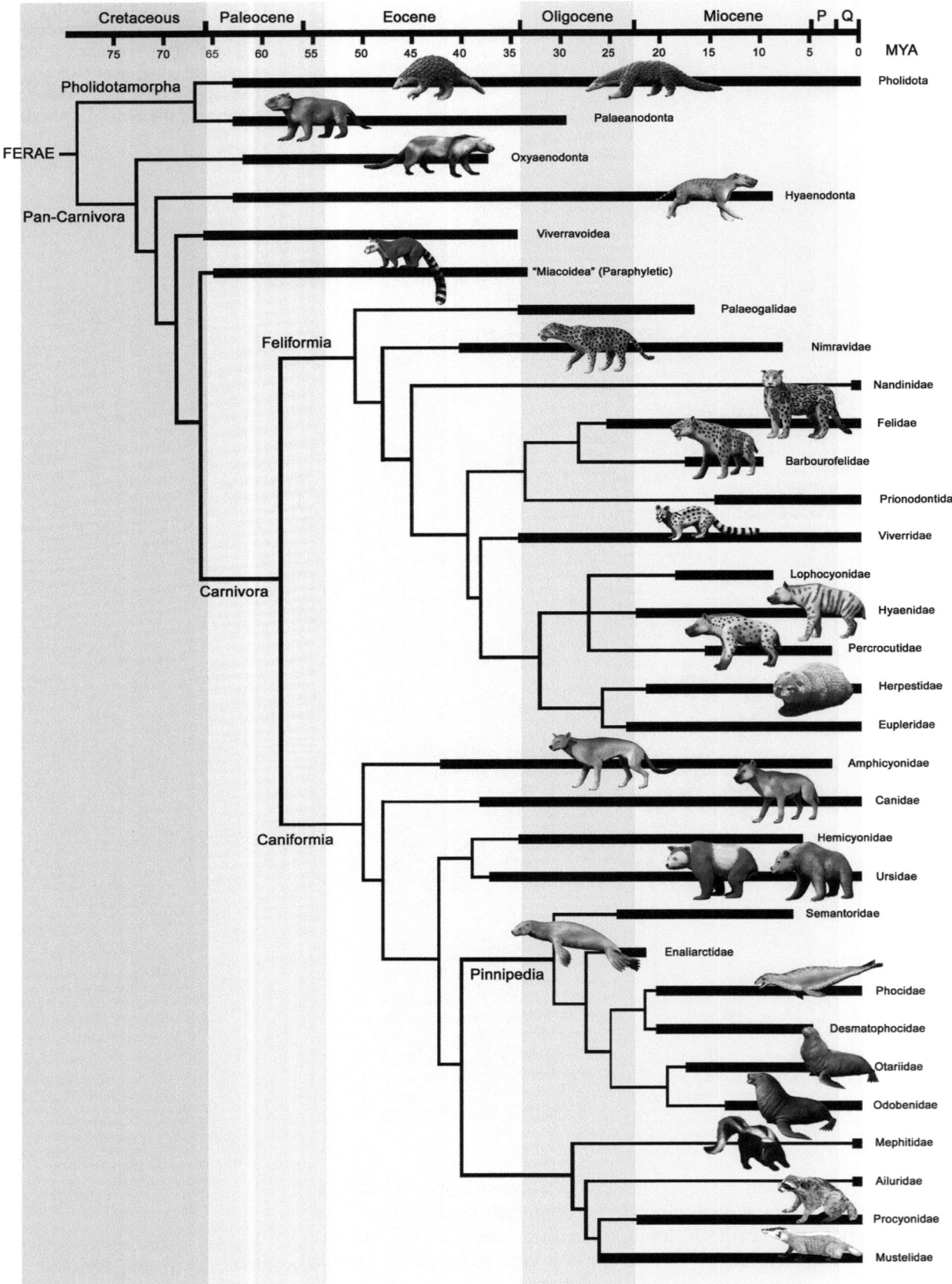

Figure 23.5 Family tree of carnivorous mammals, including the archaic Creodonta, and the radiation of the modern predators, order Carnivora.

raccoons, weasels, seals, sea lions, and many others). In discussing these animals, we must be careful with our words. The word "carnivore" refers to any meat-eater (technically including even carnivorous plants like the Venus' flytrap), but a "carnivoran" is a member of the order Carnivora. Not all carnivorans are carnivorous, either. Some are omnivores (such as bears and raccoons), and two (the giant panda and the red panda) are specialized herbivores, feeding on bamboo.

In addition to having large stabbing canine teeth, almost all predatory mammals have modified their cheek teeth for slicing meat. Thus, the teeth have evolved from simple rounded cusps of their ancestors into teeth that are sets of shearing blades that occlude precisely with the opposite blade to perform a scissor-like action. In addition, a specialization that all carnivorans and creodonts share (not seen in the same way in other carnivorous mammals) is that they develop a specialized pair of enlarged cheek teeth known as carnassials (**Figure 23.6**). This pair of upper and lower teeth serves as the main cutting and breaking tools of the jaw. They are specially adapted for slicing tough meat and tendons, and in many carnivores, for crushing and breaking bone. A glance at the mouth of any cat or dog immediately reveals these crucial teeth. If you watch a dog chew a bone, you will see them use the side of their mouth to bring these powerful carnassial teeth into action.

Carnassial teeth occur in both creodonts and carnivorans, but they key difference is their position. Except for seals and sea lions, which have modified all their teeth into simple conical pegs for fish catching, all carnivorans have their carnassial teeth involving the last, or fourth upper premolar (P4), and m1, the first lower molar (**Figure 23.6**). By contrast, the carnassial shearing pair is further back in the jaws of creodonts: either between the first upper molar and the second lower molar (in oxyaenids) or the second upper molar and the third lower molar (in hyaenodonts).

The position of these crucial teeth is key to identifying which order a fossil belongs to. It may also help explain why carnivorans became so

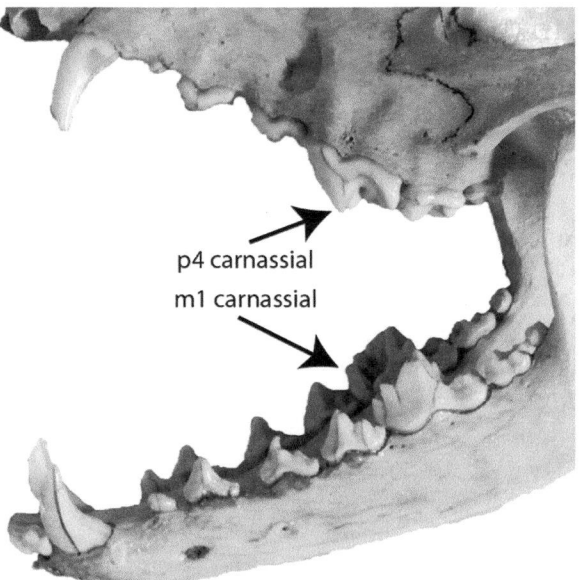

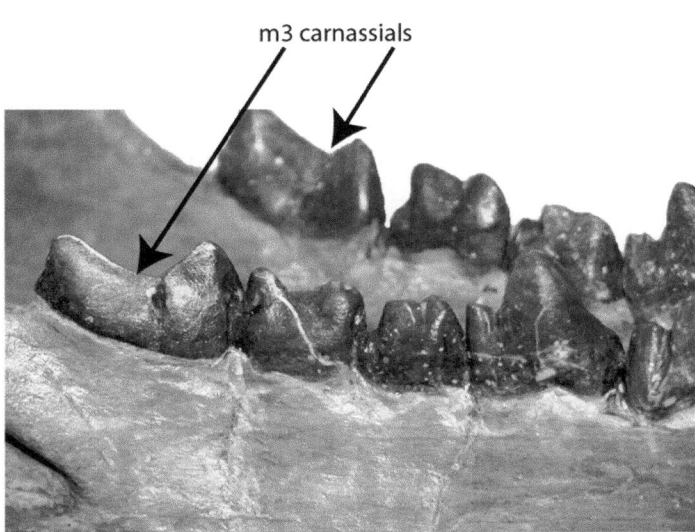

Figure 23.6 Most carnivorous mammals have an enlarged set of cheek teeth called carnassials, which are used for slicing meat and tendons, crushing bones, and breaking tough objects. (Left) In this dog, a terrestrial member of the order Carnivora (dogs, cats, weasels, hyaenas, and their kin), the carnassial teeth are the upper fourth premolar (p4) and the lower first molar (m1). (Right) In creodonts like *Hyaenodon*, the carnassial teeth are in the back of the jaw, either the last molar (m3) or the next-to-last molar (m2). (Photo by the author.)

successful and replaced creodonts. The forward position of their carnassials gave their skulls greater evolutionary flexibility in shape and feeding style, so they could shorten their faces and reduce other parts of the tooth row without affecting the crucial carnassial teeth. In other cases, some carnivorans (like bears) have modified their carnassials and post-carnassial molars into teeth specialized for crushing and other functions. By contrast, the creodonts had carnassials in the back of the mouth, which limited the options of what they could do with their teeth, so they kept the same stereotyped dentition, and never developed any true cat-like forms (which have a short snout, and almost no molars behind their carnassials), bear-like creatures, or other highly specialized groups.

Creodonts

The creodonts were the first major group of Cenozoic mammals to adopt the predatory lifestyle. Originally mistaken for carnivorans, or considered to be ancestors of carnivorans, more recent analyses show that creodonts are an early experiment in predator evolution that was eventually replaced by carnivorans. They are relatively primitive in their body form, without the wide range of different shapes (like a bear or dog or weasel or cat) seen in carnivorans.

Creodonts dominated the meat-eater niches during the Paleocene and Eocene, but by the late Oligocene had vanished from North America and Europe, and straggled on in Asia and Africa until the late Miocene. The reason for their extinction is not known. It seems likely that the more advanced carnivorans outcompeted them in many parts of the world, while creodonts were stuck without much evolutionary flexibility in their jaws, teeth, or limbs. The last known creodonts were species of *Hyaenodon* and *Dissopsalis*, which vanished from Asia and Africa about 11 Ma.

There are two families of creodonts: the Paleocene and early Eocene oxyaenids, which had unspecialized skeletons with low, massive, flat skulls, small brains, long tails, and generally short robust limbs, so they were ambush predators, not fast runners; and the hyaenodonts. The oxyaenid *Oxyaena* from the late Paleocene and early Eocene was built like a wolverine, with powerful robust limbs and a flexible body up to a meter (3.3 feet) long. *Patriofelis* from the middle Eocene of North America, had a huge robust skull; it reached up to 1.8 meters (6 feet) in length and up to 90 kg, or about the size of a modern lion or panther (**Figure 23.7[A]**). The most impressive of the oxyaenids was *Sarkastodon* from the late middle Eocene of Mongolia (**Figure 23.7[B]**). It was twice as long as that of lion-sized *Patriofelis*, so it was probably about 3 meters (10 feet) in length and weighed about 800 kg (1000 pounds), the size of a very big bear. Although most oxyaenids were stereotyped in the heavy-bodied lion-like predator shape, one lineage (*Machairoides* and *Apataelurus*) developed long saber-toothed canines, and converged on the two lineages of saber-toothed carnivorans (**Figure 23.4[A]**).

The other main lineage of creodonts was the hyaenodonts. Their name is misleading, because they were creodonts and not related to true hyaenas, which are members of the order Carnivora. Nor was their anatomy or teeth particularly hyaena-like. Instead, their teeth were adapted mostly for specialized shearing, without the robust bone-crushing cheek teeth found in true hyaenas. Hyaenodonts had more delicate wolf-like skeletons and slender limbs compared to hyaenas, or especially compared to oxyaenids. Hyaenodontids arose in Asia in the late Paleocene, and quickly spread across the northern continents in the early Eocene, where the weasel-like proviverrines were most common predators in

Figure 23.7 Creodonts were archaic carnivorous mammals that occupied the main predatory ecological niches in the Paleocene and early Eocene before the evolutionary radiation of modern carnivorans. (A) *Patriofelis*. (B) *Sarkastodon*. (C) *Hyaenodon*.

most size categories. Even bigger was the Oligocene Mongolian creodont *Hyaenodon gigas*. It was 1.4 meters (5 feet) at the shoulder, about 3 meters (10 feet) long, and weighed about 500 kg. Many different species of *Hyaenodon* evolved (**Figure 23.7[C]**), and they were highly successful predators from the middle Eocene to the late Miocene, a span of about 26 million years (longer than just about any other fossil mammal). Hyaenodontids were among the most common predators in the early Oligocene in North America but also Eurasia. By the late Oligocene, however, *Hyaenodon* had vanished from both North America and Europe, pushed out by other more advanced carnivorans. However, hyaenodontids persisted as important predators in Asia and Africa well into the late Miocene. The biggest of these was the Miocene *Hyaenaelurus*, a lion-sized predator up to 3 meters (10 feet) long and weighing about 300 kg, which hunted large prey in Africa and Eurasia during the middle Miocene, 15–11 Ma. Although oxyaenids were extinct before 37 Ma, the hyaenodontids persisted long after them, vanishing about 11 Ma in Africa and Asia.

Carnivorans

All the living flesh-eating mammals today are members of order Carnivora (**Figure 23.5**). They include about 280 living species in 13 families, and about ten times that many extinct species. They have the most extreme size range of any mammalian order, from the least weasel (only a few inches long and weighs about 25 grams), to the huge polar bears that weigh up to a ton, to the gigantic elephant seals, which can weigh 5000 kg (11,000 lb) and reach 7 meters (24 feet) in length. As pointed out earlier, carnivorans have a number of specialized adaptations for the predatory life, from the large canines for stabbing and grabbing prey, to the cheek teeth adapted for slicing meat. Most also have long sharp claws for fighting and slashing and other functions, digestive modifications for a diet of meat, and acute senses (especially smell, sight, and hearing) for detecting prey and hunting it down.

Both anatomical data and now molecular data confirm that there are two main branches (**Figure 23.5**) to the Carnivora: the Feliformia (cats, hyaenas, mongooses, and their relatives), and the Caniformia (dogs, bears, raccoons, weasels and their relatives, plus the seals and sea lions). Both groups evolved from primitive carnivorans of the Paleocene and Eocene lumped into wastebasket groups, the "miacids" (the fossils closest to the Caniformia, mostly early to middle Eocene in age) and the "viverravids" (the fossils closer to the Feliformia, mostly Paleocene in age). Most "miacids" and "viverravids" were small creatures (**Figure 23.8[A]**) about the size and proportions of weasels or mongooses. During the Paleocene and most of the Eocene, these small carnivorans were overshadowed by the much bigger creodonts, and did not begin to diversify in size or shape until most of the creodonts vanished in the late Eocene and Oligocene in North America and Eurasia.

One major branch of the carnivorans is the feliforms, the cats, hyaenas, civets, mongoose, and their relatives (**Figure 23.5**). Although they have

Figure 23.8 Reconstructions of some extinct carnivorans. (A) The primitive archaic "miacid" *Tapocyon*. (B) The saber-toothed "false cat" or nimravid *Pogonodon*. (C) The remarkable saber-toothed predator *Barbourofelis*, which might have a nimravid, or another example of independent evolution of saber-toothed dentition. (D) The huge "bone-crushing dog" or borophagine, *Epicyon*. (E) The amphicyonid ("beardog") *Daphoenocyon*.

some features in common (like retractile claws in many of them, and tendency to shorten the snout and reduce the cheek teeth behind the carnassials), the Feliformia is defined by uniquely specialized features in the skull region and braincase. The validity of this group was supported in the 1990s and the years since, when molecular analyses confirmed that all feliforms were closely related to one another.

The first cat-like carnivorans are a group sometimes called "false cats" or "paleofelids" or more properly, the nimravids. They are extremely cat-like in their body shape and teeth (some are even saber-toothed), but the evidence outside these features demonstrate that they are not cats at all, but may even be related to the dog branch of the Carnivora (**Figures 23.5** and **23.8[B,C]**). They first appeared in the middle Eocene, and flourished as the dominant cat-like predator in the late Eocene and Oligocene, vanishing near the end of the Oligocene, about 26 Ma. For the next 7–8 million years, there was a "cat gap" in North America with no cat-like forms, then about 18.5 Ma true cats (family Felidae) came over from their origins in Asia, with the earliest American form being a primitive cat named *Pseudaelurus*. For the rest of the Cenozoic, cats diversified across the Northern Hemisphere, with numerous genera (including the Ice Age North American cheetah), and also a big radiation of saber-toothed cats (machairodonts) (**Figure 23.5**). Although 12 genera and 73 species of machairodonts are known, the most famous is the late Ice Age saber-toothed cat, *Smilodon*, found in both North America and South America.

After cats, the most familiar feliforms are the hyaenas. We think of them only as skulking scavengers, shadowing lions to take over their kill, but hyaenas are skilled pack hunters who kill about 95% of their own food, but only scavenge when a carcass is available. With their robust heavy jaws and teeth, they can crush and break open bones for their marrow, and consume almost the entire carcass. They first appeared in the Miocene, about 17 Ma, and soon radiated into 20 genera and 70 species that occupied the cat-like predatory role before cats evolved and took it over; some were also long-legged predators that could run like cheetahs, and others were shaped more like dogs. Although dominant in Africa and Eurasia during the Neogene, one genus (*Chasmoporthetes*) managed to reach North America in the late Pliocene. The most spectacular hyaena was the *Dinocrocuta*, a bear-sized monster that weighed about 400 kg (880 lb), and had a massive skull with immensely powerful crushing jaws. It was common in the middle and late Miocene of Africa and Eurasia. By the Pleistocene, there was *Pachycrocuta*, a mega-scavenger weighing about 110 kg (240 lb), about the size of a lion. Its jaws and teeth were strong enough to even break elephant bones, and it is thought to be the predator that collected lots of bones of *Homo erectus* ("Peking man") in the famous Zhoukoudian caves near Beijing.

In addition to the cats and hyaenas, the rest of the feliforms include the herpestids (mongoose, meerkats, and kin), a common smaller predator in Africa and Asia with 34 different species in 14 genera, the euplerids of Madagascar (including the fossa and the Malagasy civet), and the viverrids, including 38 species of civets, genets, and binturongs, also found mainly in tropical Asia and Africa. All of these have an excellent fossil record in the Old World.

The other main branch of the Carnivora is the Caniformia, or dogs, bears, seals, raccoons, weasels, and their kin. The evolution of dogs is now very well understood due to an extensive set of studies in the 1990s based on huge new fossil collections. Dogs originated in the late Eocene of North America with a small weasel-shaped form called *Hesperocyon*. This primitive group of dogs radiated through the Oligocene and early Miocene in North America, to be replaced a group of dogs with robust jaws and teeth (like those of hyaenas) known as the borophagines, or "bone crushing dogs" (**Figure 23.8[D]**). These culminated with bear-sized dogs like *Epicyon* and *Borophagus*, which had huge crushing teeth, and apparently not only were the main predator in North America during the Miocene, but also performed the roles of hyaenas. They vanished at the end of the Ice Ages, but in the meanwhile, they had been replaced by the radiation of the modern subfamily Caninae, to which all living dogs, wolves, coyotes, foxes, dholes, dingoes, and other canines belong.

Another extinct family is the amphicyonids, or "beardogs" (which are neither bears nor dogs, but their own distinct family). These started with tiny fox-sized creatures (*Daphoenus*) in the late Eocene and Oligocene to the early Miocene *Daphoenodon*, and then to huge predators in both Eurasia and North America. The largest was the bear-sized amphicyonids *Ischyrocyon* and especially *Amphicyon*, which may have weighed almost 600 kg (1320 lb), largest predatory mammal seen up to that time in North America (**Figure 23.9[A]**). As they got larger, amphicyonids switched from the more slender dog-like build, walking on the tips of its toes, to a fully robust bear-like build and were walking on the palms of its feet and hands.

Bears (family Ursidae) is another diverse group with 8 species alive today. However, the bear family has a long history with at least 13 genera and

Figure 23.9 Photos of some extinct gigantic carnivorans. (A) Skeleton of the giant amphicyonid, or beardog, the genus *Amphicyon*. (Photos by the author.)

B

Figure 23.9 (Continued) (B) Reconstruction of the giant short-faced bear, *Arctodus*.

dozens of species going back to the late Eocene (38 Ma). The earliest bears (*Parictis*) from the late Eocene of North America were small and looked much like raccoons, with a similar omnivorous diet. By the early Oligocene, there were similar bears (*Amphicynodon*) in Eurasia, emigrants from North America. From these primitive amphicynodontines, the first bear subfamily was the hemicyonines, with the dog-like Eurasian *Cephalogale* appearing in the early Oligocene, and later genera such as *Phoberocyon* (at 20 Ma) and *Plithocyon* (at 15 Ma) migrating back from Eurasia to North America. The short-faced bears (*Arctodus*) were larger than any living bear, standing 3.7 meters (12 feet) on its hind legs, with a 4.3 meters (14 foot) arm-span, and weighing almost 1000 kg (2200 lb), the largest land carnivorans ever known (**Figure 23.9[B]**).

An important branch of the Caniformia is the Pinnipedia, or seals, sea lions, and walruses (**Figure 23.5**). For a while their origins were controversial, but now both anatomical and molecular evidence shows they are clearly related to the earliest bears. There are 33 living species in 22 genera, spread among three living families. The first is the Phocidae, or true seals. They have no external ears, and are unable to walk with their hind flippers. Second are the Otariidae, or sea lions. They have external ears and hind flippers that can rotate forward for walking. Finally, there are the Odobenidae, or walruses, as well as over 50 extinct species.

The earliest known fossil pinniped is *Pujila darwini* (**Figure 29.10[A]**), an otter-like creature from the lower Miocene lake beds of the Canadian Arctic. The next more advanced fossil is *Enaliarctos*, from the early

Figure 23.10 Reconstruction of some extinct pinnipeds. (A) The early Miocene primitive pinniped *Pujila*. (B) The early middle Miocene archaic pinniped *Enaliarctos*. (C) The four-tusked archaic walrus *Gomphotaria*.

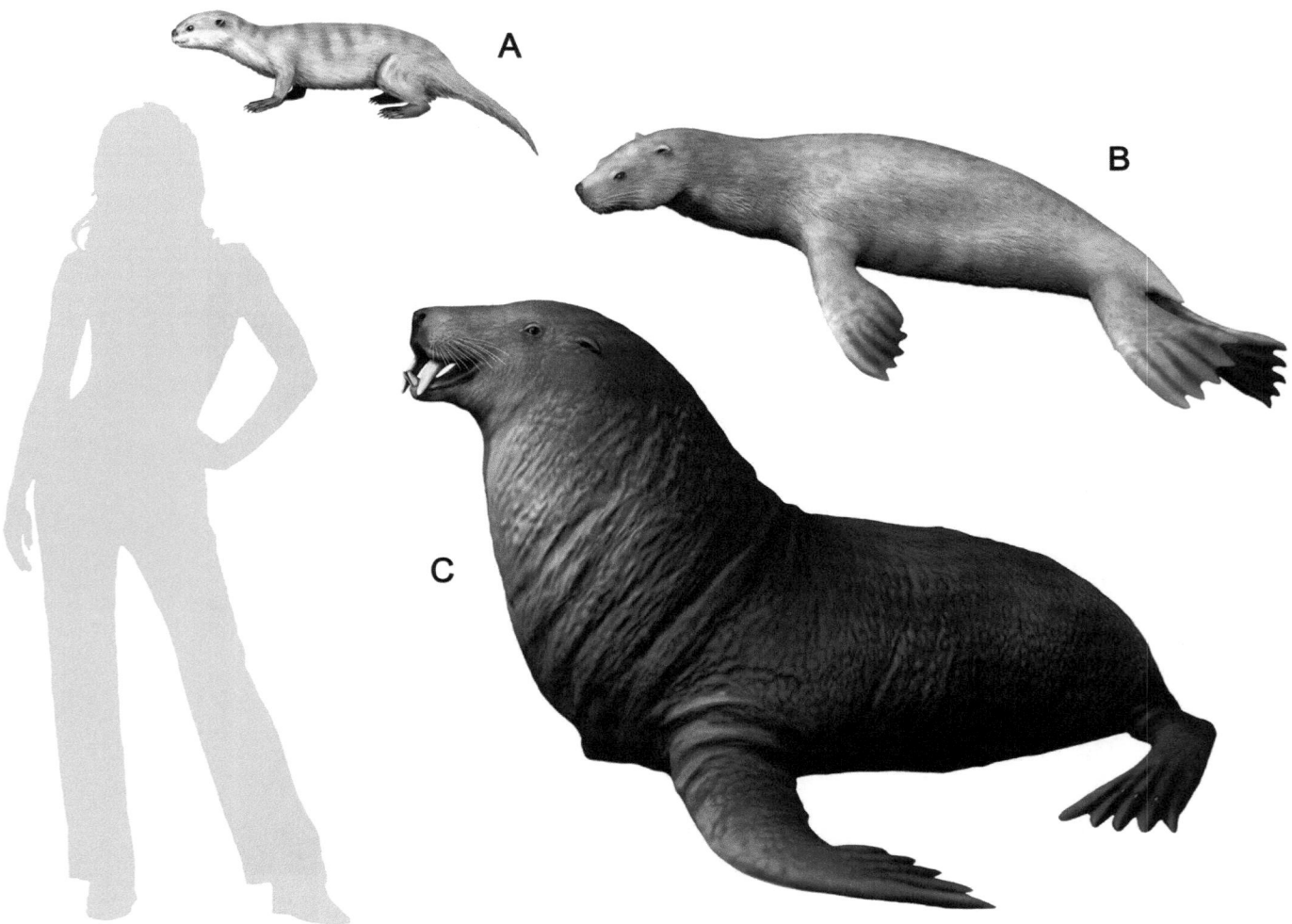

Miocene (24–22 Ma) of California and Oregon (**Figure 29.10[B]**). *Enaliarctos* superficially looks like a seal with flippers and streamlined body, as well as large eyes, whiskers, and ears suitable for hearing underwater. However, its hands and feet were not yet fully modified into the classic flipper, and its teeth and braincase were still quite bear-like. *Enaliarctos* apparently swam with both front and hind flippers, a transitional stage between the hindlimb propulsion of true seals and the forelimb propulsion of sea lions.

After *Enaliarctos*, the three pinniped families diverged. The earliest sea lions split in the middle Miocene, about 16 Ma, and most of their fossil record comes from the North Pacific, before eventually spreading to the Southern Hemisphere oceans. Walruses first appeared about 18 Ma, with fossils such as *Proneotherium* of North America and *Prototaria* of Japan, which looked like sea lions with large canines. The evolution of walruses is well documented by fossils, including the four-tusked *Gomphotaria* of the middle Miocene (**Figure 29.10[C]**). Then came the Pliocene *Valenictis*, which had intermediate-length upper tusks, a short lower jaw with no lower tusks, and a mouth that had not yet developed the peg-like teeth and suction-pump mechanism of modern walruses.

The true seals can also be traced back to the early Miocene with the extinct family Desmatophocidae, which are transitional fossils between modern seals and *Enaliarctos*. Desmatophocids still swam with both front and hind flippers, and did not yet have the skull and tooth specializations seen in modern seals.

Finally, there are a number of smaller groups in the Caniformia, including the red "panda" (family Ailuridae), which is not closely related to the familiar giant panda; the raccoons and their relatives (family Procyonidae); and the huge family Mustelidae, the weasels and their kin, including 57 living species such as the ferrets, minks, polecats, martens, fishers, and stoats, but also unusual forms like the big ferocious wolverine, the digging badgers, the unrelated honey-badgers, and a separate subfamily for the 7 genera and 12 species of otters. All are predators on small animals, mainly birds, rodents, and fish. Skunks used to be included in the Mustelidae, but are now in their own family, the Mephitidae.

FURTHER READING

Agnarsson, I.; Zambrana-Torrelio, C.M.; Flores-Saldana, N.P.; May-Collado, L.J. 2011. A time-calibrated species-level phylogeny of bats (Chiroptera, Mammalia). *PLoS Currents*. 3: RRN1212.

Agusti, J.; Anton, M. 2002. *Mammoths, Sabertooths, and Hominids: 65 Million Years of Mammalian Evolution in Europe*. Columbia University Press, New York.

Anton, M.; Turner, M. 2000. *The Big Cats and Their Fossil Relatives*. Columbia University Press, New York.

Barycka, E. 2007. Evolution and systematics of the feliform Carnivora. *Mammalian Biology*. 72 (5): 257–282.

Bishop, K.L. 2008.The evolution of flight in bats: Narrowing the field of plausible hypotheses. *The Quarterly Review of Biology*. 83 (2): 153–169.

Eiting, T.P.; Gunnell, G.F. 2009. Global completeness of the bat fossil record. *Journal of Mammalian Evolution*. 16 (3): 151–173.

Eizirik, E.; Murphy, W.J.; Koepfli, K.P.; Johnson, W.E.; Dragoo, J.W.; Wayne, R.K.; O'Brien, S.J. 2010. Pattern and timing of diversification of the mammalian order Carnivora inferred from multiple nuclear gene sequences. *Molecular Phylogenetics and Evolution*. 56 (1): 49–63.

Ewer, R.F. 1973. *The Carnivores*. Cornell University Press, Ithaca, NY.

Flynn, J.J.; Finarelli, J.A.; Zehr, S.; Hsu, J.; Nedbal, M.A. 2005. Molecular phylogeny of the Carnivora (Mammalia): Assessing the impact of increased sampling on resolving enigmatic relationships. *Systematic Biology*. 54 (2): 317–337.

Flynn, J.J.; Finarelli, J.A.; Spaulding, M. 2010. Phylogeny of the Carnivora and Carnivoramorpha, and the use of the fossil record to enhance understanding of evolutionary transformations, pp. 25–63. In Goswami, A.; Anthony, F., eds. *Carnivoran Evolution: New Views on Phylogeny, Form and Function*. Cambridge University Pres, Cambridge, UK.

Flynn, J.J.; Finarelli, J.A.; Zehr, S.; Hsu, J.; Nedbal, M.A. 2005. Molecular phylogeny of the Carnivora (Mammalia): Assessing the impact of increased sampling on resolving enigmatic relationships. *Systematic Biology*. 54 (2): 317–337.

Flynn, J.J., Neff, N.A.; Tedford, R.H. 1988. Phylogeny of the Carnivora, pp. 73–116. In Benton, M.J., ed. *The Phylogeny and Classification of the Tetrapods, vol. 2: Mammals*. Clarendon Press, Oxford.

Hallett, M., Harris, M.A. 2020. *On the Prowl: In Search of Big Cat Origins*. Colubmia University Press, New York.

Hunt, R.M., Jr.; Barnes, L.G. 1994. Basicranial evidence for ursid affinity of the oldest pinnipeds. *Proceedings of the San Diego Society of Natural History*. 29: 57–67.

Gaudin, T. 2009. The phylogeny of living and extinct pangolins (Mammalia, Pholidota) and associated taxa: A morphology-based analysis. *Journal of Mammalian Evolution*. 16 (4): 235–305.

Janis, C. 1993. Tertiary mammal evolution in the context of changing climates, vegetation, and tectonic events. *Annual Reviews of Ecology and Systematics*. 24: 467–500.

Janis, C.; Gunnell, G.F.; Uhen, M.D., eds. 2008. *Evolution of Tertiary Mammals of North America, Vol II: Small Mammals, Xenarthrans, and Marine Mammals*. Cambridge University Press, Cambridge.

Janis, C.; Scott, K.M.; Jacobs, L.L., eds. 1998. *Evolution of Tertiary Mammals of North America, vol. I: Terrestrial Carnivores, Ungulates, and Ungulate-Like Mammals*. Cambridge University Press, Cambridge.

Kurtén, B. 1968. *Pleistocene Mammals of Europe*. Columbia University Press, New York.

Kurtén, B. 1988. *Before the Indians*. Columbia University Press, New York.

Kurtén, B.; Anderson, E. 1980. *Pleistocene Mammals of North America*. Columbia University Press, New York.

McKenna, M.C.; Bell, S.K. 1997. *Classification of Mammals above the Species Level*. Columbia University Press, New York.

Novacek, M.J.; Wyss, A.R.; McKenna, M.C. 1988. The major groups of eutherian mammals, pp. 31–73. In Benton, M.J., ed. *The Phylogeny and Classification of the Tetrapods, vol. 2: Mammals*. Clarendon Press, Oxford.

Prothero, D.R. 1994. Mammalian evolution, pp. 238–270. In Prothero, D.R.; Schoch, R.M., eds. *Major Features of Vertebrate Evolution*. Paleontological Society Short Course 7. Paleontological Society, Lawrence, KS.

Prothero, D.R. 2006. *After the Dinosaurs: The Age of Mammals*. Indiana University Press, Bloomington, IN.

Prothero, D.R. 2016. *The Princeton Field Guide to Prehistoric Mammals*. Princeton University Press, Princeton, NJ.

Roca, A.L.; Bar-Gal, G.K.; Eizirik, E.; Helgen, K.M.; Maria, R.; Springer, M.S.; O'Brien, S. J.; Murphy, W.J. 2004. Mesozoic origin for West Indian insectivores. *Nature*. 429 (6992): 649–651.

Rose, K.D.; Archibald, J.D., eds. 2005. *The Rise of Placental Mammals: The Origin and Relationships of the Major Extant Clades*. Johns Hopkins University Press, Baltimore, MD.

Savage, D.E.; Russell, D.E. 1983. *Mammalian Paleofaunas of the World*. Addison Wesley, Reading, MA.

Savage, R.J.G.; Long, M.R. 1986. *Mammal Evolution: An Illustrated Guide*. Facts-on-File Publications, New York.

Schultz, N.G.; Lough-Stevens, M.; Abreu, E.; Orr, T.; Dean, M.D. 2016. The baculum was gained and lost multiple times during mammalian evolution. *Integrative and Comparative Biology*. 56 (4): 644–656.

Simmons, N.B.; Seymour, K.L.; Habersetzer, J.; Gunnell, G.F. 2008. Primitive early Eocene bat from Wyoming and the evolution of flight and echolocation. *Nature*. 451 (7180): 818–821.

Solé, F.; Ladevèze, S. 2017. Evolution of the hypercarnivorous dentition in mammals (Metatheria, Eutheria) and its bearing on the development of tribosphenic molars. *Evolution & Development*. 19 (2): 56–68.

Solé, F.; Richard, S.; Tiphaine, C.; de Eric, B.; Thierry, S. 2014. Dental and tarsal anatomy of *Miacis latouri* and a phylogenetic analysis of the earliest carnivoraforms (Mammalia, Carnivoramorpha). *Journal of Vertebrate Paleontology*. 34 (1): 1–21.

Solé, F.; Thierry, S.; De Eric, B.; Vlad, C.; Emmanuel, G. 2016. New carnivoraforms from the latest Paleocene of Europe and their bearing on the origin and radiation of Carnivoraformes (Carnivoramorpha, Mammalia). *Journal of Vertebrate Paleontology*. 36 (2): e1082480.

Springer, M.S.; Murphy, W.J.; Eizirik, E.; O'Brien, S.J. 2003. Placental mammal diversification and the Cretaceous: Tertiary boundary. *Proceedings of the National Academy of Sciences*. 100 (3): 1056–1061.

Stucky, R.K. 1990. Evolution of land mammal diversity in North America during the Cenozoic. *Current Mammalogy*. 2: 375–432.

Szalay, F.S.; Novacek, M.J.; McKenna, M.C., eds. 1993. *Mammal Phylogeny*. Springer-Verlag, Berlin.

Teeling, E.C.; Springer, M.S.; Madsen, O.; Bates, P.; O'Brien, S.J.; Murphy, W.J. 2005. A molecular phylogeny for bats illuminates biogeography and the fossil record. *Science*. 307 (5709): 580–584.

Tsagkogeorga, G.; Parker, J.; Stupka, E.; Cotton, J.A.; Rossiter, S.J. 2013. Phylogenomic analyses elucidate the evolutionary relationships of bats. *Current Biology*. 23 (22): 2262–2267.

Turner, A.; Anton, M. 2004. *National Geographic Prehistoric Mammals*. National Geographic Society, Washington, DC.

Van de Bussche, R.A.; Hoofer, S.R. 2004. Phylogenetic relationships among recent chiropteran families and the importance of choosing appropriate out-group taxa. *Journal of Mammalogy*. 85 (2): 321–330.

Waddell, P.J.; Okada, N.; Hasegawa, M. 1999. Towards resolving the interordinal relationships of placental mammals. *Systematic Biology*. 48 (1): 1–5.

Wang, X.; Tedford, R. 2010. *Dogs: Their Fossil Relatives and Evolutionary History*. Columbia University Press, New York.

Werdelin, L.; Sanders, W.L., eds. 2010. *Cenozoic Mammals of Africa*. University of California Press, Berkeley, CA.

Werdelin, L.; Yamaguchi, N.; Johnson, W.E.; O'Brien, S.J. 2010. Phylogeny and evolution of cats (Felidae), pp. 59–82. In Macdonald, D.W.; Loveridge, A.J., eds. *Biology and Conservation of Wild Felids*. Oxford University Press, Oxford, UK.

Woodburne, M.O., ed. 2004. *Late Cretaceous and Cenozoic Mammals of North America: Biostratigraphy and Geochronology*. Columbia University Press, New York.

Zhou, X.; Xu, S.; Xu, J.; Chen, B.; Zhou, K.; Yang, G. 2012. Phylogenomic analysis resolves the interordinal relationships and rapid diversification of the laurasiatherian mammals. *Systematic Biology*. 61 (1): 150–164.

LAURASIATHERIA II

THE UNGULATES

24

We live in a world where most of the attention gets grabbed by the carnivores—and mammals are no exception. We have TV shows entitled "Fangs" but none (alas) called "Molars", and the hoofed mammals are often regarded as little more than fodder.

—Christine Janis, 2003

HORNS, HOOVES, AND FLIPPERS

After rodents and bats, the third largest group of placentals is the hoofed mammals, or ungulates. Hoofed mammals make up about 33% of the living and extinct mammalian genera, and nearly all the large-bodied herbivores are ungulates. According to the anatomical and fossil-based phylogenies (**Figure 22.2**), they include the even-toed artiodactyls (pigs, peccaries, hippos, camels, deer, antelopes, giraffes, cattle, sheep, and goats, plus their descendants, the whales), and the odd-toed perissodactyls (horses, rhinos, tapirs, and their extinct kin). In the 1980s and 1990s, studies of the anatomy of mammals also suggested that the ungulates included the tethytheres (elephants, manatees, and their extinct relatives), the aardvarks, and the woodchuck-like hyraxes or conies. But molecular evidence has pushed these groups into the Afrotheria.

Ungulates not only have dominated the large herbivore niche through most of the Cenozoic, but also are the dominant aquatic predators and filter feeders, and some were even carnivorous. Some ungulates have long slender limbs for fast running (especially antelopes and horses), but others are large-bodied with robust limbs (such as rhinos, hippos, and many extinct groups). Some can even climb trees. Ungulates have occupied a wide variety of ecological niches given the constraints of their body size and diet.

Until about 35 years ago, the interrelationships of the major ungulate groups were obscured by a paraphyletic ancestral "wastebasket" group, the order "Condylarthra". "Condylarths" had nothing in common except that they were primitive ungulates that were not members of any of the living orders. As long as this "wastebasket" group obscured the evidence, there was no possibility that ungulate relationships could be deciphered. However, when cladistic analysis was applied to the "condylarths" and other ungulates, there was a clear pattern of branching among the ungulate groups that has withstood repeated testing from additional morphological and molecular analyses. It turned out that throwing taxa into the "Condylarthra" wastebasket hid a phylogenetic pattern for over a century, but by shifting to a focus on shared evolutionary novelties (plus the great increase in numbers of taxa and characters), scientists were able tease out that pattern.

The earliest ungulates are known from the early Late Cretaceous (about 85 Ma) of Uzbekistan, and show that the major placental divergences

DOI: 10.1201/9781003128205-24

must have come quite early. Better specimens of ungulates are known from the latest Cretaceous, where *Protungulatum* is among the more common taxa. Although these Mesozoic ungulates are known mostly from isolated teeth, jaws, and bones, they still have diagnostic ungulate features. Their molars are square and lower-crowned, with rounder cusps, for eating vegetation rather than insects, and they already have distinctive features of the ankle that are recognizably ungulate.

In the Paleocene, the ungulates split into a number of distinct groups. Some of these archaic ungulates (such as the arctocyonids, hyopsodonts, and periptychids) have long been lumped into the order "Condylarthra", but each is distinctive and related to a different part of the ungulate radiation. The Paleocene arctocyonids (**Figure 24.1[A]**) were the most primitive of the ungulates, about the size and shape of a raccoon, and probably with a similarly omnivorous diet. The hyopsodonts (**Figure 24.1[B]**), on the other hand, were most common in the early and middle Eocene, and were among the last of the surviving "condylarths". They were long-bodied and short-legged, and shaped somewhat like dachshunds, except that their multi-cusped teeth were clearly adapted for grinding vegetation. Another group of "condylarths", the phenacodonts (**Figure 24.1[C]**), are not closely related to the other archaic ungulates, but might be related to the perissodactyls.

The most surprising of these "condylarths" is a group of hoofed predators known as mesonychids. They were the first group of mammals to become

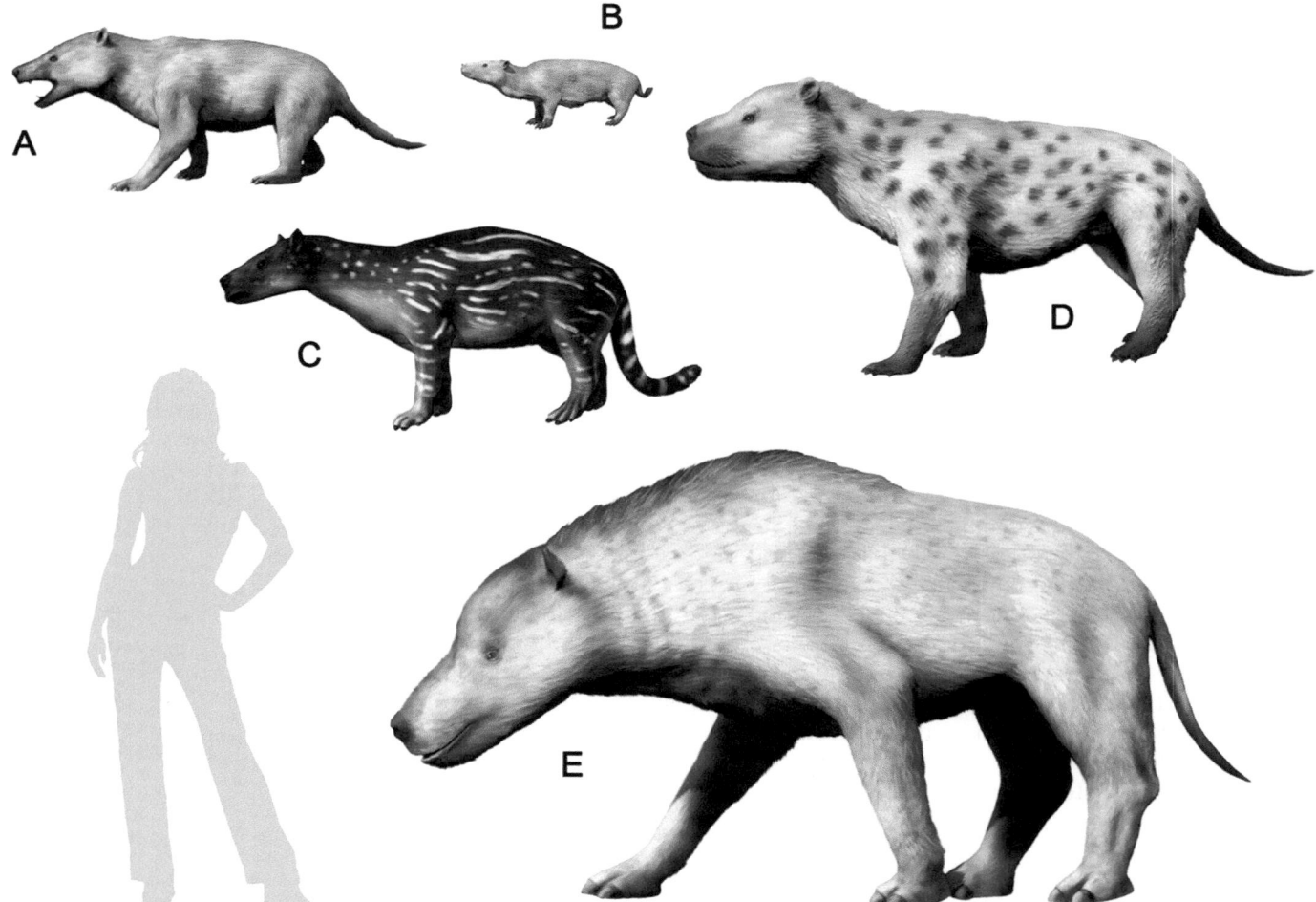

Figure 24.1 Reconstructions of some archaic ungulates. (A) *Arctocyon*, (B) *Hyopsodus*, (C) *Phenacodus*, (D) *Harpagolestes*, (E) *Andrewsarchus*.

specialized meat-eaters, appearing in the middle Paleocene before more specialized carnivorous mammals (the creodonts, and eventually the carnivorans) occupied that niche. Most mesonychids were the size and shapes of large wolves or bears (although some were as small as foxes). They had a heavy robust skull armed with sharp canine teeth, and huge round-cusped molar teeth suitable not only for eating meat, but also for bone crushing. Their body was also very wolf-like, with a long tail and limbs. Like many modern carnivores, they walked on the tips of their long toes, rather than flat-footed. Despite all these carnivorous adaptations, however, mesonychids were derived from hoofed mammals. The proof is in their toes, which had hooves rather than claws.

By the early Eocene, mesonychids had reached their peak of diversity, with wolf-sized beasts such as *Mesonyx* or *Harpagolestes* (**Figure 24.1[D]**) reigning as the largest carnivorous mammals of their time. However, they had to share their world with two other groups of carnivorous mammals: the creodonts (which soon surpassed them in size and diversity); and the true carnivorans (which were still weasel-sized, and did not become large dog-sized or cat-sized predators until the Oligocene). By the middle Eocene, the mesonychids rapidly declined in North America and Eurasia, where they had once dominated. The cause for this decline are unclear. Whatever the reason, the mesonychids were very rare in the late middle Eocene, and they disappeared from North America at the end of the middle Eocene, and from Asia during the late Eocene. The last of the Asian mesonychids, however, was a truly spectacular beast known as *Andrewsarchus* (**Figure 24.1[E]**). Only a single skull of this animal is known, almost a meter long, more than twice the size of any bear that has ever lived! If the rest of the animal were also bear-like, it would have been about 4 meters long and 2 meters high at the shoulder, and weighed almost four times as much as the largest known bear. Mesonychids were long considered related to whales until more recent evidence has placed them as a close relative of both whales plus artiodactyls.

Artiodactyls

One of the first ungulate groups to branch off was the even-toed ungulates, or artiodactyls. They are so called because the axis of symmetry in their hand and foot runs between the third and fourth digits, so they usually have an even number of toes, either two or four toes. Artiodactyls also have a very distinctive ankle bone that has a pulley-like facet on the top and bottom surface. This gives their feet very efficient movement in a fore-aft plane for rapid running, but restricts their ability to rotate their feet in a way that more generalized mammals can. Artiodactyls are the largest group of living ungulates, with over 190 living species, including most of the domesticated hoofed mammals (cattle, sheep, goats, camels, pigs) and thus they are the source of most of our meat, milk, and wool.

The earliest artiodactyls are known from the lower Eocene rocks of Pakistan, and shortly thereafter they spread to the rest of Eurasia and North America. These early forms were very delicately built, resembling a small hornless antelope, and some had such long hind legs that they may have hopped (**Figures 24.2[A]** and **24.3[A]**). During the Eocene, these archaic artiodactyls quickly diversified into a great variety of lineages—the heavy-bodied, omnivorous pigs (**Figure 24.2[B]**); the pig-like (but unrelated) American peccaries or javelinas (**Figure 23.3[C]**; the aquatic hippos; and a number of other pig-like forms.

Among the most bizarre of these pig-like forms were the entelodonts (**Figure 24.3[B,C]**). Thanks to various documentaries like *Walking with*

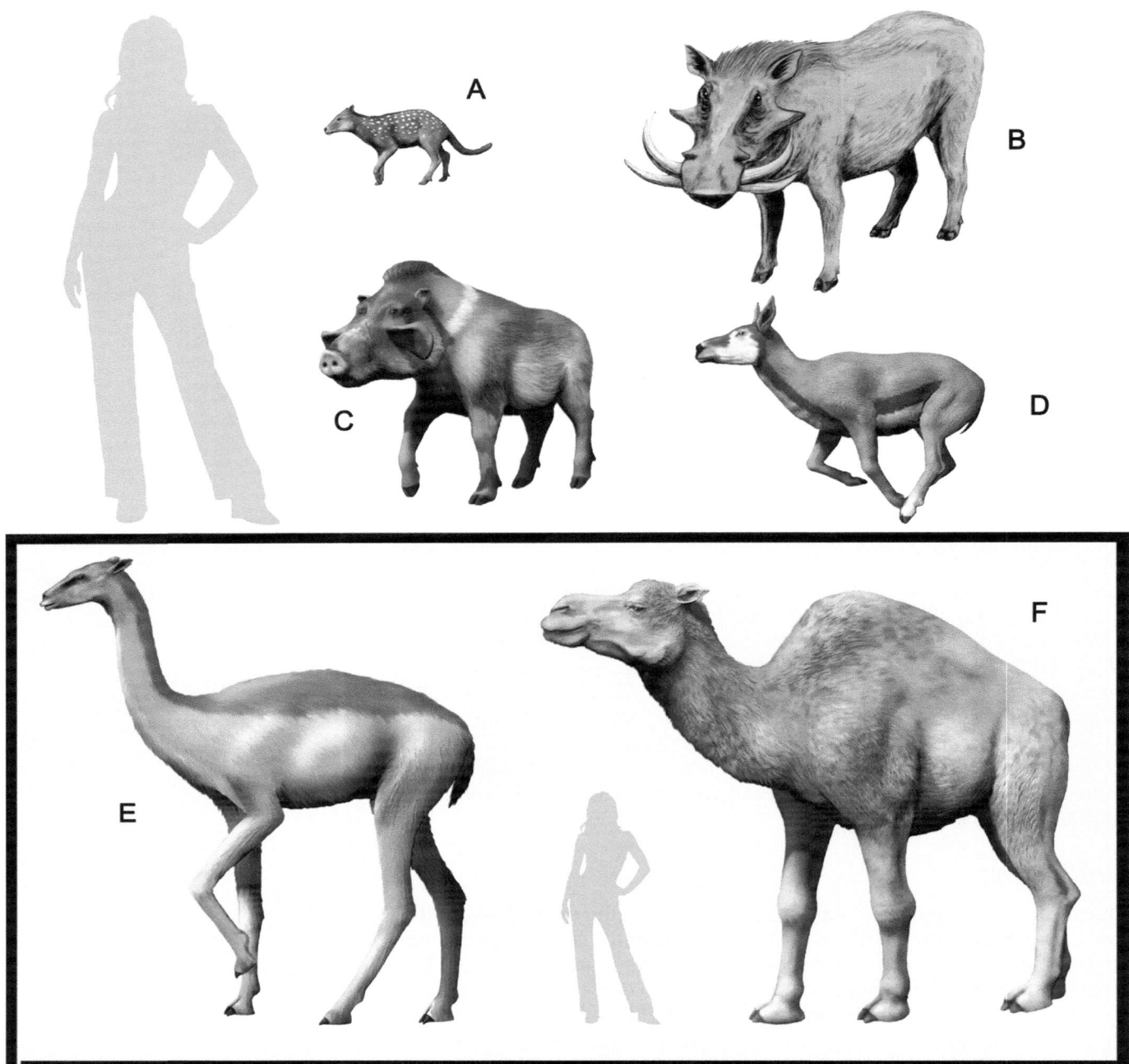

Figure 24.2 Reconstructions of some extinct artiodactyls. (A) The earliest known artiodactyl *Diacodexis*. (B) The bizarre extinct warthog *Metridiochoerus*. (C) The Miocene peccary with huge flanges on its cheekbones, known as *Skinnerhyus*. (D) The gazelle-like camel *Stenomylus*. (E) The giraffe-like camel *Aepycamelus*. (F) The gigantic camel *Gigantocamelus*.

Beasts, they have become media stars and acquired nickname such as "hell pigs", "killer pigs", or "terminator pigs". Like many pigs and their relatives, they had large heads with blunt rounded cusps on their cheek teeth and big canine tusks, along with a chunky body supported by four robust limbs with four hooved fingers and toes. However, pig-like these features were, there are no clear unique evolutionary specializations that unite entelodonts with the pigs, peccaries, hippos, and other suoid artiodactyls. Most of the features that place entelodonts close to other suoids on recent phylogenies are primitive features of the teeth and skeleton that could have evolved by convergent evolution,

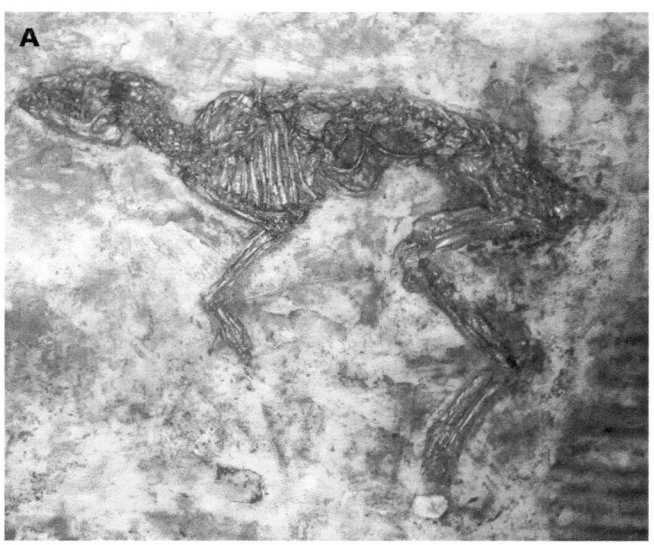

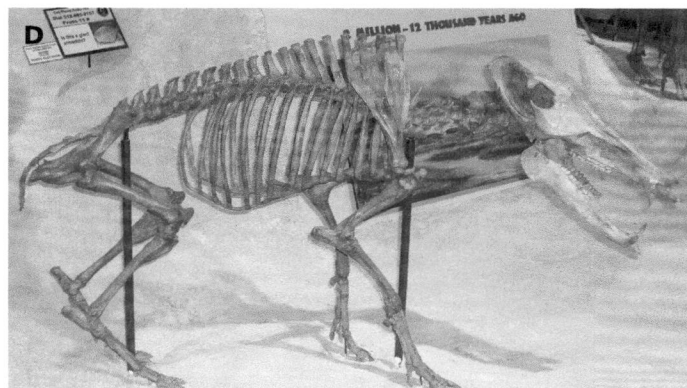

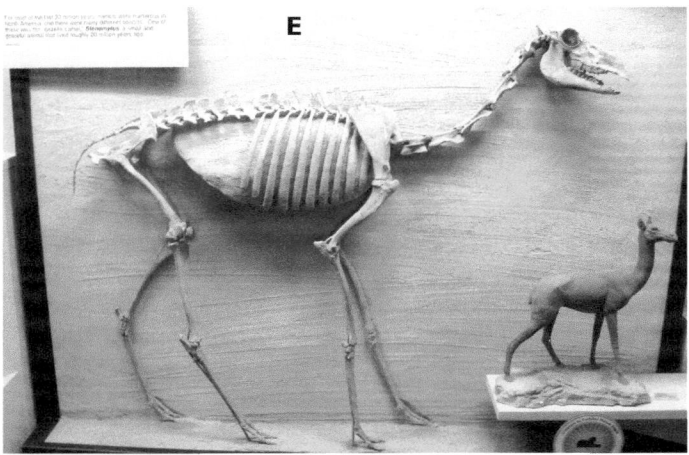

Figure 24.3 Fossils of some extinct non-ruminant artiodactyls. (A) The primitive diacodexid artiodactyl *Messelobunodon*, with long hopping legs, from the middle Eocene of the Messel lake beds, Germany. (B) The skeleton of one of the last and largest entelodonts, *Daeodon*, from the early Miocene of Nebraska. (C) Reconstruction of *Daeodon*. (D) The extinct Ice Age long-nosed peccary, *Mylohyus*, with the bizarre flanges sticking out of its cheekbones, typical of many Miocene through Pleistocene peccaries. (E) The "gazelle-camel" *Stenomylus*, know from large early Miocene beds. It had extremely high-crowned molars in its jaws, with very deep roots, presumably used to feed on gritty grasses. (Courtesy Wikimedia Commons.)

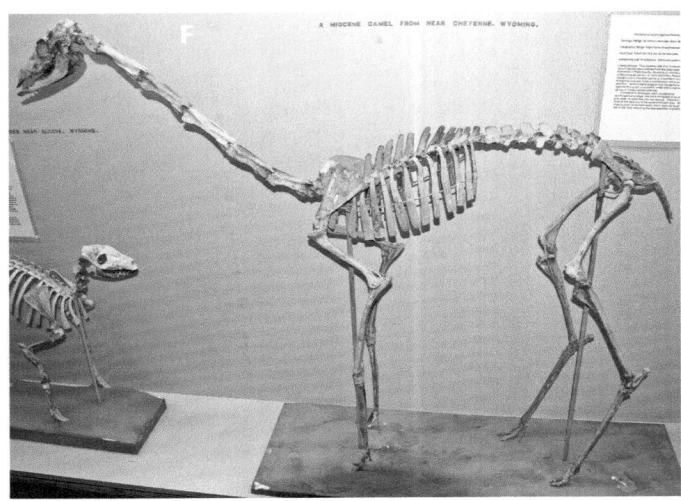

Figure 24.3 (Continued) (F) One of the "giraffe-camels", *Oxydactylus*, from the early Miocene. Later forms, like *Aepycamelus*, were very giraffe-like in build with extremely long neck and legs. (G) The enormous *Gigantocamelus*, one of the largest camels ever known.

so the true biological affinities of entelodonts are still unclear. The latest analyses place them closer to the hippo-whale branch, rather than near the pigs.

Entelodonts first show up in the late middle Eocene of China (40 Ma), with *Eoentelodon*, then spread to North America with *Brachyhyops* (about 38 Ma). During the late Eocene and early Oligocene (37–30 Ma), the dominant entelodont was *Archaeotherium*, a creature about the size of a domestic hog. Many specimens had bony knobs and widely flaring flanges on their cheekbones and around their eyes and bumps on their lower jaws. The culmination of this trend was the hippo-sized entelodont *Daeodon* (formerly called *Dinohyus*) from the early Miocene (18–19 million years old) Agate Springs fossil beds in Nebraska (**Figure 24.3[B,C]**). These monsters were almost 2.1 meters (7 feet) tall at the shoulder, and weighed up to 431 kg (930 lb). Their skull alone is almost 90 cm (3 feet) long.

Early camels did not have humps, but were built more like deer or antelopes, or like the four living species of South American camels (llamas, alpacas, guanacos, vicuñas) which lack humps as well. Camels were once a strictly North American group, playing the roles on this continent that were occupied by other groups elsewhere (**Figures 24.2[D–F]** and **24.3[E–G]**). For example, during the Miocene there were long-necked, long-legged "giraffe-camels" (**Figure 24.3[F]**), delicate "gazelle-camels" (**Figure 24.3[E]**), and yet others that paralleled the shapes of many African antelopes (since North America never hosted true antelopes). In the Pliocene, camels migrated to South America across the Panamanian land bridge, giving rise to the llamas, alpacas, guanacos, and vicuñas still living there today. In the late Miocene, they also crossed the Bering Strait to the Old World, where they evolved into dromedaries and Bactrian camels, the only groups with humps. During the Ice Ages they reached huge size with creatures like *Titanotylopus* and *Gigantocamelus*. Then about 10,000 years ago they vanished from their North American ancestral homeland.

In the late Eocene and Oligocene, another great evolutionary breakthrough occurred when a group of artiodactyls, the ruminants, developed a four-chambered stomach system. Ruminants first swallow their food and then let it ferment in the first stomach chamber, the rumen, where cellulose-digesting bacteria help break up the plant matter. When they have a chance to rest, ruminants regurgitate food from the rumen and "chew their cud", which helps break it down even further. By the time the cud is swallowed again, most of the nutrients can be absorbed by the intestines, so ruminants get the maximum nutrition out of each bite of vegetation. By contrast, most other herbivorous mammals (horses, rhinos, elephants, rabbits) are hindgut fermenters, and have no specialized foregut fermentation chamber, so they can get only a limited amount of nutrition out of the relatively indigestible cellulose in the food as it passes rapidly through their intestine and caecum. Consequently, hindgut fermenters must eat much larger quantities of food than ruminants, and are not as efficient or versatile. (Rabbits get around this by eating their feces, so the food goes through their digestive tract twice, the second time with the cellulose-digesting bacteria already working on it.) With this great innovation, the ruminants (especially the deer, giraffes, cattle, antelopes, goats, and sheep) eventually became the dominant hoofed mammals of the later Cenozoic, and may have pushed out many other groups, such as the horses. Extinct ruminants came in many different shapes and sizes, from the tiny primitive hornless *Leptomeryx* (**Figure 24.4[A]**), to a variety of extinct mouse deer (family Tragulidae) and musk deer (family Moschidae) (**Figure 24.4[B]**) to a wide variety of pronghorns (**Figures 24.4[B,C]** and **24.5[B]**) to the weird-looking horned palaeomerycids (**Figures 24.4[D]** and **24.5[C]**) to a variety of giraffes with short necks and strange horns (**Figures 24.4[E,F]** and **24.5[D]**) to gigantic deer with incredible horns (**Figures 24.4[G,H]** and **24.5[E]**) and cattle which sometimes had gigantic wide horns and huge sizes (**Figures 24.4[I]** and **24.5[F]**).

One of the most amazing stories in evolutionary biology is the origin of whales from land mammals (**Figure 24.6**). By the middle Eocene, there were archaic fossil whales with a fully whale-like body, including a horizontal tail fluke, forelimbs modified into flippers, and no hindlimbs. On the basis of their distinctive triangular teeth, paleontologists had long looked for whale origins among a group of carnivorous hoofed mammals known as the mesonychids.

For years the oldest known whales of the early middle Eocene were known only from fossils of fully aquatic animals without hindlimbs. Recently, however, numerous transitional fossils between whales and their ancestors have been found from the early Eocene of Africa and Asia. They go all the way back to *Indohyus* in the early Eocene, which was a very primitive land ungulate, but had a few whale-like features. *Pakicetus* and *Ichthyolestes* were slightly more aquatically adapted, although still mostly built like a wolf. The most impressive of these fossils transitional between hoofed land mammals and whales is *Ambulocetus* from the early Eocene of Pakistan. Although it still has a primitive whale-like skull and teeth, its front and hind feet are both adapted for swimming, yet it does not yet have a tail fluke. Other fossil whales have even more specialized front flippers (*Dorudon*, *Rodhocetus*, and *Georgiacetus*), and have reduced their hindlimbs to tiny vestiges, and a tail with a horizontal fluke. Then in 2001, two groups of paleontologists hunting for fossils in Pakistan independently discovered that the earliest whales had the distinctive "double-pulley" astragalus bone in their ankles, unique to the artiodactyls. The idea that whales were descended from artiodactyls (specifically, the hippopotamus lineage, and their extinct anthracothere ancestors) was long suggested by molecular evidence, but finally corroborated by fossils. The transformation from whale-like artiodactyls (the anthracotheres

Figure 24.4 Reconstructions of some extinct ruminant artiodactyls. (A) The tiny mouse deer relative *Leptomeryx*. (B) The Miocene pronghorn *Merriamoceros*. (C) The pronghorn *Osbornoceros*. (D) The deer-like palaeomerycid *Procranioceras*. (E) The early giraffid *Prolibytherium*. (F) The gigantic Miocene giraffid *Brahmatherium*. (G) The deer *Eucladoceros*, with multiple tines on its antlers. (H) The "Irish elk", the largest Ice Age deer known *Megaloceros*. (I) The Ice Age long-horned *Bison latifrons*.

Figure 24.5 Fossils of some of the more unusual extinct ruminants. (A) The extinct musk deer, *Micromeryx*, from the Miocene of Europe, with the prominent tusks in males instead of horns or antlers. (B) The tiny pronghorn, *Merycodus*, with the Y-shaped horns in the males (back) and hornless females (front), from the late Miocene of Nebraska. [Photos (D) and (E) by the author; the rest courtesy Wikimedia Commons.]

Figure 24.5 (Continued) (C) The three-horned skull of the deer-like palaeomerycid *Procranioceras skinneri*, from the late Miocene of Nebraska. (D) The giant moose-like giraffid *Sivatherium*, from the Miocene of Asia. (E) The "Irish elk", which is not an elk, nor is it restricted to Ireland, but found over much of northern Europe during the Ice Ages. Properly known as *Megaloceros*, it was a gigantic deer with huge antlers weighing several hundred pounds, which had to be regrown each year by the bucks.

Figure 24.5 (Continued) (F) The extinct Ice Age bison, *Bison latifrons*, with the incredibly long horns.

and their modern descendants, the hippos) to a fully aquatic whale is now one of the best-documented major evolutionary transitions in the fossil record.

By the Oligocene, the archaic archaeocete whales like *Basilosaurus* (**Figure 24.6**) were extinct, and were replaced by a radiation of the two modern groups of cetaceans, the odontocetes (toothed whales, including sperm whales, killer whales, dolphins, and porpoises) and the mysticetes (baleen whales, including the blue whale, right whale, humpback whale, gray whale, and many others). The more familiar odontocetes are predators, feeding on fish and squid with their many conical teeth. The baleen whales, on the other hand, are toothless, and their mouth is filled with screens of horny tissue called baleen, which is used to filter out small fish and krill. Baleen whales such as the blue whale swallow a large mouthful of seawater, and as they close their mouths, they force out the water through the filter, leaving all the food trapped in their mouths.

Perissodactyls

The other major group of ungulates, the perissodactyls, are the order of herbivorous "odd-toed" hoofed mammals that includes the living horses, zebras, asses, tapirs, rhinoceroses, and their extinct relatives (**Figure 24.7**). They are recognized by a number of unique specializations, but their most diagnostic feature is their hands and feet. Most perissodactyls have either one or three toes on each foot, and the axis of symmetry of the foot runs through the middle digit. They are divided into three groups: the Hippomorpha (horses and their extinct relatives); the Titanotheriomorpha (the extinct brontotheres; **Figure 24.7[A,B]**); and the Moropomorpha (tapirs, rhinoceroses, and their extinct relatives). The

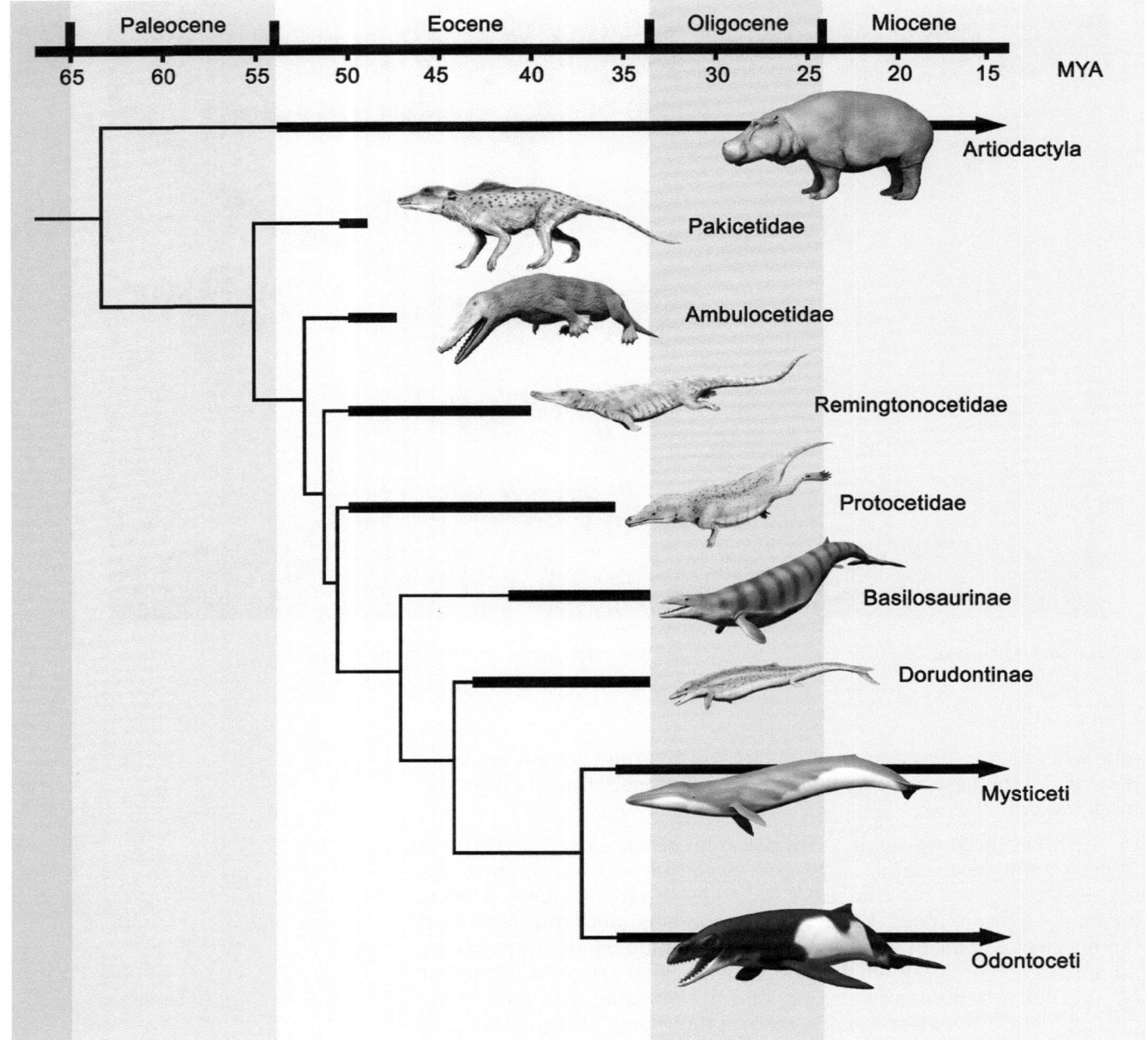

Figure 24.6 Family tree showing the evolution of whales from anthracotheres, the common ancestor between whales and hippos, and showing the stepwise evolution from a land animal to a fully aquatic animal.

Moropomorpha include not only the familiar living animals like rhinos and tapirs, but also the weird perissodactyls known as chalicotheres (**Figure 24.7[C]**).

Perissodactyls apparently evolved from ancestors in the Paleocene of Asia. This was confirmed in 1989, when a specimen recovered from upper Paleocene deposits in China was described and named *Radinskya*. This specimen shows that perissodactyls originated in Asia around 57 Ma, and since then other relatives of the perissodactyls have been found in Pakistan and India. More recently, a group of extinct mammals known as cambaytheres from the early Eocene of India seem to be even closer to perissodactyls.

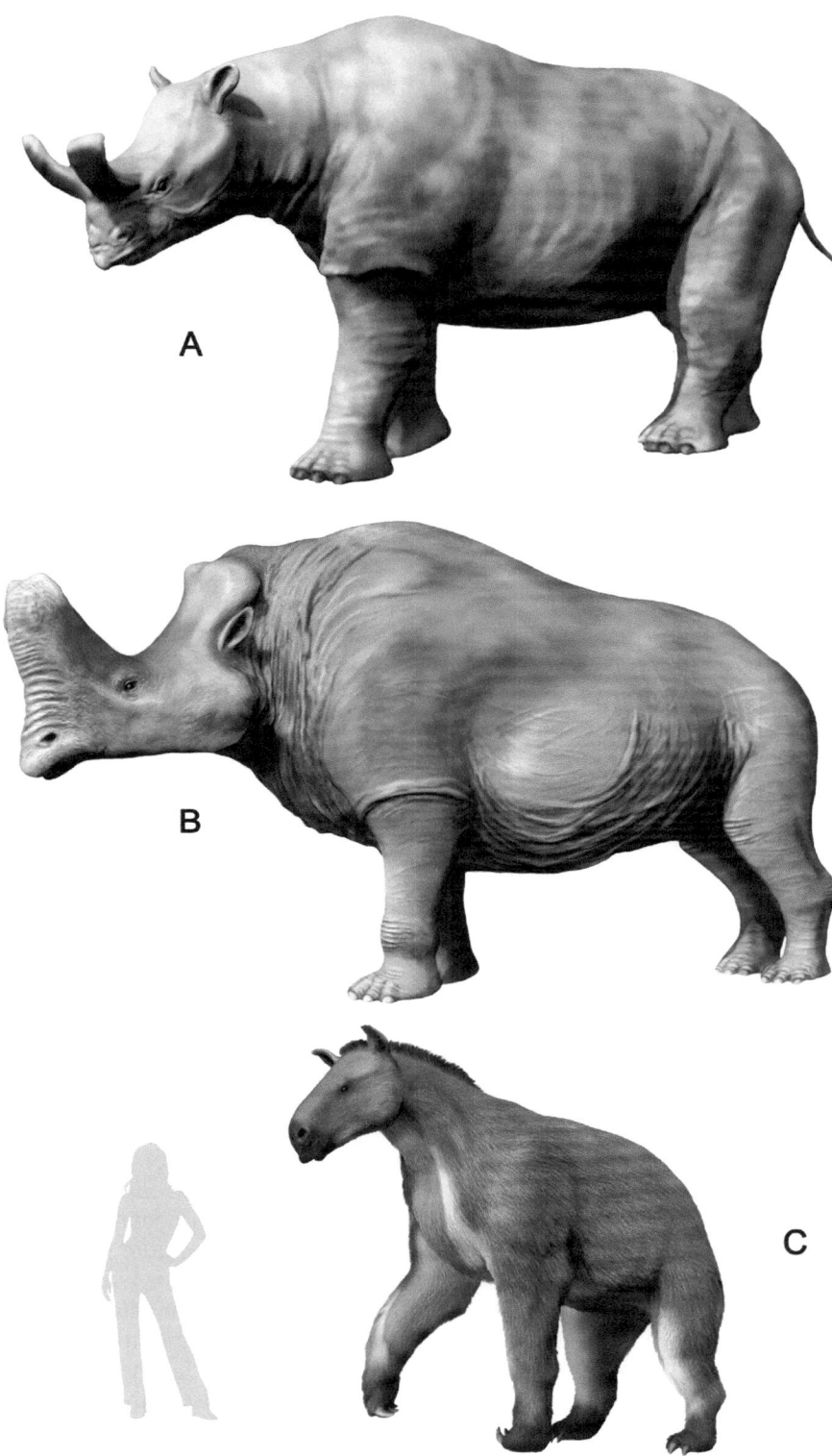

Figure 24.7 Reconstructions of some extinct perissodactyls. (A) The elephant-sized late Eocene North American brontothere *Megacerops*, the very last of this group known. (B) The late Eocene Mongolian brontothere *Embolotherium*, with the fused horns making a battering ram. (C) The weird clawed chalicothere *Moropus*.

By the early Eocene, the major groups of perissodactyls had differentiated, and migrated from Asia to Europe and North America. Before the Oligocene, the brontotheres (**Figure 24.7[A,B]**) and the archaic tapirs were the largest and most abundant hoofed mammals in Eurasia and North America. After these groups became extinct, horses and rhinoceroses were the most common perissodactyls, with a great diversity of species and body forms. Both groups were decimated during another mass extinction about 5 Ma, and today only 5 species of rhinoceros, four species of tapir, and a few species of horses, zebras, and asses cling to survival in the wild. The niches of large hoofed herbivores were taken over by the ruminant artiodactyls, such as cattle, antelopes, deer, and their relatives.

From their Asian origin, the hippomorphs spread all over the northern continents. In Europe, the horse-like palaeotheres substituted for true horses (**Figures 24.8** and **24.9[A]**). North America became the center of evolution of true horses, which occasionally migrated to other continents. *Protorohippus* (once called *Hyracotherium* or *Eohippus*) was a beagle-sized horse with four toes on the front feet that lived in the early Eocene. Its descendants evolved into many different lineages living side-by-side. The late Eocene-early Oligocene collie-sized three-toed horses *Mesohippus* and *Miohippus* were once believed to be sequential segments on the unbranched trunk of the horse evolutionary tree. However, they coexisted for millions of years, with five different species of the two genera living at the same time and place. From *Miohippus*-like ancestors, horses diversified into many different ecological niches. One major lineage, the anchitherines, retained low-crowned teeth, presumably for browsing soft leaves in the forests. Some anchitherines, such as *Megahippus*, were almost as large as the modern horse. *Anchitherium* migrated from North America to Europe in the late early Miocene, the first true horse to reach Europe.

In the middle Miocene, there were at least 12 different lineages of three-toed horses in North America, each with slightly different ecological specializations. This situation is analogous to the diversity of modern antelopes in East Africa. The ancestors of this great radiation of horses are a group of three-toed, pony-sized beasts that have long been lumped into the "wastebasket" genus "*Merychippus*". However, recent analyses have shown that the species of "*Merychippus*" are ancestral to many different lineages of horses. True *Merychippus* was a member of the hipparion lineage, a group of three-toed horses that developed highly specialized teeth, and had a distinctive concavity in the bone on the front of the face. Hipparions were a highly diverse and successful group of horses, with seven or eight different genera not only spread across North America, but also migrating to Eurasia. Merychippines were also ancestral to lineages such as *Calippus* (a tiny dwarf horse), *Protohippus*, and *Astrohippus*.

On two different occasions (*Pliohippus* and *Dinohippus*) three-toed horses evolved into lineages with a single toe on each foot. In the early Pliocene, most of these three-toed and one-toed horse lineages became extinct, leaving only *Dinohippus* to evolve into the modern horse *Equus*. The main lineage of horses that survived the latest Miocene extinctions were known as the equines. The living genus *Equus* first appeared in the Pliocene, and was widespread throughout the northern hemisphere. When the Isthmus of Panama rose about 2.5 Ma, horses also spread to South America. There they evolved into distinctive horses with a short proboscis known as the hippidions. At the end of the last Ice Age (about 10,000 years ago), horses became extinct in the New World. Columbus reintroduced horses to their ancestral homeland in 1493.

Brontotheres or titanotheres (**Figures 24.7[A,B]** and **24.9[B]**) began as pig-sized, hornless animals about 53 Ma, and quickly evolved into multiple lineages of cow-sized animals with long skulls and no horns. In the late middle Eocene (between 40 and 47 Ma), there were six different lineages of brontotheres. Some had long skulls, while others had short snouts and broad skulls. Still others had a pair of tiny blunt horns on the tip of their noses. Between 37 and 34 Ma, their evolution culminated with huge, elephant-sized beasts bearing large paired blunt horns

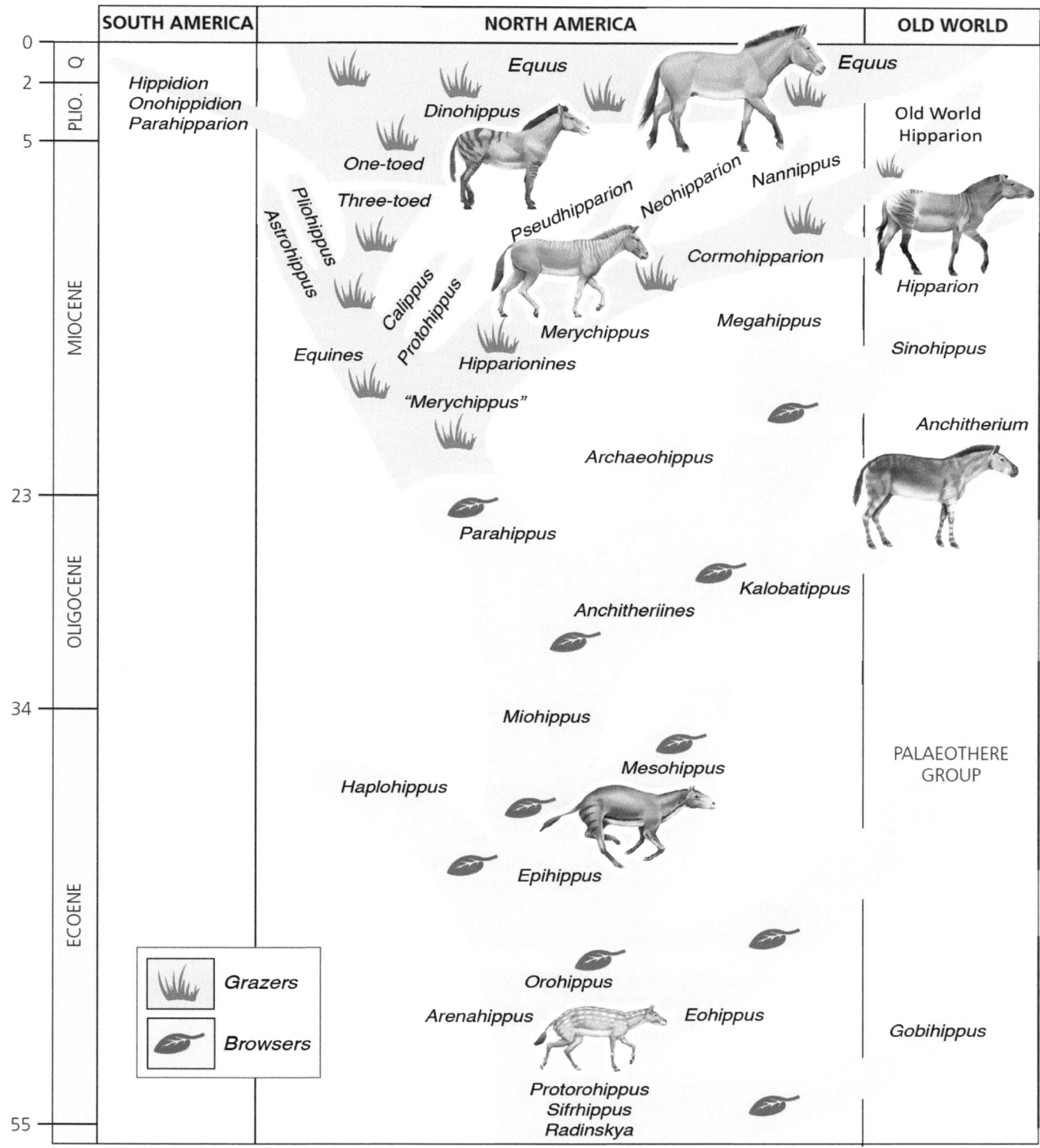

Figure 24.8 Evolution of the horses from dog-sized animals with low-crowned teeth and multiple toes, to the modern horses with high-crowed teeth and only one toe in each hand and foot.

on their noses (**Figures 24.7[A]** and **24.9[B]**). Throughout their history, brontotheres were the largest animals in North America. They also appeared in Asia in the late Eocene, where beasts such as *Embolotherium*, with a huge single "battering-ram" horn evolved (**Figure 24.7[B]**). Recent research has shown that the extinction of brontotheres about 34 Ma was due to a global climatic change (triggered by the first Antarctic

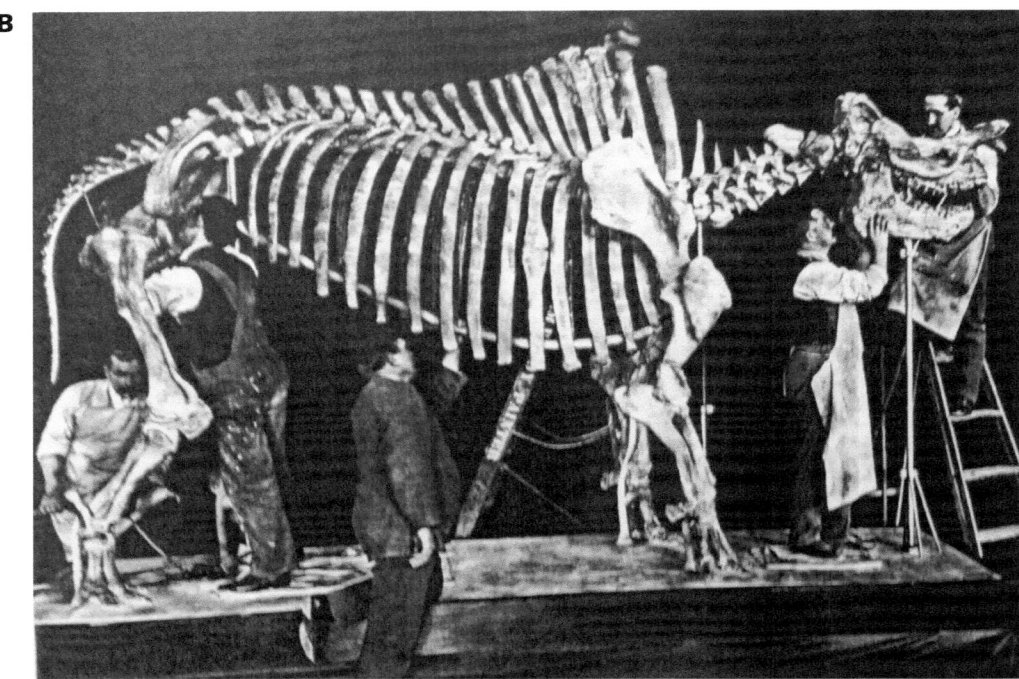

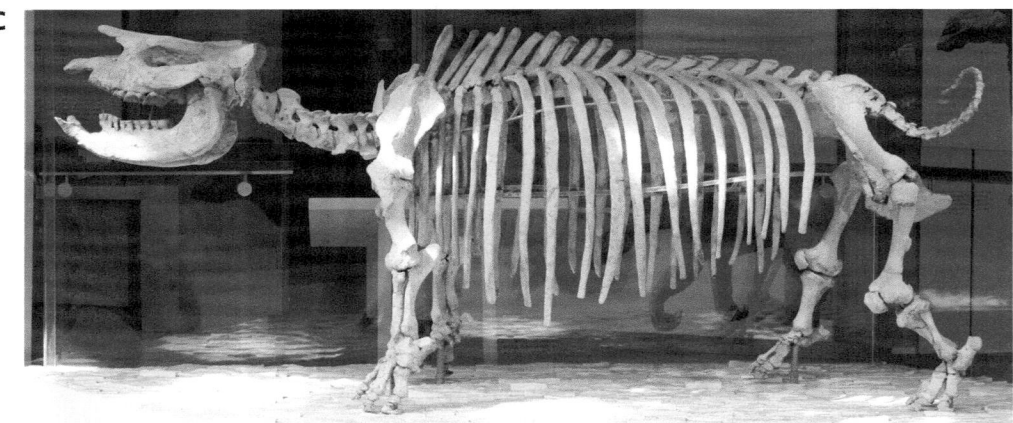

Figure 24.9 Fossils of some major groups of perissodactyls. (A) Display of the evolution of the horse skeleton from *Protorohippus* ("*Eohippus*" or "*Hyracotherium*" in the left foreground) to *Mesohippus* (right foreground) to *Merychippus* (extreme right) to *Plesippus* (center) to the living horse *Equus* (extreme left). (B) The last of giant brontotheres, *Megacerops coloradensis*. (C) The hippo-like Miocene rhinoceros, *Teleoceras fossiger*. (Courtesy Wikimedia Commons.)

glaciers) that caused worldwide cooling and drying of climates. This climatic change decimated the forests of the temperate regions and eliminated most of the soft, leafy vegetation on which brontotheres fed.

The earliest moropomorphs and tapir relatives, such as *Homogalax*, occur in lower Eocene strata. They are virtually indistinguishable from the earliest horses, such as *Protorohippus*. From this unspecialized ancestry, a variety of archaic tapir-like animals diverged. Most retained the simple leaf-cutting teeth characteristic of tapirs, and like brontotheres, they died out at the end of the Eocene when their forest habitats shrank. Only the modern tapirs, with their distinctive long proboscis, still survive in the jungles of Central and South America (three species), and Southeast Asia (one species). All are stocky, pig-like beasts with short stout legs and oval hooves, and a short tail. They have no natural defenses against large predators (such as jaguars or tigers) except fleeing through dense brush and swimming to make their escape.

The horse-like clawed chalicotheres (**Figure 24.7[C]**) are closely related to some of these archaic tapirs. When chalicotheres were first discovered, paleontologists refused to believe that among the jumble of disassociated bones, the claws belonged to a hoofed mammal related to horses and rhinos. However, many articulated specimens have clearly shown that chalicotheres are an example of a hoofed mammal that has secondarily regained its claws. There has been much speculation as to what chalicothere used their claws for. Traditionally, they were considered useful for digging up roots and tubers, except that the fossilized claws show no sign of the characteristic scratches due to digging. Instead, chalicotheres apparently used their claws to hook and haul down limbs and branches to eat leaves (much as ground sloths might have done), rather than for digging. *Chalicotherium* had such long forelimbs and short hindlimbs that it apparently knuckle-walked like a gorilla, with its claws curled inward. Chalicotheres were always rare throughout their history in North America and Eurasia, but nevertheless survived until the Ice Ages in Africa.

Rhinoceroses have been highly diverse and successful throughout the past 50 million years (**Figures 24.9[C]** and **24.10**). They have occupied nearly every niche available to a large herbivore, from dog-sized running animals, to several hippo-like forms, to the largest land mammal that ever lived, *Paraceratherium* (**Figure 24.11**). Most rhinoceroses were hornless. Unlike the horns of cattle, sheep, and goats, rhino horns are made of cemented hair fibers, and have no bony core, so they rarely fossilize. The presence and size of the horn must be inferred from the roughened area on the surface of the skull where it once attached.

The earliest rhino relatives, known as *Hyrachyus*, were widespread over Eurasia and North America in the early middle Eocene, and are even known from the Canadian Arctic. They apparently crossed back and forth between Europe and North America using a land bridge across the North Atlantic (before that ocean opened to its present width). From *Hyrachyus*, three different families of rhino diverged. One family, the amynodonts, was a hippo-like amphibious group, with stumpy legs and a barrel chest. In addition, amynodonts are usually found in river and lake deposits. They occupied this niche long before the hippo evolved. The last of the amynodonts, which had a short trunk like an elephant, died out in Asia in the middle Miocene.

The second family was known as the hyracodonts, or "running rhinos", because they had unusually long slender legs compared to other rhinos. They were particularly common in Asia and North America in the middle and late Eocene. The last of the North American forms was *Hyracodon*,

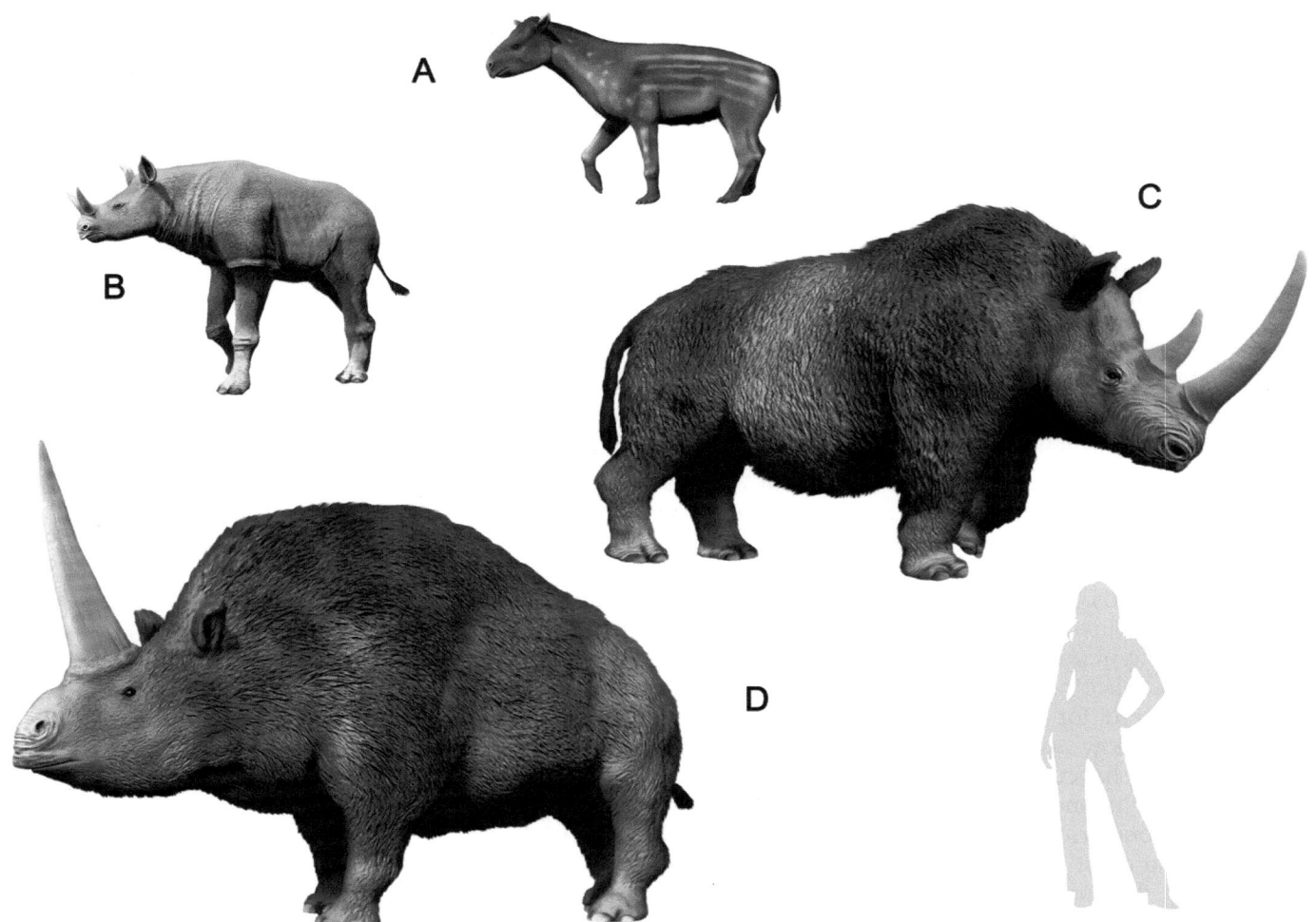

Figure 24.10 Reconstructions of some extinct rhinoceroses, including: (A) the running rhinoceros *Hyracodon*; (B) the earliest horned rhino, *Menoceras*, with paired horns on the noses of males; (C) the Ice Age woolly rhino *Coelodonta*; (D) the huge Ice Age rhino *Elasmotherium*, with a single gigantic horn on its forehead and none on its nose.

which was about the size and proportions of a Great Dane, and survived until the late Oligocene (**Figure 24.10[A]**). The second group of hyracodonts was the gigantic indricotheres (**Figure 24.11**), which were the largest mammals in Asia during the late Eocene and Oligocene (about 40 to 30 Ma). The biggest of all was *Paraceratherium* (once called *Baluchitherium* or *Indricotherium*), which was 18 feet (6 meters) tall at the shoulder and weighed 44,000 lb (20,000 kg). It was so tall that it must have browsed leaves from the tops of trees, as giraffes do today. Despite its huge bulk, it did not have the massive limbs and short, compressed toes of most giant land animals, such as sauropod dinosaurs, brontotheres, or elephants. Instead, it reveals its heritage as descended from a running rhino by retaining its long slender toes—even though it was much too large to run. Indricotheres were also the last of the hyracodonts, vanishing from Asia in the early Miocene.

The third family is the true rhinoceroses, or family Rhinocerotidae. They first appeared in Asia and North America in the late middle Eocene, and lived side-by-side with the hyracodonts and amynodonts on both continents. Up until this point, all the rhinoceroses we have mentioned were hornless. Rhinos with horns first appeared in the early late Oligocene; two different lineages independently evolved paired horns on the tip of the nose (**Figure 24.10[B]**). Both of these groups

Figure 24.11 A life-sized reproduction of the gigantic hyracodontid rhino *Paraceratherium*, compared to an elephant, and to its closest relative *Hyracodon* in front of it. (Courtesy Wikimedia Commons.)

became extinct in the late early Miocene, when two new subfamilies immigrated to North America from Asia: the browsing (leaf-eating) aceratherines, and hippo-like grazing teleoceratines. In the middle and late Miocene, browser-grazer pairs of rhinos were found all over the grasslands of Eurasia, Africa, and North America. The teleoceratine *Teleoceras* was remarkably similar to hippos in its short limbs, massive barrel-shaped body, and high-crowned teeth for eating gritty grasses (**Figure 24.9[C]**).

A mass extinction event that occurred at the end of the Miocene about 5 Ma wiped out North American rhinos, and decimated most of the archaic rhino lineages (especially the teleoceratines and aceratherines) in the Old World. The surviving lineages diversified in Eurasia and Africa, and even thrived during the Ice Ages. For example, the woolly rhinoceros (**Figure 24.10[C]**) was widespread in the glaciated regions of Eurasia, although it never crossed into North America (unlike the woolly mammoth and bison, which did). Another elephant-sized Ice Age rhino, *Elasmotherium*, had a gigantic 2-meter long horn on its forehead, not its nose (**Figure 24.10[D]**). The five living species of rhinoceros (white and black rhinos of Africa, the Indian, Sumatran, and Javan rhinos of southeast Asia) are all on the brink of extinction due to heavy poaching for their horns, which on the Asian "medicine" black market is more valuable than gold or cocaine.

Miscellaneous Mammals with Hooves

In addition to the living groups whose relationships are now well established, there are a number of extinct groups of large-bodied hoofed herbivores which may or may not be related to the true ungulates. Their relationships have been debated for a long time, but so far most of the evidence is not strong enough to convince everyone.

One such group of animals are the bizarre six-horned saber-toothed uintatheres (**Figure 24.12[A]**). The primitive ones, like *Prodinoceras* from the late Paleocene, were pig-sized creatures with no horns, but large upper canine tusks. In the early Eocene, the rhino-sized *Bathyopsis* continued the trend of large body size and huge upper canines with a big flanges of bone on the lower jaw to receive them. The largest of them, like *Uintatherium* and *Eobasileus*, were the size of elephants. They were the largest mammals known from the middle Eocene of North America and Asia. Then by the late middle and late Eocene, brontotheres took over as the largest land mammals, and uintatheres vanished. Even though they had elephant-sized bodies with broad flat feet like elephants, uintathere heads were truly weird. As already mentioned, they had three pairs of knob-like horns on the top of their skull: over their noses, over their eyes and on the back of their skulls. Even stranger are the huge upper canine tusks which protrude downward below their lower jaw, which also has paired knobs along the bottom. Finally, their teeth were surprisingly small low-crowned grinders, much too small for such a huge body, which would have needed a lot of food. These teeth could only chew soft leafy vegetation, so presumably that's what they fed on during the jungles of North America and Asia in the middle Eocene, and their disappearance by the late Eocene was apparently related to the shrinking of the tropical forests as the earth got colder and drier. And there is still no consensus as to what animals they were related to, or whether they were ungulates at all.

Another group of bizarre extinct creatures were the native ungulates of South America. This continent was isolated from the rest of the world after the Late Cretaceous, so no perissodactyls or artiodactyls or proboscideans were able to reach it until the late Miocene or later. Consequently, without competition from the ungulate groups of North America and Eurasia, South America developed a huge diversity of its own native hoofed mammal groups. Many of these developed into body forms that mimicked horses, rhinos, pigs, camels, and even rabbit-like creatures, apparently occupying those ecological niches in the absence of those groups from other continents. It's an astounding example of convergent evolution, showing how certain ecological niches can be filled by totally different groups of animals (such as the convergence of different mammal groups to form saber-toothed predators—**Figure 23.4**).

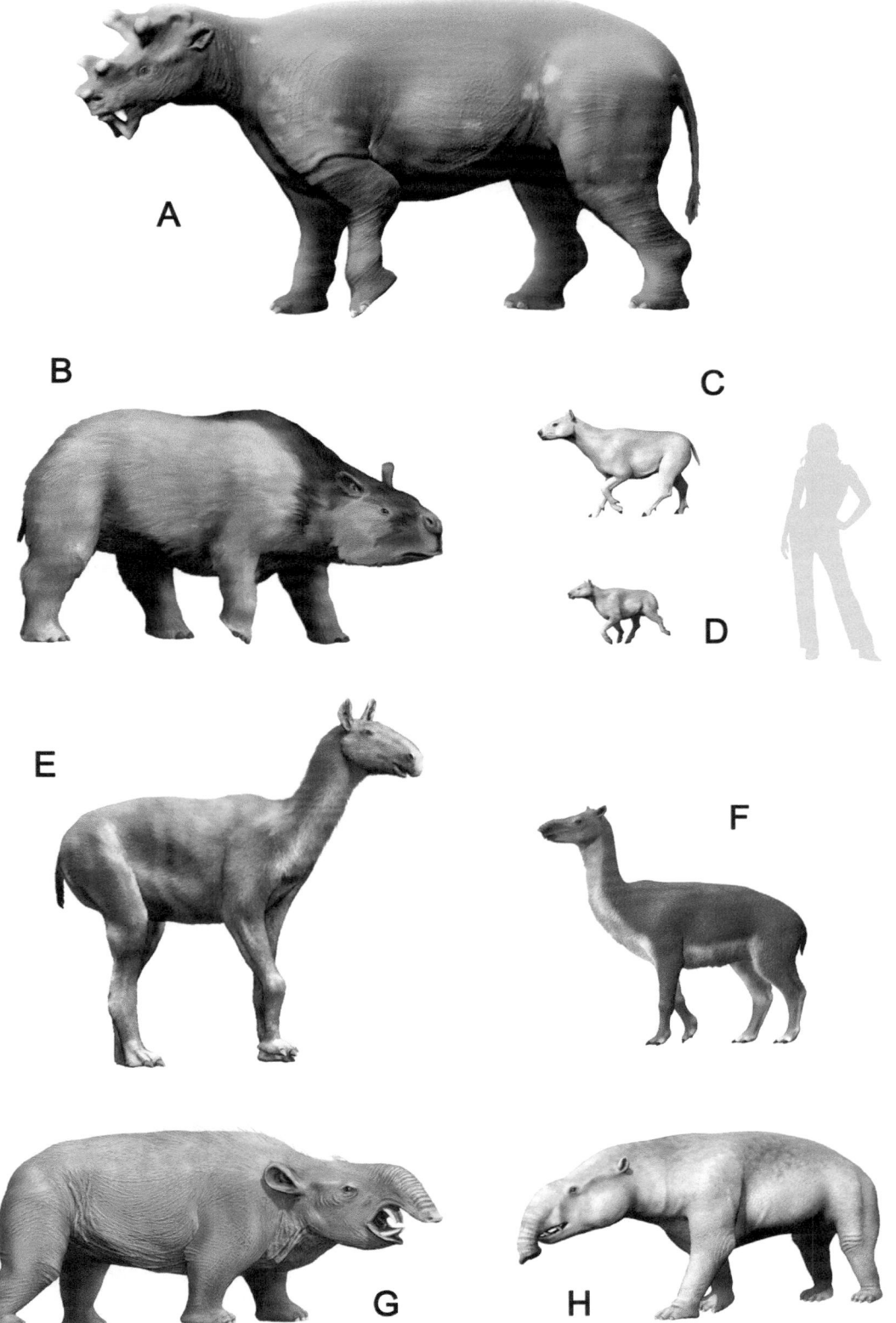

Figure 24.12 A variety of extinct mammals bearing hooves (which may or may not be related to true ungulates, like perissodactyls and artiodactyls). (A) The giant six-horned saber-toothed *Uintatherium*, from the middle Eocene of Utah, Colorado, and Wyoming. (B) The hippo-like notoungulate *Toxodon*, from the Pleistocene of South America. (C) The Miocene litoptern *Diadaphorus*, from the Miocene of South America, which superficially resembled three-toed Miocene horses of North America. (D) The giraffe and camel-like litoptern *Macrauchenia*, from the Ice Ages of South America. (E) The horse-like litoptern *Thoatherium* from South America, which was more committed to a single toe that horses were. (F) The llama-like litoptern *Theosodon*. (G) The mastodont-like *Pyrotherium*, with a short trunk like a proboscidean, but a member of a uniquely South American group. (H) The mastodont-like *Astrapotherium*, from the Miocene of South America, with a longer trunk and four large tusks in its upper and lower jaws.

The most diverse group are the Notoungulata ("southern ungulates"), with 14 families containing at least 100 genera and several hundred species. They are known from the late Paleocene right up to the mass extinctions at the end of the Pleistocene. Some were built like hippos (*Toxodon*, **Figure 24.12[B]**), while others were like rhinos with a bony horn on the forehead (*Trigodon*), others were like sheep with horns (*Adinotherium*), while *Homalodotherium* had claws and a build like the chalicotheres. *Thomashuxleya* was built much like a warthog, while the archaeohyracids resembled the true hyraxes of Africa. One of the notoungulates, *Pachyrukhos*, was a striking mimic of a rabbit. It even had large ears, long hind legs for hopping, large eyes for nocturnal vision, and chisel-like front teeth.

A second group, the Litopterna, included *Macrauchenia* (**Figure 24.12[D]**), a fossil found by Charles Darwin on the *Beagle* voyage in 1834. It was about 3 meters (10 feet) long with a camel-like body and long neck, and weighed over 1000 kg (2200 lb)—but it apparently had a long proboscis reminiscent of a tapir or mastodont. Its long legs were adapted for fast running, and had flexible ankles for rapid cornering and turning. Its teeth were relatively low-crowned compared to other South American mammals, and their geochemistry suggests that it was mostly a leaf browser, using its long neck and trunk to strip the leaves from taller trees. It first appeared about 7 Ma, and vanished near the end of the last Ice Age, about 20,000 to 10,000 years ago. The litoptern *Theosodon* (**Figure 24.12[F]**) looked much like a llama, while another group converged on horses. These included the three-toed *Diadaphorus* (**Figure 24.12[C]**), which resembled three-toed horses of North America, and the one-toed *Thoatherium* (**Figure 24.12[E]**). *Thoatherium* looked much like the late Miocene horses of North America, but it was an even more specialized runner than any northern horse. While even the most advanced true horses have tiny splints of their reduced side toes alongside their main toe, *Thoatherium* had only one central toe on each foot, with no trace of side toes. Once again, South American hoofed mammals demonstrate amazing convergence on unrelated mammals from other continents. But in this case, *Thoatherium* has reduced its side toes even more than has the living horse, so it was a better one-toed horse than any true horse ever was!

Two other strange orders of South American ungulates are known. The pyrotheres (**Figure 24.12[G]**) were built somewhat like a mastodont, with the nasal opening shifted backward suggesting it had a short trunk, and large tusks in its jaws. The astrapotheres are even more bizarre. *Astrapotherium* itself (**Figure 24.12[H]**), which was the size of a rhino or mastodont (about 3 meters or 10 feet long), with a very long trunk and spindly legs, features that suggest it was amphibious. Its skull is truly amazing, with large hippo-like curved upper and lower tusks protruding from its mouth, and a deeply retracted nasal bone suggesting a longer proboscis or trunk. Once again, we have an example of a South American hoofed mammal mimicking hippos or mastodonts, complete with the trunk and tusks.

All four of these native South American hoofed mammal groups are extinct, so it was long a mystery as to what they might be related to. But recent work on their recovered protein sequences from late Ice Age *Toxodon* and *Macrauchenia* suggest that both might be related to the base of the perissodactyl radiation. There are no proteins or other organic molecules left in the last of the Miocene pyrotheres or astrapotheres to analyze, so their relationships continue to be a mystery.

Pantodonts

One of the earliest groups of mammals on earth to evolve large body size was the extinct order known as pantodonts (**Figure 24.13**). They

Figure 24.13 The pantodonts were the first mammals to grow large in the Paleocene and early-middle Eocene of Eurasia and North America. (A) The middle Paleocene genus *Pantolambda*. (B) The late Paleocene *Barylambda*. (C) The large tusked *Titanoides*, from the late Paleocene of Alberta and North Dakota. (D) One of the last of the pantodonts, the early Eocene *Coryphodon*.

were stocky creatures with a primitive skeleton, and robust limbs, often with blunt claws or hoof-like structures. Later pantodonts had a hippo-like build that suggested they were partially aquatic. Other pantodonts had long tails and slender limbs suggesting they could have been tree dwellers.

The most distinctive feature of pantodonts is their cheek teeth, which had a distinctive "V" shape pattern of crests, and resembling the Greek letter "lambda" (Λ). This is why many pantodont genera have the root "lambda" in their names. The primitive pantodont tooth pattern is similar to that of modern tapirs, and well suited to eating soft leaves and other browse from the Paleocene and Eocene jungles where they lived. Pantodonts were never a very diverse group (fewer than two dozen genera and species are known), but they were widespread in North America and eastern Asia (China and Mongolia) in the Paleocene and early Eocene. They also lived in both the Canadian Arctic and on Svalbard Island in the European Arctic, and briefly in Europe.

From the beginning of the Paleocene, pantodonts stood out due to their relatively large body size at a time when most other mammals were rat-sized to cat-sized (**Figure 24.13**). In the early Paleocene, the earliest known pantodont in Asia (*Bemalambda*) was the size of a large dog. Sheep-sized *Pantolambda* was the first pantodont to migrate from Asia to North America during the middle Paleocene (**Figure 24.13[A]**). By the late Paleocene they evolved to pony-sized *Barylambda* (2.5 meters long, weighing about 650 kg). It had a heavy skeleton with long claws, and a robust tail like that of like a ground sloth, so it may have reared up on its hind feet and tail to reach higher vegetation. *Barylambda* was bigger than any other mammal in North America at that time (**Figure 24.13[B]**). An even larger pantodont was the bear-sized *Titanoides* of the late Paleocene of North America (**Figure 24.13[C]**). It was up to 3 meters (10 feet) long and weighed about 150 kg (330 lb). Its broad robust snout sported sharp canine tusks.

By the early Eocene, *Coryphodon* (last of the American pantodonts) was the size of a small rhinoceros (**Figure 24.13[D]**). Small species were about 2.25 meters (7.4 feet) long, and weighed about 500 kg (1100 lb). The largest species weighed as much as 700 kg (1500 lb). *Coryphodon* was significantly larger than any land mammal in the early Eocene. It migrated from Asia to North America, replacing *Barylambda*, and even spread to Europe and the Canadian Arctic, before vanishing at the beginning of the middle Eocene. *Coryphodon* had very robust stocky limbs and was built like a hippopotamus, suggesting it had an aquatic lifestyle. It had a stocky muscular neck and head and body, broad muzzle, and prominent canine tusks for battling with other animals, and rooting up vegetation. There is evidence that the tusks are much larger in males than in females. Through most of their range, *Coryphodon* lived off leaves and soft browse, but those that lived in the Canadian Arctic show evidence from their tooth wear of switching to a diet of leaf litter, twigs, and evergreen needles during the six months of darkness when plants were not growing.

The very last of the pantodonts, *Hypercoryphodon* from Mongolia, was the size of a rhinoceros, and survived until the end of the middle Eocene (about 40 Ma). It is known primarily from a huge skull found in Mongolia in the 1920s by Walter Granger, and described by Granger and Henry Fairfield Osborn in 1932.

The relationships of pantodonts are still controversial. Many classifications have combined them with other heavy-bodied early Cenozoic mammals, especially hoofed mammals, based on their primitive skeletal features, but this is not strong evidence of relationship based on shared evolutionary novelties. Some have noted the similarity of the "V"-shaped crests on their cheek teeth, and suggested that they are related to primitive insectivorous mammals or possibly rabbit relatives of the Cretaceous, but such a simple tooth pattern could easily be due to convergent evolution. For now, their relationships to the other groups of mammals are unresolved, since there is no strong evidence or consensus that links them with any particular group.

FURTHER READING

Agusti, J.; Anton, M. 2002. *Mammoths, Sabertooths, and Hominids: 65 Million Years of Mammalian Evolution in Europe.* Columbia University Press, New York.

Archibald, J.D.; Zhang, Y.; Harper, T.; Cifelli, R.L. 2011. *Protungulatum*, confirmed Cretaceous occurrence of an otherwise Paleocene eutherian (placental?) mammal. *Journal of Mammalian Evolution.* 18 (3): 153–161.

Asher, R.J.; Bennet, N.; Lehmann, T. 2009. The new framework for understanding placental mammal evolution. *BioEssays.* 31 (8): 853–864.

Boisserie, J.-R.; Lihoreau, F.; Brunet, M. 2005. The position of Hippopotamidae within Cetartiodactyla. *Proceedings of the National Academy of Sciences.* 102 (5): 1537–1541.

DeMiguel, D.; Azanza, B.; Morales, J. 2014. Key innovations in ruminant evolution: A paleontological perspective. *Integrative Zoology.* 9 (4): 412–433.

Franzen, J.-L. 2010. *The Rise of Horses.* Johns Hopkins University Press, Baltimore, MD.

Froehlich, D. 2002. The systematics and taxonomy of the early Eocene equids (Perissodactyla). *Zoological Journal of the Linnean Society.* 134 (2): 141–256.

Gatesy, J.; Milinkovitch, M.; Waddell, V.; Stanhope, M. 1999. Stability of cladistic relationships between Cetacea and higher-level artiodactyl taxa. *Systematic Biology.* 48 (1): 6–20.

Gingerich, P.D.; ul Haq, M.; Zalmout, I.S.; Khan, I.H.; Malkani, M.S. 2001. Origin of whales from early artiodactyls: Hands and feet of Eocene Protocetidae from Pakistan. *Science*. 293 (5538): 2239–2242.

Groves, C.P.; Grubb, P. 2011. *Ungulate Taxonomy*. Johns Hopkins University Press, Baltimore, MD.

Janis, C. 1993. Tertiary mammal evolution in the context of changing climates, vegetation, and tectonic events. *Annual Reviews of Ecology and Systematics*. 24: 467–500.

Janis, C. 2008. An evolutionary history of browsing and grazing ungulates, pp. 21–45. In Iain, J.G.; Herbert, H.T.P., eds. *The Ecology of Browsing and Grazing*. Ecological Studies, 195. Springer, Berlin.

Janis, C.M.; Scott, K.M. 1987. The interrelationships of higher ruminant families with special emphasis on the members of the Cervoidea. *American Museum Novitates*. 2893: 1–85.

Janis, C., Scott, K.M., Jacobs, L.L., eds. 1998. *Evolution of Tertiary Mammals of North America, vol. I: Terrestrial Carnivores, Ungulates, and Ungulate-like Mammals*. Cambridge University Press, Cambridge.

Kurtén, B. 1968. *Pleistocene Mammals of Europe*. Columbia University Press, New York.

Kurtén, B. 1988. *Before the Indians*. Columbia University Press, New York.

Kurtén, B.; Anderson, E. 1980. *Pleistocene Mammals of North America*. Columbia University Press, New York.

MacFadden, B.J. 1992. *Fossil Horses: Systematics, Paleobiology, and Evolution of the Family Equidae*. Cambridge University Press, Cambridge.

McKenna, M.C. 1975. Toward a phylogenetic classification of the Mammalia, pp. 21–46. In Luckett, W.P.; Szalay, F.S., eds. *Phylogeny of the Primates*. Plenum Press, New York.

McKenna, M.C.; Bell, S.K. 1997. *Classification of Mammals above the Species Level*. Columbia University Press, New York.

McKenna, M. C.; Chow, M.; Suyin, T.; Luo, Z. (1989). *Radinskya yupingae*, a perissodactyl-like mammal from the late Palaeocene of China, pp. 24–36. In Prothero, D.R.; Schoch, R.M., eds. *The Evolution of Perissodactyls*. Oxford University Press, Oxford, UK.

Montgelard, C.; Catzeflis, F. M.; Douzery, E. 1997. Phylogenetic relationships of artiodactyls and cetaceans as deduced from the comparison of cytochrome b and 12S rRNA mitochondrial sequences. *Molecular Biology and Evolution*. 14 (5): 550–559.

Novacek, M.J. 1992. Mammalian phylogeny: Shaking the tree. *Nature*. 356: 121–125.

Novacek, M.J. 1994. The radiation of placental mammals, pp. 220–237. In Prothero, D.R., and R.M. Schoch, eds. *Major Features of Vertebrate Evolution*. Paleontological Society Short Course 7. Paleontological Society, Lawrence, KS.

Novacek, M.J.; Wyss, A.R. 1986. Higher-level relationships of the recent eutherian orders: Morphological evidence. *Cladistics*. 2: 257–287.

Novacek, M.J., Wyss, A.R.; McKenna, M.C. 1988. The major groups of eutherian mammals, pp. 31–73. In Benton, M.J., ed. *The Phylogeny and Classification of the Tetrapods, vol. 2: Mammals*. Clarendon Press, Oxford.

O'Leary, M.A.; Bloch, J.I.; Flynn, J.J.; Gaudin, T.J.; Giallombardo, A.; Giannini, N.P.; Goldberg, S.L.; Kraatz, B.P.; Luo, Z.-X.; Meng, J.; Ni, X.; Novacek, M.J.; Perini, F.A.; Randall, Z.S.; Rougier, G.W.; Sargis, E.J.; Silcox, M.T.; Simmons, N.B.; Spaulding, M.; Velazco, P.M.; Weksler, M.; Wible, J.R.; Cirranello, A.L. 2013. The placental mammal ancestor and the post-K-Pg radiation of placentals. *Science*. 339 (6120): 662–667.

Prothero, D.R. (1994). Mammalian evolution, pp. 238–270. In Prothero, D.R., and R.M. Schoch, eds. *Major Features of Vertebrate Evolution*. Paleontological Society Short Course 7. Paleontological Society, Lawrence, KS.

Prothero, D.R. 2005. *The Evolution of North American Rhinoceroses*. Cambridge University Press, Cambridge, UK.

Prothero, D.R. 2006. *After the Dinosaurs: The Age of Mammals*. Indiana University Press, Bloomington, IN.

Prothero, D.R. 2009. Evolutionary transitions in the fossil record of terrestrial hoofed mammals. *Evolution: Education and Outreach*. 2 (2): 289–302.

Prothero, D.R. 2016. *The Princeton Field Guide to Prehistoric Mammals*. Princeton University Press, Princeton, NJ.

Prothero, D.R.; Foss, S., eds. 2007. *The Evolution of Artiodactyls*. Johns Hopkins University Press, Baltimore, MD.

Prothero, D.R.; Manning, E.M.; Fischer, M. 1988. The phylogeny of the ungulates, pp. 201–235. In Benton, M.J., ed. *The Phylogeny and Classification of the Tetrapods, vol. 2: Mammals*. Clarendon Press, Oxford.

Prothero, D.R.; Schoch, R.M., eds. 1989. *The Evolution of Perissodactyls*. Oxford University Press, New York.

Prothero, D.R.; Schoch, R.M. 2002. *Horns, Tusks, and Flippers: The Evolution of Hoofed Mammals and Their Relatives*. Johns Hopkins University Press, Baltimore, MD.

Rose, K.D. 2006. *The Beginning of the Age of Mammals*. Johns Hopkins University Press, Baltimore, MD.

Rose, K.D.; Archibald, J.D., eds. 2005. *The Rise of Placental Mammals: The Origin and Relationships of the Major Extant Clades*. Johns Hopkins University Press, Baltimore, MD.

Rose, K. D.; Holbrook, L. T.; Kumar, K.; Rana, R. S.; Ahrens, H. E.; Dunn, R. H.; Folie, A.; Jones, K. E.; Smith, T. 2019. Anatomy, relationships, and paleobiology of *Cambaytherium* (Mammalia, Perissodactylamorpha, Anthracobunia) from the lower Eocene of western India. *Journal of Vertebrate Paleontology*. 39: 1–147.

Savage, D.E.; Russell, D.E. 1983. *Mammalian Paleofaunas of the World*. Addison Wesley, Reading, MA.

Savage, R.J.G.; Long, M.R. 1986. *Mammal Evolution: An Illustrated Guide*. Facts-on-File Publications, New York.

Spaulding, M. O'Leary, M.A., Gatesy, J. 2009. Relationships of Cetacea (Artiodactyla) among mammals: Increased taxon sampling alters interpretations of key fossils and character evolution. *PLoS One*. 4 (9): e7062.

Springer, M. S.; Burk-Herrick, A.; Meredith, R.; Eizirik, E.; Teeling, E.; O'Brien, S. J.; Murphy, W. J. 2007. The adequacy of morphology for reconstructing the early history of placental mammals. *Systematic Biology*. 56: 673–684.

Springer, M.S.; Meredith, R.W.; Janecka, J.E.; Murphy, W.J. 2011. A historical biogeography of Mammalia. *Proceedings of the Royal Society B*. 366: 2478–2502.

Springer, M. S.; Stanhope, M. J.; Madsen, O.; de Jong, W. W. 2004. Molecules consolidate the placental mammal tree. *Trends in Ecology and Evolution*. 19: 430–438.

Stucky, R.K. 1990. Evolution of land mammal diversity in North America during the Cenozoic. *Current Mammalogy*. 2: 375–432.

Szalay, F. S.; Novacek, M. J.; McKenna, M. C. (eds.). 1993. *Mammal Phylogeny*. Springer-Verlag, Berlin.

Thewissen, J.G.M.; Cooper, L. N.; Clementz, M. T.; Bajpai, S.; Tiwari, B. N. 2007. Whales originated from aquatic artiodactyls in the Eocene epoch of India. *Nature*. 450 (7173): 1190–1194.

Turner, A.; Anton, M. 2004. *National Geographic Prehistoric Mammals*. National Geographic Society, Washington, DC.

Welker, F.; Collins, M. J.; Thomas, J. A.; Wadsley, M.; Brace, S.; Cappellini, E.; Turvey, S. T.; Reguero, M.; Gelfo, J. N.; Kramarz, A.; Burger, J.; Thomas-Oates, J.; Ashford, D. A.; Ashton, P. D.; Rowsell, K.; Porter, D. M.; Kessler, B.; Fischer, R.; Baessmann, C.; Kaspar, S.; Olsen, J. V.; Kiley, P.; Elliott, J. A.; Kelstrup, C. D.; Mullin, V.; Hofreiter, M.; Willerslev, E.; Hublin, J. J.; Orlando, L.; Barnes, I.; MacPhee, R. D. 2015. Ancient proteins resolve the evolutionary history of Darwin's South American ungulates. *Nature*. 522 (7554): 81–84.

Werdelin, L.; Sanders, W.L., eds. 2010. *Cenozoic Mammals of Africa*. University of California Press, Berkeley, CA.

Woodburne, M.O., ed. 2004. *Late Cretaceous and Cenozoic Mammals of North America: Biostratigraphy and Geochronology*. Columbia University Press, New York.

Zhou, X.; Xu, S.; Xu, J.; Chen, B.; Zhou, K.; Yang, G.; et al. 2011. Phylogenomic analysis resolves the interordinal relationships and rapid diversification of the Laurasiatherian mammals. *Systematic Biology*. 61 (1): 150–164.

EUARCHONTO-GLIRES

RODENTS, RABBITS, PRIMATES—AND HUMANS

25

We must, however, acknowledge, as it seems to me, that man with all his noble qualities, still bears in his bodily frame the indelible stamp of his lowly origin.

—Charles Darwin, *Descent of Man*

We're all one dysfunctional family
No matter where we nomads roam
Rift Valley Drifters, drifting home genome by genome
Take a look inside your genes, pardner, then you will see
We've all got a birth certificate from Kenya
—Roy Zimmerman, "Rift Valley Drifters"

THE EUARCHONTOGLIRES

In the 1990s, another discovery that emerged from the molecular studies of the relationships of mammals was that the rodents and rabbits formed a natural group—but even more remarkable, that their closest relatives were the primates (**Figure 22.2**). This may come as a surprise or even a shock to some people but our own closest kin after all the various types of apes, monkeys, lemurs, and other primates are the rodents and rabbits. Before this molecular evidence was discovered, there was a general consensus that primates were all closely related, and that they might be related not only to colugos and tree shrews, but also to bats and elephant shrews. (Bats are now Laurasiatheria, elephant shrews are now Afrotheria.) In 1910, William King Gregory group called this assemblage the "Archonta". Likewise, zoologists and paleontologists long assumed that the small body size and chisel-like incisors of rodents and rabbits were just convergent evolution from groups with different origins. Nevertheless, Gregory in 1910 put these two gnawing groups in a taxon he called Glires.

GLIRES

Many people are surprised to hear that rabbits *aren't* rodents. But the differences between them were noticed as far back as the 1850s, when zoologists placed them in separate orders. Although both groups are small-bodied herbivores or omnivores with ever-growing chisel-like front incisors for gnawing, they are very different in anatomical details. In particular, rodents have only one pair of upper chisel-like incisors (**Figure 25.1[A]**), but rabbits have two pairs (**Figure 25.1[B]**). The snout region of the skull in rabbits is made of porous spongy bone. Functional

DOI: 10.1201/9781003128205-25

Figure 25.1 Comparison of the anatomical differences between rodents and rabbits (lagomorphs).
(A) A rodent skull, showing the single pair of ever-growing chisel-like upper incisors that are rooted deeply in the skull, and curve around as they come into occlusion with the lower incisors, where they are worn down to an edge with the self-sharpening mechanism of a hard band of enamel on the front edge, and softer dentin on the inside. (B) A rabbit skull, showing the two pairs of upper incisors. The spongy bone in the snout is there to conserve weight and bone mass, because all the stresses are taken up by the triangular framework of bone around the edge, and not by the middle.

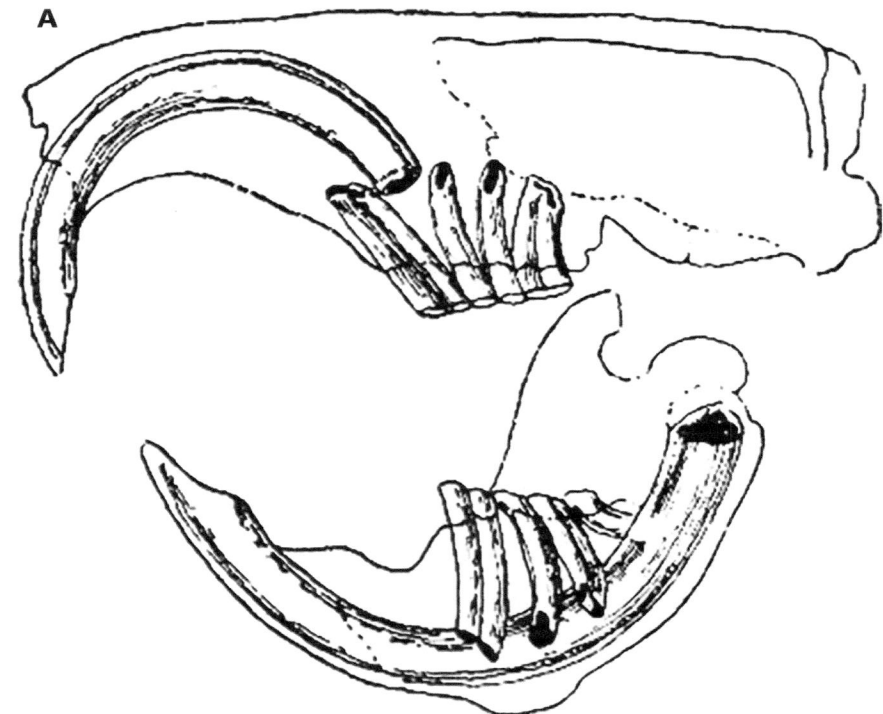

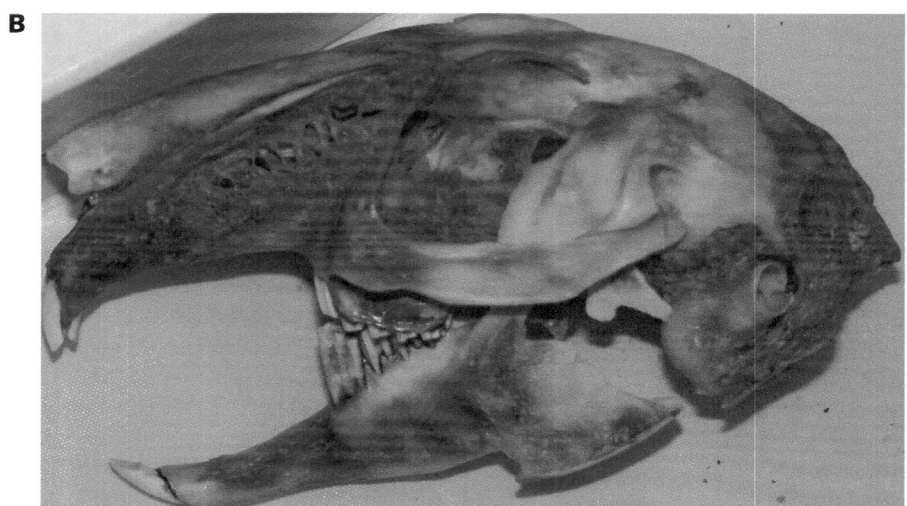

morphology studies have suggested that rabbits have minimized the amount of dense bone needed in their snouts, since the stress is taken up by the triangle of solid bones around the edge of the snout.

There are also numerous detailed differences in the skull and skeleton. Rabbits tend to have short tails, and longer hind legs to move by jumping, while rodents are mostly runners and climbers (although kangaroo mice, kangaroo rats, gerbils, jerboas, and a few others have convergently evolved to the jumping locomotion of rabbits). For much of the twentieth century, biologists and paleontologists regarded rodents and lagomorphs as unrelated groups that had independently developed the gnawing lifestyle in different ways. Only in the 1980s and later has opinion swung the opposite way with new anatomical analyses, new fossils, and finally molecular data that was previously unknown.

The best evidence for their close relationship is fossils of primitive ancestral Glires from the early Cenozoic of Asia that were closely related to rodents plus lagomorphs, but not a member of either group. Often known by the name "anagalids" and "eurymylids" (**Figure 25.2[A]**), these fossils are often placed in their own wastebasket group, with some of them (such as *Tribosphenomys* and *Rhombomylus*) being closer to rodents, while others (such as *Mimotona* and *Gomphos*) closer to rabbits. Some paleontologists trace the origin of these animals back to a well-known group of Cretaceous Mongolian fossils known as zalambdalestids. By the Paleocene of China, there are numerous fossils of a group called

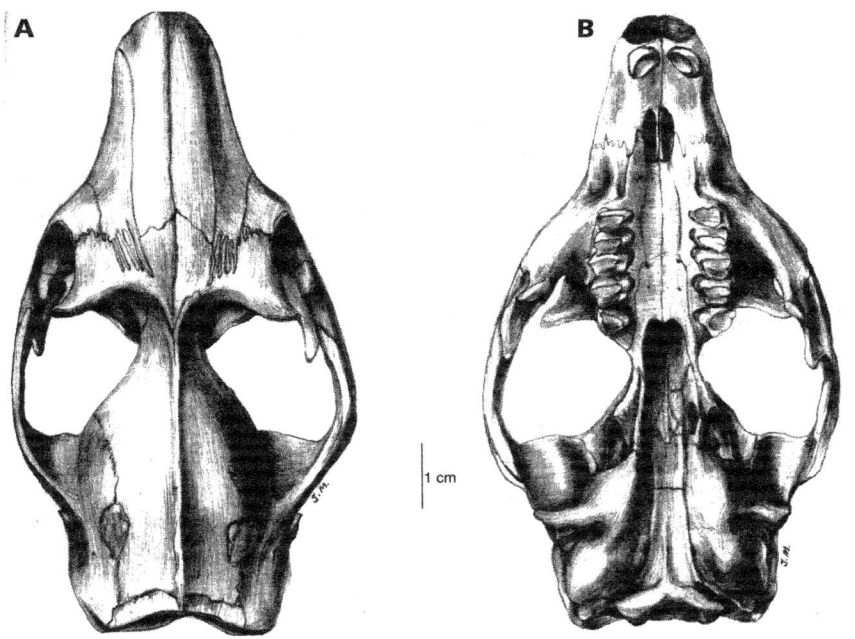

Figure 25.2 Some more remarkable fossil rodents. (A) The skull of the primitive glirid *Rhombomylus*, from the Paleocene of China. Creatures like this were ancestral to the entire radiation of rodents and lagomorphs. (B) The skeleton of the burrowing rodent *Epigaulus*, with the remarkable horns on its snout. (C) The bear-sized Ice Age beaver *Castoroides*. Scale bar in cm. [(A–C) Courtesy Wikimedia Commons. (D) By the author.]

Figure 25.2 (Continued) (D) Comparison of the skulls of *Castoroides* (large dark skull) and the modern beaver, *Castor canadensis* (smaller white skull).

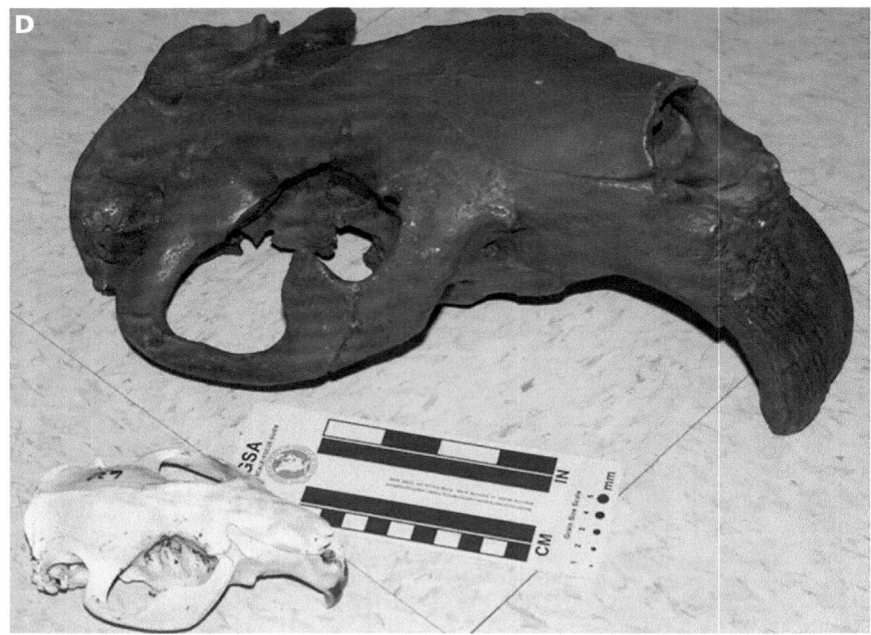

pseudictopids, which are increasingly more like rodents and lagomorphs. These fossils are the size of rodents and rabbits, with many of the specializations of the skeleton as well, but do not yet have the fully developed ever-growing gnawing incisors. *Mimatona* from the early Paleocene of China had the lagomorph-like double pair of upper chisel incisors, but the crown pattern of the molar teeth is very primitive and not yet truly lagomorph. *Eurymylus*, *Rhombomylus*, and other fossils from the Paleocene and early Eocene of China have the classic single pair of gnawing incisors like rodents, but do not yet have the cheek tooth specializations that separate rodents from more primitive forms. Thus, we have many primitive fossils that not only link rodents and lagomorphs, but also allow us to trace their origins back into the Paleocene and even Late Cretaceous.

Rodents

By far the most abundant, diverse, and successful group of placental mammals is the order Rodentia. They are incredibly diverse (over 40% of the the species of living mammals, or at least 350 genera and 1700 species, are rodents), disparate (occupying body forms from the pig-sized capybara, to aquatic beavers and muskrats, gliding, tree-climbing, and burrowing squirrels, spiny porcupines, subterranean gophers and naked mole rats, and hundreds of different kinds of rats and mice), and they are also incredibly abundant. One only needs to think about the reproductive ability of rats or mice or hamsters to realize why they are by far the most common mammals on the planet. If it were not for predators, the earth would be a planet of rodents.

One of the most unusual rodents was a group called the mylagaulids, distantly related to beavers. Many of the had a pair of horns on their noses (**Figure 25.2[B]**), but the function of these horns is still controversial. They also dug deep corkscrew-shaped burrows that looked like a spiral staircase, with a living chamber at the very bottom. When these corkscrews of sediment filling the burrow were first found, they were given the name *Daemonelix* ("devil's corkscrews").

Rodents are usually the dominant group of mammals in the small-body-size niche, but occasionally they become huge. The largest living rodent, the capybara, weighs about 40 kg, but the Pleistocene beaver *Castoroides* weighed about 200 kg and reach 2.5 meters in length, as large as a bear (**Figures 25.2[C,D]** and **25.3[D]**). There was a gigantic Ice Age capybara *Neochoerus*, which at 113 kg (250 pounds) was about the size of a cow (**Figure 25.3[B]**). Meanwhile, South America had even larger giant rodents, including the extinct rhinoceros-sized Plio-Pleistocene pacarana, *Josephoartigasia* (**Figure 25.3[A]**), which was 3 meters (10 feet) long and weighed up to 1500 kg (3382 lb). It was discovered only a few years ago; the previous record-holders were the bison-sized Miocene pacarana *Phoberomys* ("terror mouse"), which was 3 meters (10 feet) long, but weighed only about 700 kg (1500 lb), and the slightly smaller late Miocene-Pleistocene cow-sized fossil *Telicomys*, which was about 2.7 meters (7 feet) long.

Rodents have a number of unique features, but their most obvious is their pair of chisel-like (gliriform) upper and lower incisors, which are used to gnaw their hard-shelled food and vegetation, and in some groups, to cut down trees or dig tunnels or burrows. These incisors are constantly growing, with open roots, and must be continuously worn down into a sharp point by abrading them together (**Figure 25.1[A]**). They are self-sharpening, since the hard enamel band on the front edge of the incisor wears down much more slowly than the soft dentin on the back of the tooth. If there is a problem with occlusion so the incisors are not sharpened down, they will continue growing in a curve until they curl around and puncture the top of the skull. There is a toothless gap (diastema) behind the incisors, and then a row of premolars and molars that are adapted for grinding their diet of seeds, nuts, and vegetation.

This small-bodied, seeds/nuts/vegetation-gnawing diet and lifestyle was very successful, as demonstrated by the fact that multituberculates occupied this niche for most of the Mesozoic, and several groups of

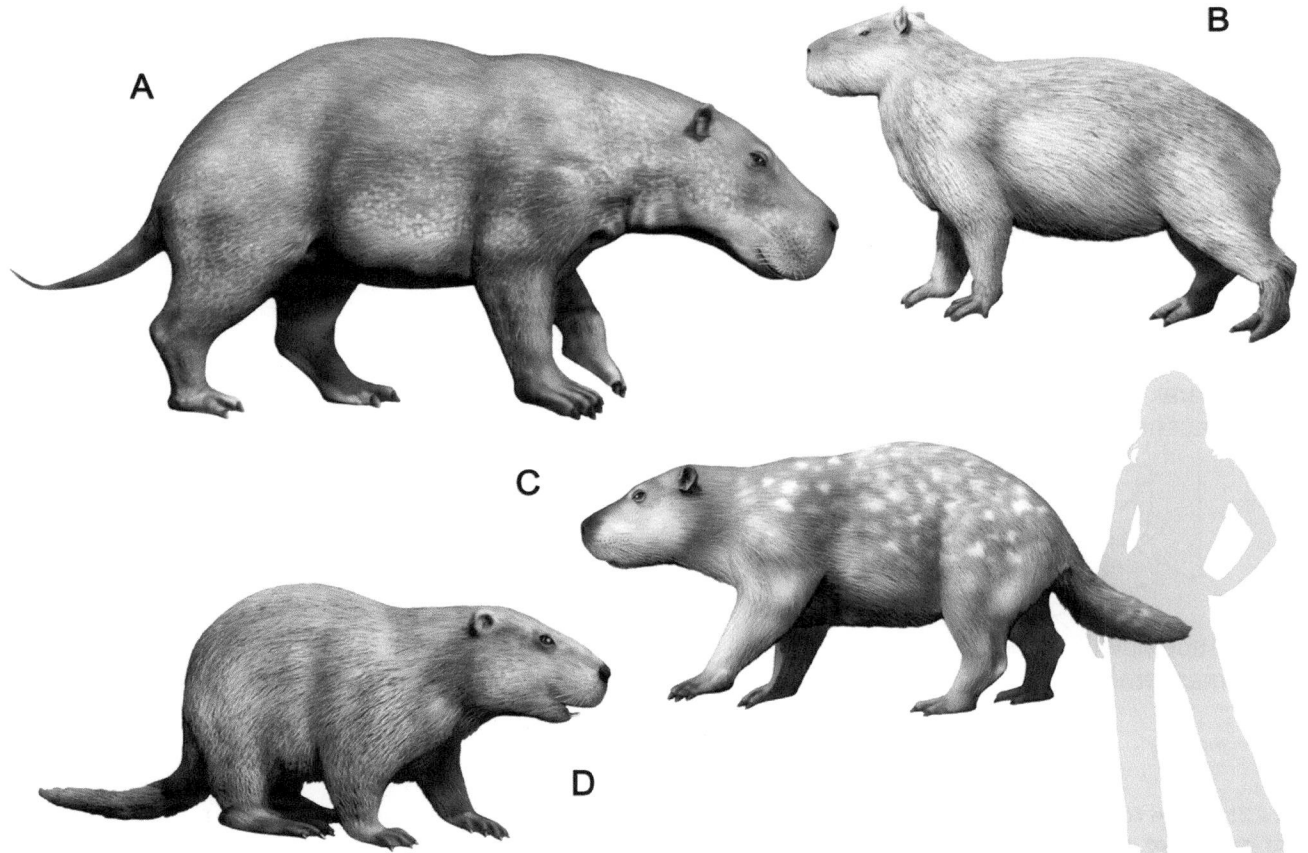

Figure 25.3 Reconstructions of some Rodents of Unusual Size. (A) The extinct rhino-sized pacarana from South America, *Josephoartigasia*. (B) The giant Ice Age capybara from South America, *Neochoerus*. (C) The Miocene giant South American rodent *Telicomys*. (D) The bear-sized giant Ice Age beaver *Castoroides*.

primates also were built like rodents in the Paleocene. When rodents spread from Asia to North America and Europe in the early Eocene, however, they began to displace the earlier occupants, so that by the Oligocene, multituberculates and rodent-like primates were extinct. The early rodents had very primitive protrogomorph skulls (**Figure 25.4[A]**), with the masseteric muscles attached only to a limited area along the base of the zygomatic arch (as in many other mammals). By the late Eocene, they had diversified into three main lineages. The sciuromorphs are only slightly more specialized than the ancestral protrogomorphs, with the masseter muscles extending up along the front of the zyomatic arch to the side of the snout. Sciuromorphs include the squirrels and all their relatives, including chipmunks, woodchucks and marmots, and the beavers. In the second condition, known as hystricomorph, the masseter muscle passes up through the zygomatic arch and onto the snout through a hole for the passage of nerves called the infraorbital foramen. Hystricomorphs include not only the porcupines (both North American and African), and some other African rodents, but also the incredible radiation of native South American rodents, the caviomorphs (including the guinea pigs, capybaras, chinchillas, agoutis, and many less familiar animals). The caviomorphs first arrived in South America in the late

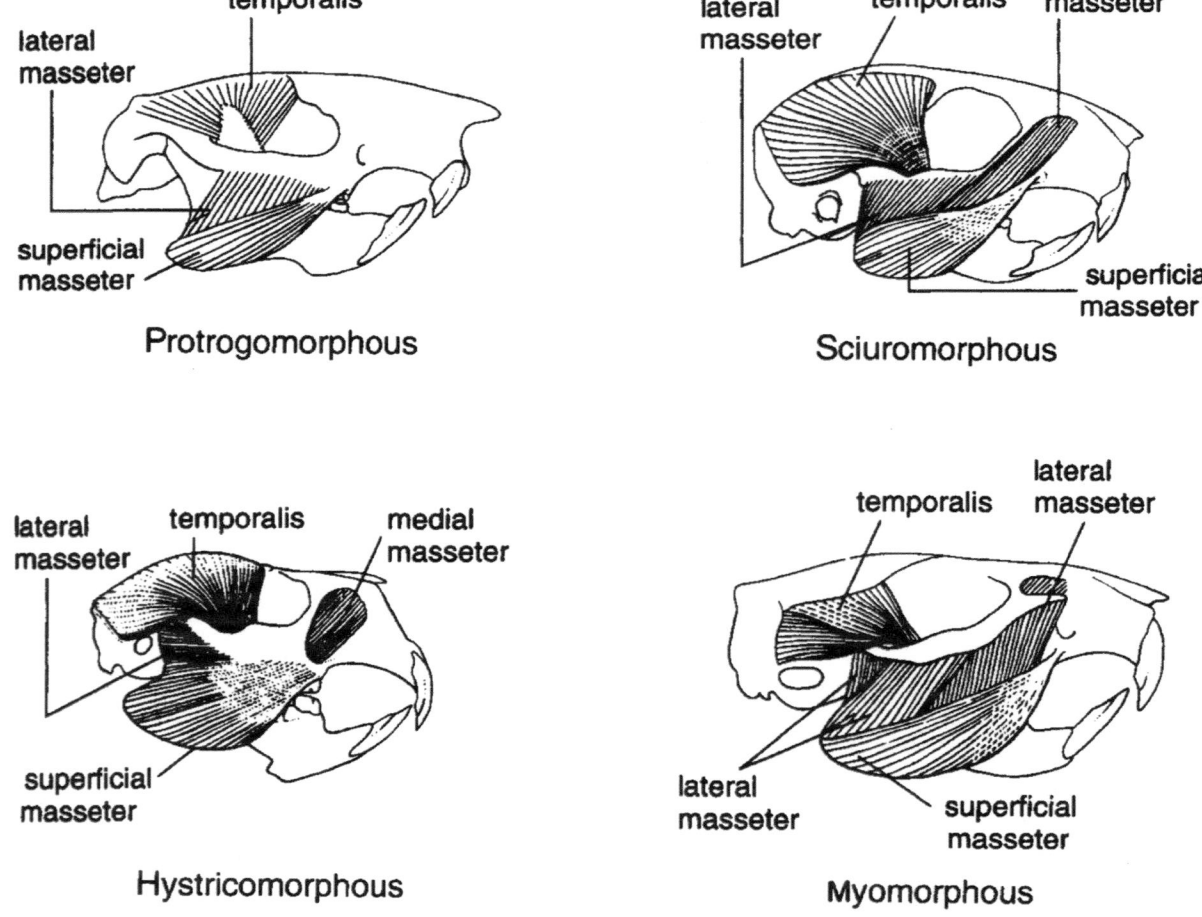

Figure 25.4 The four different configurations of jaw muscles and holes in the skull (infraorbital foramen) of the major groups of rodents. The most primitive rodents are protrogomorphs, with masseter muscles below the zygomatic arch. In sciuromorphs, the masseter muscles extending up along the front of the zyomatic arch to the side of the snout. In the hystricomorphs, the masseter muscle passes up through the zygomatic arch and onto the snout through a hole for the passage of nerves called the infraorbital foramen. The most common condition is known as myomorph; it combines a strand of the masseter passing along the front of the zygomatic arch with another passing through the infraorbital foramen. (Redrawn from several sources.)

Eocene, presumably from African hystricomorph ancestors that rafted there across the Atlantic. (The same scenario applies to the New World monkeys as well.) The most specialized condition is known as myomorph, and it combines a strand of the masseter passing along the front of the zygomatic arch with another passing through the infraorbital foramen. The vast majority of rodents, including the rats, mice, hamsters, voles, lemmings, and their kin, exhibit this condition.

Lagomorpha

Rabbits, hares, and their relatives have always been placed in their own order Lagomorpha, along with the hamster-like pikas. Lagomorphs have two pairs of chisel-like incisors, in contrast to the single pair in rodents, and a number of other unique specializations. They include over 92 species: 62 species of rabbits in 12 genera, and 10 species of pika in one genus. As discussed earlier, lagomorphs are recognized by their unique combination of two pairs of gnawing incisors in their upper jaw, compared to only a single pair in rodents (**Figure 25.1[B]**). Rabbits and hares have short tails compared to most rodents, and long ears that help them hear predators and also to radiate excess body heat, and long hind legs for their jumping gait.

Rabbits are also unusual in that they eat large quantities of vegetation without a foregut fermentation chamber like ruminants have, so their cellulose-digesting bacteria are located near the end of their digestive tract in the caecum. To compensate, most rabbits will eat their feces and run their food through their gut a second time, so the bacteria from the first passage through their gut will have time to break down the cellulose and make more efficient digestion.

Lagomorphs are also famous for their rapid breeding rates. Typically, the female rabbit can bear a large litter of babies, leave them in the burrow for safety to feed, wean them in about a month, and then immediately become pregnant again. This allows them to have several large litters in a year, and in good times with abundant food or few predators, they can multiply at incredible rates.

Only two families of lagomorphs survive today: the Leporidae (hares and rabbits), and the Ochotonidae (pikas). Leporids can be traced back to the early Eocene of Asia, where fossils such as *Shamolagus*, *Lushilagus*, *Dituberolagus*, and *Strenulalagus* represent extremely primitive rabbits just slightly more advanced than archaic Glires like *Gomphos* (**Figure 25.5[A]**) and *Mimotona* from the Paleocene and early Eocene. In the late middle Eocene, leporids crossed the Bering land bridge and began to evolve in North America as well. Some of them, like *Palaeolagus* (**Figure 25.5[B]**) from the Big Badlands of South Dakota, are extremely common fossils, and known from many complete skeletons as well as thousands of jaws. Leporids continued in low diversity during the Miocene, apparently outcompeted by ochotonids. When the spread of cold grasslands occurred in the Pliocene and Pleistocene, leporids underwent a huge explosive radiation in diversity. The also spread to Africa in the late Miocene, and were among the immigrants from North America to South America during the Pliocene when they crossed the Panama land bridge and evolved new genera there.

There are also some remarkable extinct rabbits, such as *Nuralagus rex* (**Figure 25.5[C]**), from deposits about 3–5 Ma on the Mediterranean island of Minorca, near Spain. It was about 12 kg (26 pounds) in weight, about six times as heavy as a normal rabbit! It was bulky and heavy, not slim and fast like modern rabbits. This is typical of many islands, where there are few predators, so normally small and fast animals, like rabbits and hedgehogs (such as *Deinogalerix*—see **Figure 23.1**), can afford to grow gigantic in the absence of strong selection pressure by predators.

Figure 25.5 Reconstruction of some extinct lagomorphs. (A) The Asian early Eocene archaic lagomorph *Gomphos*. (B) The late Eocene-Oligocene lagomorph *Palaeolagus*, common in the Big Badlands of South Dakota. (C) The giant rabbit *Nuralagus rex*, from the late Cenozoic of the island of Minorca. Scale bar is 20 cm.

EUARCHONTA

As discussed earlier, in 1910 William King Gregory first proposed a taxon called Archonta ("ruling beings") for the primates, tree shrews, colugos, elephant shrews, and bats. That taxon fell out of use until the late 1980s and 1990s, when it re-emerged in the phylogenies based on anatomical evidence. It was modified in the late 1990s, when molecular analyses showed that primates, colugos, and tree shrews are indeed closely related (so they were renamed the "Euarchonta"), but bats and elephant shrews were part of Laurasiatheria and Afrotheria, respectively.

Tree Shrews

Of those three groups of Euarchonta, tree shrews, or family Tupaiidae, have long had a place in anthropology textbooks as the nearest relative of true primates. Today there are twenty species in five genera in two families found entirely in the jungles of Southeast Asia. They are shrew-like in build (although usually larger in size), living in the trees and ground during the daytime, and feeding on a wide variety of foods (insects, small vertebrates, fruit, nuts, and seeds). Originally, tupaiids were considered "insectivores" and thrown into a taxonomic wastebasket with other insectivorous mammals. But thanks to molecular data, they are now considered euarchontans, closest relatives of primates.

Colugos

The most unusual group of living euarchontans is the dermopterans ("skin wing" in Greek), or colugos. They are sometimes called "flying lemurs", although that is a misnomer, because they are not lemurs, nor do they fly. Only two species survive today in the jungles of the Philippines and Southeast Asia. They vaguely resemble flying squirrels in shape and size, and they have a large furry membrane (patagium) between their front and hind limbs, and between the hind limbs and tail, that they use to glide from tree to tree. However, their heads are more like those of primates, with large forward-facing eyes for good stereovision, especially at night when they feed on leaves, sap, flowers, fruits and seeds. Their teeth bear some resemblance to those of some very primitive primates, and their lower front incisors are modified into a comb-like device for grooming their fur, convergently evolved with the "tooth combs" seen in some early primates. Colugos are known from a number of fossil teeth and jaws of two groups of common Paleocene and early Eocene fossils, the Mixodectidae and the Plagiomenidae. Then colugos virtually vanish from the fossil record as the jungles of the Eocene disappeared, and only those two surviving lineages in Asia remain.

PRIMATES

True primates have a number of unique and distinctive specializations, including shorter snouts, forward-pointing eyes for stereoscopic vision with a bar of bone behind the eye socket, nails instead of claws on most of their fingers and toes, relatively large brains, and many features of the teeth and the skull region.

One of the oldest groups of close primate relatives were the plesiadapids (**Figures 25.6[A,B]** and **25.7**). The earliest known fossil is *Purgatorius*, a shrew-like creature (**Figure 25.6[A]**) from the latest Cretaceous and early Paleocene, which was very primitive and insectivore-like but still had primate-like features in its teeth and ankles. Most plesiadapids (**Figure 25.6[D]**) were built much like squirrels or lemurs, although some were the size and shape of a woodchuck (*Platychoerops*) or like a tarsier, with huge forward-facing eyes (like *Carpolestes*). Plesiadapids had some similarities to true primates in their cheek teeth, and a few of them had nails instead of claws, opposable thumbs, and bodies like lemurs or tarsiers (like *Carpolestes*) but otherwise they lacked the primate specializations, such as fully forward-facing eyes for binocular vision, and a fully primate cheek tooth pattern. The more squirrel-like plesiadapids (**Figure 25.6[D]**) had large forward-pointing incisors with a big gap (diastema) between the incisors and the cheek teeth. It is somewhat like the condition found in rodents, although plesiadapids did not develop the full chisel mechanism or ever-growing incisors found in all rodents and rabbits. Given their huge diversity and abundant fossils, they apparently thrived in the Paleocene jungles of Montana, Wyoming, and western Europe. They probably ate a variety of fruits, flowers, seeds, and fruits.

Dozens of species in 38 genera in 9 families of plesiadapids are known from the fossil record. Plesiadapids were most abundant in the Paleocene of North America and Europe, where their species are so common and rapidly evolving, that they are used to tell time in this interval. Most plesiadapids vanished in the early Eocene with the invasion of true primates (adapids and omomyids), but a few genera straggled on into the middle Eocene, and a few even lasted until the end of the middle Eocene (40 Ma). Since most were tree dwellers, their extinction was probably due to a combination of the loss of the tropical forests during the late Eocene, plus the competition from true primates.

The living true primates (excluding plesiadapids) are extremely successful with 13 families, 71 genera and 424 species of lemurs, lorises, pottos, tarsiers, monkeys, apes, and humans (**Figure 25.7**). Hundreds of more extinct species and genera are known from dozens of families. Most are specialized tree-dwellers, living largely on fruits, seeds, and maybe leaves, although some (like baboons, gorillas, chimps, and humans) spend most or all their time on the ground. They are typically split into two larger groups, the strepsirhines (once called "prosimians"), which include lemurs, lorises, and galagos; and the haplorhines, which include the tarsiers plus anthropoids (or Simiiformes), which include the monkeys, apes, and humans.

Strepsirhini

The earliest fossil true primate is known as *Altiatlasius* from the late Paleocene of Morocco, and by the early Eocene true primates had spread widely

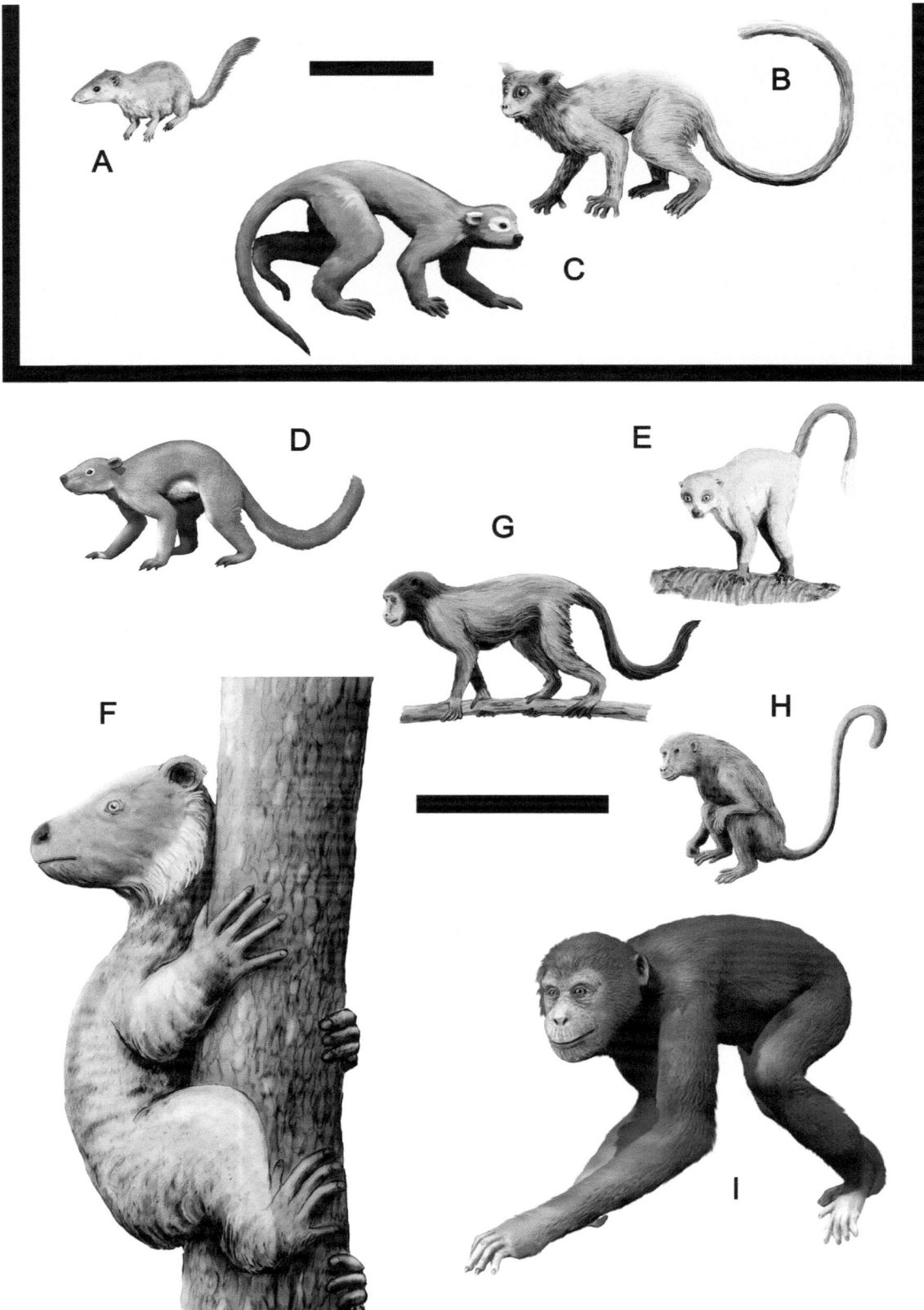

Figure 25.6 Reconstructions of some extinct primates. (A) The earliest known primate from the latest Cretaceous and early Paleocene, *Purgatorius*. (B) The Eocene omomyid primate *Necrolemur*. (C) The earliest South American monkey, known as *Branisella*. (D) The early Paleocene to early Eocene plesiadapid *Plesiadapis*. (E) The middle Eocene adapid primate *Darwinius*, from the famous Messel lake shales in Germany. (F) The giant gorilla-sized lemur *Megaladapis* from Quaternary deposits in Madagascar. (G) The late Eocene anthropoid *Aegyptopithecus*, from the Fayûm beds of Egypt. (H) The late Oligocene anthropoid *Saadanius*. (I) The ape *Oreopithecus*, from the Miocene of Europe. Scale bar for (A)–(C): 10 cm; scale bar for (D)–(I): 50 cm.

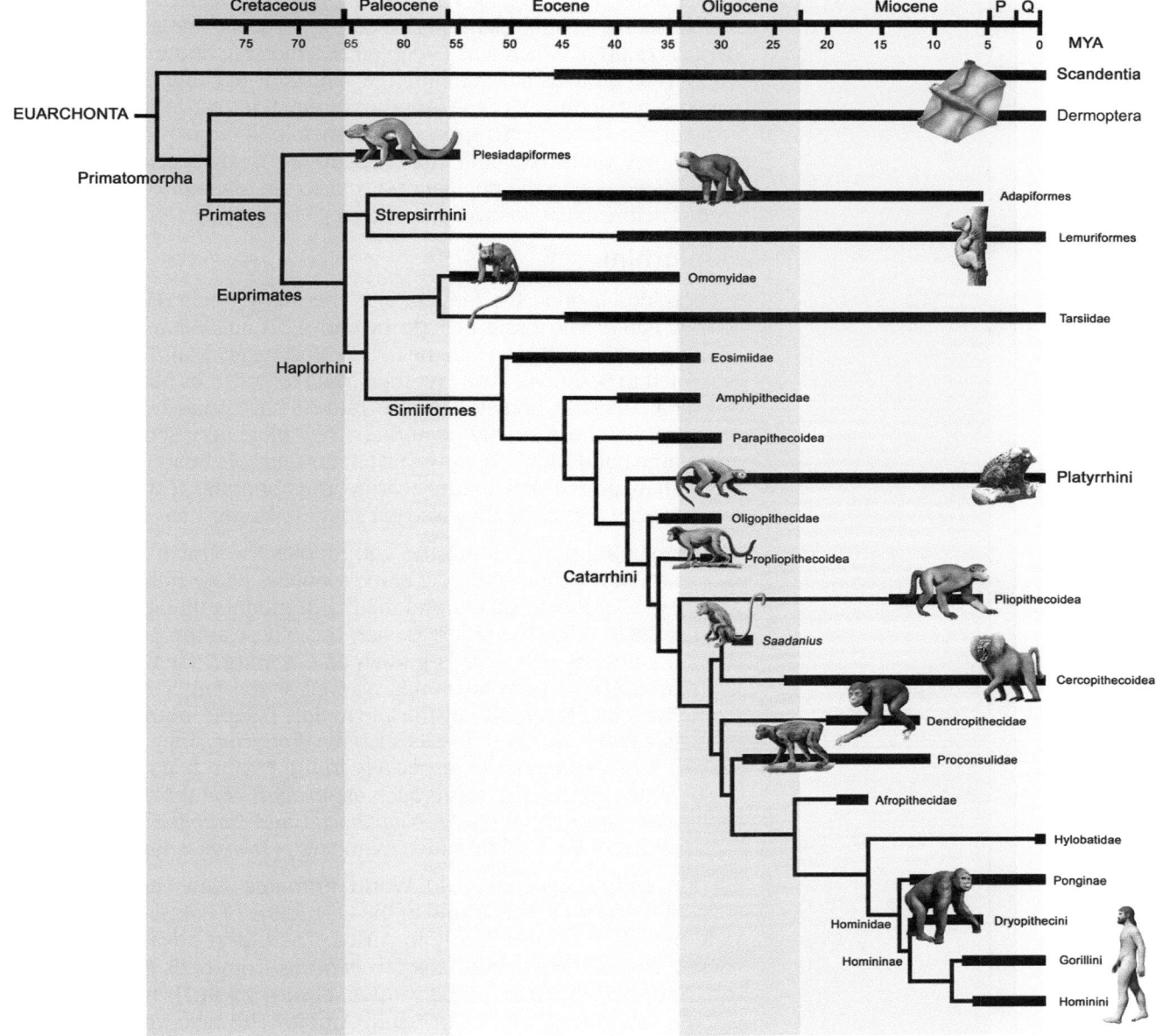

Figure 25.7 Family tree of the primates and their nearest relatives.

to Eurasia and North America. The early Eocene *Teilhardina* first appears in Asia, then in Europe, and finally in North America over a few thousand years. Two main families dominate the early Eocene of the northern continents (**Figure 25.5**): the larger, more lemur-like adapids (**Figure 25.6[E]**) (with longer snouts and smaller eyes facing more sideways), and the smaller, more tarsier-like omomyids (**Figure 25.6[B]**), with shorter snouts and large forward-facing eyes. Both groups were incredibly abundant and diverse through the early and middle Eocene of North America, but vanished completely by the end of the middle Eocene (40 Ma).

The survivors of this early radiation of strepsirhines are the living lemurs, found entirely in Madagascar, plus the lorises and galagos of Asia and Africa, with over 100 species in 15 genera and 5 families. Lemurs apparently evolved from adapiforms that reached Madagascar in the early Eocene (54 Ma), and there are fossils of close relatives of lemurs from the Eocene of Algeria and Tunisia. Lemurs evolved in isolation on Madagascar since that time. They developed into many different sizes and shapes, from the tiny mouse lemur (only 30 g, or 1.1 ounces) to the gorilla-sized ground-dwelling *Archaeoindris*, which weighed 440 lb (200 kg). The Ice Age lemur *Megaladapis* (**Figure 25.6[F]**), which was 5 feet tall (1.5 meters), and may have reached 50 kg (110 lb), was still roaming the forests of Madagascar until humans wiped it out about 1420 A.D. Another group, the long-armed sloth lemurs, hung upside-down from branches with their long-curved finger like modern sloths; they vanished about 1620 A.D. Today, many species of lemurs are endangered due to the population explosion in Madagascar and the widespread deforestation, as well as poachers who kill them for bushmeat.

Haplorhini

Haplorhines include the tarsiers plus the anthropoids (**Figure 25.7**). The earliest haplorhine fossils include the early Eocene (55 Ma) *Archicebus* from China, and *Xanthorhysis* from the middle Eocene (45 Ma). Today tarsiers are restricted to Southeast Asia, but their fossil record goes back to the Eocene of China, Thailand, and even Africa. Tarsiers have huge eyes that dominate their face, and are strictly nocturnal. They have very short snouts, small ears, large hands and feet with long fingers and toes bearing nails, and long tails. They are the only entirely carnivorous primates, catching insects by leaping from branches; they also eat snakes, lizards, bats, and birds.

The remaining primates are the anthropoids, or simian primates (Simiiformes). These include the New World monkeys (Platyrrhini), the Old World monkeys (Cercopithecidae), and their descendants the apes and humans (**Figure 25.7**). Like the oldest tarsier, the oldest anthropoids come from the early Eocene with fossils known as *Eosimias* ("dawn monkey") from China and *Afrasia* from Myanmar, as well as the larger Amphipithecidae from China and Myanmar. By the late middle Eocene, anthropoids are well established in Africa with fossils such as *Afrotarsius*. They are much better known in the late Eocene, especially in the Fayûm beds of Egypt. These include the squirrel-size forms such as *Apidium*, and the dog-sized *Aegyptopithecus* (**Figure 25.6[F]**), *Propliopithecus*, and *Parapithecus*, which might be considered the most primitive members of the catarrhines.

By the early Oligocene, Old World primates vanished from Eurasia (except for a few fossils found in the Oligocene of Pakistan), and anthropoid evolution occurred only in Africa. The oldest advanced catarrhine fossils are late Oligocene: *Nsungwepithecus* from beds 26 Ma in Tanzania, and the gibbon-sized *Saadanius* (**Figure 25.6[G]**) from beds about 29 Ma, found near Mecca in Saudi Arabia. The next youngest fossil is *Kamoyapithecus* from beds dated at 24 Ma in Kenya. By the Miocene, there were many primitive catarrhines such as *Victoriapithecus* (20 Ma), the earliest Old World monkey, and *Prohylobates* at about 17 Ma. There was a big evolutionary radiation in the middle Miocene of Old World not only in Africa but also in Eurasia, where Old World monkeys are known from hundreds of sites. Today, they are spread across this entire region, with mainly baboons and colobus monkeys in Africa, and macaques and rhesus monkeys more common in Eurasia and North Africa.

The New World monkeys, or Platyrrhini (**Figure 25.7**), today are found only in Central and South America. They include over 64 species and 17 genera in five families, including the marmosets and tamarins, spider monkeys, squirrel monkeys, howler monkeys, capuchins and uakaris, woolly monkeys, and sakis. They nearly all have prehensile tails. In fact, they are the only primates that can grab branches with their tails. Also, they have relatively flat noses ("platyrrhine" means "flat nosed") contrasted with the narrower noses of Old World monkeys ("catarrhine" means "narrow nosed").

The Platyrrhini have an excellent fossil record in South America, starting with the earliest fossil New World monkeys, reported from isolated teeth in beds about 36 Ma (late Eocene) in the Peruvian Amazon. The oldest relatively complete platyrrhine fossil is *Branisella* (**Figure 25.6[C]**) from late

Oligocene beds about 26 Ma in Salla, Bolivia. It is very similar to the late Eocene African fossil *Proteopithecus*, suggesting that platyrrhines rafted from Africa to South America on floating vegetation about 36 Ma, at a time when the Atlantic was 1000 km narrower than it is now. (The New World rodents, or caviomorphs, did the same thing, and at about the same time.) By the Miocene, the radiation of New World monkeys was in full swing, with a least 20 different genera known from Argentina to Bolivia to Colombia.

Hominoidea (Apes and Humans)

In addition to the Old World monkeys, the catarrhines include the apes and humans (now placed in the family Hominidae) (**Figures 25.7** and **25.8**). Apes differ from monkeys in that they have lost their tails, and usually have a wider degree of motion in their shoulder joint, allowing some of them to swing through branches hand over hand (brachiation).

Although there are only a few types of living apes (two species of chimpanzee, plus gorillas, orangutans, and several species of gibbons), apes had a much greater diversity in the geologic past. The oldest known ape fossil is *Rukwapithecus*, from beds about 25 Ma (late Oligocene) in Tanzania. During the Miocene, apes underwent a spectacular evolutionary radiation in Africa and Eurasia, and were much more common and diverse than monkeys. Over 40 fossil genera and over 100 species are known (14 genera in just the Miocene of Africa). They ranged from the size of a housecat (3 kg) to the size of a gorilla (80 kg), and ate a wide variety of foods, from leaves and fruit to more omnivorous diets. At about 16.5 Ma, *Afropithecus* escaped Africa and its descendants

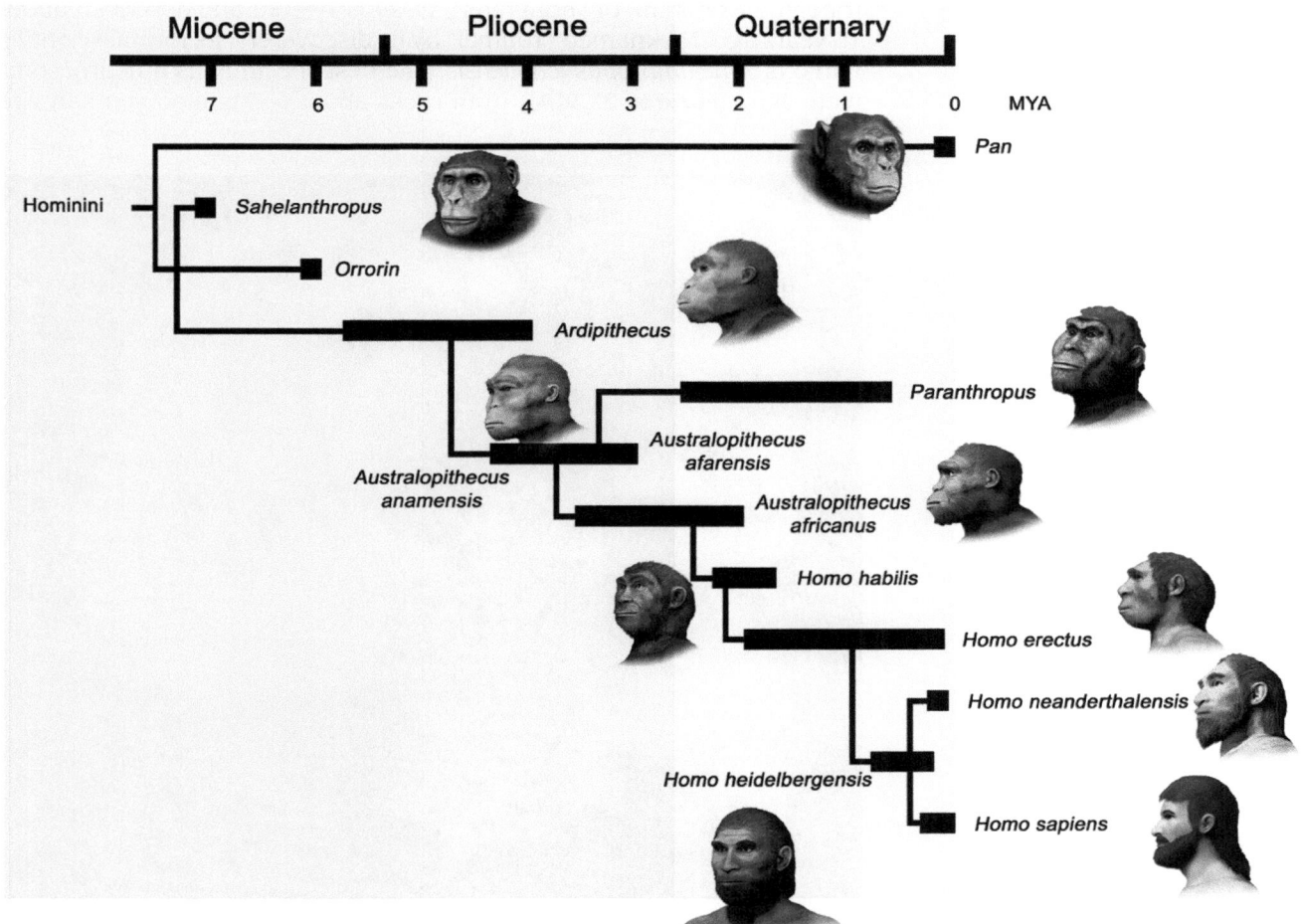

Figure 25.8 Family tree of the major species of humans (tribe Hominini) and their nearest relatives.

spread across Eurasia, leading to a whole new evolutionary explosion of apes. Some are particularly well known, such as *Sivapithecus* (including *Ramapithecus*) from 12 Ma in Pakistan, once suggested as a human ancestor but now recognized as a primitive orangutan. The dryopithecine apes were widespread across much of Eurasia. *Oreopithecus* was a famous fossil from Europe (**Figure 25.6[I]**), first described in 1872. Living about 7 Ma, *Oreopithecus* was much more specialized for leaf eating than most apes. By the end of the Miocene, climate began to dry up and the great ape radiation was decimated. Only a few survived into the Pliocene, while the Old World monkeys diversified instead.

HUMAN EVOLUTION

The topic of human evolution is often contentious and controversial, because there are many people who do not accept the overwhelming evidence that humans are part of the animal kingdom, and also members of the primates, and especially closely related to apes such as the chimps and gorillas. But to scientists, there is no question that we are classified with the other great apes in the Hominini. The fact that we share over 98% of our DNA with the chimpanzees and gorillas is just further confirmation of our close affinities. In addition, for a long time the fossil record of human evolution was very incomplete, so not much could be said about how humans evolved. But in the twenty-first century, we now have thousands of good specimens of fossil humans and closely related species, going all the way back to 7 Ma. The fact of human evolution is something that can no longer be denied except by those who refuse to look at evidence (**Figure 25.8**).

The oldest fossil that can be truly described as a member of our own tribe, the Hominini, or "hominins", was discovered and described about 15 years ago. Nicknamed "Toumai" by its discoverers, its formal scientific name is *Sahelanthropus tchadensis*. The best specimen is a nearly complete skull (**Figures 25.9[A]**) from rocks about 6–7 million years in age

Figure 25.9 Some examples of the more complete hominin fossils. (A) The distorted skull of *Sahelanthropus*, the oldest fossil in the human lineage, from beds in Chad about 6–7 Ma. [Photos (A, E, H) courtesy Wikimedia Commons. Photos (B) and (C) courtesy T. White. Photos (D, F, G) by the author.]

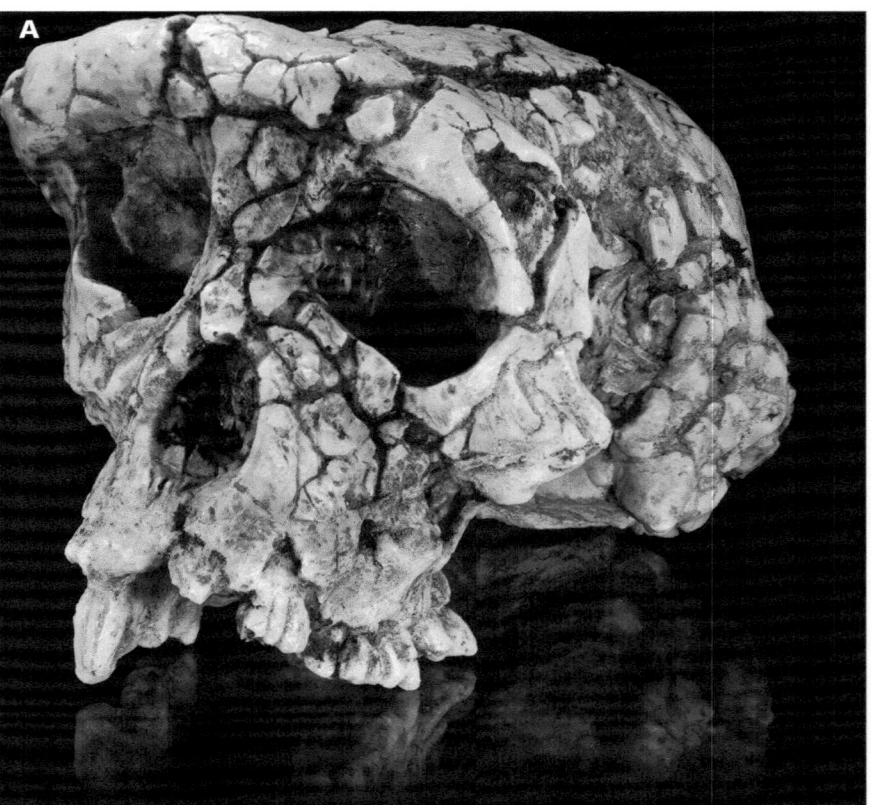

B

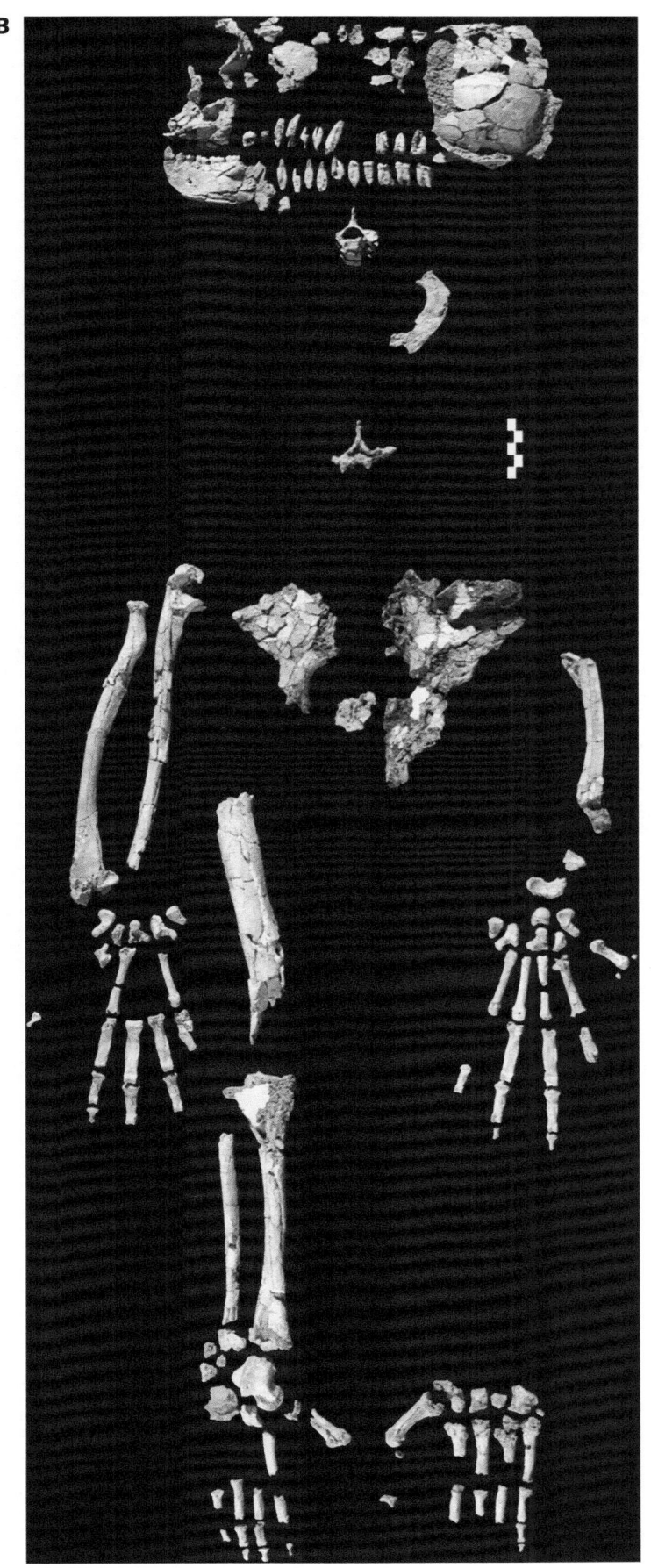

Figure 25.9 (Continued) (B) The nearly complete skeleton of *Ardipithecus ramidus*, from rocks in Ethiopia dated at 4.4 Ma.

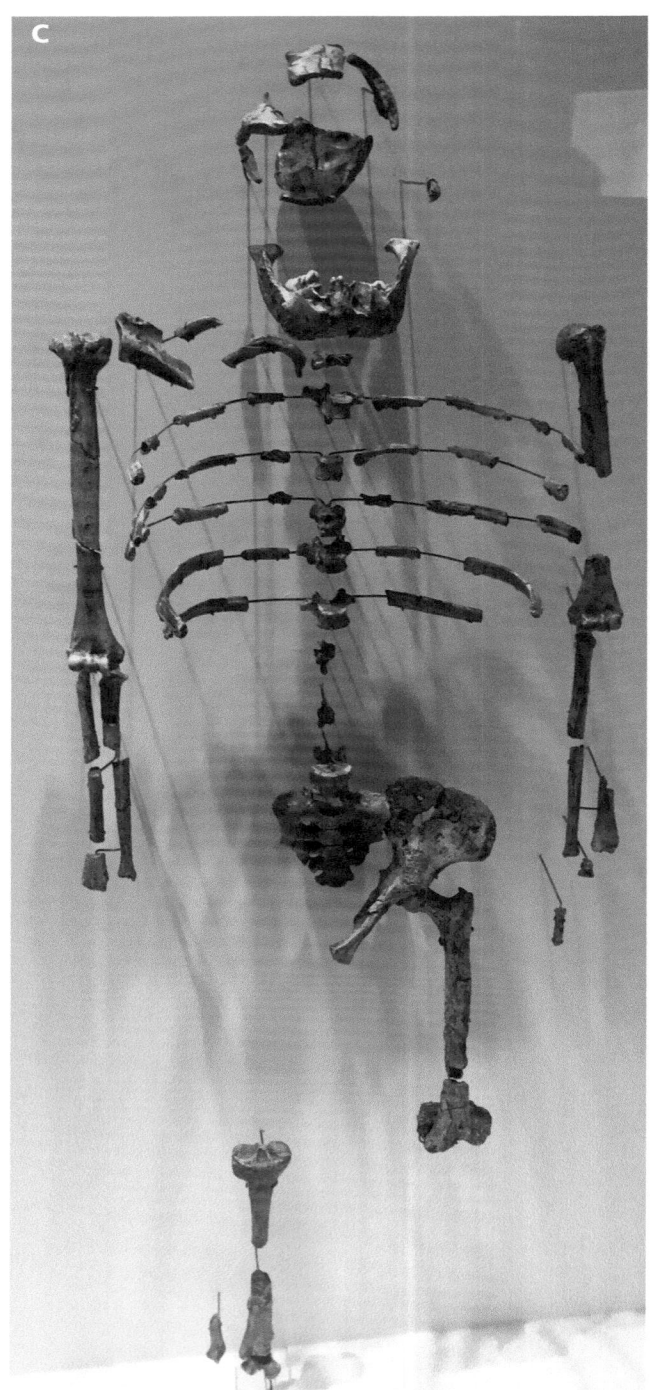

Figure 25.9 (Continued) (C) Skeleton of *Australopithecus afarensis*, or "Lucy", from beds in Ethiopia dated 2.95–3.85 Ma. (D) Reconstruction of *A. afarensis*.

from the Sub-Saharan Sahel region of Chad (hence the scientific name, which translates to "Sahel man of Chad"). Although the skull is very chimp-like with its small size, small brain, and large brow ridges, it had remarkably human-like features, with a flattened face, reduced canine teeth, enlarged cheek teeth with heavy crown wear, and an upright posture—all of this at the very beginning of human evolution. Just slightly younger is *Ororrin tugenensis*, from the upper Miocene Lukeino Formation in the Tugen Hills in Kenya dated between 5.72 and 5.88 Ma. *Ororrin* is known mainly from fragmentary remains, but the teeth have the thick

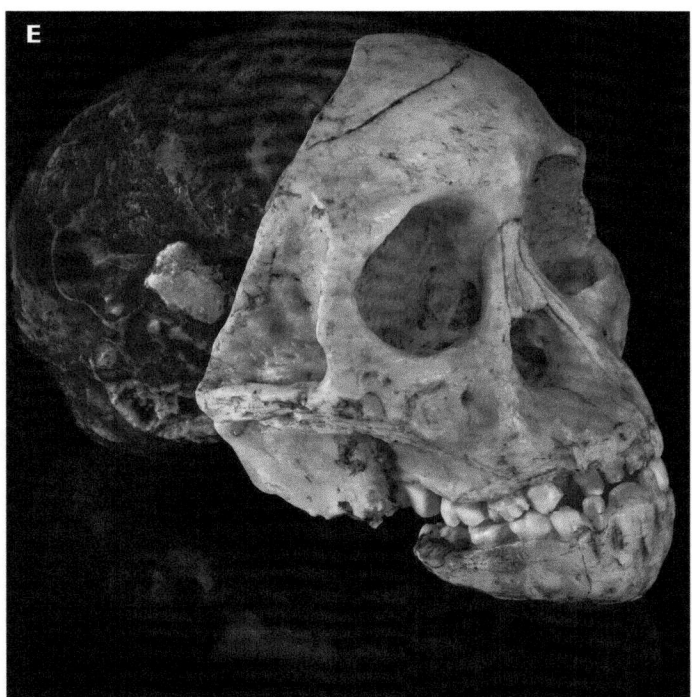

Figure 25.9 (Continued) (E) The famous "Taung child" skull of *Australopithecus africanus*, with a small ape-like brain exposed in the back, and the relatively flat face with small brow ridges, typical of more advanced hominins. (F) Display of late Pliocene hominins. In the middle row are skulls of the robust genus *Paranthropus*, including (left to right), the primitive *P. aethiopicus* (the "Black Skull"), the robust "Nutcracker Man" *P. boisei*, and on the right the original species of the genus *Paranthropus*, *P. robustus*. In the bottom row are their lower jaws, showing how enormous and distinctive their lower molars were, especially as the crowns were ground flat by a gritty diet. In the top row are more gracile *Australopithecus africanus*. (G) Side-by-side comparison of the skulls of *Homo habilis* (right) and *Homo rudolfensis* (left), the earliest species in our genus.

enamel typical of early hominins, and the thigh bones clearly show that it walked upright. Slightly younger still are the remains of *Ardipithecus kadabba*, found in Ethiopian rocks dated between 5.2 and 5.8 Ma. These consist of a number of fragmentary fossils, but the foot bones show that hominins used the "toe off" manner of upright walking as early as 5.2 Ma. Our human lineage was well established by the latest Miocene and fully upright in posture, even though our brains were still small and our body size not much different than that of contemporary apes.

H

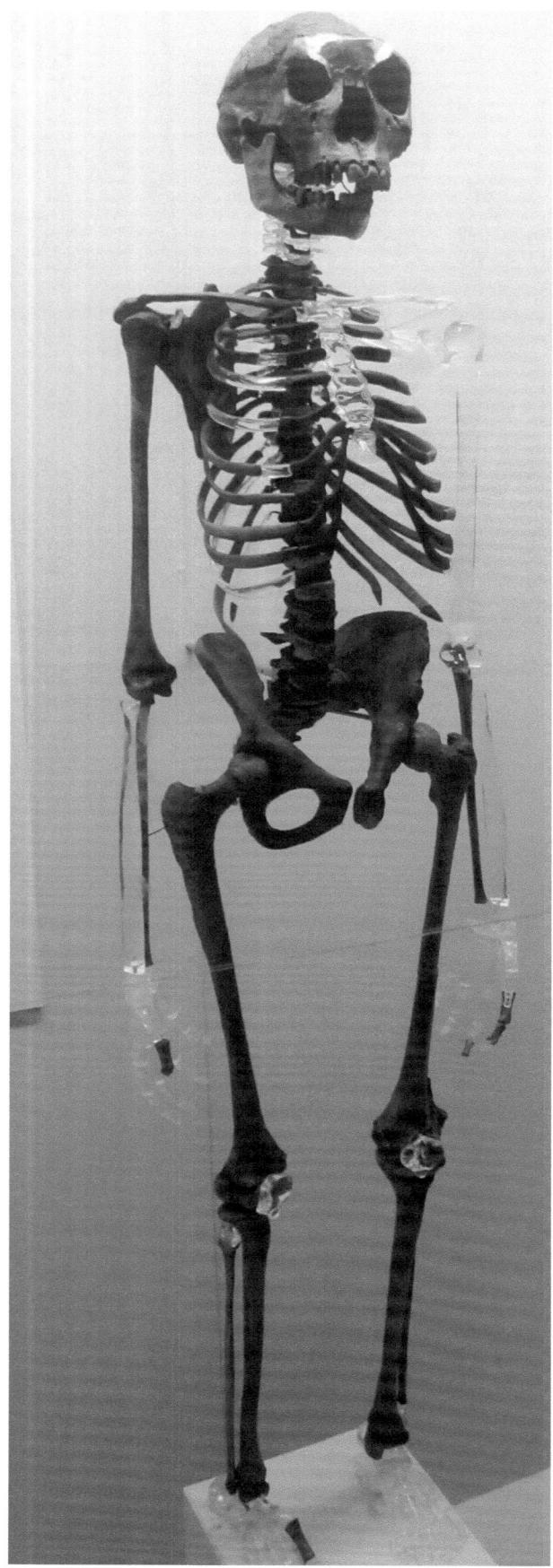

Figure 25.9 (Continued) (H) The nearly complete skeleton of *Homo ergaster*, known as the "Nariokotome boy", found by Alan Walker and crew on the shores of West Turkana in 1984. It dates to about 1.7 Ma.

The Pliocene saw an even greater diversity of hominins (**Figure 25.8**), with a number of archaic species overlapping in time with the radiation of more advanced hominins. Archaic relics of the Miocene included *Ardipithecus ramidus*, found in Ethiopia in 1992 from rocks 4.4 million years in age, which had human-like reduced canine teeth and a U-shaped lower jaw (instead of the V-shaped lower jaw of the apes). *Ardipithecus ramidus* is now known from nearly complete skeletal material (**Figure 25.9[B]**), making it the oldest hominin skeleton known. Rocks in Kenya about 3.5 million years in age also yield other more primitive forms like *Kenyapithecus platyops*.

By 4.2 Ma, however, the first members of the advanced genus *Australopithecus*, the most diverse genus of our family in the Pliocene, are also found. The oldest of these fossils is *Australopithecus anamensis* from rocks near Lake Turkana in Kenya ranging from 3.9 to 4.2 million years in age. These creatures were fully bipedal, as shown not only by their bones but also by hominin trackways near Laetoli, Tanzania. In 2019, a complete skull of this species was reported for the first time, which made its anatomy and relationships much better understood.

The most famous of these early australopithecines is *A. afarensis* (from rocks 2.95–3.85 Ma near Hadar, Ethiopia), better known as "Lucy" by its discoverers Don Johanson and Tim White. Celebrating by the campfire the night after they made the discovery, they were singing along with a tape of the Beatles' "Lucy in the Sky with Diamonds" and decided to nickname the fossil "Lucy" (**Figure 25.9[C,D]**). When it was discovered in 1974, *Australopithecus afarensis* was the first early hominin to clearly show a bipedal posture (based on the knee joint and pelvic bones) but was not as upright as later hominins. These were still small creatures (about 3 feet, or 1 meter tall) with small brains, and very ape-like in having large canine teeth and a large protruding jaw.

By the late Pliocene, hominins had become very diverse in Africa (**Figure 25.8**). These included not only the primitive forms *Australopithecus garhi* (dated at 2.5 million years) and *A. bahrelghazali* (dated at 3.5 million years) but the one of the best-known australopithecines, *Australopithecus africanus* (**Figure 25.9[E]**). Originally described by Raymond Dart in 1924 based on a juvenile skull (the "Taung child"), for decades the Eurocentric anthropology community refused to accept it as ancestral to humans. But as more South African caves yielded better specimens to paleontologists like Robert Broom (especially the adult skull nicknamed "Mrs. Ples"), it became clear that *Australopithecus africanus* was a bipedal, small-brained African hominin, not an ape. *Australopithecus africanus* was a rather small, gracile creature, with a dainty jaw, small cheek teeth, no skull crest, and a brain only 450 cc in volume. On the basis of its gracile and very human-like features, *Australopithecus africanus* is often considered the best candidate for ancestry of our own genus *Homo*.

In addition to *Australopithecus africanus*, the late Pliocene of Africa also yields a number of highly robust hominins. For a long time, they were lumped into a very broad concept of the genus *Australopithecus*, either as distinct species or even dismissed as robust males of *Australopithecus africanus*. In recent years, however, anthropologists have come to regard them as a separate robust lineage, now placed in the genus *Paranthropus*. The oldest of these is the curious "Black Skull" (so called because of the black color of the bone), discovered in 1975 on the shores of West Lake Turkana, Kenya, from rocks about 2.5 million years in age (**Figure 25.9[F]**). Although it is small in brain size, the skull is robust with large bony ridge along the top midline of the skull (called a sagittal crest), massive molars, and a dish-shaped face. Currently, scientific opinion places the "Black Skull" as the earliest member of *Paranthropus*, *P. aethiopicus*. It was followed by the most robust of all hominins, *P. boisei*, from rocks in East Africa ranging from 2.3 to 1.2 million years in age (**Figure 25.9[F]**). The first specimen found of this species was nicknamed "Nutcracker Man" for its huge thick-enameled molars, robust jaws, wide flaring cheekbones, and strong crest on the top of its head, suggesting a diet of nuts or seeds or even bone cracking. Discovered by Mary Leakey at Olduvai Gorge in 1959, it was originally named "*Zinjanthropus boisei*" by Louis Leakey, who made his reputation from it. The rocks of South Africa between 1.6 and 1.9 million years in age yield the original species of *Paranthropus*, *P. robustus* (**Figure 25.9[F]**). These too had massive jaws, large molars, and large skull crests but were not as robust as *P. boisei*. *Paranthropus robustus* lived side by side in the same South African caves as *A. africanus*. It is not only more robust but also larger than that species as well, with some individuals weighing as much as 120 pounds.

Finally, the early Pleistocene produces the first fossils of our own genus *Homo*, which are easily distinguished from contemporary *Australopithecus* and *Paranthropus* by a larger brain size, flatter face, no skull crest, reduced brow ridges, smaller cheek teeth, and reduced canine teeth. The first of these to be described was *Homo habilis* (whose name literally means "handy man"), discovered in the 1960s by Louis and Mary Leakey in Olduvai Gorge, Tanzania, from beds about 1.75 million years in age (**Figures 25.8** and **25.9[G]**).

Originally, all of the earliest *Homo* specimens were shoehorned into the species *H. habilis*, but now paleoanthropologists recognize that this material is too disparate to belong to one species, so several are now recognized. These include the a more advanced-looking skull (**Figure 25.8[H]**) now known as *H. rudolfensis* (dated to about 1.9 Ma), which made Richard Leakey's reputation, and the very advanced but short-lived *Homo ergaster* (**Figure 25.8[H]**), from beds 1.6–1.8 million years in age. These species are known not only from bones but also from their primitive stone tools, especially choppers and hand axes of the "Oldowan" technology.

Many of the archaic Pliocene taxa persisted into the early Pleistocene (as recently as 1.6 Ma), including *Paranthropus robustus* and *P. boisei*, *Homo ergaster*, and *Homo habilis*. The best-known fossil of *Homo ergaster* is a nearly complete skeleton of a boy who died when he was about 8–9 years old, found on the shores of West Lake Turkana by Alan Walker and his crew in 1984. Nicknamed "Nariokotome Boy" (**Figure 25.8[H]**), it is estimated that he would have been 2 meters tall if fully grown, which is taller than most modern humans.

By 1.9 Ma, however, a new species had appeared: *Homo erectus* (**Figure 25.10**). This human was not only bipedal and stood erect (as its species name implies) but was also almost as large in body size as we are. Its brain capacity was about 1 liter (1000 cc), only slightly less than ours. *Homo erectus* made crude choppers and hand axes ("Acheulean culture" tools) and was the first species to make and use fire. Originally, *Homo erectus* was confined to Africa, where all of our other ancestors had long lived. By around 1.8 Ma, we have evidence that *Homo erectus* migrated outside our African homeland, as specimens from Indonesia (originally described as "*Pithecanthropus erectus*" or "Java man") have been dated at that age. In addition, specimens are known from elsewhere in Eurasia, such as Romania and the Republic of Georgia, that are almost as old. By about 500,000 years ago, we have abundant fossils of *H. erectus* in many parts of Eurasia, including the famous specimens from the Chinese caves at Zhoukoudian, originally called "Peking Man" and dated as old as 460,000 years ago. The latest dating suggests that *H. erectus*

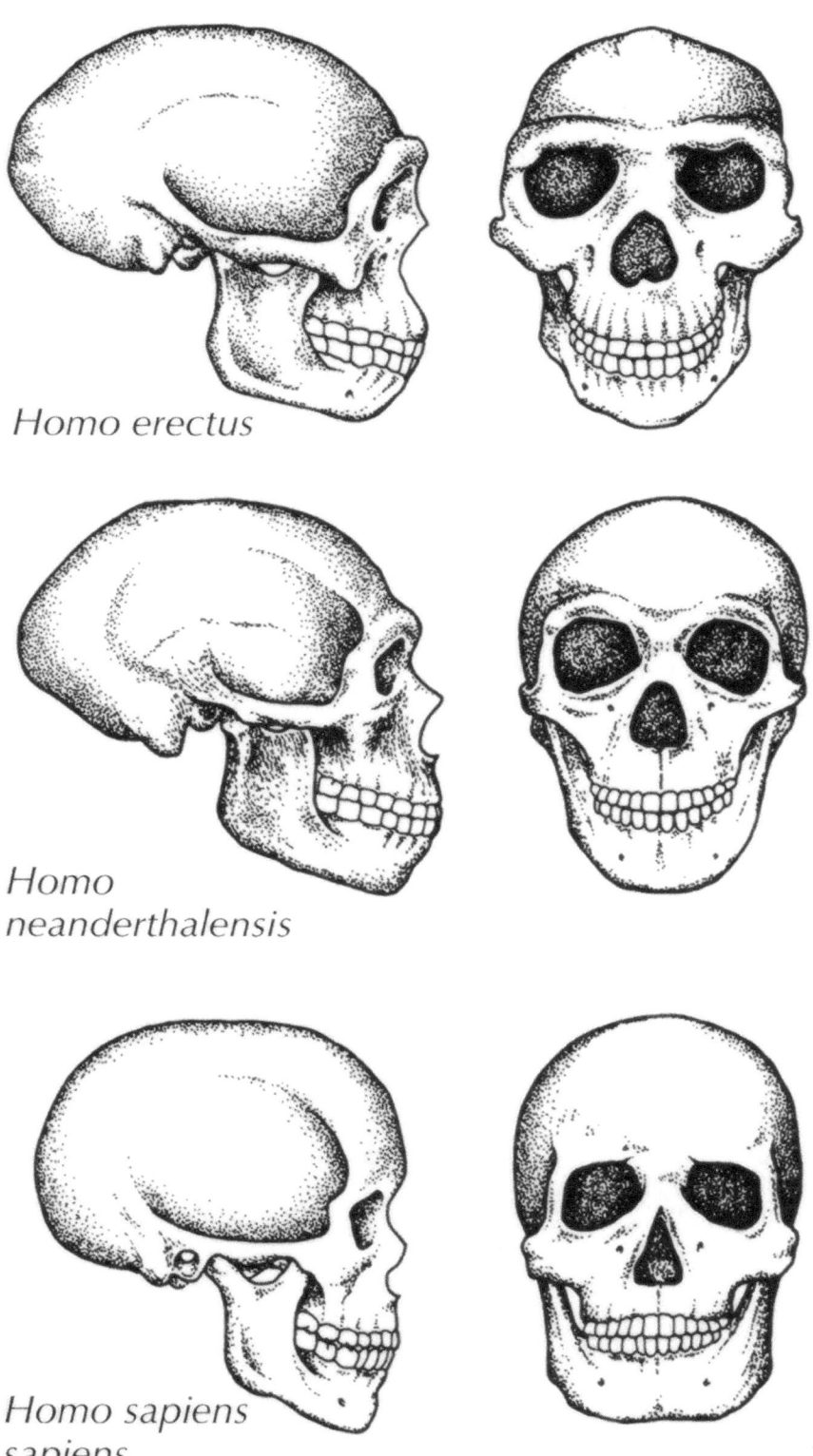

Homo erectus

Homo
neanderthalensis

Homo sapiens
sapiens

Figure 25.10 Comparisons of the skulls of *Homo erectus* (top), *Homo neanderthalensis* (center) and modern *Homo sapiens* (bottom). All have relatively large brains, with about 1000 cc capacity in *H. erectus*, while both Neanderthals and modern humans have brains in the 1500–1700 cc range. The two extinct species have strong brow ridges, a protruding snout without much of a chin, and broader, heavier cheekbones compared to modern humans. Neanderthals have about the same brain size as modern humans, but their skull is a bit flatter with a point on the back end. (Redrawn from several sources.)

may have persisted as recently as 143,000 years ago and possibly 74,000 years ago, overlapping with modern *H. sapiens. Homo erectus* was not only the first widespread hominin species but also one of the most successful and long-lived species, spanning more than 1.8 million years in duration between 1.9 and 0.143 Ma. During much of that long time, it was the only species of *Homo* on the planet and changed very little in brain size or body proportions. If longevity is a measure of success, then it could be argued that *Homo erectus* was even more successful than we are.

By about 400,000 years ago, another species was established in western Europe and the Near East: the Neanderthals (**Figures 25.10** and **25.11**). In 1857, these were the first fossil humans ever to be discovered, although their fragmentary fossils were originally dismissed as the remains of diseased Cossacks that had died in caves. The first complete descriptions of skeletons were based on an specimen from a cave at La Chapelle aux Saints in France that suffered from old age and rickets, so for decades Neanderthals were thought to be stoop-shouldered, bow-legged, and primitive, the classic stereotypical grunting "cave men".

Modern research has shown that Neanderthals were very different from this outdated image. Although their skulls are distinct from ours in having a protruding face, large brow ridges, no chin, and a flatter skull that sticks out in the back, they had, on average, a slightly *larger* brain capacity than we do, and they practiced a complex culture that included ceremonial burials suggesting religious beliefs. Their bones (and presumably bodies) were robust and muscular and slightly shorter than the average modern human, because they lived exclusively in the cold climates of the glacial margin of Europe and the Middle East, where their short stocky build (like a modern Inuit or Laplander) would have been an advantage. Their tool kits and culture were also more complex, with Mousterian hand axes, spear points, and other complex devices, as well as bone and wooden tools. Some of these tools show complex working and simple carving, so they were artistic as no hominin before had ever been. The famous discoveries at Shanidar Cave in Iraq showed that Neanderthals buried their dead with multiple kinds of colorful flowers, suggesting that they may have had at least some kind of religious beliefs and possibly belief in an afterlife.

Figure 25.11 Reconstructed skeleton and life-sized model of a Neanderthal. (Photo courtesy Wikimedia Commons.)

For decades, anthropologists treated Neanderthals as a subspecies of *Homo sapiens*, but recent work suggests that they were a distinct species. The best fossil evidence of this comes from Skhul and Qafzeh caves on Mt. Carmel in Israel, where layers bearing Neanderthal remains are interbedded and alternate with layers containing early modern humans. In 1997 Neanderthal DNA was sequenced and they are clearly not *Homo sapiens*, but genetically distinct as well. However, their DNA shows evidence that all modern humans of non-African descent have a bit of Neanderthal DNA in them, so there must have been some interbreeding between them in Eurasia where they overlapped.

Neanderthals were the only extinct species of human known in from DNA sequencing until 2010, when molecular biology shocked the world with the announcement that there was yet another species of human during the last 40,000 years. Digging in Denisova Cave in the Altai Mountains of Siberia near the Mongolian-Chinese border, Russian archeologists found a juvenile finger bone, a toe bone, and a few isolated teeth of a hominin mixed with artifacts including a bracelet. The artifacts gave a radiocarbon date of 41,000 years ago, so the age was well established. But when the molecular biology lab of Svante Pääbo and Johannes Krause at the Max Planck Institute in Leipzig, Germany (who first sequenced Neanderthal DNA), analyzed the mitochondrial DNA of the finger bone, they found it had a unique genetic sequence that was distinct from both Neanderthals and modern humans. The nuclear DNA was also distinct, but suggested that these people were closely related to the Neanderthals. They may also have interacted with modern humans, because they share about 3–5% of their DNA with Melanesians and Australian Aborigines. The mitochondrial DNA data suggest that they branched off from the human lineage about 600,000 years ago, and represent a separate "out of Africa" migration distinct from the much earlier (1.8 Ma) *Homo erectus* exodus, or the much younger (300,000 years ago) emigration of *H. rhodesiensis-H. heidelbergensis* from Africa to Eurasia.

These mysterious people whose DNA was so distinctive are now called the "Denisovans". Since there are so few fossils, we cannot say much about their physical appearance or anything else other than that they have distinctive DNA that is found in no other human species. In fact, scientists are still reluctant to give the Denisovans a formal scientific name, because there is not enough fossil material to describe the anatomy of the species in any useful sense. So the Denisovans are mysterious, showing us that the bones don't tell the whole tale, but that there may have been numerous other human species on this planet that haven't left a fossil record.

Almost as surprising as the 2010 discovery of the Denisovans was the 2003 announcement of a primitive dwarfed species of humans found only on the island of Flores in Indonesia. Found at a site called Liang Bua Cave, their fossils and artifacts are dated between 1 million and 74,000 years ago. The most striking feature of these people is their tiny size, only about 1.1 meters (3 feet 7 inches) tall in a fully grown adult, so they have been nicknamed the "hobbits". Yet these are not modern African pygmies (which are tiny but fully modern humans), but an entire population of dwarfed people that appear to have been descended from a *Homo erectus* ancestry (or possibly even from *Homo habilis* ancestry) about a million years ago, then became dwarfed. Size reduction is a common effect on oceanic islands, with many types of animals (especially elephants, mammoths and hippos) undergoing dwarfing on islands ranging from Malta to Crete to Cyprus to Madagascar. The reason for this dwarfing is clear: they are living on the smaller food resource base of an island, so cannot get the kind of nutrition needed to grow to normal sizes. In addition, on most islands they are typically they are not

Figure 25.12 A museum display of replicas of the skulls of the many fossil humans brings home the point that there is now an excellent fossil record of human evolution, with many transitional fossils from those that are very ape-like to those that are only slightly different from modern humans. (Photo by the author.)

under pressure from large predators on islands as well, or competing with the same large herbivores. Although the interpretation of these fossils is controversial, most anthropologists agree that they were a distinct species, which has been formally named *Homo floresiensis*.

Finally, we find the first fossil skulls and skeletons that look almost indistinguishable from our own species. Some of these, dubbed "archaic *Homo sapiens*" or more formally, *Homo heidelbergensis*, are known from a few deposits in Africa dating as old as 300,000 years (**Figure 25.12**). About 90,000 years ago, skulls from Africa (such as from Klasies Mouth Cave in South Africa) look almost completely modern in appearance and are universally regarded as *Homo sapiens* (our species). Like *Homo erectus*, early *Homo sapiens* spent most of its history in Africa, and migrated to Eurasia about 200,000 years ago, and then about 70,000 years ago. There these people came into contact with Neanderthals, and for about 30,000 years they coexisted. Mysteriously, Neanderthals vanished 40,000 years ago. Whether they were wiped out by *Homo sapiens* or by some other cause is not clear. The subject has been one of endless debate and speculation. Pat Shipman argues that modern humans had an advantage in domesticated dogs, which helped them overcome Neanderthals in hunting and in warfare. Whatever happened, modern *Homo sapiens* soon took over the entire Old World, developing complex cultures (the "Cro-Magnon people") including famous cave paintings of Europe, and many kinds of weapons and tools.

This brief review of the hominin fossil record hardly does justice to the richness and quality of the specimens or to the incredible amount of anatomical detail that has been deciphered. If it all seems like too much to absorb, just gaze at the fossils in **Figure 25.12**. They look vaguely like modern human skulls, but they definitely show the change from more primitive hominins that some people see as "mere apes" (even though they were all completely bipedal and had many other human characteristics) up through forms that everyone would agree look much like "modern humans" (even though they had many distinctive anatomical features, like those found in Neanderthals, that make them a distinct species). Even non-scientists can glance at these fossils and see the hallmarks of their own ancestry.

FURTHER READING

Agusti, J.; Anton, M. 2002. *Mammoths, Sabertooths, and Hominids: 65 Million Years of Mammalian Evolution in Europe.* Columbia University Press, New York.

Asher R. J.; Geisler J. H.; Sánchez-Villagra, M. R. 2008. Morphology, paleontology, and placental mammal phylogeny. *Systematic Biology* 57: 311–317.

Asher, R.J.; Meng, J.; Wible, J.R.; McKenna, M.C.; Rougier, G.W.; Dashzeveg, D.; Novacek, M.J. 2005. Stem Lagomorpha and the antiquity of Glires. *Science.* 307 (5712): 1091–1094.

Beard, K.C. 2004. *The Hunt for the Dawn Monkey: Unearthing the Origin of Monkeys, Apes, and Humans.* University of California Press, Berkeley.

Conroy, G.C. 1990. *Primate Evolution.* W. W. Norton, New York.

Delson, E.C. 1985. *Ancestors: The Hard Evidence.* Liss, New York.

Diamond, J. 1992. *The Third Chimpanzee: The Evolution and Future of the Human Animal.* HarperCollins, New York.

Douzery, E.J.P.; Delsuc, F.; Stanhope, M.J.; Huchon, D. 2003. Local molecular clocks in three nuclear genes: Divergence times for rodents and other mammals and incompatibility among fossil calibrations. *Journal of Molecular Evolution.* 57: S201–S213.

Fabre, P.H.; et al. 2012 .A glimpse on the pattern of rodent diversification: A phylogenetic approach. *BMC Evolutionary Biology.* 12: 88.

Foley, R.A.; Lewin, R. 2003. *Principles of Human Evolution.* Wiley-Blackwell, New York.

Ge, D.; Wen, Z.; Xia, L.; Zhang, Z.; Erbajeva, M.; Huang, C.; Yang, Q. 2013. Evolutionary history of Lagomorphs in response to global environmental change. *PLoS ONE.* 8 (4): e59668.

Gregory, W.K. 1910. The orders of mammals. *Bulletin of the American Museum of Natural History.* 27: 1–525.

Harari, Y.N. 2015. *Sapiens: A Brief History of Humankind.* Harper, New York.

Harris, E.E. 2015. *Ancestors in Our Genome: The New Science of Human Evolution.* Oxford University Press, Oxford, UK.

Hopkins, S.S.B. 2005. The evolution of fossoriality and the adaptive role of horns in the Mylagaulidae (Mammalia: Rodentia). *Proceedings of the Royal Society B.* 272 (1573): 1705–1713.

Janis, C., Gunnell, G.F.; Uhen, M.D., eds. 2008. *Evolution of Tertiary Mammals of North America, Vol II: Small Mammals, Xenarthrans, and Marine Mammals.* Cambridge University Press, Cambridge.

Johanson, D.; Edey, M. 1981. *Lucy: The Beginnings of Humankind.* Simon & Schuster, New York.

Johanson, D.; Edgar, B. 1996. *From Lucy to Language.* Simon & Schuster, New York.

Johanson, D.; Johanson, L.; Edgar, B. 1994. *Ancestors: In Search of Human Origins.* Villard, New York.

Johanson, D.; Wong, K. 2009. *Lucy's Legacy: The Quest for Human Origins.* Crown, New York.

Jurmain, R.; Kilgore, L.; Trevathan, D.; Ciochon, R. 2014. *Introduction to Physical Anthropology.* Cengage, New York.

Krause, D.W. 1986. Competitive exclusion and taxonomic displacement in the fossil record: The case of rodents and multituberculates in North America. *Contributions to Geology, University of Wyoming, Special Paper.* 3: 95–117.

Kurtén, B. 1968. *Pleistocene Mammals of Europe.* Columbia University Press, New York.

Kurtén, B. 1988. *Before the Indians.* Columbia University Press, New York.

Kurtén, B.; Anderson, E. 1980. *Pleistocene Mammals of North America.* Columbia University Press, New York.

Larsen, C.S. 2014. *Our Origins: Discovering Physical Anthropology.* W.W. Norton, New York.

Lewin, R. 1988. *In the Age of Mankind: A Smithsonian Book on Human Evolution.* Smithsonian Institution Press, Washington, DC.

Li, C.K.; Wilson, R.W.; Dawson, M.R. 1987. The origin of rodents and lagomorphs. *Current Mammalogy.* 1: 97–108.

Luckett, W.P.; Hartenberger, J.-L., eds. 1985. *Evolutionary Relationships among Rodents.* Plenum Press, New York.

Marks, J. 2002. *What It Means to Be 98% Chimpanzee.* University of California Press, Berkeley.

McKenna, M.C. 1975. Toward a phylogenetic classification of the Mammalia, pp. 21–46. In Luckett, W.P.; Szalay, F.S., eds. *Phylogeny of the Primates.* Plenum Press, New York.

McKenna, M.C.; Bell, S.K. 1997. *Classification of Mammals above the Species Level.* Columbia University Press, New York.

Novacek, M.J. 1992. Mammalian phylogeny: Shaking the tree. *Nature.* 356: 121–125.

Novacek, M.J. 1994. The radiation of placental mammals, pp. 220–237. In Prothero, D.R.; Schoch, R.M., eds. *Major Features of Vertebrate Evolution.* Paleontological Society Short Course 7. Paleontological Society, Lawrence, KS.

Novacek, M.J.; Wyss, A.R. 1986. Higher-level relationships of the recent eutherian orders: Morphological evidence. *Cladistics.* 2: 257–287.

Novacek, M.J.; Wyss, A.R.; McKenna, M.C. 1988. The major groups of eutherian mammals, pp. 31–73. In Benton, M.J., ed. *The Phylogeny and Classification of the Tetrapods, vol. 2: Mammals.* Clarendon Press, Oxford.

Pääbo, S. 2014. *Neanderthal Man: In Search of Lost Genomes.* Basic Books, New York.

Potts, R.; Sloan, C. 2010. *What Does It Mean to Be Human?* National Geographic Society, Washington, DC.

Prothero, D.R. 1994. Mammalian evolution, pp. 238–270. In Prothero, D.R.; Schoch, R.M., eds. *Major Features of Vertebrate Evolution.* Paleontological Society Short Course 7. Paleontological Society, Lawrence, KS.

Prothero, D.R. 2006. *After the Dinosaurs: The Age of Mammals.* Indiana University Press, Bloomington, IN.

Prothero, D.R. 2016. *The Princeton Field Guide to Prehistoric Mammals.* Princeton University Press, Princeton, NJ.

Prothero, D.R. 2018. *When Humans Nearly Vanished: The Eruption of Toba Volcano.* Smithsonian Books, Washington, DC.

Reich, D. 2018. *Who We Are and How We Got Here: Ancient DNA and the New Science of the Human Past.* Pantheon, New York.

Roberts, A. 2018. *Evolution: The Human Story.* DK Books, London.

Rose, KD. 2006. *The Beginning of the Age of Mammals.* Johns Hopkins University Press, Baltimore, MD.

Rose, K.D.; Archibald, J. D., eds. 2005. *The Rise of Placental Mammals: The Origin and Relationships of the Major Extant Clades.* Johns Hopkins University Press, Baltimore, MD.

Rose K. D.; Deleon, V. B.; Missian, P.; Rana, R. S.; Sahni, A.; Singh, L.; Smith, T. 2008. Early Eocene lagomorph (Mammalia) from western India and the early diversification of Lagomorpha. *Proceedings of the Royal Society B.* 2007: 1661.

Rutherford, A. 2016. *A Brief History of Everyone Who Ever Lived: The Human Story Retold through Our Genes.* The Experiment LLC, New York.

Savage, DE.; Russell, D.E. 1983. *Mammalian Paleofaunas of the World.* Addison Wesley, Reading, MA.

Savage, R.J.G.; Long, M. 1986. *Mammal Evolution: An Illustrated Guide.* Facts-on-File Publications, New York.

Sawyer, G.J.; Deak, V.; Sarmiento, E.; Milner, R. 2007. *The Last Human: A Guide to the Twenty-Two Species of Extinct Humans.* Yale University Press, New Haven, CT.

Shipman, P. 2015. *The Invaders: How Humans and Their Dogs Drove Neanderthals to Extinction.* Belknap Press, Cambridge, MA.

Sibley, C.G.; Ahlquist, J.E. 1984. The phylogeny of hominoid primates, as indicated by DNA-DNA hybridization. *Journal of Molecular Evolution.* 20: 2–15.

Stringer, C. 2012. *Lone Survivors: How We Came to Be the Only Humans on Earth.* Times Books, London.

Stringer, C.; Andrews, P. 2005. *The Complete World of Human Evolution.* Thames & Hudson, London.

Stringer, C.; Gamble, C. 1993. *In Search of Neanderthals: Solving the Puzzle of Human Origins.* Thames & Hudson, London.

Stucky, R.K. 1990. Evolution of land mammal diversity in North America during the Cenozoic. *Current Mammalogy.* 2: 375–432.

Szalay, F.S.; Novacek, M.J.; McKenna, M.C., eds. (1993). *Mammal Phylogeny.* Springer-Verlag, Berlin.

Tattersall, I. 1993. *The Human Odyssey: Four Million Years of Human Evolution.* Prentice Hall, Upper Saddle River, NJ.

Tattersall, I. 2012. *Masters of the Planet: The Search for Our Human Origins.* St. Martin's Griffin, London.

Tattersall, I. 2015. *The Strange Case of the Rickety Cossack and Other Cautionary Tales from Human Evolution.* St. Martin's Press, New York.

Tattersall, I.; Schwartz, J. 2000. *Extinct Humans.* Westview, New York.

Turner, A.; Anton, M. 2004. *National Geographic Prehistoric Mammals.* National Geographic Society, Washington, DC.

Walker, A.; Shipman, P. 1996. *The Wisdom of the Bones: In Search of Human Origins.* Knopf, New York.

Werdelin, L.; Sanders, W.L., eds. 2010. *Cenozoic Mammals of Africa.* University of California Press, Berkeley, CA.

Wible, J.R. 2007. On the cranial osteology of the Lagomorpha. *Bulletin of Carnegie Museum of Natural History.* 2007 (39): 213–234.

Wood, A.E. 1955. A revised classification of the rodents. *Journal of Mammalogy.* 36 (2): 165–187.

Woodburne, M.O., ed. 2004. *Late Cretaceous and Cenozoic Mammals of North America: Biostratigraphy and Geochronology.* Columbia University Press, New York.

Wu, S.; Wu, W.; Zhang, F.; Ye, J.; Ni, X.; Sun, J.; Edwards, S. V.; Meng, J.; Organ, C. L. 2012. Molecular and paleontological evidence for a post-Cretaceous origin of rodents. *PLoS ONE.* 7 (10): e46445.

INDEX

T - #0558 - 071024 - C464 - 280/210/20 - PB - 9780367473167 - Gloss Lamination